Nimesulide – Actions and Uses

Edited by K. D. Rainsford

Birkhäuser Verlag
Basel · Boston · Berlin

Editor

K.D. Rainsford
Biomedical Research Centre
Sheffield Hallam University
Howard Street
Sheffield, S1 1WB
UK

A CIP catalogue record for this book is available from the Library of Congress, Washington D.C., USA

Bibliographic information published by Die Deutsche Bibliothek
Die Deutsche Bibliothek lists this publication in the Deutsche Nationalbibliografie;
detailed bibliographic data is available in the Internet at <http://dnb.ddb.de>.

ISBN 10: 3-7643- 7068-8 Birkhäuser Verlag, Basel – Boston – Berlin
ISBN 13: 978-3-7643-7068-8

Part of Springer Science+Business Media
Printed on acid-free paper produced from chlorine-free pulp. TCF ∞

Printed in Germany
Typesetting: Fotosatz-Service Köhler GmbH, Würzburg
Cover design: Micha Lotrovsky, Therwil
Cover illustration: 3D nimesulide molecule
ISBN 10: 3-7643-7068-8
ISBN 13: 978-3-7643-7068-8

9 8 7 6 5 4 3 2 1

www.birkhauser.ch

Contents

A. Bernareggi and K. D. Rainsford
Pharmacokinetics of nimesulide

A. Maroni and A. Gazzaniga
Pharmaceutical formulations of nimesulide

K.D. Rainsford, M. Bevilacqua, F. Dallegri, F. Gago, L. Ottonello, G. Sandrini, C. Tassorelli, and I.G. Tavares
Pharmacological properties of nimesulide

M. Bianchi, G.E. Ehrlich, F. Facchinetti, E.C. Huskisson, P. Jenoure, A. La Marca, K.D. Rainsford

Clinical applications of nimesulide in pain, arthritic conditions and fever

List of contributors

A. Bernareggi, Cell Therapeutics Inc., Europe, Via Ariosto 23, 20091 Bresso, Italy; e-mail: alberto.bernareggi@ctimilano.com

M. Bevilacqua, U O Endocrinologia e Diabetologia, Ospedale L Sacco-Polo Universitario, 20157 Milano, Italy;
e-mail: m.bevilacqua@hsacco.it; mauriziobevilacqua@fastwebnet.it

M. Bianchi, Department of Pharmacology, Faculty of Medicine, University of Milan, Via Vanvitelli 32, 20129 Milano, Italy; e-mail: mauro.Bianchi@unimi.it

F. Bissoli, Clinica S Gaudenzio, Divisione Medicina, Via Enrico Bottini 3, 20100 Novara, Italy; e-mail: bissolifranco@hotmail.com

I. Bjarnason, Department of Medicine, Guy's, King's and St Thomas' Medical School, University of London, Bessemer Road, London SE5 9PJ, UK; e-mail: ingvar.bjarnason@kcl.ac.uk; ingvar.bjarnason@virgin.net

A. Conforti, Università di Verona, Istituto di Farmacologia, Policlinico Borgo Roma, 37134 Verona, Italy

F. Dallegri, First Clinic of Internal Medicine, Department of Internal Medicine, University of Genova Medical School, 16132 Genova, Italy;
e-mail: dalle@unige.it

G. E. Ehrlich, University of Pennsylvania, 1 Independence Place 1101, 241 South Sixth Street, Philadelphia, PA 19106-3731, USA; e-mail: ge2@mindspring.com

F. Facchinetti, Clinica Ostetrica & Ginecologia, Via del Pozzo 71, 41100 Modena, Italy, e-mail: facchinetti.fabio@unimore.it

F. Gago, Departamento de Farmacologia, Universidad de Alcalá, E-28871, Alcalá de Henares, Madrid, Spain; e-mail: federico.gago@uah.es

A. Gazzaniga, Università degli Studi di Milano, Istituto di Chimica Farmaceutica e Tossicologia, Viale Abruzzi, 42, 20131 Milano, Italy;
e-mail: Andrea.Gazzaniga@unimi.it

E. C. Huskisson, 14A Milford House, 7 Queen Anne Street, London W1M 9FD, UK; e-mail: edwardhuskisson@aol.com

P. Jenoure, crossklinik am Merian Iselin Spital, Föhrenstrasse 2, 4009 Basel, Switzerland; e-mail: jenoure@swissonline.ch

L. Maiden, Department of Medicine, Guy's, King's and St Thomas' Medical School, University of London, Bessemer Road, London SE5 9PJ, UK

A. La Marca, Mother Infant Department and UCADH Unit of Reproduction, University of Modena & Reggio Emilia, Via del Pozzo 71, 41100 Modena, Italy; e-mail: antlamarca@libero.it

A. Maroni, Università degli Studi di Milano, Istituto di Chimica Farmaceutica e Tossicologia, Viale Abruzzi, 42, 20131 Milano, Italy;
e-mail: alessandra.maroni@unimi.it

N. Moore, Department of Pharmacology, Université Victor Segalen, Bordeaux, France; e-mail: nicholas.moore@pharmaco.u-bordeaux2.fr

U. Moretti, Clinical Pharmacology Unit, Department of Medicine and Public Health, Section of Pharmacology, University of Verona, 37134 Verona, Italy.
e-mail: umoretti@sfm.univr.it

L. Ottonello, First Clinic of Internal Medicine, Department of Internal Medicine, University of Genova Medical School, 16132, Genova, Italy;
e-mail: otto@unige.it

K.D. Rainsford, Biomedical Research Centre, Sheffield Hallam University, Howard Street, Sheffield S1 1WB, UK; e-mail: k.d.rainsford@shu.ac.uk

G. Sandrini, IRCCS Fondazione "Istituto Neurologico C. Mondino", Dipartimento di Scienze Neurologiche, Univerità di Pavia, Via Mondino 2, 27100 Pavia, Italy; e-mail: gsandrin@unipv.it and giorgio.sandrini@unipv.it

K. Takeuchi, Department of Medicine, Guy's, King's and St Thomas' Medical School, University of London, Bessemer Road, London SE5 9PJ, UK

C. Tassorelli, IIRCCS Fondazione "Istituto Neurologico C. Mondino", Dipartimento di Scienze Neurologiche, Univerità di Pavia, Via Mondino 2, 27100 Pavia, Italy; e-mail: cristina.tassorelli@mondino.it

I.G. Taveres, Academic Department of Surgery, Guy's, King's and St Thomas' School of Medicine, The Rayne Institute, London, SE5 9NU, UK;
e-mail: ignatius.tavares@kcl.ac.uk

G.P. Velo, Ospedale Policlinico, Via delle Menegone 10, 37134 Verona, Italy;
e-mail: gpvelo@sfm.univr.it

Preface

There can be few drugs used to treat pain and inflammation that have came from such modest and inauspicious beginnings to be so widely accepted in the world today as the title drug for this book, nimesulide. Originally it was developed in the mid-late 1960's by Riker Laboratories (USA) as part of a programme of drug discovery in new non-steroidal anti-inflammatory drugs (NSAID) and pesticides. Helsinn Healthcare SA (Lugano, Switzerland) obtained the world-wide rights for this drug in the 1980's and this company has been the prime mover responsible for its subsequent development. This has involved extensive clinical studies in various arthritic and pain states as well as investigations into the mode of action of nimesulide. From the latter studies it emerged that the drug has selectivity for inhibition of the cyclo-oxygenase-2 (COX-2) enzyme that is responsible for prostaglandins involved in the development of inflammation. This discovery made during the early 1990's led to the recognition that nimesulide was probably the first drug among those NSAIDs used clinically to have COX-2 selectivity. Recently, there has been considerable debate about the degree of COX-2 selectivity shown by the coxibs and other NSAIDs. Nimesulide is classified as a *preferential* COX-2 inhibitor, due to the small degree of inhibition of COX-1 observed in many studies.

It has become clear in recent years that inhibition of COX-2 while significant is not the sole basis for controlling pain and inflammatory conditions. Furthermore, it has also emerged since the discovery of its COX-2 effects that the actions of nimesulide have been found to be more extensive than were originally envisaged in its early stages of development (i.e. inhibition of prostaglandin production and anti-oxidant activities). In addition, it is a potent inhibition of histamine release, modulator of cytokines, steroid receptor mimicry and range of enzymatic activities that underlie degradation of cartilage and bone in osteoarthritis and other joint diseases. Some of the actions of nimesulide may be important in understanding why this drug has low gastrointestinal (GI) side effects along with its proven ability to spare production of GI-protective prostaglandins. Thus, the broad-based biochemical and cellular actions of nimesulide along with its pharmacokinetic properties (rapid absorption, short-lived plasma half life) appear to underlie its reputation for being a very effective drug in controlling a variety of painful and inflammatory states while having low GI and some of the common side effects in comparison with other NSAIDs.

This book represents the first comprehensive monograph on nimesulide covering all aspects relating to its chemical and biological developments, pharmacokinetics, pharmaceutical properties, basic and clinical pharmacodynamics, clinical uses in various pain and inflammatory conditions as well as the evaluation, assessment and mechanisms underlying adverse side-effects from nimesulide.

The book would not have been possible without the valuable contributions of the leading experts in the field who have made significant contributions to understanding of the actions, uses and safety of the drug. The invaluable help and advice provided by the medical and scientific staff at Helsinn Healthcare including access to their scientific databases is also most gratefully acknowledged.

This book represents the original work of the authors and editor who are totally responsible for its contents. The opinions and views of these contributors are theirs alone. Thus, this book is an independent assessment of the state of art of knowledge on the drug.

I should like to acknowledge the valuable secretarial and administrative help of Mrs Marguerite Lyons of the Biomedical Research Centre at Sheffield Hallam University as well that of Mrs Veronica Rainsford-Koechli, the assistance in preparing a computer-based literature retrieval system proposed by Mr Alexander Rainsford, and the ever-willing help and assistance of the Library Staff of the Adsetts Learning Centre at Sheffield Hallam University and the Royal Society of Medicine Library, London.

Finally, but not last, my sincere thanks to Dr Hans-Detlef Klüber, Mrs Karin Neidhart and staff at Birkhäuser Verlag, Basel, for their help in the preparation and production of this book.

April 2005

K.D. Rainsford
Sheffield Hallam University
Sheffield UK

The discovery, development and novel actions of nimesulide

K. D. Rainsford

Biomedical Research Centre, Sheffield Hallam University, Howard Street, Sheffield, S1 1WB, UK

Introduction

The historical development of the non-steroidal anti-inflammatory drugs (NSAIDs) has had several different phases. The use in the pre-nineteenth century period of various plant extracts for the treatment of pain classically culminated in the isolation and later synthesis, by Kolbe and Lautermann in 1874, of salicylic acid, probably the first synthetic NSAID [1, 2]. From this came the acetylated salicylate, aspirin, supposedly safer and more effective than salicylic acid at the end of that century [2]. The pyrazolones, antipyrine and aminopyrine, acetanilide and phenacetin were developed in the latter part of the nineteenth century as fever-reducing and pain-relieving agents [1, 3]. Today, these are described as non-narcotic analgesics as they do not have the anti-inflammatory properties of NSAIDs such as aspirin. The development of the analgesics, like that of other drugs to control infections in the nineteenth and early part of the twentieth centuries grew out of expansion of the dyestuff and other chemical industries in Germany, Britain and Switzerland at that time. The success in Germany of the chemical industry in the latter part of the nineteenth century was achieved from close collaborations with scientists and physicians in universities and research institutes. The German chemical industry was conscientiously scientific and highly commercial [3]. The chemical science of compound development was often based on concepts, and little basic biological information was available to enable development of targets as we know them today. Moreover, formal preclinical safety and efficacy studies, along with controlled clinical trials, were not undertaken with the new chemical derivatives – many of them derived from aniline, phenols, naphthalene and other members of the coal tar family of compounds. Clinical studies consisted of simple trials on a few patients. Full-scale toxicity studies were unheard of, although there was appreciation of the need to recognise toxic effects. Indeed with some drugs, such as aspirin, simple studies were undertaken to show that this drug caused less epithelial injury to the skin of fish than that produced by salicylic acid [2]. This period has been described as the age of 'empiricism' [1].

The serendipitous discovery by Landé and Forrestier of the antirheumatic effects of parenteral gold salts (originally discovered by Robert Koch in the 1890s

Nimesulide – Actions and Uses, edited by K. D. Rainsford

to have antitubercular activity) which led Landé in 1927 to observe that aurothioglucose in various non-tubercular conditions produced marked relief from joint symptoms [1]. Empiricism and serendipity also played a part in the applications of D-penicillamine, anti-malarials, corticosteroids, sulphasalazine and methotrexate in the pre- and post-World War II period for the treatment of rheumatoid and related arthritic conditions [1].

In 1948–1949, Brodie and Axelrod discovered that paracetamol was the main metabolite of phenacetin in humans, which was then coming under serious criticism because of methaemoglobinaemia, hepatic and renal problems. Hinsberg and Treupel had found, in 1894, that paracetamol had antipyretic activity like that of phenacetin and antipyrine, although the effects were evident at higher doses of the latter two drugs than with paracetamol [4]. Because of the advent of aspirin and other analgesics paracetamol was forgotten until the observations of Brodie and Axelrod, after which it was marketed in the 1950s in the US in combination with aspirin and caffeine and in the UK on its own in 1956 and thereafter had a slow introduction in other countries. Again serendipity played a considerable part in the discovery and development of paracetamol.

In the late 1940s phenylbutazone was discovered by Stenzel at J R Geigy Pharmaceuticals in Basel, Switzerland, looking for acidic compounds to solubilise the basic compound, aminopyrine, in attempts to use it as an injectable form and improve the latter's effectiveness for arthritic conditions [1]. Studies soon established that the combination was more effective and had a longer duration of effect than aminopyrine from which it emerged that phenylbutazone was the more active of the two components. The key to the discovery of phenylbutazone was undoubtedly the animal assays for anti-inflammatory activity pioneered by Gerhard Wilhelmi at J R Geigy Pharmaceuticals, notably the ultraviolet (UV) light-induced erythema in guinea pigs [1, 5].

Animal assays for anti-inflammatory activity (including the cotton pellet granuloma and carrageenan-induced paw oedema in rats) and the beginnings of structure-activity determinations in empirical screening played a major part in the discovery of indomethacin, an indole, which was based on an idea by T-Y Shen and Charlie Winter that 5-hydroxytryptamine (serotonin) was important in inflammation [1]. The UV erythema assay in guinea pigs was employed by Stewart Adams in the discovery of ibuprofen in the early 1960s but significantly he employed assays for analgesic activity (the Randall-Selitto test in rats) and gastrointestinal toxicity in dogs, as well as detailed investigations on the absorption and distribution of radiolabeled drugs to discriminate those which had low liver accumulation [5].

Knowledge of the mechanisms underlying the development of inflammation in the pre-prostaglandin era [6] and of the actions of aspirin, phenylbutazone, indomethacin and ibuprofen were rudimentary at the time of the discovery of the newer drugs in the 1950s–1960s. Histamine, kinins, possibly 5-hydroxytrypta-

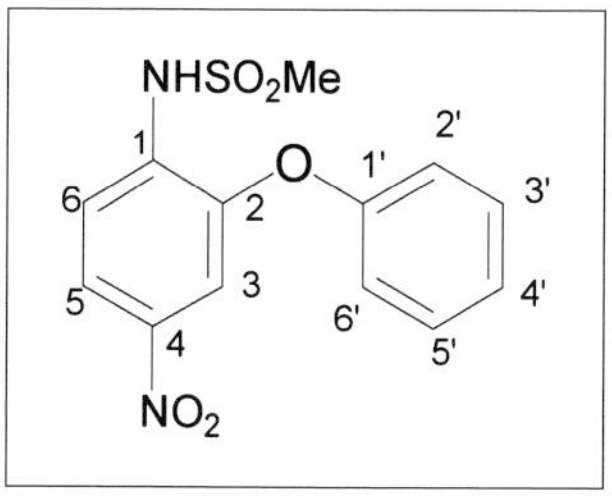

Figure 1
Chemical structure of nimesulide [CA Registry 51803-78-2] known systematically as: Methanesulfonamide, N-(4-nitro-2-phenoxyphenyl)-, or 2-phenoxy-4-nitromethanesulfonanilide, or 4-nitro-2-phenoxymethanesulfonanilide.

mine and a range of metabolic effects involving mitochondrial production of adenosine triphosphate (ATP) and the connective tissue components, as well as effects on leucocytes were considered possible targets for the action of these drugs [7–9] – later to be known as non-steroidal anti-inflammatory agents (NSAIDs) to distinguish them from anti-inflammatory corticosteroids. The pioneering studies of the late Professor Derek Willoughby, Professor Gerald Weissman, Dr Anthony Allison, Dr Philip Davies and many others in the period of the late 1950s to the 1970s saw recognition of a whole range of cellular inflammatory events that are regulated by leucocytes and various plasma – and tissue – derived factors, the interferons, lymphokines and other progenitors of the cytokines heralded the broader and more complex view of inflammation [10, 11]. It was only later after the discovery in 1971 by Professor Sir John Vane, FRS, Nobel Laureate, and his colleagues that the inhibition of the production of prostaglandins in inflammation and platelet functions represented a mechanism for the actions of aspirin and related drugs [1, 6].

In this historical setting the discovery of nimesulide (4-nitro-2-phenoxymethane sulphonanilide; Fig. 1) took place before the period when the prostaglandins were being first found to have roles in inflammation, pain, fever and thrombosis*. Since inevitably the state of the science underlying disease processes serves as the basis for drug discovery at any one period in time it is to the period of the 1960s that we look to understand the biochemical and cellular responses involved in the development of inflammation and pain. The concepts of inflammation and pain at

* *The US patent granted to Moore et al. [14] cites continuation-in-part or abandoned applications dating back to 13 April 1970. Thus, it can be assumed that the concept development of R-805 and others in this series took place in the period before the discovery by Vane (1971) and others of the effects of aspirin and other analgesics on inhibiting production of prostaglandins as a basis to their action in inflammation and other therapeutic actions.*

that time centred on the roles of (a) histamine, kinins and slow reacting substance in anaphylaxis and other systemic mediators of pain and acute inflammatory reactions, (b) the emerging involvement of polymorpho-neutrophil leucocytes (PMNs), monocytes/macrophages and lymphocytes in regulating the major inflammatory reactions, and (c) the changes in the cartilage, synovial and bone metabolism of collagen, glycosaminoglycans/proteoglycans, glucose, fatty acid and in mitochondria [7–9]. Pain was considered to be linked to inflammation [8]. Most of the anti-inflammatory drugs were discovered in this period by testing of compounds *in vivo* in animal models.

Discovery of R-805 – nimesulide

The development of nimesulide arose from investigations by Dr George (GGI) Moore (a medicinal–organic chemist; Fig. 2), Dr Karl F Swingle (a pharmacologist), Dr Bob (RA) Scherrer (a medicinal chemist) and their colleagues at Riker Laboratories Inc (Northridge, California, US, later part of the 3M Company at St Paul, Minnesota, US). They had the idea that since the evidence in the late 1960s suggested that free radicals were important in chronic inflammatory diseases then drugs which scavenge these radicals might have novel anti-inflammatory mechanisms to control chronic inflammation. They undertook a detailed structure-activity analysis and determined the pharmacological properties of the sulphonamides [12]. This class of agents had previously been considered in the 1940s to have antirheumatic activity as a consequence of their antibiotic effect by Svartz and her colleagues at Pharmacia in Sweden and this culminated in the development of the sulphonamide–salicylate conjugate, sulphasalazine [13].

Dr Moore has kindly provided a statement about the thinking and important aspects concerning the concepts that underlay the development of the methane sulphonanilides leading to the identification of nimesulide:

> My name is George G. I. Moore, and I am the inventor of nimesulide, originally R-805. I am currently a Corporate Scientist at 3M Co., working at the St Paul (MN) main campus. Following a BA (Honors in Chemistry) from Cornell University in 1962 and a PhD in Organofluorine Chemistry from University of Colorado in 1965, I joined 3M's fledgling pharmaceuticals project. Our synthetic group included several noteworthy chemists, such as John Gerster, who was to invent the first fluoroquinolone antibacterial and later the immune response modifier imiquimod, and Bob Scherrer, inventor of Parke-Davis' meclofenamic and mefenamic anti-inflammatory agents. At that time, our main approach was application of 3M fluorochemistry to pharmaceutical and agrochemical syntheses. In the antiinflammatory area, two fluoroalkanesulfonanilides (triflumidate and diflumidone) had been identified for clinical

Figure 2
Dr George Moore, the chemist who discovered nimesulide (originally coded R-805). He was born in Boston (USA) in 1941, graduated BA (Honors) in chemistry at Cornell University in 1962, then PhD in organofluorine chemistry at University of Colorado in 1965. He then joined the 3M Company (St Paul, MN), which was subsequently incorporated into Riker Laboratories and then moved from Northridge, CA, to St Paul, MN. He is now a Corporate Scientist in the Industrial Business Laboratory at the 3M Company. Thanks to Dr Moore for providing this photo and biographical details.

trials. My role was synthetic expansion of this series, but by late 1969, it seemed that no improvement in the acute therapeutic ratio was forthcoming, and management decided to curtail syntheses. My young assistant, Larry Lappi, and I had just made a 'final' series which included 4-nitro-2-phenoxy trifluoromethanesulfonanilide, the CF_3-analog of what would become R-805. Karl Swingle, our chief anti-inflammatory pharmacologist, found this exceptionally potent in rat paw carrageenan and other models. With renewed management support, we developed a selective nitration process which allowed us to rapidly make a series of analogs. R-805 was synthesised in early 1971, and it unexpectedly had, by far, the best acute therapeutic ratio. (In both anti-inflammatory and herbicidal activities until this point, the order of activity for

> RSO_2NH-Ar had been CF_3>CF_2H>CFH_2>CH_3; in R-805, the 4-NO_2 offset the usual decrease in acidity.) Scherrer, who had been mentoring this synthetic program, used his study of R-805 partitioning into octanol–water to develop his concept of physiological distribution of ionisable drugs. Following secondary evaluations, the material was designated as R-805 for clinical trials in our newly-acquired subsidiary, Riker Laboratories.
>
> This work led to a broader discovery. I focused on the special role of the 4-NO_2 group in this material and several related sulphonanilides. Screening of a variety of materials and use of the emerging science of QSAR showed no correlation with acidity or lipophilicity. There was a weak correlation with the radical stabilisation parameter, 'E_R', (weak in that E_R values were available for only four substituents), with nitro by far the best. This, and the recently-published involvement of PGs in inflammation, led me to hypothesise in late 1971 that free radical scavenging might be involved. I made many types of modified antioxidants, primarily phenolic but including N- and C-based radicals. Swingle found an exceptionally high percentage of these series was effective in his models, strengthening an antioxidant–anti-inflammatory association. We went on to identify one of these for topical trials.
>
> At about this time, Riker reassessed its business plan and decided to discontinue the anti-inflammatory area. R-805 was made available for license, and we all went on to other things, but we still take pride in the fact that nimesulide is used today.

Moore and co-workers recognised from their structure-activity analyses that the anti-inflammatory properties of trifluoro-alkane-sulphonamides are related to the powerful lipophilic properties of the CF_3SO_2 group which serves as a powerful electron attractor (Hammet coefficient, s = 1.3) and their acidic properties [12]. The development of nimesulide (R-805) was to some extent an extension of the recognition of the acidic properties of the nitro-group which is located at the *para*-position of the methyl-sulphonamido-moiety [14] (Fig. 1).

In the structure-activity analysis of this series the anti-inflammatory activities were determined using the UV erythema assay in guinea pigs and the rat paw carrageenan assay, while the analgesic activity was determined in the Randall-Selitto in rats and the phenylquinone writhing test in mice [12, 14–17]. Assays of prostaglandin synthesis inhibition were later performed using the bovine seminal vesicle microsomal preparation *in vitro* [15], which was a standard preparation employed at that stage (containing what is now known to be COX-1). Studies by Rufer and colleagues [18] discovered the basis of the oxy-radical scavenging effects of nimesulide during prostaglandin endoperoxide metabolism were similar to those of the phenolic compound, MK-886, which had been previously shown by Kuehl and co-workers [19] to stimulate prostaglandin production *in vitro* as a result of scavenging the peroxy-radical formed during the oxygenation of the 15-carbon moiety of arachidonic acid. This formed one basis in support of the free radical concept being a basis for the therapeutic target set by Moore and his colleagues in their development of the methane sulphonanilides. Later studies [16] (also reviewed in Chapter 4) have subsequently shown that there are other antioxidant mechanisms involved in the anti-inflammatory activity of nimesulide.

Chemical synthesis

The synthesis mentioned in the original patent [14] (Fig. 3) involved dissolving 2-phenoxymethanesulphonanilide (initially prepared by treating 2-phenoxyaniline with methyl-sulphonyl chloride) in glacial acetic acid with warming, then mixing in 70% nitric acid (Fig. 4). After heating, the mixture is poured onto water and the precipitate collected by filtration. Following recrystallisation from ethanol, a light tan solid is recovered with MPt 143–144.5 °C which is 4-nitro-2-phenoxymethane-sulphonanilide. Several other synthetic procedures for the synthesis of nimesulide, its intermediates and analogues have been subsequently reported [20–26] (Fig 5).

Of the efforts to produce other methane sulphonanilides only diflumidone [15] appears to have proven to be a clinical candidate, but is no longer under development.

Development of nimesulide

Following the initial discovery, and the pharmacological and toxicological studies of R-805 it was investigated for clinical efficacy and safety in patients with rheumatoid arthritis [27]. These studies showed that the drug was effective in controlling pain and inflammation. Some of these studies were performed at what is now regarded as very high doses (up to 800 mg/d) and it was not surprising that some liver enzymes were elevated in these patients.

In 1980, Helsinn Healthcare SA of Lugano, Switzerland, acquired the worldwide licensing rights for nimesulide and proceeded to invest in extensive clinical and basic studies on the actions of the drug. The production by Helsinn of nimesulide was first commenced in 1985. The first certificate of analysis released is reported in Figure 6. It was first introduced in Italy in 1985. Nimesulide is now marketed in over 50 countries worldwide [27, 28], through partnerships with leading pharmaceutical companies in most of these countries [27]. The countries where it is marketed by Helsinn and its partners include many in continental Europe, Central and South America, and the Far East. For commercial reasons the drug has not been marketed by Helsinn or others in the US, UK or Australia [27]. The various trade mark names for nimesulide registered worldwide and originated by Helsinn are shown in Appendix A. Nimesulide is produced and sold

United States Patent Office

3,840,597
Patented Oct. 8, 1974

1

3,840,597
SUBSTITUTED 2-PHENOXY ALKANE-SULFONANILIDES

George G. I. Moore, Birchwood, and Joseph Kenneth Harrington, Edina, Minn., assignors to Riker Laboratories, Inc., Northridge, Calif.

No Drawing. Continuation-in-part of application Ser. No. 118,476, Feb. 24, 1971, which is a continuation-in-part of abandoned application Ser. No. 28,148, Apr. 13, 1970. This application July 3, 1972, Ser. No. 268,606

Int. Cl. C07c *143/74*

U.S. Cl. 260—556 F **31 Claims**

ABSTRACT OF THE DISCLOSURE

Diphenyl ethers wherein an alkyl- or haloalkylsulfonamido substituent group is oriented ortho to the ether linkage and a nitro or amino substituent is oriented in the 4 or 5 positions with respect to the sulfonamido group are active anti-inflammatory agents.

This is a continuation-in-part of the copending application, Ser. No. 118,476 filed Feb. 24, 1971, and now abandoned, which is a continuation-in-part of application Ser. No. 28,148 filed Apr. 13, 1970, now abandoned.

This invention relates to diphenyl ethers substituted by an alkyl- or haloalkylsulfonamido group and a nitro or amino group (as defined herein) wherein the orientation of the groups is critical. In particular the invention relates to such compounds wherein the alkyl- or haloalkylsulfonamido group is oriented in the 2 position (ortho) with respect to the ether linkage and the nitro or amino group is oriented in the 4 or 5 position with respect to the alkyl- or haloalkylsulfonamido group, and to salts thereof. The rings and the sulfonamido nitrogen are optionally substituted. The compounds are anti-inflammatory agents. Methods for the preparation and use of the compounds are also described.

Alkysulfonamido and haloalkylsulfonamido substituted diphenyl ethers have ben alluded to heretofore. Thus, see British patents 738,758, 854,956 and 856,452, French patent 1,188,591 and U.S. Pat. 3,223,582. However, none of these patents disclose or suggest the compounds of the present invention wherein a nitro and amino group must be present, nor do they suggest the critical nature of the orientation of the substituent groups to obtain high activity. Furthermore, the pharmacological activity of the compounds of the invention is not suggested by the prior art.

Many non-steroidal anti-inflammatory agents have been discovered in recent years, and some are currently marketed for the treatment of various conditions treated by anti-inflammatory, analgetic and antipyretic agents. However, these agents have significant side-effects which prevent their use in many patients. The search for anti-inflammatory agents with reduced side effects and improved therapeutic ratio is continuing. The compounds of the present invention are effective anti-inflammatory agents with excellent therapeutic ratios.

It is therefore an object of the present invention to provide compounds which are anti-inflammatory agents.

It is another object of the invention to provide anti-inflammatory compositions containing one or more haloalkyl- or alkylsulfonamidoaryl compounds as active ingredients therein.

It is a further object of the invention to provide a method for controlling inflammation in mammalian tissue.

Still other objects will be made apparent by the following specification.

2

DETAILED DESCRIPTION OF THE INVENTION

According to the present invention there is provided a class of compounds of the formula

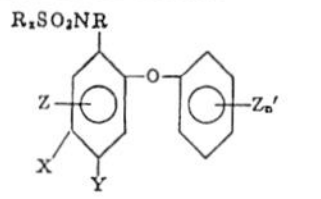

I

wherein R_x is an optionally halogenated lower alkyl radical, R is hydrogen, cyano, alkyl, alkylsulfonyl, haloalkylsulfonyl, a cation or

$$-\overset{\overset{\displaystyle O}{\|}}{C}-A-R^1,$$

where R^1 is alkyl and A is oxygen or a carbon-carbon bond, X is alkoxy, alkyl, halogen, acetamido, nitro, hydrogen, amino, alkoxycarbamoyl or dialkylamino, Y is nitro, amino, alkoxycarbamoyl, dialkylamino or hydrogen, provided that one of X and Y is nitro, amino, alkoxycarbamoyl, or dialkylamino, Z is halogen, nitro or hydrogen, Z′ is halogen, alkyl, alkoxy, nitro, amino alkanamido, haloalkyl, hydroxy, dialkylamino, alkoxycarbamoyl, alkylthio, alkylsulfonyl, alkanoyl, or alkylsulfinyl and n is 0–2 (zero, one or two), provided that the individual aliphatic groups appearing in the R_x, R, X, Y and Z moieties, including those characterized as lower alkyl, contain from one to four carbon atoms each. By alkanamido herein is meant the group

$$\text{alkyl-}\overset{\overset{\displaystyle O}{\|}}{C}-NH-$$

and by alkoxycarbamoyl is meant

$$\text{alkyl-O}-\overset{\overset{\displaystyle O}{\|}}{C}NH-.$$

Compounds of the invention wherein R is hydrogen or a cation are presently most preferred. The compounds in which R is alkyl or alkylsulfonyl are preferred to those in which R is haloalkylsulfonyl, cyano or

$$\overset{\overset{\displaystyle O}{\|}}{C}AR'.$$

When R is alkyl, alkylsulfonyl or haloalkylsulfonyl it preferably contains one carbon atom. The preferred halogens in the haloalkylsulfonyl R moieties are fluorine and chlorine. When R is

$$-\overset{\overset{\displaystyle O}{\|}}{C}-A-R',$$

A is preferably oxygen, and R′ preferably contains one or two carbon atoms.

R_x may be straight or branched chain when it contains three or four carbon atoms. R_x preferably contains one carbon atom. R_x is preferably methyl, chloromethyl, fluoromethyl, difluoromethyl or trifluoromethyl, and most preferred is methyl.

It is preferred that n is zero or one. Most preferred is n equal to zero. When n is one, Z′ is preferably oriented para or ortho, and most preferably Z′ is halogen oriented para. Orientation is relative to the diphenyl ether oxygen.

It is presently preferred that Z is hydrogen. When Z is halogen it is preferably chlorine.

Most preferably X is hydrogen and Y is nitro. Other preferred combinations are those in which X is amino and Y is hydrogen; X is ethoxycarbamoyl and Y is hydrogen; X is dimethylamino and Y is hydrogen; and X is acetamido and Y is nitro.

Figure 3
US Patent number 3,840,597 issued to George GI Moore and JK Harrington from earlier applications [continuation in part] of February 24 1971 and April 13 1970 and assigned to Riker Laboratories Inc., Northridge, CA, USA [14]. The initial date of the application (1970) clearly antedates the first report of Vane and colleagues of the discovery of action of NSAIDs in controlling prostaglandin production.

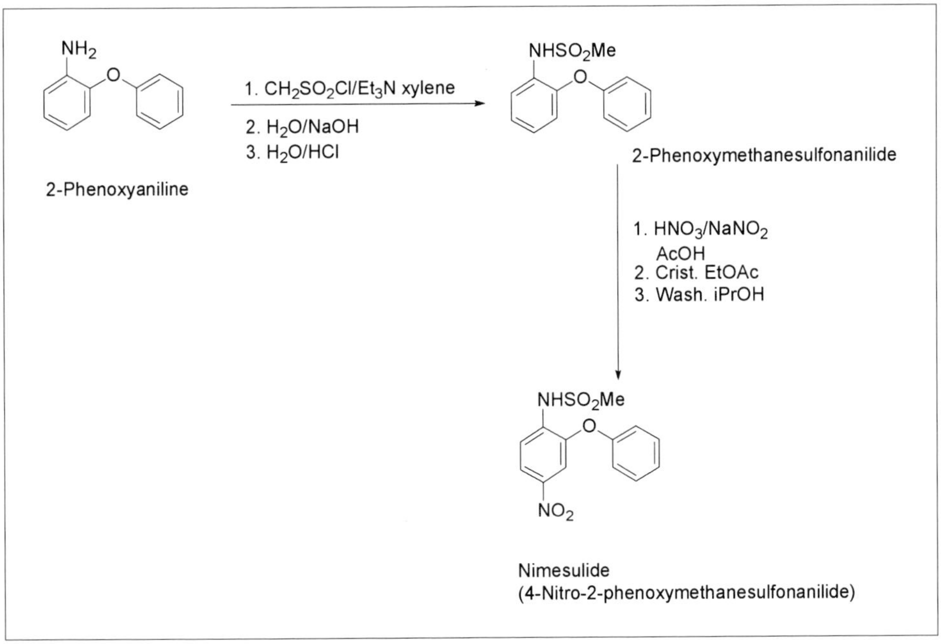

Figure 4
Scheme for the synthesis of nimesulide [14].

by a considerable number of generics manufacturers in Italy, India, China and South America, which is a reflection on its widespread acceptance as an effective pain-relieving and anti-inflammatory agent.

The principal indications for the drug in most countries are for the relief of pain, symptomatic treatment of painful osteoarthritis, extra-articular disorders including tendinitis, bursitis, post-surgical pain including that from dental surgery, ear, nose and throat conditions, dysmenorrhoea and other acute pain states [28]. The most recent Summary of Product Characteristics in force in the EU countries and showing the endorsed indications of the drug as approved in 2003 by the European Medicines Evaluation Agency (EMEA) is shown in Appendix B. This has been prepared and approved from the most up-to-date information on the safety and efficacy of nimesulide and must be regarded as an international standard for recommendations for the use of this drug.

Clinical studies supporting therapeutic claims have been undertaken by Helsinn worldwide in over 90,000 patients [28]. To date over 346 million treatment courses have been employed using the product from Helsinn [28].

After acquiring the licence worldwide, Helsinn then licensed the product for veterinary indications to the French pharmaceutical company, Virbac S.A. [27].

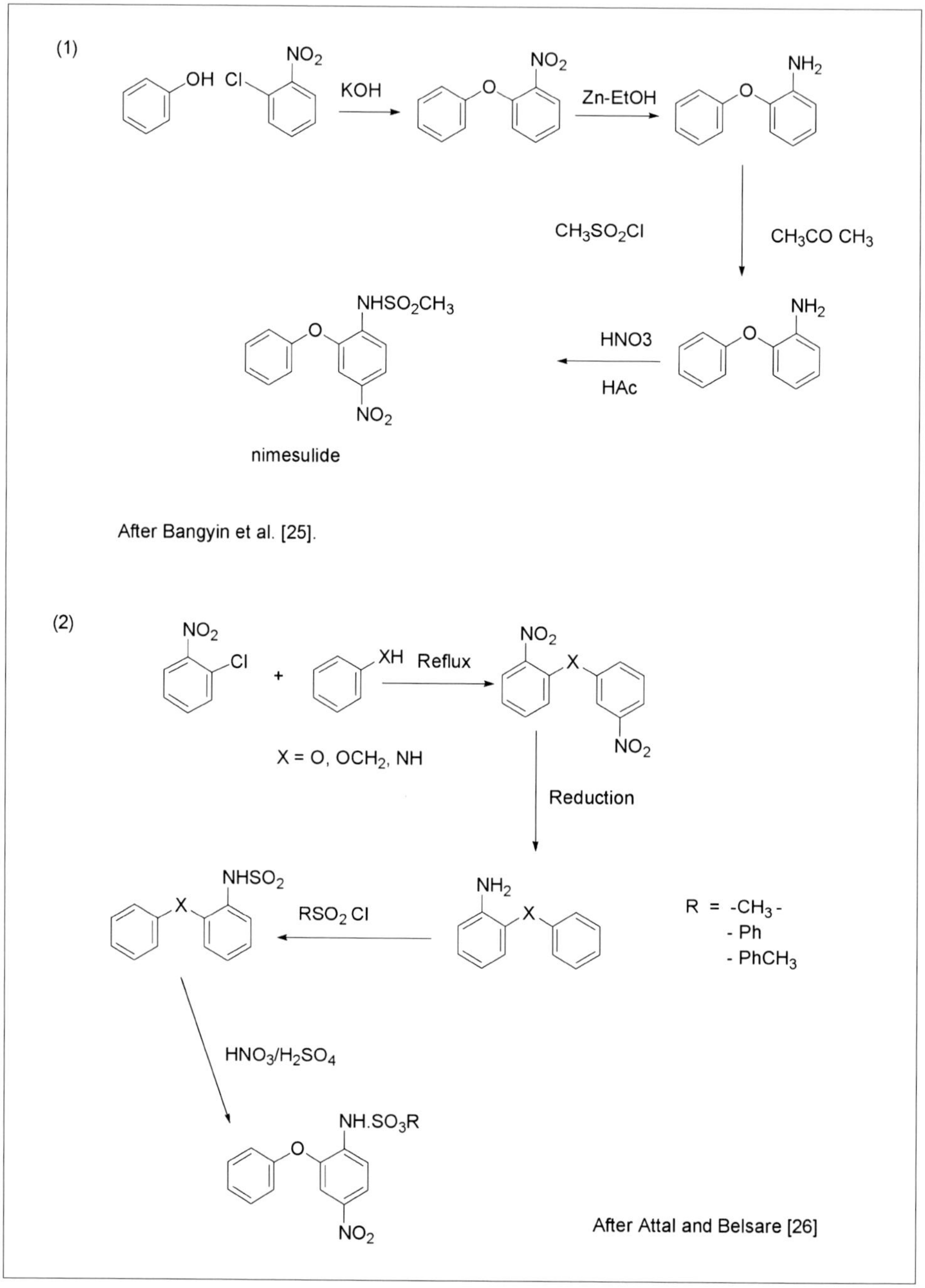

Figure 5
Some schemes for the synthesis of nimesulide, intermediates and analogues.

HELSINN

HELSINN SA
Via Livio 14 6830 Chiasso Tel. (091) 448057
P.O. Box 876 Switzerland Telex 842374 HLSN CH

A N A L Y T I C A L   C E R T I F I C A T E

PRODUCT : NIMESULIDE   BATCH No. :85-1/1

FORMULA : $C_{13}H_{12}N_2O_5S$ MW:308.31   ANALYSIS No.:90/85

Description	:	Yellow powder, odorless
Identification	:	Conform (TLC, UV, IR)
Assay (HPLC)	:	---
(UV)	:	99.30 % on dry substance
(potentiometric)	:	99.95 % on dry substance
Melting point	:	149.4 °C
TLC	:	Single spot at UV 254 nm
Color	:	Est. at 450 nm < 0.600
Loss on drying	:	< 1.00 %
Residual solvents	:	< 500 ppm (Ethyl Acetate)
Solubility	:	Conform
IR-Spectrum	:	Conform
Particle size	:	95 % < 5 microns

COMMENTS: DATE: March 1, 1985

SIGNATURES
Analytical Dept.: R. Bosio
Techn. Director : Dr. R. Wandeler

Figure 6
The first analytical certificate for production of nimesulide (of Helsinn's origin). The drug was first marketed in Italy in 1985.

Physical and chemical properties

Recently, a monograph for nimesulide was included in the European Pharmacopoeia (Ph. Eur mon. 01/2002:1548).

Nimesulide is a pale white-yellowish crystalline powder with a melting point of 147–149 °C and a molecular weight of 308.31 [29, 30]. It is a weak acid having a pKa of 6.4–6.8 [18, 30–32]. It has poor aqueous solubility but is soluble in

Table 1 – Solubility of nimesulide in various solvents

Solvent(s)			Solubility mg/ml	Dielectric Constant (ε) of Solvent(s)
Water			0.014	78.36
Glycerol			0.218	42.5
Methanol			8.812	32.63
Ethanol			3.320	24.3
Butanol			2.120	17.1
n-Octanol			0.970	9.72
Ethylene Glycol			0.510	37.7
Propylene Glycol			1.760	32.0
Polyethylene Glycol (PEG) 400			63.120	12.4
Glycerol	80% + Ethanol	20%	0.691	38.86
	60%	40%	1.693	35.22
	10%	90%	4.040	26.12
PEG 400	80%	20%	9.900	21.92
	60%	40%	24.640	19.54
	90%	10%	65.600	13.59
Water	80%	20%	0.101	67.55
	60%	40%	0.125	56.74
	90%	10%	3.320	24.30
Glycine-NaOH buffer	pH	7	0.034	–
		7.9	0.081	–
		8.84	0.807	–
		9.42	3.886	–
		9.52	6.914	–
		10.17	34.639	–

Partition coefficient in n-octanol/water = 1.788, pKa = 6.4–6.8 [18, 30–32]. The pKa varies according to different solvents/system. From: Seedher & Bhatia (2003) [34].

acetone, chloroform and ethyl acetate and is slightly soluble in ethanol [29, 30]. Details of the solubility in various solvents and solvent mixtures are shown in Table 1 [34]. Of the alcohols the drug is most soluble in methanol with progressive decrease in solubility with increase in carbon length of the respective alcohol and decrease in dielectric constant of the solvent (Tab. 1). The drug is most soluble in polyethylene glycol (PEG) 400 and this is a potentially useful solvent system for oral dosing of laboratory animals. The amount of PEG employed in oral dosing can be reduced by adding ethanol (Tab. 1). No doubt the addition

Table 2 – Crystal and Molecular Properties of Nimesulide

$C_{13}H_{12}N_2O_5S$ $M_r = 308.31$

Crystal form Monoclinic $C2/_c$

Dimensions

a = 33.657 Å (3)	θ = 28.31–32.35°
b = 5.1305 Å (3)	μ = 2.310 mm^{-1}
c = 16.0816 Å (10)	T = 293 K (2)
β = 92.368° (8)	Prism 0.30 × 0.30 × 0.27
V = 2774.5 Å (3)	
Z = 8	
D_x = 1.476 mg m^{-3}	

Molecular structure with atom-labelling scheme.

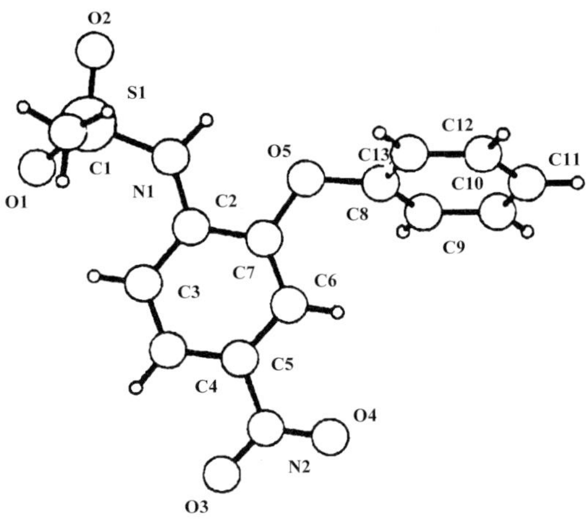

From: Dupont et al. [35].

of water to PEG-ethanol systems would ensure relatively high solubility so reducing the mass of the organic solvents added to an oral dosage form. Of particular utility are the observations that the poor water solubility of nimesulide is overcome when the drug is dissolved in relatively small amounts (10%) of added ethanol (Tab. 1) and this may be an advantage when preparing mixtures of the drug for tissue culture. There is a pronounced increase in aqueous solubility when the drug is dissolved in glycine-NaOH buffer at pH >7.2 (Tab. 1). Some COX-2 selective inhibitors (meloxicam, celecoxib, rofecoxib) also show similar trends in solvent and solution properties to nimesulide although there are quantitative differences [33].

The liposolubility of nimesulide as determined by its partition coefficient, Log P, in *n*-octanol/water is 1.788 [34].

The crystal structure of nimesulide has been reported by Dupont et al. [35] and the details of this are shown in Table 2. The stereochemical structure (Tab. 2) reveals that the O^5 phenyl moiety is out of plane by about 75° with respect to the nitro-sulphonanilide [34]. The molecular conformation is stabilised by intramolecular NH···O hydrogen bond [35]. The cohesion of the nimesulide crystal is the result of the NH···O and van der Waal's interactions [35].

Acid-base hydrolysis of *N*-amido-methyl-sulphonamides at high temperatures (50°C) has been reported by Iley et al. [36]. The acid-catalysed pathway involves protonation of the amide followed by expulsion of a neutral amide and formation of a sulphonyliminium ion. The base-catalysed hydrolysis by nucleophilic attack of the hydroxide ion at the amide carbonyl carbon atom forms benzamide and sulphonamide by an Elcbrev mechanism involving ionisation of the sulphonamide.

Chemical analysis

Analysis in plasma and other biological fluids as well as in solids of nimesulide and its metabolites can be performed by high performance liquid chromatography (HPLC) using reverse phase columns and UV detection [29, 30, 37] (see also Chapter 2; Bernareggi and Rainsford), and HPLC combined with mass spectrometry [38–41]. The HPLC methods mostly employ either aqueous (with or without buffers such as phosphate) acetonitrile or methanol mixtures. In water based systems there will be two ionised states of nimesulide (with and without protonation of the amino group) present whereas the use of acidic phosphate buffers will control this and enable the non-ionised form to be determined [41].

A comprehensive determination of all the major metabolites of nimesulide present in urine and faeces, including phenolic glucuronides and sulphates, has recently been reported [40]. Determination of nimesulide in solid dosage forms has been undertaken by reverse-phase HPLC using electrochemical detection [41], or by fluorimetry using diazotisation of the drug with *N*-(1-naphthyl) ethylene [42], or by second order derivative UV spectrophotometry [43].

UV spectrophotometric analyses of pure and solid dosage forms have been applied using 50% v/v and 100% acetonitrile as solvents [44]. The limits of detection in these solvent systems were 0.46 µg/ml and 1.04 µg/ml respectively, and high precision and accuracy was claimed for these methods. The advantage of employing acetonitrile as the solvent is that this can be used to extract the drug from various matrices. Also subsequent HPLC can be performed following initial UV spectrophotometry of the samples by directly injecting the acetonitrile extract onto the HPLC column without further purification, or if necessary using reverse-phase mini-columns.

A rapid, sensitive and specific method has been reported by Patravale et al. [45] for the quantitative analysis of nimesulide and degradation products in solid dosage forms using high performance thin layer chromatography (HPTLC). Quantification was achieved using UV scanning densitometry. Using methanolic extraction recovery of nimesulide was found to be 99.5% with the limits of detection and quantitation being 60 and 100 ng respectively. This technique, while requiring some fastidiousness, offers considerable potential for routine laboratory analysis of solid nimesulide.

The extraction of nimesulide, like that of some other NSAIDs, from solid matrix forms may be achieved using supercritical CO_2 fluid extraction [46]. The reported method [46] applied to solubilisation of nimesulide achieved dynamic saturation at pressures between 100–220 bar at temperatures of 312.5 K and 331.5 K. Nimesulide and some other NSAIDs had relatively high solubilities with nimesulide having solubility of $0.85–9.85 \times 10^5$ mole fraction [46]. With automation and method development supercritical fluid extraction could be applied to extraction of the drug from complex biological matrices or fluids (e.g., frozen and crushed brain, bone marrow and bone, urine) where conventional solvent extraction methods may be more difficult.

Electrochemical detection applied to HPLC analysis of drugs, including the NSAIDs, often proves difficult because of the problem of poisoning occurring frequently of the electrode. Catarino et al. [47] developed a technique to overcome this problem by employing a twin channel system with passage of a regenerating solvent over the surface of the electrode. The method was applied to the amperometric determination of nimesulide in pharmaceutical preparations.

Chemical reactions of nimesulide

A key chemical property of nimesulide is its antioxidant potential and this has been investigated using a number of different chemical and biochemical procedures [48–52].

Direct evidence of oxy-radical scavenging activity of nimesulide and its 4-hydroxy-metabolite was shown by Maffei Facino and co-workers using electron spin resonance spectroscopy (ESR) [48]. Using 5,5-dimethyl-1-pyrroline-*N*-oxide (DMPO) in chloroform as a spin trapping agent and ultrasonic irradiation of water (sonolysis) to generate hydroxyl-radicals ($OH^•$) these authors observed that 1–50 µmol/L nimesulide caused a concentration-dependent reduction in the DMPO-OH adduct observed by ESR; at the highest concentration the signal was almost completely inhibited (Fig. 7). 4-Hydroxy-nimesulide was appreciably less active in trapping the $OH^•$ radicals since the concentration required for 50%

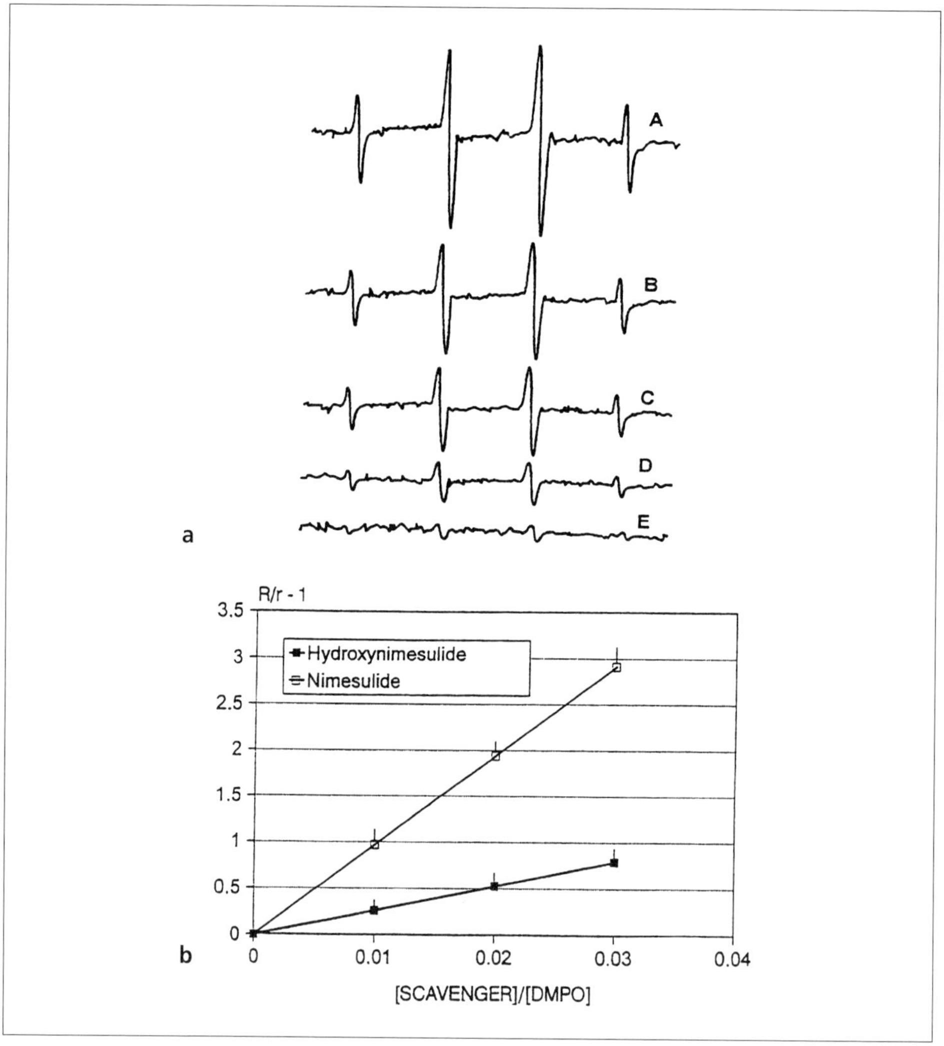

Figure 7
The ESR Spectra (upper panel; Figure 7a) and graph of the kinetic reactions (lower panel; Figure 7b) showing the hydroxyl scavenging activity of nimesulide and 4-hydroxynimesulide. (a) The ESR spectra were of the DMPO-OH spin adduct in the absence (A) and in the presence of increasing concentrations of nimesulide (B = 1 μmol/L; C = 10 μmol/L; D = 50 μmol/L; E = 100 μmol/L). These were recorded after 15 mins of ultrasound radiation. (b) Kinetic reactions of hydroxyl radicals with DMPO and nimesulide or 4-hydroxy-nimeslude. R and r are the initial rates of formation of DMPO-OH in the absence and presence of the two compounds. Data are means ± standard deviation of 5 determinations. From: Maffei Facino et al. [48]; reproduced with permission of the publishers of Arzneimittelforschung.

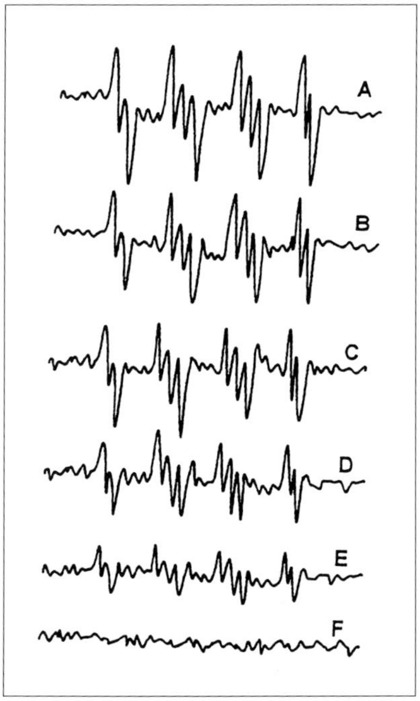

Figure 8
ESR spectra of the $O_2^{\bullet -}$ scavenging effect of 4-hydroxy-nimesulide. A = DMPO-OOH spin adduct (control); B = 1 µmol/L, C = 10 µmol/L, D = 50 µmol/L, E = 100 µmol/L and F = 200 µmol/L of 4-hydroxy-nimesulide.

quenching of the DMPO-OH spin adduct was seven times greater than that observed with nimesulide (Fig. 7). Using the xanthine–xanthine oxidase system for generating superoxide ($O_2^{\bullet}$-) and DMPO as a spin trap to yield the DMPO-OOH radical, Maffei Facino et al. found that 4-hydroxy-nimesulide, but not nimesulide itself, was effective in inhibiting the formation of the DMPO-OOH adduct with an IC_{50} of 40 µ mol/L (Fig. 8). These results are particularly interesting since they show differential effects of nimesulide and its 4-hydroxy-metabolite as oxyradical scavengers. Since the cell or tissue damaging effects of $OH^{\bullet}$ radicals are greater than those of $O_2^{\bullet}$- it would appear that the pharmacologically important total oxyradical scavenging activity might be due to nimesulide rather than its 4-hydroxy-metabolite as based on these chemical reactions. This was supported by studies showing that the oxyradical chain initiation in lipid peroxidation was more inhibited by the nimesulide than by its 4-hydroxy-metabolite [48] (Fig. 9). The system employed the water sonolysis procedure to generate oxyradicals as described above and the lipid peroxidation of phosphatidyl-choline liposomes

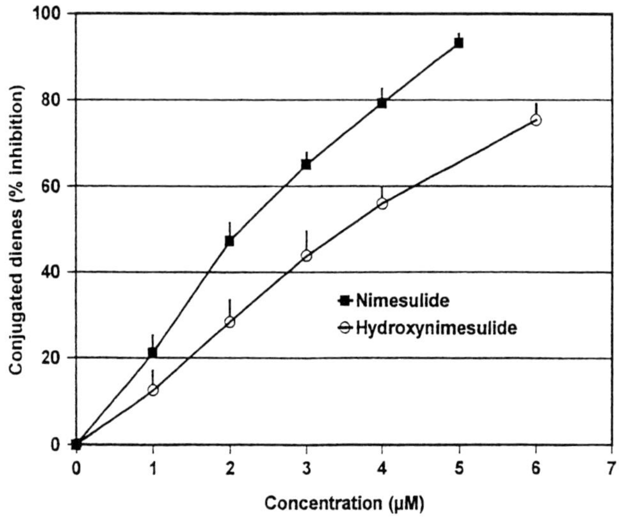

Figure 9
Effects of nimesulide and 4-hydroxy-nimesulide on formation of conjugated dienes. Values are means ± standard deviation of 5 determinations. All values were statistically significant from control ($p < 0.001$). From: Maffei Facino et al. [48]; reproduced with permission of the publishers of Arzneimmittelforschung.

was measured by the simultaneous assay of the oxidation of the conjugate dienes using absorbance and second-derivative UV spectrophometry (at a wave length of 233 nm). In a later study [49] they also observed reduction in the lipid substrate determined by HPLC, and the production of carbonyl breakdown products as 2,4-dinitro-phenyl-hydrazones. Their results showed that in the initiation phase of lipid peroxidation where OH• is generated 4-hydroxynimesulide is less effective as an oxyradical scavenger than nimesulide. However, at the post-initiation phase the decomposition of conjugated dienes (which leads via formation of alkoxyradicals to formation of secondary aldehydes) was potently inhibited by 4-hydroxy-nimesulide added at the beginning of the propagation phase with an IC_{50} of 2.67 μmol/L [48].

The iron-catalysed Fenton reaction (R-COOH + $Fe^{2+} \rightarrow$ RO• + Fe^{3+} + OH•) employed in the lipid peroxidation of phosphatidyl choline liposomes observed by ESR was also found to be inhibited by 4-hydroxynimesulide [48]. Overall these studies have been considered to form the basis of a chain-breaking antioxidant reaction by 4-hydroxy-nimesulide whose mechanism is shown in Figure 10.

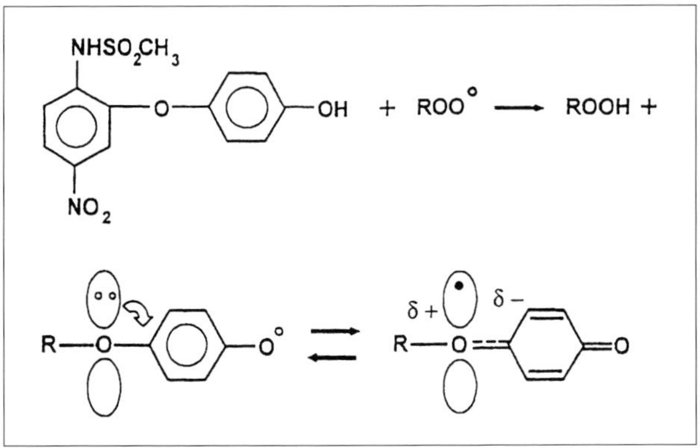

Figure 10
Postulated chain-breaking reactions of 4-hydroxy-nimesulide accounting for the mechanism its anti-oxidant activity. From: Maffei Facino et al. [48]; reproduced with permission of the publishers of Arzneimittelforschung.

In comparison with the other NSAIDs, diclofenac and indomethacin, these drugs also exhibited oxyradical and lipid peroxy-radical scavenging effects [50]. However, the hydroxyl-radical scavenging effects were most potent with nimesulide which had an IC_{50} = 1.85 µmol/L while indomethacin and diclofenac had IC_{50} values of 6.85 and 2.5 µmol/L respectively [50].

An HPLC method using the stable free radical generator, α,α-diphenyl-β-picrylhydrazyl (DPPH) radical in methanol was employed by Karunankar et al. [51] to compare the antioxidant effects of nimesulide with a diverse range of drugs. The change from the deep purple colour of DPPH was monitored by HPLC at wave lengths of 256 and 517 nm [51]. Nimesulide (1.0–5.0 ng/mL), like that of aspirin in the same concentration range, or some other drugs reduced DPPH showing that these drugs have free radical scavenging activity. Unfortunately, the free radical scavenging effects of 4′-hydroxy nimesulide was not studied by these authors.

A number of biochemical methods have also been employed to demonstrate the relative antioxidant activities of nimesulide, and its 4′-hydroxy and N-acetylamino-metabolites using enzymic or cell-based systems [49, 50, 52]. These confirm the selective effects of nimesulide and its metabolites as chain-breaking antioxidants.

Using HPLC and TLC, Kovarikova et al. [53] investigated the photochemical reactions of the sodium salt of nimesulide upon exposure to UV light at 2-phenoxy-4-nitrosanilide and methane sulphonic acid. Thus, monitoring for the presence of these products can be employed for pharmaceutical analysis of nimesulide to detect photodegradation. Photochemical reactions can be of major concern with NSAIDs both from the point of view of pharmaceutical stability but also as a potential for producing skin reactions [54, 55]. While the latter is not a likely consequence with nimesulide because of the relatively low frequency of skin reactions and there being no evidence for skin exposure to UV light, as with some NSAIDs, to skin reactions, there is the possibility that photochemical reactions may be of importance for pharmaceutical stability of the drug.

An electrochemical reaction involving the nitro radical anion produced by nimesulide has been investigated [56]. This property of nimesulide is a further aspect of novel chemistry of this NSAID. Hypochlorous acid (HOCl) is a product of neutrophil activation, which may have anti-infective effects and at high levels may lead to initiation or contribution to inflammatory reactions. Part of the anti-inflammatory effects of NSAIDs like that of nimesulide may be due to their actions on production of this oxygen radical species (Chapter 4; Rainsford et al.). Using a combined HPLC separation procedure, which fluorometric detection, Van Antwerpen et al. [57] used p-amino-benzoic acid (PABA) oxidation induced by HOCl to assay the effects of nimesulide and some other NSAIDs on this system. The PABA chlorination was inhibited by meloxicam and some other oxicams, the effects of which were more potent than that of nimesulide. The rate constants for effects of nimesulide were $2.3 \pm 0.6 \times 10^2$ contrasted with that of meloxicam $1.7 \pm 0.3 \times 10^4$ and other oxicams which had values of around 10^3.

In conclusion, nimesulide has some unique chemical properties in some respect related to the presence of the nitro-group and the phenolic group of the 4-hydroxyl-metabolite which underlies its antioxidant activity. In other respects the pKa and liposolubility separate nimesulide from other NSAIDs.

Versatile formulations

Nimesulide has been formulated into a wide range of pharmaceutical forms. However those registered and available in most of the countries worldwide are tablets, granules for oral suspension and suppositories. The pharmacokinetic and pharmaceutical properties of some of these are discussed in Chapters 2 and 3. Here, some aspects of the chemistry of these are discussed. Of particular interest are the attempts to develop formulations of nimesulide with the aim to enhance its absorption and minimise the contact of crystals or particles with the gastric mucosa and so reduce the gastrointestinal irritancy of the drug (e.g., cyclodextrin inclusion formulations), those developed to enable transcutaneous delivery so that the drug may be applied to the skin, and parenteral formulations to enable the drug to be given by intramuscular or intravenous injection (though the latter is not particularly favoured at present).

A considerable number and type of cyclodextrin (CD) inclusion formulations or complexes with NSAIDs have been developed [58–65]. Of these, relatively few have been used clinically with any therapeutic success. Among these developments that have been investigated clinically are the oral CD formulations of piroxicam [60–63], an ophthalmic CD formulation of diclofenac [64], and as well some oral CD formulations of nimesulide [65–68]. While the clinical benefits involving possible fast onset of action and/or better gastrointestinal (GI) tolerance have yet to be studied in detail with these formulations, especially with a drug that intrinsically has fast onset of analgesia and low GI adverse reactions (Chapters 3 and 5), nonetheless the development of these CD formulations is of interest chemically and pharmaceutically.

Several CD formulations of nimesulide have been prepared and reported in the patent literature [69–71] as well as in journal articles [72–79]. The physicochemical properties of a number of cyclodextrin (CD) inclusion formulations of nimesulide have been described [72–79].

Among these studies with various CD formulations was one reported by Vavia and Adhage [72, 73] who described the use of a standard freeze-drying method with complexation being determined by differential scanning calorimetry (DSC), Fourier transform infrared (FTIR) spectroscopy and x-ray diffractometry (XRD). The dissolution rate of the hydroxypropyl CD-drug complex was faster than that of the β-CD-drug complex or nimesulide alone [72]. Using DSC and XRD, these authors established that ball-milling of the freeze-dried β-CD-nimesulide in the ratio of 4:1 by weight produced superior solubilisation than other ratios down to 1:1. Greater absorption and bioavailability was observed by the 4:1 inclusion complex.

Chowdary and Nalluri [74] prepared solid inclusion complexes by needing of nimesulide and β-CD in molar ratios of 1:2 respectively and observed higher dissolution rates with this preparation compared with those made by co-evaporation. A subsequent study by this group confirmed these results and showed the formation of the 1:2 molar complex by DSC, XRD, ^{1}H-nuclear magnetic resonance spectroscopy, mass spectrometry and scanning electron microscopy. Braga et al. [79] employed the sodium salt of nimesulide in preparing the crystalline β-CD inclusion complex with this drug from co-precipitation in aqueous media. The use of sodium hydroxide or salts of this or other alkalis of nimesulide with various CDs has been reported in the patent literature [71] or the drug is solubilised by addition of an organic solvent [70]. The use of organic solvents may not always be acceptable pharmaceutically [71] so the preparation of sodium or other salts may be more effective especially in view of the improvement in the solubility of nimesulide (Tab. 1).

Higher rates of association and dissolution of nimesulide have been reported with hydroxypropyl-CD than with β-CD, reflecting the significant hydrophobic effect between the drug and flexible hydroxypropyl moieties [72].

Currently one formulation of nimesulide with cyclodextrins is commercialised. Aspects concerning the physicochemical properties of nimesulide in solvent systems (see Tab. 1) are important for exploiting development of novel formulations of a sparingly soluble drug such as nimesulide especially those for injectable use [80–83]. The partitioning kinetics of nimesulide has been investigated using an aqueous buffer/*n*-octanol systems [84]. The partitioning kinetics appears to be directly related to the aqueous solubility. In phospholipid liposomes nimesulide, like many other NSAIDs, binds to the lipid bilayer by hydrophobic interactions [85]. These features will be important in relation to penetration by the drug through cellular membranes (see also Chapter 4; K. D. Rainsford et al.).

The use of the sodium salt of nimesulide (usually prepared by solubilising the drug in acetone and sodium carbonate) was exploited in preparation of CD inclusion compounds [71]. Solubilisation with fatty acids has also been exploited for preparing CD complexes [71] and this could be a useful means for preparing the drug for biological assays. Sodium salts of nimesulide could be prepared for formulating the drug for parenteral administration. However, there are limits to solubility of the drug with sodium salts for parenteral use even though this and other alkali metal ions have been shown to be useful for preparing micronised oral formulations with improved bioavailability and pharmacokinetics [82]. Solubilisation of nimesulide using sodium salts of bicarbonate, saccharinate and benzoate together with sodium hydroxide and ethanol have been formulated to be used as a mouthwash or tincture [83].

An injectable formulation of nimesulide, principally for intramuscular use, has been described in a series of patents by Jain and Singh [80]. The formulation comprises dimethyl acetamide, benzyl benzoate, benzyl alcohol and ethyl oleate in quantities ranging from 5–65% each. This is a substantial quantity of solvent excipients and raises issues about the bulk and irritant or other toxic activities of such complex formulations. It is, however, claimed that in preclinical toxicology studies the formulations have a favourable therapeutic index.

Water soluble formulations of nimesulide have been prepared using the lysine salt for injection [81]. Solubilised forms of nimesulide for oral use have been developed including effervescent carbonate preparations [86], those with various surfactants (e.g., Tween 80, Cremophore EL) with or without CDs [82]. Hydroalcoholic formulations of nimesulide have been prepared using various alcohols (e.g., ethanol, glycerol) and buffered to pH 8 with sodium salts of bicarbonate, saccharinate, benzoate and sodium hydroxide [87]. These have been claimed to be useful for mouthwashes for the treatment of inflammation of the rhinopharyngeal or oral (presumably buccal) mucosae [87]. At such high pH values (around 8.0) these solutions may be irritants. It might be possible to employ these preparations for the treatment of periodontal disease though this has not been demonstrated yet.

The use of alkaline salts, especially sodium salts, of nimesulide for the preparation of micronised formulations has been described [88]. The combined properties of salifaction and micronising give these formulations particular advantages to enable rapid absorption and low mucosal irritancy.

Some oral formulations of nimesulide have been developed to control the release of the drug [89–97], enhance its gastric absorption [98–101] or reduce the possibility of causing gastric irritancy [102]. Of these polylactate microparticles have been shown to affect the crystalline state of nimesulide [97]. An interesting system has been developed to control release of nimesulide by a 'multiple unit' system with pellets of polysorbate, cellulose, sodium carmellose, maltodextrin, pregelatinised starch with lactose and with an inner coating of magnesium stearate, talc and Eudragit and an outer coating of Methocel, talc and water to enable both immediate and extended release of the drug [89]. This is obviously a very complex collection of excipients and it will be interesting to see if this proves to be a viable and cost-effective means of having extended release of the drug.

Recrystallising nimesulide and solubilising in Tween 80 with polyvinyl pyrrolide has been found to increase the analgesic activity of the drug [103]. The use of hydrotopes (e.g., sodium salts of ascorbate, benzoate or salicylate, or piperazine or nicotinamide) has been found to improve the solubilisation of nimesulide [104, 105]. Of these the piperine hydrotope was found most suitable for use as an injectable formulation [104]. Piperine is obtained from the black pepper, *Piper nigrum*, long pepper, *P. longum* or related species and can be prepared synthetically [104]. Improved analgesic activity and pharmacokinetics was demonstrated in piperine-containing preparations of nimesulide in rodent models [104].

Liposome delivery systems (as noted earlier) have been developed for investigating lipid–drug interactions. A formulation of 800 mg cholesterol, 800 mg hydrogenated lecithin and 1.25 g nimesulide by weight was prepared into liposomes that were then freeze-dried and 114 mg added to hard gelatine capsules which were then coated with Eudragit [106]. The blood levels of nimesulide were found of peak at 5 h showing that this formulation was effectively delaying the release of the drug. Liposome delivery or formulations containing liposomes are attractive for enabling sustained release and reducing the propensity for gastric irritation and associated dyspeptic symptoms.

Much attention has been devoted to the development of topical or transdermal preparations of nimesulide for percutaneous delivery of the drug. Among the various formulations that have been developed there are essentially two groups – those where solubilising agents and skin penetrants have been incorporated into the formulation [107–117], or gel formulations [118–126]. Of these the gel formulations have probably proven the most successful [118–121]. Indeed, the 3.0% gel formulation of nimesulide is successfully marketed to date in nine countries and has found wide acceptance for the relief of pain in acute musculoskeletal conditions [121]. A number of the formulations featuring solubilising agents/skin penetrants have focussed on claims concerning potential to develop yellow skin colouration. Most gel formulations do not present with this problem. The Helsinn formulations have carboxyvinylpolymer as a gel-forming agent and/or a solvent comprising either ethanol, isopropanol or polyacrylamide – isoparaffin and diethylene glycol monoethyl ether. A particular feature of these gels is their stability, its non-alcoholic base, and this contrasts with some of the other topical preparations employing organic solvents and skin penetrants [120]. Its low systemic bioavailability means that it has a low risk for gastrointestinal or other major organ system toxicities, while at the same time the gel formulation has been found to have good anti-inflammatory activities in animal models and humans [120].

Description of the pharmacokinetic and pharmaceutical properties of the gel formulations will be found in Chapters 2 and 3 respectively, while the relevant clinical efficacy in control of musculoskeletal pain is discussed in Chapter 5.

Recently, novel unilammellar or multilammellar lipid film systems, known as niosomes, have been employed to prepare encapsulated formulations of nimesulide [127, 128]. The theory behind the development of niosomally entrapped drugs is that they have improved interactions with the dermal layers of skin both by reducing trans-epidermal water loss and increasing smoothness through replenishment of skin lipids [127]. In one niosomal system non-ionic surfactants (e.g., Tween 80 or Span 20) and cholesterol, dissolved in chloroform/methanol, solvents evaporated and then prepared as Carbopol 934 gels using an aqueous polypropylene glycol–glycerine system to which the drug was added [127]. The penetration of this through human cadaver skin and effects of topically applied preparations in the rat carrageenan paw oedema assay were compared with some gel formulations of nimesulide [127]. Aside from showing good skin penetration through human skin in vitro, the niosomal-nimesulide was found to have about 3–4 times the anti-inflammatory activity compared with plain drug in gel or a marketed gel formulation (Panacea) [127]. These results suggest it may be worth investigating the potential of gel formulations incorporating niosomes to enhance absorption of nimesulide through the skin. Some simplification of the preparation of niosomes may be advantageous from the point of view of production of the gel formulations.

Novel 'non-pain' uses of nimesulide

The uses of nimesulide in controlling pain, inflammation and fever are well known and are discussed in Chapter 5; and their adverse effects are discussed in Chapter 6 of this book. Here, some potentially novel applications of the drug are reviewed in preventing cancers, Alzheimer's disease, neurodegenerative and related dementias, immunodeficiency disorders, cataract formation and in some gynaecological conditions.

Nimesulide in cancer

Since the findings by Bennett et al. in 1975–1977 [129–131] that cancerous tissues of the human colon have markedly increased output of PGE_2, there has been much interest in the possibility that NSAIDs may reduce the growth and proliferation of colorectal and other cancers [132]. The role of PGE_2 and COX-2 upregulation in the proliferation of cancers and the case for using selective COX-2 inhibitors and conventional NSAIDs in prevention of cancers of the gastrointestinal tract, breast, prostate have been investigated. NSAIDs have been shown to have protective or inhibitory effects against experimentally-induced cancers in rodents [133–135]. There is also epidemiological, case-controlled and cohort studies in various populations showing that the risk of developing colorectal cancer can be reduced by about one-half following long-term intake of aspirin and other NSAIDs [132].

The case has been made for targeting COX-2 as a means of controlling the proliferation of cancer cells and angiogenesis stimulated by these cells [133–141]. However, there are some notable exceptions to this concept among them that some NSAIDs that are not COX-2 inhibitors have weak effects on PG production, e.g., sulindac, sulindac sulphoxide, R-flurbiprofen [141–145]. Moreover, recent studies suggest that the signal transduction pathway involving NFκB–IκB regulation in both the main target for the actions of NSAIDs [133–135] controlling proliferation of cancer cells and apoptosis, although COX-2-prostaglandin production may have ancillary effects. Inhibition of NFκB signalling will lead to reduction in the synthesis of COX-2 and other proteins including metalloproteinases that are responsible for aiding and abetting tumour growth and proliferation. Recent studies [146] suggest that inhibition of the NFκB pathway enhances TNFα-related apoptosis-inducing-ligand (TRAIL) to induce death in tumour cells. This may, in part, explain the apoptosis of tumour cells observed with several NSAIDs [133–135]. Moreover, the overexpression of the promotor driving the expression of the death receptor 4 (DR4) [147] may also drive apoptosis in cells that are TRAIL-resistant or expression of a key enzyme determining one of the apoptosis-inducing pathways, caspase-3.

Recent studies also suggest that 15-lipoxygenase may represent an additional target for NSAIDs [148–151]. The apoptosis induced by NSAIDs in melanoma cells is also shown to be independent of direct effect on COX-2 [151] providing further support for the view that NSAIDs act in controlling growth and apoptosis of cancer cells by indirectly affecting the regulation of the production of this enzyme protein as well as those involving metalloproteinases and components of apoptosis pathways [146–152].

Against the background of these studies on the comparative effects of the NSAIDs on tumour growth, proliferation and apoptosis there is impressive literature showing that nimesulide has multiple sites of action on these components of tumourogenesis. No studies appear to have been undertaken to examine the effects of this drug on the prevention of tumour growth and proliferation in human cancers although other NSAIDs have been found effective in preventing colon, breast and possibly prostate cancers [153–156].

Of the *in vivo* studies undertaken in rodent models of carcinogenesis and tumour growth and proliferation, nimesulide has been found to inhibit rat bladder carcinogenesis induced by *N*-butyl-*N*-(4-hydroxybutyl)nitrosamine [157], coincident with increased COX-2 expression [158], mouse colon carcinogenesis induced by azoxymethane [159], rat mammary carcinogenesis induced by 2-amino-1-methyl-6-phenyl-imidazo[4,5-*b*]pyridine [160], rat chemical induced tongue carcinogenesis coincident with increased expression of COX-2 and iNOS [161], mice infected with *Helicobacter pylori* and exposed to the chemical carcinogen, MNU, to produce gastric adenocarcinomas [162], the N-nitroso-bis(2-oxopropyl)amine induced pancreatic cancer induced in hamsters [163], mouse hepatomas coincidently treated with 5-fluorouracil [164, 165], 4-nitroquinoline 1-oxide induced dysplasia and carcinomas of the tongue in rats [166], and intestinal polyposis induced in *Apc* gene deficient mice [167] and in *Min*-mice [168]. One study suggests the drug has no effect on polyposis in *Apc* mice [169].

In vitro studies have shown that nimesulide has multiple modes of action in controlling growth, proliferation and apoptosis of cancer cell lines and tumours in addition to inhibiting COX-2 regulated prostaglandin production [170–185]. Among the targets for effects of nimesulide on gene regulation and intracellular signalling are the pro-apoptotic gene, *Par-4* [174], the *Bax*- regulated apoptosis [175, 179, 186], cell cycle arrest [187], VEGF cell surface receptor expression [188], expression of *c-Jun* [181], and suppression of telomerase activity via blockade of *Aκ/PκB* activation [185]. Further aspects of the molecular actions of nimesulide in relation to cell growth and differentiation are discussed in Chapter 4.

Of particular interest for cancer therapy are recent *in vitro* studies suggesting that nimesulide may act synergistically to increase the cytotoxicity of doxorubicin in the human lung adenocarcinoma cell line, A549 coincident with caspase-3 induction and apoptosis; the effects of the combination of the two drugs being greater than that individually [189]. Using the same cell line subcutaneously implanted in nude mice it has been found that nimesulide acted synergistically with cisplatin to inhibit tumour growth and *in vitro* the combination was found also to cause additive or synergistic effects, depending on the drug concentrations, of apoptosis [190].

The administration of nimesulide, as well as some other COX-2 inhibitors prior to photodynamic therapy of implanted C-26 cells in mice resulted in marked potentiation of the anti-tumour effects of the latter treatment [191]. Similar inhibitory effects of combined photodynamic therapy and nimesulide on the inhibition of tumour growth were found in a wide variety of oral and skin tumour explants showing high COX-2 expression [192]. Moreover, in two oral squamous cell carcinoma cell lines either expressing COX-2 (HSC-2) or not expressing this enzyme (HSC-4), only nimesulide-inhibited growth in the HSC-2 cells in combination with the photodynamic therapy [192]. These effects of nimesulide may only be a reflection of the genetic system or their controls (e.g., NFκB, *Cjun*) not being present in the cell line (HSC-4) that expresses COX-2, and not the expression of COX-2 *per se*.

Alzheimer's disease and neurodegenerative disorders

Alzheimer's disease (AD) has all the hallmarks for being a chronic inflammatory condition that is probably initiated by pathogenic β-amyloid deposition in plaques in certain regions of the central nervous system [193–195]. The activation of microglial cells by amyloid leads to local production of proinflammatory cytokines, oxyradicals, eicosanoids (principally COX-2 derived prostanoids) with infiltration and activation of lymphocytes and expression of cell surface receptors involved in ligand interactions with inflammatory cells or molecules [193]. COX-2 activation occurs by cytokines (e.g., IL-1β, TNFα and IL-6) and so represents a target for the actions of NSAIDs [193, 195]. Likewise, oxyradicals and production of IL-1β, TNFα and IL-6 as well as the NFκB signalling pathway are potential targets for the effects of those NSAIDs that affect their production [196] (see also Chapter 4; Rainsford et al.).

Early epidemiological studies, especially in arthritic patients taking NSAIDs long-term, suggested that there may be improvements in cognitive function or preventative effects of these drugs on the symptoms of AD [195–199]. More convincing data came from the longitudinal study in 1,686 patients by Stewart et al. [200] who showed that the risk of developing AD was reduced by 60% following use of NSAIDs for two or more years; that by aspirin users over the same period was associated with risk reduction of 36%, while there was no significant benefits from use of paracetamol over the same period. Another epidemiological study involving 1,648 patients showed that concurrent use of anti-inflammatory agents (and oestrogens in women) was associated in improvements in mental functions and cognition [201, 202]. A smaller scale clinical trial in 41 patients with mild-to-moderate AD treated for 25 weeks with a combination of diclofenac and misoprostol (as a gastroprotective agent) did not show any benefits over placebo [203]. However, an open label study in 73 patients with vascular dementia showed that treatment with the salicylate platelet aggregation inhibitor, triflusal, for 12 months did result in improved cognitive functions compared with control [204]. While vascular dementia and AD may be dissimilar in pathology it is interesting that this and other studies have shown benefits of NSAIDs in vascular dementia.

A randomised 'pilot' parallel group study of nimesulide 100 mg twice daily for 12 weeks in 40 AD patients with mild–moderate disease who were taking cholinesterase inhibitors showed little if any benefits of this NSAID on cognitive scores, clinical status or activities of daily living and behaviour compared with controls [205].

Unfortunately, the trials with nimesulide and the diclofenac/misoprostol combination were in small number of patients [203, 205] who were treated for relatively short periods of time (being 12 weeks [205] or 25 [203] weeks, respectively) so this is hardly a basis for giving definitive answers to the question of whether or not individual NSAIDs have benefits in AD.

More recent prospective studies in larger groups of AD patients with mild–moderate disease failed to show any benefits of 12 months treatment with the COX-2 selective inhibitor rofecoxib, 25 mg once daily, or naproxen sodium, 220 mg twice daily, compared with placebo [206, 207]. These latter trials show that selective COX-2 inhibition from rofecoxib treatment is unlikely to confer any benefits in AD patients. The results with naproxen sodium may reflect on the low dose of this drug or other features.

The studies with rofecoxib and naproxen [206, 207] were at least in trials that were probably adequately powered and possibly of sufficient duration (1 yr) to permit determination of trends for therapeutic benefit. However, it should be noted that the epidemiological study by Stewart and co-workers [200] that did show risk reduction by NSAIDs in AD extended for two or more years of use of these drugs. It is, therefore, possible that longer-term treatments may be required in any prospective, controlled trials. This may present a problem for the ethics of a study involving a placebo treatment arm to the study.

The results of the rofecoxib studies [206, 207] may, however, prove instructive. Perhaps selective COX-2 blockade is not alone sufficient for controlling the progression of a complex chronic inflammatory condition with such severe and serious irreversible neurodegenerative changes as in AD. Thus, application of nimesulide (or even other NSAIDs) that have multiple modes of action on eicosanoid metabolism the production of oxyradicals and proinflammatory cytokines (e.g., TNFα, IL-6), intracellular signalling and cell surface expression on leucocytes and endothelial cells might be expected to have greater potential protective or therapeutic benefits in AD than observed with a selective COX-2 inhibitor.

The studies in experimental models in rodents and *in vitro* in inflammatory cellular systems would appear to give some support for nimesulide being of potential use in prevention or treatment of AD. Of the non-prostaglandin mechanisms that may be involved in the actions of nimesulide in the pathogenesis of AD, the studies of Avramovich et al. [208] are of particular interest in showing that nimesulide (like ibuprofen, indomethacin and thalidomide) can stimulate the neural cell secretion of the non-amyloidogenic α-secretase form of the soluble amyloid precursor protein (sAPPα). These authors used the rat phaechromocytoma PC12 and human SH-SY5Y neuroblastoma cells and they found that nimesulide 0.1–1.0 µmol/l stimulated secretion of sAPPα into the culture media [208]. Ibuprofen 0.1–1.0 µmol/l, thalidomide, or its non-teratogenic analogue supidimide, and higher concentrations of indomethacin also stimulated release of sAPPα from these cells. Inhibitors of protein kinase C (PKC) and mitogen-activated protein kinase (MAPK) pathways partially blocked the stimulatory effect of nimesulide 1 µmol/l suggesting that PKC/MAPK signalling pathways are involved in the nimesulide-induced stimulation of sAPPα secretion [208]. The effects of nimesulide appear to be mediated by a metallo-proteinase that is also sensitive to the hydroxamate, Ro-31-9770, which inhibits this enzyme [208].

In other models of neuronal injury somewhat variable results have been obtained with nimesulide. Thus, in a model of closed head injury in rats nimesulide 30 mg/kg i.p. decreased cortical and hippocampal PGE_2 but, like other NSAIDs did not improve cerebral oedema or Neurological Severity Scores [209]. In many respects this is a severe model of brain injury and so it is not surprising that NSAIDs had no therapeutic benefits even though PGE_2 concentrations in the brain were reduced by these drugs [209]. In contrast, nimesulide 6 mg/kg i.p. at 30 min after injury and thereafter for 10 days improved cognitive deficit (in the Barnes circular maze) and motor dysfunction in rats exposed to 2 m impact acceleration model of diffuse traumatic injury [210, 211]. This model of brain trauma, while having marked effects on brain functions, is probably not as severe as the closed head injury model [209].

In a model of epilepsy induced in mice by administration of haloperidol, nimesulide and naproxen, but not the COX-2 specific drug rofecoxib, reduced the catalepsy score [212].

The link between LPS, proinflammatory cytokines (e.g., TNFα) or neurokinin and the expression of COX-2 and iNOS and products of these enzymes in neural cells has been shown in a variety of cellular systems to be inhibited by nimesulide [213–216]. In transgenic mice that have over expression of neuronal COX-2 there is induction of complement component, C1qB [217]. Since complement is deposited in AD brain cells this could represent a component of the inflammatory response in AD. Nimesulide has been found to reduce the mRNA coding for C1qB implying that the COX-2 inhibition by nimesulide may protect against inflammatory changes involving complement deposition that is regulated or influenced by COX-2.

Nimesulide has been found to protect against the decrease in the expression of the mRNA coding for a key cortical protein, p18, that leads to COX-2 over expression which also leads to acceleration of glutamate-mediated apoptosis coincident with pRb phosphorylation [218].

In models of cognitive dysfunction in mice, employing scopolamine or lipopolysaccharide treatments or aged animals, nimesulide, rofecoxib and naproxen given repeatedly each day for 7 days significantly reversed the cognitive retention deficits [219]. In conditioned place preference tests in rats, nimesulide at the low doses of 0.1 to 1.0 mg/kg induced place preference [220] inferring that some influence on reward or other behavioural influences may be affected by the drug.

Kainic acid-induced seizures in rats lead to enhanced expression of COX-2 in the hippocampus and cortex, which are reduced by therapeutic doses of nimesulide after application of the neurotoxin but not by prior treatment [221, 222].

Overall these studies suggests that nimesulide may have generalised neuroprotective effects as a consequence of inhibition of COX-2, NO and oxyradicals. There may also be influences of nimesulide on neural excitability or plasticity [223] although the exact basis of this is unclear.

A number of patents (preliminary or granted) exist claiming benefits from the use or application of nimesulide in preventing AD, cognitive impairment, Parkinson's disease, amyotrophic lateral sclerosis or amyloid- or generalised neurodegenerative disorders [224–230].

Miscellaneous uses

There have been a number of studies reported in which nimesulide has been used to induce uterine relaxation (as a tocolytic agent), managing labour, inducing closure of the patent ductus arteriosus and some other states. While the drug is probably relatively safe to use in these indications it is worth cautioning that to the author's knowledge no formal safety investigations both preclinical and clinical have been performed to form a sound basis for evaluating the clinical toxicity of the drug. The application of nimesulide in these gynaecological and obstetric conditions must be regarded as 'off label' and experimental.

The pharmacological basis for employing nimesulide, and other NSAIDs, for inducing uterine relaxation is based on their effects *in vitro* in relaxing myometrial contractivity. Recent studies for instance in the micromolar range (1–100 μmol/l) shows that nimesulide, celecoxib and meloxicam all produce myometrial relaxation in pregnant (before and after labour) and non-pregnant human myometrial tissues [231]. Glucocorticoid-induced premature labour in sheep has been found to be prolonged by nimesulide 20 mg/kg/d coincident with reduction in the maternal and foetal plasma levels of 13, 14-dihydro-15-ketoprostaglandin $F_{2\alpha}$ and prostaglandin E_2 and reduced uterine myometrial activity [232, 233]. The effects of nimesulide were more pronounced when the drug was given in combination with the oxytocin antagonist, atosiban [232, 233].

In a small study in five women in pre-term labour who were resistant to i.v. ritodrine 8 days treatment with nimesulide 100 mg b.i.d. resulted in prolongation of pregnancy for a mean of 27 days (range 6–69 days) [234]. Oligohydramnios occurred in all foetuses after 3–9 days therapy but resolved upon discontinuation of the drug in most patients. This condition has been reported in other patients [235] and must, therefore, be a cause for concern in applying nimesulide for premature labour, although the risks in these patients must be balanced against the benefits of therapy and the better safety profile of nimesulide compared with use of indomethacin, the drug most frequently employed for this condition.

In a randomized double-blind study in 30 pre-term patients who were of 28–32 weeks gestation, the physiological and tocolytic effects of nimesulide 200 mg b.i.d. were compared with indomethacin 100 mg b.i.d. and sulindac 200 mg b.i.d. initially for 48 h, then followed-up for 72 h thereafter [236]. All the drug treatments reduced foetal urine output, amniotic fluid index and ductal blood flow over the 48 h treatment period, which then returned to normal in the following 72 h. The authors concluded that nimesulide causes similar short-term foetal effects to the other two drugs.

It is of interest that a patent has been claimed for use of nimesulide, preferably in conjunction with progestins, for substantially preventing or reducing at least one of the changes associated in the female reproductive system associated with the onset or continuation of labour [237].

The question whether nimesulide should be employed as a uterine contractile agent requires further toxicological evaluation in order to determine its relative safety compared with indomethacin.

Patents have been granted claiming the use of COX-2 inhibitors, including nimesulide, for overcoming the immunodeficiency of agents such as the HIV infection [238]. The rationale for this treatment is that COX-2 activity (which is increased *inter alia* in lymph nodes and associated T cells) leads to increased PGE_2, which in turn increases the levels of cAMP leading to protein kinase A signalling and impaired lymphocyte functions. In mice with the mouse equivalent of AIDS it was found that T cells were impaired and that administration of COX-2 inhibitors overcame the immune deficiency in lymph node cells. Studies with HIV patients' $CD3^+$ T cells showed they also responded to treatment with COX-2 inhibitors to overcome the COX-2 derived increase in PGE_2 and consequent immune deficiency. The effects of the COX-2 inhibitors were superior to treatment with indomethacin.

NSAIDs have been used for the treatment of Bartter's syndrome, an inherited condition that results in excess renal induction of PGE_2 coincident with renal salt loss, hypercalcuria, nephro-calcinosis and secondary hyperaldosteronism [239]. With identification of increased expression of COX-2 in the macula densa leading to hyperreninalnia in these patients, trials of COX-2 inhibitors, including nimesulide have shown benefit in restoring renin-aldosterone and other renal functions in Bartter's syndrome patients [240, 241].

A patent claiming benefits of nimesulide as an 'anti-cataract agent' has been reported [242]. The evidence was based on inhibition by nimesulide of depolymerisation of hyaluronic acid and the development of opacity of rat lens incubated *in vitro* for 4 days in the presence of glucose and foetal calf serum, the treatment of which leads to lens protein denaturation [242]. No *in vivo* evidence appears to have been reported to support these claims, although other NSAIDs have been found to suppress formations of cataract and reduce inflammation during and following cataract surgery [243–248]. The actions of nimesulide in preventing these conditions may be related to its antioxidant as well as COX-2 inhibitory effects.

Conclusions

Nimesulide has a variety of potentially novel, non-pain, effects some of which may be related to its known pharmacological actions relating to its anti-inflammatory effects. The effects of the drug on intracellular signalling pathways that regulate cell growth and other cellular controls may represent some unique sites of action of the drug.

References

1. Otterness IG (1995) The discovery of drugs to treat arthritis: A historical view. In: VJ Merluzzi, J Adams (eds): *The search for anti-inflammatory Drugs*. Birkhäuser, Boston, 1–26
2. Rainsford KD (2004) History and development of the salicylates. In: K D Rainsford (ed): *Aspirin and related drugs*. Taylor and Francis, London, 1–23
3. McTavish JR (2004) The industrial history of analgesics: the evolution of analgesics and antipyretics. In: K D Rainsford (ed): *Aspirin and related drugs*. Taylor and Francis, London, 25–43
4. Prescott LF (2001) *Paracetamol (Acetaminophen). A Critical Bibliographic Review*. Taylor and Francis, London
5. Rainsford KD (1999) History and development of ibuprofen. In: KD Rainsford (ed): *Ibuprofen. A critical bibliographic review*. Taylor and Francis, London, 1–24
6. Vane JR (1971) Inhibition of prostaglandin synthesis as a mechanism of action for aspirin-like drugs. *Nature* 231: 232–235
7. Müller W, Harwerth H-G, Fehr K (eds) (1971) *Rheumatoid Arthritis. Pathogenic Mechanisms and Consequences in Therapeutics*. Academic Press, London & New York
8. Smith MJH, Smith PK (eds) (1966) *The salicylates. A critical bibliographic review*. Wiley Interscience, New York, London & Sydney
9. Whitehouse MW (1968). The molecular pharmacology of anti-inflammatory drugs: some possible mechanisms of action at the biochemical level. *Biochem Pharmacol* Suppl: 293–307
10. Giroud JP, Willoughby DA, Velo, GP (eds) (1975) *Future trends in inflammation II*. Bikhäuser Verlag, Basel & Stuttgart
11. Weissman G (1992) Inflammation. Historical perspective. In: Gallin J, Goldstein IM, Snyderman R (eds): *Inflammation: Basic principles and clinical correlates. 2nd Edition*. Ravel Press, New York, 5–9

12. Moore, GGI (1974) Sulfonamides with anti-inflammatory activity. In: Scherrer, RA, Whitehouse MW (eds): *Antiinflammatory agents. Chemistry and pharmacology, Volume 1.* Academic Press, New York, San Francisco & London, 160–177
13. Rainsford KD (2004) Occurrence, properties and synthetic developments of the salicylates. In: Rainsford KD (ed): *Aspirin and related drugs.* Taylor and Francis, London, 45–96
14. Moore GGI, Harrington JK (1974) Substituted 3-phenoxy alkane sulphonanilides. US Patent No. 3,840,597, October 8, 1974
15. Vigdahl RL, Tukey RH (1977) Mechanism of action of novel anti-inflammatory drugs diflumidone and R-805. *Biochem Pharmacol* 26: 307–311
16. Swingle KF, Hamilton RR, Harrington JK, Kvam DC (1971) 3-Benzoyldifluoromethanesulfonanilide, sodium salt (diflumidone sodium, MBR 416408): a new anti-inflammatory agent. *Arch Int Pharmacodyn Ther* 189: 129–144
17. Swingle KF, Moore GGI, Grant TJ (1976) 4-Nitro-2-phenoxymethanesulfonanilide (R-805): a chemically novel anti-inflammatory agent. *Arch Int Pharmacodyn Ther* 221: 132–139
18. Rufer C, Schillinger E, Böttcher I, Repenthin W, Herrmann CH (1982) Non-steroidal anti-inflammatories-XII. Mode of action of anti-inflammatory methane sulfonanilides. *Biochem Pharmacol* 31: 3591–3596
19. Kuehl FA, Humes JL, Egan RW, Ham EA, Beveridge GC, van Arman CG (1977). Role of prostaglandin endoperoxide PGG_2 in inflammatory processes. *Nature* 265: 170–173
20. Riker Lab Inc (1975). *p*-Nitration of alkane sulfonanilides, by careful treatment with nitration agent in at least equimolar amt. WPI Acc No. 1975-01873W/197501
21. Cignarella G, Vianello P, Berti F, Rossoni G (1996) Synthesis and pharmacological evaluation of derivatives structurally related to nimesulide. *Eur J Med Chem* 31: 359–364
22. Kensai Y, Saito M, Masami G, (with Taisho Pharmaceutical Co Ltd) (1997) Japanese PCT Application No. WO 9746520
23. Yoshinari Y, Yutaka O, Kazuto S, Hideji S, Katsuo H (1998) Preparation of sulfonanilide derivatives as anti-inflammatory agents. Japanese Kokai Tokkyo Koho – Japanese Patent JP 02022260
24. Fernandes AC daS, Borges JER, Pereira MFBM da S, Romao CJC, Correia LMB, Correia PB (1999) Method for the preparation of aryl ethers using silver salt as catalyst under ultrasound irradiation. Portugese Patent Application No. PT 1999-102315
25. Bangyin C, Hanping Z, Lihong Z, Xinfen Z, Zhenguo Z, Shugang G, Hong L (2000) Synthesis and structural identification of nonsteroidal anti-inflammatory drug nimesulide. *Tongji Yike Daxue Xuebao* [Acta Univ Med Tongji] 29: 67–68
26. Attal V, Belsare DP (2003) Synthesis and anti-inflammatory activity of arylsulfonanilides structurally related to nimesulide. *Ind J Pharmaceut Sci* 65: 135–138
27. Personal Communication from Helsinn Healthcare SA
28. www.nimesulide.net (accessed 14 September 2004)
29. Singla AK, Chawla M, Singh A (2000) Review. Nimesulide: some pharmaceutical and pharmacological aspects – An update. *J Pharm Pharmacol* 52: 467–486

30. *British Pharmacopoeia*, Vol II (2004). Nimesulide. The Stationery Office, London, 1378–1379
31. Fallavena PRB, Schapovral EES (1997) pKa determination of nimesulide in methanol-water mixtures by potentiometric titration. *Int J Pharm* 158: 109–112
32. Singh S, Sharda N, Mahajan L (1999) Spectrophotometric determination of pKa of nimesulide. *Int J Pharm* 176: 261–264
33. Piel G, Pirotte I, Delneuville I, Neven P, Llabres G, Delarge J, Delattre L (1997) Study of the influence of both cyclodextrins and L-lysine on the aqueous solubility of nimesulide: isolation and characterization of nimesulide-L-lysine-cyclodextrin complexes. *J Pharm Sci* 86: 475–480
34. Seedher N, Bhatia S (2003) Solubility enhancement of Cox-2 inhibitors using various solvent systems. *AAPS PharmSciTech* 4: 1–9
35. Dupont L, Pirotte B, Masereel B, Delarge J, Geczy J (1995) Nimesulide. *Acta Cryst* C51: 507–509
36. Iley J, Lopes F, Moreira R (2001) Kinetics and mechanism of hydrolysis of N-amido-methylsulfonamides. *J Chem Soc, Perkin Trans* 2: 749–753
37. Chang SF, Miller AM, Ober RE (1977) Determination of an anti-inflammatory methane-sulfonanilide in plasma by high-speed liquid chromatography. *J Pharm Sci* 66: 1700–1703
38. Carini M, Aldini G, Stefani R, Marinello C, Facino RM (1998) Mass-spectrometric characterization and HPLC determination of the main urinary metabolites of nimesulide in man. *J Pharm Biomed Anal* 18: 201–211
39. Barrientos-Astigarraga RE, Vannuchi YB, Sucupira M, Moreno RA, Muscara MN, De Nucci G (2001). Quantification of nimesulide in human plasma by high-performance liquid chromatography/tandem mass spectrometry. Application to bioequivalence studies. *J Mass Spectrom* 36: 1281–1286
40. Macpherson D, Best SA, Gedik L, Hewson AT, Rainsford KD, Parisi S (2004) The biotransformation and pharmacokinetics in humans of nimesulide. Submitted
41. Alvarez-Lueje A, Vásquez-Vergara P. Núñez-Vergara LJ, Squella JA (1998) HPLC determination of nimesulide in tablets by electrochemical detection. *Anal Lett* 31: 1173–1184
42. Lakshmi CSR, Reddy MN, Naidu PY (1998) Fluorimetric determination of nimesulide with N-(1-naphthyl) ethylene. *Indian Drugs* 35: 519–520
43. Altinöz S, Dursun ÖÖ (2000) Determination of nimesulide in pharmaceutical dosage forms by second order derivative UV spectrometry. *J Pharm Biomed Anal* 22: 175–182
44. Chandran S, Saggar S, Priya KP, Saha RN (2000) New ultraviolet spectrophotometric method for the estimation of nimesulide. *Drug Dev Ind Pharm* 26: 229–234
45. Patravale VB, D'Souza S, Narkar Y (2001) HPTLC determination of nimesulide from pharmaceutical dosage forms. *J Pharm Biomed Anal* 25: 685–688
46. Macnaughton SJ, Kikic I, Foster NR, Alessi P, Cortesi A, Colombo I (1996) Solubility of anti-inflammatory drugs in supercritical carbon dioxide. *J Chem Eng Data* 41: 1083–1086

47. Catarino RI, Conceicao AC, Garcia MB, Goncalves ML, Lima JL, dos Santos MM (2003) Flow amperometric determination of pharmaceuticals with on-line electrode surface renewal. *J Pharm Biomed Anal* 33: 571–580
48. Maffei Facino R, Carini M, Aldini G, Saibene L, Morelli R (1995) Differential inhibition of superoxide, hydroxyl and peroxyl radicals by nimesulide and its main metabolite 4-hydroxynimesulide. *Arzneim-Forsch* 45(II): 10–17
49. Facino RM, Carini M, Aldini G (1993) Antioxidant activity of nimesulide and its main metabolites. *Drugs* 46 (Suppl 1): 15–21
50. Maffei Facino R, Carini M, Aldini G, Saibene L, Macciocchi A (1993) Antioxidant profile of nimesulide, indomethacin and diclofenac in phosphatidylcholine liposomes (PCL) as membrane model. *Int J Tissue React* 15: 225–234
51. Karunakar N, Prabhakar MC, Krishnar DR (2003) Determination of the antioxidant activity of some drugs using high-pressure liquid chromatography. *Arzneim-Forsch* 53: 254–259
52. Mouithys-Mickalad AM, Zheng SX, Deby-Dupont GP, Deby CM, Lamy MM, Reginster JY, Henrotin YE (2000) *In vitro* study of the antioxidant properties of non-steroidal anti-inflammatory drugs by chemiluminescence and electron spin resonance (ESR). *Free Radic Res* 33: 607–621
53. Kovarikova P, Mokry M, Klimes J (2003) Photochemical stability of nimesulide. *J Pharm Biomed Anal* 26: 827–832
54. Rainsford KD (1992) Mechanisms of rash formation and related skin conditions induced by non-steroidal anti-inflammatory drugs. In: Rainsford KD, Velo GP (eds): *Side-effects of Anti-inflammatory Drugs 3*. Kluwer Academic Publishers, Lancaster, 287–301
55. Moore DE (2002) Drug-induced cutaneous photosensitivity: incidence, mechanism, prevention and management. *Drug Saf* 25: 345–372
56. Squella JA, Gonzalez P, Bollo S, Nunez-Vergara LJ (1999) Electrochemical generation and interaction study of the nitro radical anion from nimesulide. *Pharm Res* 16: 161–164
57. Van Antwerpen P, Dubois J, Gelbeke M, Neve J (2004) The reactions of oxicam and sulfonanilide non steroidal anti-inflammatory drugs with hypochlorous acid: determination of the rate constants with an assay based on the competition with para-aminobenzoic acid chlorination and identification of some oxidation products. *Free Radic Res* 38: 251–258
58. Florence AT, Jani PU (1994) Novel oral drug formulations. Their potential in modulating adverse effects. *Drug Saf* 10: 233–266
59. Redenti E, Szente L, Szejtli J (2001) Cyclodextrin complexes of salts of acidic drugs. Thermodynamic properties, structural features, and pharmaceutical applications. *J Pharm Sci* 90: 979–986
60. Bonardelli P, Oliani C, Monici Preti PA, Pellicano P, Quattrocchi G (1990) Efficacy and gastrointestinal tolerability of beta-cyclodextrin-piroxicam and tenoxicam in the treatment of chronic osteoarthritis. *Clin Ther* 12: 547–555
61. Santucci L, Fiorucci S, Chiucchi S, Sicilia A, Bufalino L, Morelli A (1992) Placebo-controlled comparison of piroxicam-beta-cyclodextrin, piroxicam, and indomethacin on

gastric potential difference and mucosal injury in humans. *Dig Dis Sci* 37: 1825–1832

62. Wang D, Miller R, Zheng J, Hu C (2000) Comparative population pharmacokinetic pharmacodynamic analysis for piroxicam-beta-cyclodextrin and piroxicam. *J Clin Pharmacol* 40: 1257–1266
63. Pijak MR, Turcani P, Turcaniova Z, Buran I, Gogolak I, Mihal A, Gazdik F (2002) Efficacy and tolerability of piroxicam-beta-cyclodextrin in the outpatient management of chronic back pain. *Bratisl Lek Listy* 103: 467–472
64. Mester U, Lohmann C, Pleyer U, Steinkamp G, Volcker E, Kruger H, Raj PS (2002) A comparison of two different formulations of diclofenac sodium 0.1% in the treatment of inflammation following cataract-intraocular lens surgery. *Drugs RD* 3: 143–151
65. Scolari G, Lazzarin F, Fornaseri C, Carbone V, Rengo S, Amato M, Cicciu D, Braione D, Argentino S, Morgantini A et al (1999) A comparison of nimesulide beta cyclodextrin and nimesulide in postoperative dental pain. *Int J Clin Pract* 53: 345–348
66. Raja M, Sivaseelam A, Subbiah R, Yuvaraj NR, Chandran K, Annamalai K (2001) Evaluation of efficacy and safety of nimesulide with betacyclodextrin vs nimesulide tablets in osteoarthritis. *J Indian Med Assoc* 99: 451–452
67. Fioravanti A, Storri L, Di Martino S, Bisogno S, Oldani V, Scotti A, Marcolongo R (2002) A randomized, double-blind, multicentre trial of nimesulide-beta-cyclodextrin versus naproxen in patients with osteoarthritis. *Clin Ther* 24: 504–519
68. Dasgupta KS, Deshpande AS, Vedi JN, Patel S (2002) Evaluation of efficacy of nizer *versus* nimesulide tablets in otitis media. *J Indian Med Assoc* 100: 619
69. Dubini E, Vidi A, Bietti G, Maffione G (1991) Inclusion compounds of nimesulide with cyclodextrins. Boehringer Ingelheim Italia (IT). Patent No. WO 9117774
70. Sicart Girona I (1991) Inclusion complexes with cyclodextrins. Leetrim Ltd., German Offen., German Patent, DE 1991-4116659
71. Geczy J (1998) Nimesulide salt cyclodextrin inclusion complexes. Europharmaceutical S.A., Cyclolab Cyclodextrin Research. US Patent No. US 5744165
72. Vavia PR, Adhage NA (1999) Inclusion complexation of nimesulide with beta-cyclodextrins. *Drug Dev Ind Pharm* 25: 543–545
73. Adhage NA, Vavia PR (2000) β-Cyclodextrin inclusion complexation by milling. *Pharm Pharmacol Commun* 6: 13–17
74. Chowdary KP, Nalluri BN (2000) Nimesulide and beta-cyclodextrin inclusion complexes: physicochemical characterization and dissolution rate complexes. *Drug Dev Ind Pharm* 26: 1217–1220
75. Nalluri BN, Chowdary KP, Murthy KV, Hayman AR, Becket G (2003) Physicochemical characterization and dissolution properties of nimesulide-cyclodextrin binary systems. *AAPS PharmSciTech* 4: E2
76. Margarotto L, Berini S, Cosentino C, Torri G (2000–2001) Characterization of nimesulide/beta-cyclodextrin composite obtained by solid state activation. *Metastable Mech Alloyed Nanocryst Materials* 360-3: 643–648

77. Miro A, Quaglia F, Calignano A, Barbato F, Cappello B, La Rotonda MI (2000) Physicochemical and pharmacological properties of nimesulide/beta-cyclodextrin formulations. *STP Pharma Sci* 10: 157–164
78. Ravelet C, Geze A, Villet A, Grosset C, Ravel A, Wouessidjewe D, Peyrin E (2002) Chromatographic determination of the association constants between nimesulide and native and modified beta-cyclodextrins. *J Pharm Biomed Anal* 29: 425–430
79. Braga SS, Ribeiro-Claro P, Pillinger M, Goncalves IS, Pereira F, Fernades AC, Correia PB, Teixeira-Dias JJ (2003) Encapsulation of sodium nimesulide and precursors in beta-cyclodextrin. *Org Biomol Chem* 1: 873–878
80. Jain R, Singh A (1997) Pharmaceutical injectable analgesic composition containing nimesulide. Panacea Biotec Ltd, India. US Patent No. US 5,688,829. European Patent No. EP 0812591(1997). Australia Patent No. AU 69,320 (1998). Canadian Patent No. CN 1,189,336 (1998)
81. Geczy J, Neven P, Piel G, Pirotte B, Delneuville I (1998) Water-soluble nimesulide salt and its preparation, aqueous dilution containing it, nimesulide-based combinations and their uses. US Patent No. US 5756546
82. De Tommaso V (1999) A water soluble nimesulide adduct also for injectable use. Micio Pharma Chemica Aktienge. Patent No. WO 9941233
83. Singh A, Jain R (1999) Water-miscible nonsteroidal antiinflammatory injections. Panacea Biotech Ltd, India. Patent No. JP 11228448
84. Grassi M, Coceani N, Mararotto L (2002) Modelling partitioning of sparingly soluble drugs in a two-phase system. *Int J Pharm* 239: 157–169
85. Ferreira H, Lucio M, de Catro B, Gameiro P, Lima JL, Reis S (2003) Partition and location of nimesulide in EPC liposomes: a spectrophotometric and fluorescence study. *Anal Bioanal Chem* 377: 293–298
86. Jain R, Singh A (2001) Effervescent compositions comprising nimesulide. Panacea Biotec Ltd, India. Patent No. WO0122917
87. Giorgetti PLM (2001) Pharmaceutical preparation containing nimesulide for oral administration. Errekappa Euroterapici SPA. Patent No. US 6194462
88. Monti T, Mossi W (1999) Nimesulide micronized salts. Helsinn Healthcare SA. Patent No. EP 0937709
89. Skinhoj A, Bertelsen P (1998) Modified-release multiple-units compositions of non-steroid anti-inflammatory drugs. Nycomed Danmark A/S, Denmark. WO 9912524
90. Jain R, Singh A (2001) Controlled release compositions comprising nimesulide. Panacea Biotec Ltd, India. Patent No. WO 0122791
91. Uhrich K, Macedo B (2001) Therapeutic compositions containing anti-inflammatory agents and biodegradable polyanhydrides. Rutgers, The State University of New Jersey. Patent No. WO 2001041753
92. Lai C-S, Wang T (2002) Modified forms of pharmacologically active agents for use as nonsteroidal anti-inflammatory drugs (NSAIDs). Medinox Inc, USA. Patent No. WO 2002000167

93. Martino AC, Noack RM, Pierman SA (2004) Coated pharmaceutical tablets with speckled appearance. Pharmacia Corporation, USA. Patent No. WO 2003066030
94. Garg S, Verma RK, Kaul CL (2003) Pharmaceutical composition for extended/sustained release of therapeutically active ingredient. Council of Scientific and Industrial Research, India. Patent No. WO 2003063825
95. Tandon P, Mishra B (2004) Development of multiple w/o/w emulsions showing prolonged antiinflammatory activity of nimesulide. Department of Pharmaceutics, Institute of Technology, Banaras Hindu University, Varanasi, 221005, India
96. Biyani MK, Jathar SR, Acharya MM (2001) A controlled release anti-inflammatory nimesulide formulation. Ajanta Pharma Limited, India. Patent No. WO 2001097775
97. Vandelli MA, Ruozi B, Forni F (1999) PLA microparticles for the prologed release of nimesulide: effect of preparative variables. *STP Pharma Sci* 9: 567–572
98. De Angelis P, Olivieri A (2002) Method of increasing the bioavailability of nimesulide. Bioprogress Spa, Italy. Patent No. EP 1258241
99. Ishii K, Yamada R, Ito J, Nemoto M (2004) Solid oral preparations containing sulfonamide-based anti-inflammatory agents with improved absorbability. Taisho Pharma Co Ltd, Japan. Patent No. JP 08127533
100. Borsa M (1997) Oral pharmaceutical composition having antipyretic, analgesic and anti-inflammatory activity. Leetrim Ltd, UK. Patent No. EP 843998
101. Sato M, Hizaki M, Tada Y, Yamada M (2004) Solid preparations of anti-inflammatory nimesulide or its analogs with improved absorption in digestive tract. Hisamitsu Pharmaceutical Co., Japan. Patent No. JP 09202728
102. Moroni E (1997) Non-stomach-harming and/or controlled release pharmaceutical composition based on nimesulide which can be administered orally. Aesculapius Farma SRL. Patent No. IT 1270958
103. Mutalik S, Venkates, Udupa N (2004) Fast analgesic activity from recrystallized nimesulide and its solid dispersion. *Ind J Physiol Pharmacol* 46: 115–118
104. Jain R, Singh A (1999) Pharmaceutical compositions containing NSAIDs and piperine. Panacea Biotec Ltd, India. Patent No. EP 0935964
105. Agrawal S, Pancholi SS, Jain NK, Agrawal GP (2004) Hydrotropic solubilization of nimesulide for parenteral administration. *Int J Pharm* 274: 149–155
106. Garces Garces J, Bonilla Munoz A, Parente Duena A (1998) Pharmaceutical preparation comprising coated capsules or tablets containing a liposome powder encapsulating a drug. European Patent No EP 855179
107. Miyata S, Taniguchi Y, Masuda K, Kawamura Y (1995) Antiinflammatory agent for external use. Hisamitsu Pharmaceutical Co Inc, Japan. Patent No. WO 9611002
108. Jain R, Singh A (1996) Therapeutic anti-inflammatory and analgesic composition containing nimesulide for transdermal use. Panacea Biotec Ltd, India. Patent No. US 5716609
109. Jain R, Singh A (1997) Transdermal compositions containing nimesulide. Panacea Biotec Ltd, India. Patent No. EP 0812587

110. Taniguchi Y, Miyata S, Kawamura Y, Mauda K (1997) Antiinflammatory agent for external use. Helsinn Healthcare SA, Switzerland. Patent No. EP 0782855
111. Jain R, Singh A (1997) Transdermal compositions containing nimesulide. Panacea Biotec Ltd, India. Patent No. EP 0812587
112. Valenti M, Fabiani F, Frimonti EA (2000) Topical compositions containing a non-steroidal antiinflammatory drug. Farmaceutici Formenti SpA, Italy. Patent No. WO 2000048588
113. Di Schiena MG (1998) Topical pharmaceutical formulations containing nimesulide. Dompe Int Sam (MC). Patent No. EP 0880965
114. Giorgetti PLM (1999) Pharmaceutical preparation containing nimesulide for topical use. Errekappa Eurotherapici SPA (IT). Patent No. US 5998480
115. Embil K, Figueroa R (2000) Nimesulide containing topical pharmaceutical compositions. Edko Trading Representation. Patent No. WO 0009117
116. Pedrani M, Ajani M, Villa R (2000) Pharmaceutical compositions for the topical administration in the oral cavity of on-steroid anti-inflammatory drugs useful for the stomatologic, mouth and oral cavity anti-inflammatory and analgesic therapies. Farmatron Limited, UK. Patent No. WO 2001049276
117. Giannaccini B, Saettone MF, Monti D, Boldrini E, Farmigea S, Bianchini P (2000) Use of niaouli essential oil as transdermal permeation enhancer. Patent No. WO 0053228
118. Sengupta S, Velpandian T, Sapra P, Mathur P, Gupta S (1998) Comparative analgesic efficacy of nimesulide and diclofenac gels after topical application on the skin. *Skin Pharmacol Appl Skin Physiol* 11: 273–278
119. Jain NK, Kulkarni SK (2001) Pharmacological and pharmacokinetic studies on marketed gel formulations of nimesulide. *Indian Drugs* 38: 63–66
120. Anonymous (2000) *Nimesulide Gel.* Adis International Ltd, Milano
121. Personal Communication from Helsinn Healthcare SA
122. Jain R, Singh A (1998) Novel therapeutic anti-inflammatory and analgesic composition containing nimesulide for use transdermally and process for manufacture thereof. Panancea Biotec Ltd, India. Canadian Patent No. CN 1189337
123. Giorgetti PLM (1999) Pharmaceutical preparation containing nimesulide for topical use. Errekappa Eurotherapici SPA, Italy. US Patent No. US 5998480
124. Bader S, Monti T, Hausermann E (2001) Nimesulide topical formulations in the form of liquid crystals. Helsinn Healthcare SA, Switzerland. US Patent No. US 6288121
125. Bader S, Monti T, Hausermann E (2003) Nimesulide gel systems for topical use. Helsinn Healthcare SA, Switzerland. US Patent No. US 2003036563
126. Monti T, Bader S, Hausermann E (2002) Nimesulide gel systems for topical use. US Patent No. US 2002119997
127. Shahiwala A, Misra A (2002) Studies in topical application of niosomally entrapped nimesulide. *J Pharm Pharmaceut Sci* 5: 220–225
128. Khandare JN, Hemant JB, Ramesh UR (2004) Preparation and evaluation of nimesulide niosomes for topical application. *Indian Drugs* 38: 197–202
129. Bennett A, Charlier EM, McDonald AM, Simpson JS, Stamford IF, Zebro T (1977) Prostaglandins and breast cancer. *Lancet* 2: 624–626

130. Bennett A, Tacca MD, Stamford IF, Zebro T (1977) Prostaglandins from tumours of human large bowel. *Br J Cancer* 35: 881–884
131. Bennett A (1976) Prostaglandins as factors in diseases of the alimentary tract. *Adv Prostagl Thromb Res* 2: 547–555
132. Thun MJ, Henley SJ (2003) Epidemiology of nonsteroidal anti-inflammatory drugs and colorectal cancer. In: Harris RE (ed): *COX-2 Blockade in Cancer Prevention and Therapy.* Human Press, Totawa, New Jersey, 35–55
133. Rüegg C, Zaric J, Stupp R (2003) Non steroidal anti-inflammatory drugs and COX-2 inhibitors as anti-cancer therapeutics: hypes, hopes and reality. *Ann Med* 35: 476–487
134. Harris RE (Ed) (2003) *COX-2 Blockade in Cancer Prevention and Therapy.* Humana Press, Totowa, New Jersey
135. Rainsford KD (2004) Aspirin and NSAIDs in the prevention of cancer, Alzheimer's disease and other novel therapeutic actions. In: Rainsford KD (ed): *Aspirin and Related Drugs.* CRC Press, Boca Raton (Florida), 707–755
136. Roche-Nagle G, Connolly EM, Eng M, Bouchier-Hayes DJ, Harmey JH (2004) Antimetastatic activity of a cyclooxygenase-2 inhibitor. *Br J Cancer* 91: 359–365
137. Koehne CH, Dubois RN (2004) COX-2 inhibition and colorectal cancer. *Semin Oncol* 31(Suppl 7): 12–21
138. Altorki NK, Subaramaiah K, Dannenberg AJ (2004) COX-2 inhibition in upper aerodigestive tract tumours. *Semin Oncol* 31 (Suppl 7): 30–36
139. Liao Z, Milas L (2004) COX-2 and its inhibition as a molecular target in the prevention and treatment of lung cancer. *Expert Rev Anticancer Ther* 4: 543–560
140. Mann JR, DuBois RN (2004) Cyclooxygenase-2 and gastrointestinal cancer. *Cancer J* 10: 145–152
141. Peek RM Jr (2004) Prevention of colorectal cancer through the use of COX-2 selective inhibitors. *Cancer Chemother Pharmacol* 54 (Suppl 1): S50–S56
142. Charalambous D, O'Brien PE (1996) Inhibition of colon cancer precursors in the rat by sulindac sulphone is not dependent on inhibition of prostaglandin synthesis. *J Gastroenterol Hepatol* 11: 307–310
143. Schnitzler M, Dwight T, Robinson BG (1996) Sulindac increases the expression of APC mRNA is malignant colonic epithelial cells: an *in vitro* study. *Gut* 38: 707–713
144. Duffy CP, Elliott CJ, O'Connor RA, Heenan MM, Coyle S, Cleary IM, Kavanagh K, Verhaegen S, O'Loughlin CM, NicAmlaoibh R, Clynes M (1998) Enhancement of chemotherapeutic drug toxicity to human tumour cells *in vitro* by a subset of non-steroidal anti-inflammatory drugs (NSAIDs). *Eur J Cancer* 34: 1250–1259
145. Zhang Z, DuBois RN (2000) Par-4, a proapoptotic gene, is regulated by NSAIDs in human colon carcinoma cells. *Gastroenterology* 118: 1012–1017
146. Karacay B, Sanlioglu S, Griffith TS, Sandler A, Bonthius DJ (2004) Inhibition of the NF-κB pathway enhances TRAIL-mediated apoptosis in neuroblastoma cells. *Cancer Gene Therapy* 11: 681–690

147. Kazhdan I, Marciniak RA (2004) Death receptor 4 (DR4) efficiently kills breast cancer cells irrespective of their sensitivity to tumor necrosis factor-related apoptosis-inducing ligand (TRAIL). *Gene Therapy* 11: 691–698
148. Ding X-Z, Adrian TE (2001) Role of lipoxygenase pathways in the regulation of pancreatic cancer cell proliferation and survival. *Inflammopharmacology* 9: 157–164
149. Shureiqi I, Chen D, Lee JJ, Yang P, Newman RA, Brenner DE, Lotan R, Fischer SM, Lippman SM (2000) 15-LOX-1: a novel molecular target of nonsteroidal anti-inflammatory drug-induced apoptosis in colorectal cancer cells. *J Natl Cancer Inst* 92: 1136–1142
150. Shureiqi I, Chen D, Lotan R, Yang P, Newman RA, Fischer SM, Lippman SM (2000) 15-lipoxygenase-1 mediates nonsteroidal anti-inflammatory drug-induced apoptosis independenty of cyclooxygenase-2 in colon cancer cells. *Cancer Res* 60: 6846–6850
151. Vogt T, McClelland M, Jung B, Popova S, Bogenrieder T, Becker B, Rumpler G (2001) Progression and NSAID-induced apoptosis in malignant melanomas are independent of cyclooxygenase II. *Melanoma Res* 11: 587–599
152. He Q, Luo X, Huang Y, Sheikh MS (2002) Apo2L/TRAIL differentially modulates the apoptotic effects of sulindac and a COX-2 selective non-steroidal anti-inflammatory agent in BAX-deficient cells. *Oncogene* 21: 6032–6040
153. Harris RE, Beebe-Donk J, Namboodiri KK (2001) Inverse association of non-steroidal anti-inflammatory drugs and malignant melanoma among women. *Oncol Rep* 8: 655–657
154. Garcia-Rodriguez LA, Huerta-Alvarez C (2001) Reduced risk of colorectal cancer among long-term users of aspirin and nonaspirin nonsteroidal anti-inflammatory drugs. *Epidemiology* 12: 88–93
155. Harris RE, Beebe-Donk J, Schuller HM (2002) Chemoprevention of lung cancer by nonsteroidal anti-inflammatory drugs among cigarette smokers. *Oncol Rep* 9: 693–695
156. Harris RE, Chlebowski RT, Jackson RD, Frid DJ, Ascenseo JL, Anderson G, Rodabough RJ, White E, McTiernan A, Women's Health Initiative (2003) Breast cancer and nonsteroidal anti-inflammatory drugs: prospective results from the Women's Health Initiative. *Cancer Res* 63: 6096-6101
157. Okajima E, Denda A, Ozono S, Takahama M, Akai H, Sasaki Y, Kitayama W, Wakabayashi K, Konishi Y (1998) Chemopreventive effects of nimesulide, a selective cyclooxygenase-2 inhibitor, on the development of rat bladder carcinomas initiated by *N*-butyl-*N*-(hydroxybutyl) nitrosamine. *Cancer Res* 58: 3028–3031
158. Kitayama W, Denda A, Okajima E, Tsujiuchi T, Konishi Y (1999) Increased expression of cyclooxygenase-2 protein in rat urinary bladder tomors induced by *N*-butyl-*N*-(hydroxybutyl) nitrosamine. *Carcinogenesis* 20: 2305–2310
159. Fukutake M, Nakatsugi S, Isoi T, Takahashi M, Ohta T, Mamiya S, Taniguchi Y, Sato H, Fukuda K, Sugimura T et al (1998) Suppressive effects of nimesulide, a selective inhibitor of cyclooxygenase-2, on azoxymethane-induced colon carcinogenesis. *Carcinogenesis* 19: 1939–1942

160. Nakatsugi S, Ohta T, Kawamori T, Mutoh M, Tanigawa T, Watanabe K, Sugie S, Sugimura T, Wakabayashi K (2000) Chemoprevention by nimesulide, a selective cyclooxygenase-2 inhibitor, of 2-amino-1-methyl-6-phenylimidazo[4,5-b]pyridine (PhIP)-induced mammary gland carcinogenesis in rats. *Jpn J Cancer Res* 91: 886–892
161. Yoshida K, Tanaka T, Kohno H, Sakata K, Kawamori T, Mori H, Wakabayashi K (2003) A COX-2 inhibitor, nimesulide, inhibits chemically-induced rat tongue carcinogenesis through suppression of cell proliferation activity and COX-2 and iNOS expression. *Histol Histopath* 18: 39–48
162. Hahm KB, Song YJ, Oh TY, Lee JS, Surh YJ, Kim YB, Yoo BM, Kim JH, Han SU, Nahm KT et al (2003) Chemoprevention of Helicobacter pylori-associated gastric carcinogenesis in a mouse model: is it possible? *J Biochem Mol Biol* 36: 82–94
163. Furukawa F, Nishikawa A, Lee IS, Kanki K, Umemura T, Okazaki K, Kawamori T, Wakabayashi K, Hirose M (2003) A cyclooxygenase-2 inhibitor, nimesulide, inhibits postinitiation phase of *N*-nitrosobis(2-oxopropyl)amine-induced pancreatic carcinogenesis in hamsters. *Int J Cancer* 104: 269–273
164. Li XH, Li XK, Cai SH, Tang FX, Zhong XY, Ren XD (2003) Synergistic effects of nimesulide and 5-fluorouracil on tumor growth and apoptosis in the implanted hepatoma in mice. *World J Gastroenterol* 9: 936–940
165. Li XH, Li JJ, Zhang HW, Sun P, Zhang YL, Cai SH, Ren XD (2003) Nimesulide inhibits tumor growth in mice implanted hepatoma: overexpression of Bax over Bcl-2. *Acta Pharmacol Sin* 24: 1045–1050
166. Yamamoto K, Kitayama W, Denda A, Morisaki A, Kuniyasu H, Kirita T (2003) Inhibitory effects of selective cyclooxygenase-2 inhibitors, nimesulidee and etodolac, on the development of squamous cell dysplasias and carcinomas of the tongue in rats initiated with 4-nitroquinoline 1-oxide. *Cancer Lett* 199: 121–129
167. Kitamura T, Itoh M, Noda T, Matsuura M, Wakabayashi K (2004) Combined effects of cyclooxygenase-1 and cyclooxygenase-2 selective inhibitors on intestinal tumorigenesis in adenomatous polyposis coli gene knockout mice. *Int J Cancer* 109: 576–580
168. Nakasugi S, Fukutake M, Takahashi M, Fukuda K, Isoi T, Taniguchi Y, Sugimura T, Wakabayashi K (1997) Suppression of intestinal polyp development by nimesulide, a selective cyclooxygenase-2 inhibitor, in Min mice. *Jpn J Cancer Res* 88: 1117–1120
169. Dolara P, Caderni G, Tonelli F (1999) Nimesulide, a selective anti-inflammatory cyclooxygenase-2 inhibitor, does not affect polyp number and mucosal proliferation in familial adenomatous polyposis. *Scan J Gastroenterol* 34: 1168
170. Seed MP, Freemantle CN, Alam CA, Colville-Nash PR, Brown JR, Papworth JL, Somerville KW, Willoughby DA (1997) Apoptosis induction and inhibition of colon-26 timour growth and angiogenesis: findings on COX-1 and COX-2 inhibitors *in vitro* and *in vivo* and topical disclofenac in hyaluronan. *Adv Exp Med Biol* 433: 339–342
171. Giuliano F, Warner TD (1999) *Ex vivo* assay to determine the cyclooxygenase selectivity of non-steroidal anti-inflammatory drugs. *Br J Pharmacol* 126: 1824–1830
172. Tardieu D, Jaeg JP, Deloly A, Corpet DE, Cadet J, Petit CR (2000) The COX-2 inhibitor nimesulide suppresses superoxide and 8-hydroxy-deoxyguanosine formation,

and stimulates apoptosis in mucosa during early colonic inflammation in rats. *Carcinogenesis* 21: 973–976

173. Hida T, Kozaki K, Muramatsu H, Masuda A, Shimizu S, Mitsudomi T, Sugiura T, Ogawa M, Takahashi T (2000) Cyclooxygenase-2 inhibitor induces apoptosis and enhances cytotoxicity of various anticancer agents in non-small cell cancer cell lines. *Clin Cancer Res* 6: 2006–2011
174. Zhang Z, DuBois RN (2000) Par-4, a proaptotic gene, is regulated by NSAIDs in human colon carcinoma cells. *Gastroenterology* 118: 1012–1017
175. Godlewski MM, Motyl MA, Gajkowska B, Wareski P, Koronkiewicz M, Motyl T (2001) Subcellular redistribution of BAX during apoptosis induced by anticancer drugs. *Anticancer Drugs* 12: 607–617
176. Tian G, Yu JP, Lou HS, Yu BP, Yue H, Li JY, Mei Q (2002) Effect of nimesulide on proliferation and apoptosis of human hepatoma SMMC-7721 cells. *World J Gastroenterol* 8: 483–487
177. Lin DT, Subbaramaiah K, Shah JP, Dannenberg AJ, Boyle JO (2002) Cyclooxygenase-2: a novel molecular target for the prevention and treatment of head and neck cancer. *Head Neck* 24: 792–799
178. Yamazaki R, Kusonoki N, Matsuzaki T, Hashimoto S, Kawai S (2002) Selective cyclooxygenase-2 inhibitors show a differential ability to inhibit proliferation and induce apoptosis of colon adenocarcinoma cells. *FEBS Lett* 531: 278–284
179. Godlewski MM, Gajkowski B, Lamparaska-Przybysz M, Motyl T (2002) Colocalization of BAX with BID and VDAC-1 in nimesulide-induced apoptosis of human colon adenocarcinoma COLO 205 cells. *Anticancer Drugs* 13: 1017–1029
180. Eibl G, Reber HA, Wente MN, Hines OJ (2003) The selective cyclooxygenase-2 inhibitor nimesulide induces apoptosis in pancreatic cancer cells independent of COX-2. *Pancreas* 26: 33–41
181. Pan Y, Zhang JS, Gazi MH, Young CY (2003) The cyclooxygenase 2-specific nonsteroidal anti-inflammatory drugs celecoxib and nimesulide inhibit androgen receptor activity via induction of c-Jun in prostate cells. *Cancer Epidemiol Biomarkers Prev* 12: 769–774
182. Totzke G, Schulze-Osthoff K, Janicke RU (2003) Cyclooxygenase-2 (COX-2) inhibitors sensitize tumor cells specifically to death receptor-induced apoptosis independently of COX-2 inhibition. *Oncogene* 22: 8021–8030
183. Buecher B, Broquet A, Bouancheau D, Heymann MF, Jany A, Denis MG, Bonnet C, Galmiche JP, Blottiere HM (2003) Molecular mechanisms involved in the antiproliferative effect of two COX-2 inhibitors, nimesulide and NS-398, on colorectal cancer cell lines. *Dig Liver Dis* 35: 557–565
184. Chen PY, Long QC (2004) Effects of cyclooxygenase 2 inhibitors on biological traits of nasopharyngeal carcinoma cells. *Acta Pharmacol Sin* 25: 943–949
185. Baoping Y, Guoyong H, Jieping Y, Zongxue R, Hesheng L (2004) Cyclooxygenase-2 inhibitor nimesulide suppresses telomerase activity by blocking Akt/PKB activation in gastric cancer cell line. *Dig Dis Sci* 49: 948–953

186. Okajima E, Uemura H, Ohnishi S, Tanaka M, Ohta M, Tani M, Fujimoto K, Ozono S, Okajima E, Hirao Y (2003) Expression of cyclooxygenase-2 in primary superficial bladder cancer tissue may predict risk of its recurrence after complete transurethral resection. *Aktuelle Urol* 34: 256–258
187. Li JY, Wang XZ, Chen FL, Yu JP, Luo HS (2003) Nimesulide inhibits proliferation via induction of apoptosis and cell cycle arrest in human gastric carcinoma cell line. *World J Gastroenterol* 9: 915–920
188. Eibl G, Bruemmer D, Okada Y, Duffy JP, Law RE, Reber HA, Hines OJ (2003) PGE_2 is generated by a specific COX-2 activity and increases VEGF production in COX-2 – expressing human pancreatic cancer cells. *Biochem Biophys Res Commun* 306: 887–897
189. Haynes A, Shaik MS, Chatterjee A, Singh M (2003) Evaluation of an aerosolized selective COX-2 inhibitor as a potentiator of doxorubicin in non-small-cell lung cancer cell line. *Pharm Res* 20: 1485–1495
190. Xing L, Zhang Z, Xu Y, Zhang H, Liu J (2004) The effects of nimesulide combined with cisplatin on lung cancer. *J Huazhong Univ Sci Technolog Med Sci* 24: 120–123
191. Makowski M, Grzela T, Niderla J, Lazarczyk M, Mroz P, Kopee M, Legat M, Strusinska K, Koziak K, Nowis D et al. (2003) Inhibition of cyclooxygenase-2 indirectly potentiates antitumor effects of photodynamic therapy in mice. *Clin Cancer Res* 9: 5419–5422
192. Akita Y, Kozaki K, Nakagawa A, Saito T, Ito S, Tamada Y, Fujiwara S, Nishikawa N, Uchida K, Yoshikawa K et al (2004) Cyclooxygenase-2 is a possible target of treatment approach in conjunction with photodynamic therapy for various disorders in skin and oral cavity. Br J Dermatol 151: 472–480
193. Selko DJ (1994) Amyloid beta-protein precursor: new clues to the genesis of Alzheimer's disease. Curr Opin Neurobiol 4: 708–716
194. Tuppo EE, Arias HR (2005) The role of inflammation in Alzheimer's disease. Int J Biochem Cell Biol 37: 289-305
195. Breitner JCS, Welsh KA, Helms MJ, Gaskell PC, Gau BA, Roses AD, Pericakvance MA (1995) Delayed onset of Alzheimer's disease with nonsteroidal anti-inflammatory and histamine H2 drugs. Neurobiology of Aging 16: 523-530
196. Andersen K, Launer LJ, Ott A, Hoes AW, Breteler MMB, Hofman A (1995) Do nonsteroidal anti-inflammatory drugs decrease the risk for Alzheimer's disease? The Rotterdam Study. Neurology 45: 1441–1445
197. McGeer PL, Schulzer M, McGeer EG (1996) Arthritis and anti-inflammatory agents as possible protective factors for Alzeimer's disease: a review of 17 epidemiological studies. Neurology 47: 425–432
198. Etminan M, Gill S, Samii A (2003) Effect of non-steroidal anti-inflammatory drug on risk of Alzheimer's disease: systematic review and meta-analysis of observational studies. Br Med J 327: 128
199. Szekely CA, Thorne JE, Zandi PP, Ek M, Messias E, Breitner JC, Goodman SN (2004) Nonsteroidal anti-inflammatory drugs for the prevention of Alzheimer's disease: a systematic review. Neuroepidemiology 23: 159–169

200. Stewart WF, Kawas C, Corrada M, Metter EJ (1997) Risk of Alzheimer's disease and duration of NSAID use. Neurology 48: 626–632
203. Scharf S, Mander A, Ugoni A, Vajda F, Christophidis N (1999) A double-blind, placebo-controlled trial of diclofenac/misoprostol in Alzheimer's disease. Neurology 53: 197–201
204. Lopez-Pousa S, Mercadal-Dalmau J, Marti-Cuadros AM, Vilalta-Franch J, Lozano-Gallego M (1997) Triflusal in the prevention of vascular dementia. Rev Neurol 25: 1525–1528 (in Spanish)
205. Aisen PS, Schmeidler J, Pasinetti GM (2002) Randomized pilot study of nimesulide treatment of Alzheimer's disease. Neurology 58: 1050–1054
206. Aisen PS, Schafer KA, Grundman M, Pfeiffer E, Sano M, Davis KL, Farlow MR, Jin S, Thomas RG, Thal LJ; Alzheimer's Disease Cooperative Study (2003) Effects of rofecoxib or naproxen vs placebo on Alzheimer's disease progression: a randomized controlled trial. J Am Med Assoc 289: 2819–2826
207. Reines SA, Block GA, Morris JC, Liu G, Nessly ML, Lines CR, Norman BA, Baranak CC; Rofecoxib Protocol 091 Study Group (2004) Rofecoxib: no effect on Alzheimer's disease in a 1-year, randomized, blinded, controlled study. Neurology 62: 66–71
208. Avramovich Y, Amit T, Youdim MBH (2002) Non-steroidal anti-inflammatory drugs stimulate secretion of non-amyloidogenic precursor protein. J Biol Chem 277: 31466–31473
209. Koyfman L, Kaplanski J, Artru AA, Talmor D, Rubin M, Shapira Y (2000) Inhibition of cyclooxygenase 2 by nimesulide decreases prostaglandin E2 formation but does not alter brain edema or clinical recovery after closed head injury in rats. J Neurosurg Anesthesiol 12: 44–50
210. Cernak I, O'Connor C, Vink R (2002) Inhibition of cyclooxygenase 2 by nimesulide improves cognitive outcome more than motor outcome following diffuse traumatic injury in rats. Exp Brain Res 147: 193–199
211. Cernak I, O'Connor C, Vink R (2001) Activation of cyclo-oxygenase-2 contributes to motor and cognitive dysfunction following diffuse traumatic brain injury in rats. Clin Exp Pharmacol Physiol 28: 922–925
212. Naidu PS, Kulkarni SK (2002) Differential effects of cyclooxygenase inhibitors on haloperidol-induced catalepsy. *Prog Neuropsychopharmacol Biol Psychiatry* 26: 819–822
213. Le Filliatre G, Sayah S, Latournerie V, Renaud JF, Finet M, Hanf R (2001) Cyclo-oxygenase and lipoxygenase pathways in mast cell dependent-neurogenic inflammation induced by electrical stimulation of the rat saphenous nerve. *Br J Pharmacol* 132: 1581–1589
214. Shemi D, Azab AN, Kaplanski J (2000) Time-dependent effect of LPS on PGE2 and TNF-alpha production by rat glial brain culture: influence of COX and cytokine inhibitors. *J Endotoxin Res* 6: 377–381
215. Jain NK, Kulkarni SK, Singh A (2001) Lipopolysaccharide-mediated immobility in mice: reversal by cyclooxygenase enzyme inhibitors. *Methods Find Exp Clin Pharmacol* 23: 441–444

216. Boje KM, Jaworowicz D Jr, Raybon JJ (2003) Neuroinflammatory role of prostaglandins during experimental meningitis: evidence suggestive of an *in vivo* relationship between nitric oxide and prostaglndins. *J Pharmacol Exp Ther* 304: 319–325
217. Spielman L, Winger D, Ho L, Aisen PS, Shoharmi E, Pasinetti GM (2002) Induction of the complement component ClB in brain of transgenic mice with neuronal overexpression of human cyclooxygenase-2. *Acta Neuropathol* 103: 157–162
218. Mirjany M, Ho L, Pasinetti GM (2002) Role of cyclooxygenase-2 in neuronal cell cycle activity ad glutamate-mediated excitotoxicity. *J Pharmacol Exp Ther* 301: 494–500
219. Jain NK, Patil CS, Kulkarni SK, Singh A (2002) Modulatory role of cyclooxygenase inhibitors in aging- and scopolamine or lipopolysaccharide-induced cognitive dysfunction in mice. *Behav Brain Res* 133: 369–376
220. Fattore L, Melis M, Diana M, Fratt W, Gessa G (2000) The cyclo-oxygenase inhibitor nimesulide induces conditioned place preference in rats. *Eur J Pharmacol* 406: 75–77
221. Kunz T, Oliw EH (2001) Nimesulide aggravates kainic acid-induced seizures in the rat. *Pharmacol Toxicol* 88: 271–276
222. Candelario-Jalil E, Ajamieh HH, Sam S,Martinez G, Leon Fernandez OS (2000) Nimesulide limits kainate-induced oxidative damage in the rat hippocampus. *Eur J Pharmacol* 390: 295–298
223. Chen C, Magee JC, Bazan NG (2002) Cyclooxygenase-2 regulates prostaglandin E2 signaling in hippocampal long-term synaptic plasticity. *J Neurophysiol* 87: 2851–2857
224. Grilli M, Pizzi M, Memo M, Spano P (1997) Use of selected nonsteroidal antiinflammatory compounds for the prevention and the treatment of neurodegenerative diseases. Patent No. WO 9820864
225. Pasinetti GM, Aisen PS (2004) Treatment of neurodegenerative conditions with nimesulide. Patent No. WO 9822104
226. Aisen PS, Pasinetti GM (1999) Treatment of neurodegenerative conditions with nimesulide. Patent No. US19970831402
227. Koo EHM, Golde TE, Galsko DR (2001) Nonsteroidal antiinflammatory dug (NSAID) and NSAID derivative Alzheimer's. Patent No. WO 2001078721
228. Aisen P, Pasinetti GM (2000) Treating of neurodegenerative conditions use of nimesulide for the preparation of pharmaceutical compositions. Patent No. HU 9904544
229. Pasinetti GM (2003) Inhibiting progressive cognitive impairment. Patent No. WO 2003105820
230. Isakson PC (2004) Monotherapy for the treatment of amyotrophic lateral sclerosis with cyclooxygenase-2 (COX-2) inhibitor(s). Patent No. WO 2003101441
231. Slattery MM, Friel AM, Healy DG, Morrison JJ (2001) Uterine relaxant effects of cyclooxygenase-2 inhibitors *in vitro*. *Obstet Gynecol* 98: 563–569
232. Grigsby PL, Poore KR, Hirst JJ, Jenkin G (2000) Inhibition of premature labor in sheep by a combined treatment of nimesulide. A prostaglandin synthase type 2 inhibitor, and atosiban, an oxytocin receptor antagonist. *Am J Obstet Gynecol* 183: 649–657

233. Scott JE, Grigsby PL, Hirst JJ, Jenkin G (2001) Inhibition of prostaglandin synthesis and its effect on uterine activity during established premature labor in sheep. *J Soc Gynecol Investig* 8: 266–276
234. Locatelli A, Vergani P, Bellini P, Strobelt N, Ghidini A (2001) Can a cyclo-oxygenase type-2 selective tocolytic agent avoid the fetal side effects of indomethacin? *BJOG* 108: 325–326
235. Holmes RP, Stone PR (2000) Severe oligohydramnios induced by cyclooxygenase-2 inhibitor nimesulide. *Obstet Gynecol* 96(5 Pt 2): 810–811
236. Sawdy RJ, Lye S, Fisk NM, Bennett PR (2003) A double-blind randomized study of fetal side effects during and after the short-term maternal administration of indomethacin, sulindac, and nimesulide for the treatment of preterm labor. *Am J Obstet Gynecol* 188: 1046–1051
237. Bennett PR (2004) Cyclooxygenase-2 (COX-2) selective inhibitors for managing labour and uterine contractions. Patent No. WO 9731631
238. Rahmouni-Piette S, Mutschen M, Aukurst PAL, Johansson C, Hansson V, Tasken K, Froeland SS, Klaveness J, Aandahl EM (2003) Use of COX-2 inhibitors for preventing immunodeficiency. Patent No. US2004082640
239. Rodriguez-Soriano J (1999) Bartter's syndrome comes of age. *Pediatrics* 103(3): 663–664
240. Nusing RM, Reinalter SC, Peters M, Komhoff M, Seyberth HW (2001) Pathogenetic role of cyclooxygenase-2 in hyperprostaglandin E syndrome/antenatal Bartter syndrome: therapeutic use of the cyclooxygenase-2 inhibitor nimesulide. *Clin Pharmacol Ther* 70: 384–390
241. Reinalter SC, Jeck N, Brochhausen C, Watzer B, Nusing RM, Seyberth HW, Komhoff M (2002) Role of cyclooxygenase-2 in hyperprostaglandin E syndrome/antenatal Bartter syndrome. *Kidney Int* 62: 253–260
242. Filippo D (1992) The use of nimesulide in the treatment of cataract. Patent No. EP0532900
243. Matsuo K, Hojou H, Honbou M, Miyata N (1995) Clinical efficacy of diclofenac sodium on postsurgical inflammation after intraocular lens implantation. *J Cataract Refract Surg* 21: 309–312
244. Italian Diclofenac Study Group. Efficacy of diclofenac eyedrops in preventing postoperative inflammation and long-term cystoid macular edema. *Br J Cataract Refract Surg* 23: 183–189
245. Gupta SK, Joshi S, Tandon R, Mathur P (1997) Topical aspirin provides protection against galactosemic cataract. *Indian J Ophthalmol* 45: 221–225
246. Christen WG, Manson JE, Glynn RJ, Ajani UA, Schaumberg DA, Sperduto RD, Buring JE, Hennekens CH (1998) Low-dose aspirin and risk of cataract and subtypes in a randomized trial of US physicians. *Ophthalmic Epidemiol* 5: 133–142
247. Gaynes BI, Fiscella R (2002) Topical nonsteroidal anti-inflammatory drugs for ophthalmic use: a safety review. *Drug Saf* 25: 233–250
248. Schalnus R (2003) Topical nonsteroidal anti-inflammatory therapy in ophthalmology. *Ophthalmlogica* 217: 89–98

APPENDIX A: Trademark names for nimesulide

Helsinn trademark of original nimesulide worldwide (name of Helsinn's partners – marketing authorization holders – & country):

Aulin® (Sulkaj/Albania; CSC/Austria, Bosnia, Bulgaria, Czech Republic, Slovac Republic, Poland, Romania Serbia & Montenegro, Slovenia; Schering Plough/Chile, Ecuador, Philippines, Venezuela; Gala/Indonesia; Helsinn Birex Therapeutics/Ireland; Roche/Italy; Ergo Maroc/Morocco; Angelini/Portugal; Vifor/Switzerland), **Mesulid®** (Sanofi-Aventis/Latvia, Lithuania, Belarus, Hungary, Ukraine, Georgia, Armenia, Moldavia; Therabel/Belgium, Luxemburg; Grünenthal/Columbia, Ecuador; CSC/Czech Republic; Boehringer-Ingelheim/ Greece; Schering Plough/HongKong, Philippines, Vietman, Ergha/Ireland, Rafa/Israel, Novartis/Italy, Roche/Mexico, Atco/Pakistan, Choongwae/South Korea, Harvester/Taiwan), **Nimed®** (CSC/Czech Republic, Slovac Republic, Schering Plough/Indonesia, Sanofi-Aventis/Portugal), **Nexen®** (Thérabel/France), Guaxan®(Helsinn Birex Pharmaceuticals/Spain), **Donulide®** (Wyeth-Lederle/ Portugal), **Nisulid®** (Aché/Brazil, Grünenthal/Chile, Robapharm/Switzerland), **Ainex®** (Schering Plough/Chile, Columbia, Peru, Venezuela), **Scaflan®** (Schering Plough/Venezuela), **Scaflam®** (Schering Plough/Brazil, Columbia; Lavipharm/ Greece), **Nimedex®**(Italfarmaco/Italy), **Eskaflam®** (GSK/Mexico), **Plarium®** (India), **Heugan®** (Schering Plough/Costa Rica, Dominican Republic, El Salvador, Guatemala, Panama), **Edrigyl®** (Gerolymatos/Greece) **Sulidene®** (Virbac/France), **Nimecox®** (Grünenthal/Ecuador).

(from [28] and information provided by Helsinn Healthcare SA)

Other Companies (by name):

Auroni® (Aurobindo Pharma), Flexulid® (Wander), Maxiflam® (Karnataka Antibiotics), Maxulide® (Max), Mesulid® (Stadmed), Myonal® (Uni-Sankyo), Nelsid® (Ind-Swift), Neosaid® (Blue Cross), Nilide® (Le Sante), Nimbid® (Astra IDL), Nimegesic® (Alembic), Nimesel® (Wave Pharma), Nimesul® (Albert David), Nimfast® (Indon), Nimind® (Indoco), Nimobid® (Mapra), Nimodol® (Aristo), Nimoran® (Perch), Nimsaid® (Medley), Nimuflam® (JK Drugs), Nimulid® (Panacea), Nimuspa® (Indoco), Nimusyp® (Centaur), Nise® (Dr Reddy's), Novogesic® (Glenmark), Novolid® (Brown & Burk), Orthobid® (Nicholas Piramal), Pirodol® (Menarini), Pronim® (Unichem), Pyrnim® (Saga Labs), Relisulide® (Jaggat Pharma), Remulide® (Recon), Slide® (Dee-Pharma).

(from www.webhealthcentre.com, accessed on 14/09/2004)

APPENDIX B: Summary of Product Characteristics for nimesulide as approved by the European Medicines Agency (formerly the European Medicines Evaluation Agency) in 2003

NIMESULIDE 100 MG TABLETS, SOLUBLE TABLETS, EFFERVESCENT TABLETS, COATED TABLETS, CAPSULES, HARD CAPSULES,
NIMESULIDE 50/100 MG GRANULES OR POWDER FOR ORAL SUSPENSION
NIMESULIDE 1%, 2% OR 5% ORAL SUSPENSION

1. NAME OF THE MEDICINAL PRODUCT

<TRADENAME>

2. QUALITATIVE AND QUANTITATIVE COMPOSITION

Each tablet, soluble tablet, effervescent tablet, coated tablet, capsule, hard capsule contains 100 mg nimesulide.
Each sachet contains 50 or 100 mg nimesulide.
Oral suspension containing 10 mg, 20 mg or 50 mg per ml.

For excipients, see section 6.1

3. PHARMACEUTICAL FORM

Tablet, soluble tablet, effervescent tablet or coated tablet: <Company-specific>
Granules or powder for oral suspension: <Company-specific>
Capsule, hard capsule: <Company-specific>
Oral suspension: <Company-specific>

4. CLINICAL PARTICULARS

4.1 Therapeutic indications

Treatment of acute pain.
Symptomatic treatment of painful osteoarthritis.
Primary dysmenorrhoea.

4.2 Posology and method of administration

Nimesulide-containing medicinal products should be used for the shortest possible duration, as required by the clinical situation.

Adults:
100 mg nimesulide tablets, soluble tablets, effervescent tablets, coated tablets, capsules, hard capsules, 50 mg and 100 mg granules or powder, 1%, 2% and 5% oral suspension: 100 mg bid after meal.

Elderly: in elderly patients there is no need to reduce the daily dosage (see section 5.2).

Children (< 12 years): Nimesulide containing medicinal products are contraindicated in these patients (see also section 4.3).

Adolescents (from 12 to 18 years): on the basis of the kinetic profile in adults and on the pharmacodynamic characteristics of nimesulide, no dosage adjustment in these patients is necessary.

Impaired renal function: on the basis of pharmacokinetics, no dosage adjustment is necessary in patients with mild to moderate renal impairment (creatinine clearance of 30–80 ml/min), while Nimesulide containing medicinal products are contraindicated in case of severe renal impairment (creatinine clearance < 30 ml/min) (see sections 4.3 and 5.2).

Hepatic impairment: the use of Nimesulide containing medicinal products is contraindicated in patients with hepatic impairment (see section 5.2).

4.3 Contraindications

Known hypersensitivity to nimesulide or to any of the excipients of the products.
History of hypersensitivity reactions (e.g., bronchospasm, rhinitis, urticaria) in response to acetylsalicylic acid or other non-steroidal anti-inflammatory drugs.
History of hepatotoxic reactions to nimesulide.
Active gastric or duodenal ulcer, a history of recurrent ulceration or gastrointestinal bleeding, cerebrovascular bleeding or other active bleeding or bleeding disorders.
Severe coagulation disorders.
Severe heart failure.
Severe renal impairment.
Hepatic impairment.

Children under 12 years.
The third trimester of pregnancy and breastfeeding (see sections 4.6 and 5.3).

4.4 Special warnings and special precautions for use

The risk of undesirable effects may be reduced by using Nimesulide-containing medicinal products for the shortest possible duration.
Treatment should be discontinued if no benefit is seen.

Rarely Nimesulide-containing medicinal products have been reported to be associated with serious hepatic reactions, including very rare fatal cases (see also section 4.8). Patients who experience symptoms compatible with hepatic injury during treatment with Nimesulide-containing medicinal products (e.g., anorexia, nausea, vomiting, abdominal pain, fatigue, dark urine) or patients who develop abnormal liver function tests should have treatment discontinued. These patients should not be rechallenged with nimesulide. Liver damage, in most cases reversible, has been reported following short exposure to the drug.

Concomitant administration with known hepatotoxic drugs, and alcohol abuse must be avoided during treatment with Nimesulide-containing medicinal products treatment, since either may increase the risk of hepatic reactions.

During therapy with Nimesulide-containing medicinal products, patients should be advised to refrain from other analgesics. Simultaneous use of different NSAIDs is not recommended.

Gastrointestinal bleeding or ulceration/perforation can occur at any time during treatment with or without warning symptoms or a previous history of gastrointestinal events. If gastrointestinal bleeding or ulceration occurs, nimesulide should be discontinued. Nimesulide should be used with caution in patients with gastrointestinal disorders, including history of peptic ulceration, history of gastrointestinal haemorrhage, ulcerative colitis or Crohn's disease.

In patients with renal or cardiac impairment, caution is required since the use of Nimesulide-containing medicinal products may result in deterioration of renal function. In the event of deterioration, the treatment should be discontinued (see also section 4.5).

Elderly patients are particularly susceptible to the adverse effects of NSAIDs, including gastrointestinal haemorrhage and perforation, impaired renal, cardiac and hepatic function. Therefore, appropriate clinical monitoring is advisable.

As nimesulide can interfere with platelet function, it should be used with caution in patients with bleeding diathesis (see also section 4.3). However, Nimesulide-containing medicinal products is not a substitute for acetylsalicylic acid for cardiovascular prophylaxis.

NSAIDs may mask the fever related to an underlying bacterial infection.

The use of Nimesulide-containing medicinal products may impair female fertility and is not recommended in women attempting to conceive. In women who have difficulties conceiving or who are undergoing investigation of infertility, withdrawal of Nimesulide-containing medicinal products should be considered (see section 4.6).

4.5 Interaction with other medicinal products and other forms of interaction

Pharmacodynamic interactions
Patients receiving warfarin or similar anticoagulant agents or acetylsalicylic acid have an increased risk of bleeding complications, when treated with Nimesulide-containing medicinal products. Therefore this combination is not recommended (see also section 4.4.) and is contraindicated in patients with severe coagulation disorders (see also section 4.3). If the combination cannot be avoided, anticoagulant activity should be monitored closely.

Pharmacodynamic/pharmacokinetic interactions with diuretics
In healthy subjects, nimesulide transiently decreases the effect of furosemide on sodium excretion and, to a lesser extent, on potassium excretion and reduces the diuretic response.
Co-administration of nimesulide and furosemide results in a decrease (of about 20%) of the AUC and cumulative excretion of furosemide, without affecting its renal clearance.
The concomitant use of furosemide and Nimesulide containing medicinal products requires caution in susceptible renal or cardiac patients, as described under section 4.4.

Pharmacokinetic interactions with other drugs:
Non-steroidal anti-inflammatory drugs have been reported to reduce the clearance of lithium, resulting in elevated plasma levels and lithium toxicity. If Nimesulide containing medicinal products are prescribed for a patient receiving lithium therapy, lithium levels should be monitored closely.

Potential pharmacokinetic interactions with glibenclamide, theophylline, warfarin, digoxin, cimetidine and an antacid preparation (i.e., a combination of aluminium and magnesium hydroxide) were also studied *in vivo*. No clinically significant interactions were observed.

Nimesulide inhibits CYP2C9. The plasma concentrations of drugs that are substrates of this enzyme may be increased when Nimesulide containing medicinal products are used concomitantly.

Caution is required if nimesulide is used less than 24 h before or after treatment with methotrexate because the serum level of methotrexate might increase and therefore, the toxicity of this drug might increase.
Due to their effect on renal prostaglandins, prostaglandin synthetase inhibitors like nimesulide may increase the nephrotoxicity of cyclosporins.

Effects of other drugs on nimesulide:
In vitro studies have shown displacement of nimesulide from binding sites by tolbutamide, salicylic acid and valproic acid. However, despite a possible effect on plasma levels, these interactions have not demonstrated clinical significance.

4.6 Pregnancy and lactation

The use of Nimesulide containing medicinal products is contraindicated in the third trimester of pregnancy (see section 4.3).
Like other NSAIDs Nimesulide containing medicinal products is not recommended in women attempting to conceive (see section 4.4). As with other NSAIDs, known to inhibit prostaglandin synthesis, nimesulide may cause premature closure of the ductus arteriosus, pulmonary hypertension, oliguria, oligoamnios, increased risk of bleeding, uterine inertia and peripheral oedema. There have been isolated reports of renal failure in neonates born to women taking nimesulide in late pregnancy.

Studies in rabbits have shown an atypical reproductive toxicity (see section 5.3) and no adequate data from the use of nimesulide-containing medicinal products in pregnant women are available. Therefore, the potential risk for humans is unknown and prescribing the drug during the first two trimesters of pregnancy is not recommended.

Lactation:
It is not known whether nimesulide is excreted in human milk. Nimesulide containing medicinal products are contraindicated when breastfeeding (see sections 4.3 and 5.3).

4.7 Effects on ability to drive and use machines

No studies on the effect of Nimesulide containing medicinal products on the ability to drive or use machines have been performed. However, patients who experience dizziness, vertigo or somnolence after receiving Nimesulide containing medicinal products should refrain from driving or operating machines.

4.8 Undesirable effects

The following listing of undesirable effects is based on data from controlled clinical trials* (approximately 7,800 patients) and from post marketing surveillance with reporting rates classified as: very common (>1/10); common (>1/100, <1/10), uncommon (>1/1,000, <1/100); rare (>1/10,000, <1/1,000); very rare (<1/10,000), including isolated cases.

Blood disorders	Rare	Anaemia* Eosinophilia*
	Very rare	Thrombocytopenia Pancytopenia Purpura
Immune system disorders	Rare	Hypersensitivity*
	Very rare	Anaphylaxis
Metabolism and nutrition disorders	Rare	Hyperkalaemia*
Psychiatric disorders	Rare	Anxiety* Nervousness* Nightmare*
Nervous system disorders	Uncommon	Dizziness*
	Very rare	Headache Somnolence Encephalopathy (Reye's syndrome)
Eye disorders	Rare	Vision blurred*
	Very rare	Visual disturbance
Ear and labyrinth disorders	Very rare	Vertigo
Cardiac disorders	Rare	Tachycardia*
Vascular disorders	Uncommon	Hypertension*

Vascular disorders	Rare	Haemorrhage* Blood pressure fluctuation* Hot flushes*
Respiratory disorders	Uncommon	Dyspnoea*
	Very rare	Asthma Bronchospasm
Gastrointestinal disorders	Common	Diarrhoea* Nausea* Vomiting*
	Uncommon	Constipation* Flatulence* Gastritis*
	Very rare	Abdominal pain Dyspepsia Stomatitis Melaena Gastrointestinal bleeding Duodenal ulcer and perforation Gastric ulcer and perforation
Hepato-biliary disorders (see section 4.4. Special warnings and special precautions for use)	Very rare	Hepatitis Fulminant hepatitis (including fatal cases) Jaundice Cholestasis
Skin and subcutaneous tissue disorders	Uncommon	Pruritus* Rash* Sweating increased*
	Rare	Erythema* Dermatitis*
	Very rare	Urticaria Angioneurotic oedema Face oedema Erythema multiforme Stevens Johnson syndrome

Skin and subcutaneous tissue disorders	Very rare	Toxic epidermal necrolysis
Renal and urinary disorders	Rare	Dysuria* Haematuria* Urinary retention*
	Very rare	Renal failure Oliguria Interstitial nephritis
General disorders	Uncommon	Oedema*
	Rare	Malaise* Asthenia*
	Very rare	Hypothermia
Investigations	Common	Hepatic enzymes increased*

*frequency based on clinical trial

4.9 Overdose

Symptoms following acute NSAID overdoses are usually limited to lethargy, drowsiness, nausea, vomiting and epigastric pain, which are generally reversible with supportive care. Gastrointestinal bleeding can occur. Hypertension, acute renal failure, respiratory depression and coma may occur, but are rare. Anaphylactoid reactions have been reported with therapeutic ingestion of NSAIDs, and may occur following an overdose.

Patients should be managed by symptomatic and supportive care following an NSAID overdose. There are no specific antidotes. No information is available regarding the removal of nimesulide by haemodialysis, but based on its high degree of plasma protein binding (up to 97.5%) dialysis is unlikely to be useful in overdose. Emesis and/or activated charcoal (60–100 g in adults) and/or osmotic cathartic may be indicated in patients seen within 4 h of ingestion with symptoms or following a large overdose. Forced diuresis, alkalinization of urine, haemodialysis, or haemoperfusion may not be useful due to high protein binding. Renal and hepatic function should be monitored.

5. PHARMACOLOGICAL PROPERTIES

5.1 Pharmacodynamic properties

Pharmacotherapeutic group:
ATC code: M01AX17

Nimesulide is a non-steroidal anti-inflammatory drug with analgesic and antipyretic properties which acts as an inhibitor of prostaglandin synthesis enzyme cyclooxygenase.

5.2 Pharmacokinetic properties

Nimesulide is well absorbed when given by mouth. After a single dose of 100 mg nimesulide a peak plasma level of 3–4 mg/l is reached in adults after 2–3 h. AUC = 20–35 mg h/l. No statistically significant difference has been found between these figures and those seen after 100 mg given twice daily for 7 days.

Up to 97.5% binds to plasma proteins.

Nimesulide is extensively metabolised in the liver following multiple pathways, including cytochrome P450 (CYP) 2C9 isoenzymes. Therefore, there is the potential for a drug interaction with concomitant administration of drugs which are metabolised by CYP2C9 (see under section 4.5). The main metabolite is the parahydroxy derivative which is also pharmacologically active. The lag time before the appearance of this metabolite in the circulation is short (about 0.8 h) but its formation constant is not high and is considerably lower than the absorption constant of nimesulide. Hydroxynimesulide is the only metabolite found in plasma and it is almost completely conjugated. $T_{1/2}$ is between 3.2 and 6 h.

Nimesulide is excreted mainly in the urine (approximately 50% of the administered dose).
Only 1–3% is excreted as the unmodified compound. Hydroxynimesulide, the main metabolite is found only as a glucuronate. Approximately 29% of the dose is excreted after metabolism in the faeces.

The kinetic profile of nimesulide was unchanged in the elderly after acute and repeated doses.

In an acute experimental study carried out in patients with mild to moderate renal impairment (creatinine clearance 30–80 ml/min) *versus* healthy volunteers, peak plasma levels of nimesulide and its main metabolite were not higher than in healthy

volunteers. AUC and t1/2 beta were 50% higher, but were always within the range of kinetic values observed with nimesulide in healthy volunteers. Repeated administration did not cause accumulation.
Nimesulide is contra-indicated in patients with hepatic impairment (see section 4.3).

5.3 Preclinical safety data

Preclinical data reveal no special hazards for humans based on conventional studies of safety pharmacology, repeated dose toxicity, genotoxicity and carcinogenic potential.
In repeated dose toxicity studies, nimesulide showed gastrointestinal, renal and hepatic toxicity.
In reproductive toxicity studies, embryotoxic and teratogenic effects (skeletal malformations, dilatation of cerebral ventricles) were observed in rabbits, but not in rats, at maternally non-toxic dose levels. In rats, increased mortality of offspring was observed in the early postnatal period and nimesulide showed adverse effects on fertility.

SUMMARY OF PRODUCT CHARACTERISTICS

NIMESULIDE 3% GEL/CREAM

1. NAME OF THE MEDICINAL PRODUCT

<TRADENAME>

2. QUALITATIVE AND QUANTITATIVE COMPOSITION

Nimesulide 3% gel/cream contains 3% w/w nimesulide (1 g of gel/cream contains 30 mg of nimesulide)
For excipients, see section 6.1

3. PHARMACEUTICAL FORM

Gel: <Company-specific>
Cream: <Company-specific>

4. CLINICAL PARTICULARS

4.1 Therapeutic indications

Symptomatic relief of pain associated with sprains and acute traumatic tendinitis.

4.2 Posology and method of administration

Adults: Nimesulide 3% gel/cream (usually 3 g, corresponding to a line 6–7 cm long) should be applied in a thin layer to the affected area 2–3 times daily and massaged until it is completely absorbed.
Duration of treatment: 7–15 days.
Children under 12 years: Nimesulide 3% gel/cream has not been studied in children. Therefore, safety and efficacy have not been established and the product should not be used in children (see section 4.3).

4.3 Contraindications

Known hypersensitivity to nimesulide or to any other excipients in the gel/cream.
Use in patients in whom aspirin, or other medicinal products inhibiting prostaglandin synthesis, induced allergic reactions such as rhinitis, urticaria or bronchospasm.
Use on broken or denuded skin or in the presence of local infection.
Simultaneous use with other topical creams.
Use in children under 12 years.

4.4 Special warnings and special precautions for use

Nimesulide 3% gel/cream should not be applied to skin wounds or open injuries.
Nimesulide 3% gel/cream should not be allowed to come into contact with the eyes or mucous membranes; in case of accidental contact, wash immediately with water.
The product should never be taken by mouth. Hands should be washed after applying the product.
Nimesulide 3% gel/cream should not be used with occlusive dressings.
Nimesulide 3% gel/cream is not recommended for use in children under 12 years (see section 4.3).
Undesirable effects may be reduced by using the minimum effective dose for the shortest possible duration.

Patients with gastrointestinal bleeding, active or suspected peptic ulcer, severe renal or hepatic dysfunction, severe coagulation disorders or severe/non-controlled heart failure should be treated with caution.
Since nimesulide gel 3%/cream has not been studied in hypersensitive subjects, particular caution should be used when treating patients with known hypersensitivity to other NSAIDs. The possibility of developing hypersensitivity in the course of therapy cannot be excluded.
Since with other topical NSAIDs burning sensation and exceptionally photodermatitis can occur, care should be taken during treatment with Nimesulide 3% gel/cream.
To reduce the risk of photosensitivity, patients should be warned against exposure to direct and solarium sunlight.

If symptoms persist or the condition is aggravated medical advice should be sought.

4.5 Interaction with other medicinal products and other forms of interaction

No interactions of Nimesulide 3% gel/cream with other medicinal products are known or to be expected via the topical route.

4.6 Pregnancy and lactation

There are no data relevant to the topical use of <nimesulide containing medicinal product> in pregnant women or during breastfeeding. Therefore, nimesulide 3% gel/cream should not be used during pregnancy or lactation unless clearly necessary.

4.7 Effects on ability to drive and use machines

No studies on the effect of nimesulide 3% gel/cream on the ability to drive and use machines have been performed.

4.8 Undesirable effects

The following side effects listing is based on reports from clinical studies, in a limited number of patients, where mild local reactions have been reported. The reporting rates are classified as: very common (>1/10); common (>1/100, <1/10), uncommon (>1/1,000, <1/100); rare (>1/10,000, <1/1,000); very rare (<1/10,000), including isolated cases.

Skin and subcutaneous tissue disorders (see also section 4.4)	Common	Itching Erythema

4.9 Overdose

Intoxication with nimesulide as a result of topical application of Nimesulide 3% gel or cream is not to be expected since the highest plasma levels of nimesulide following application of Nimesulide 3% gel/cream are far below those found following systemic administration.

5. PHARMACOLOGICAL PROPERTIES

5.1 Pharmacodynamic properties

Pharmacotherapeutic group: ATC code: M02AA.
Non-steroidal anti-inflammatory drug (NSAID) for topical use.
Nimesulide is an inhibitor of the prostaglandin synthesis enzyme cyclooxygenase.

Cyclooxygenase produces prostaglandins, some of them being implicated in the development and maintenance of inflammation.

5.2 Pharmacokinetic properties

When Nimesulide 3% is applied topically, plasma concentrations of nimesulide are very low in comparison with those achieved following oral intake. After a single application of 200 mg of nimesulide, in the gel form, the highest plasma level of 9.77 ng/ml was noted after 24 hours. No trace of the main metabolite 4-hydroxy-nimesulide, was detected. At steady-state (day 8) peak plasma concentrations were higher (37.25 ± 13.25 ng/ml, but almost 100 times lower than those measured following repeated oral administration.

5.3 Preclinical safety data

The local tolerance and the irritation and sensitisation potential of Nimesulide 3% have been tested in several recognised animal models. The results of these studies indicate that Nimesulide 3% is well tolerated.

Preclinical data for systemically administered nimesulide reveal no special hazards for humans based on conventional studies of safety pharmacology, repeated dose toxicity, genotoxicity and carcinogenic potential. In repeated dose toxicity studies, nimesulide showed gastrointestinal, renal and hepatic toxicity. In reproductive toxicity studies, embryotoxic and teratogenic effects (skeletal malformations, dilatation of cerebral ventricles) were observed in rabbits, but not in rats, at maternally non-toxic dose levels. In rats, increased mortality of offspring was observed in the early postnatal period and nimesulide showed adverse effects on fertility.

Pharmacokinetics of nimesulide

A. Bernareggi[1] *and K.D. Rainsford*[2]

[1]Cell Therapeutics Inc., Europe, Bresso, Italy; [2]Biomedical Research Centre, Sheffield Hallam University, Howard Street, Sheffield S1 1WB, UK

Introduction

Nimesulide (4-nitro-2-phenoxymethanesulfonanilide) is a non-acidic, non-steroidal anti-inflammatory drug (NSAID) usually administered orally in the form of tablets and granules (sachets). The normal dosage is 100 mg twice daily. Rectal administration of suppositories is also employed, although to a minor extent. Normal dosage for rectal administration is 200 mg twice daily. In some countries, a suspension of nimesulide and a topical formulation (gel) are also commercially available. Pharmacokinetic studies have been performed with all these formulations.

Comprehensive reviews have been published on nimesulide pharmacokinetics in healthy volunteers after single and multiple administration [1–3], and on the effect of age and disease on the pharmacokinetic variables [4, 5].

This chapter provides an up-to-date description of the processes related to nimesulide absorption, distribution, metabolism and elimination in animal species and in humans.

Physicochemical factors governing the oral bioavailability of nimesulide

The ability of nimesulide to cross the intestinal barrier was evaluated by the Ussing chamber [6]. This simple *in vitro* method is based on the assessment of the drug permeability through a portion of intact colon mucosa from the rabbit. The rate of drug penetration is parameterised through the apparent permeability coefficient, P_{app}, calculated according to the following equation:

$$P_{app} = \frac{V \cdot dC/dt}{A \cdot C_o}$$

Where: dC/dt = concentration change in the receiver with time, V = volume of the receiving chamber, A = exposed tissue area, C_o = concentration of the test compound in the donor chamber.

According to the P_{app} value, permeability of drugs is classified as follows:

Low permeability:	$P_{app} < 3.3 \times 10^{-6}$ cm/sec
Intermediate permeability:	3.3×10^{-6} cm/sec $< P_{app} < 9 \times 10^{-6}$ cm/sec
High permeability:	$P_{app} > 9 \times 10^{-6}$ cm/sec

Nimesulide shows a high permeability coefficient (Tab. 1), with a value of P_{app} = 48.04×10^{-6} cm/sec. In Table 1, the P_{app} values of compounds with different molecular weight and hydro-lipophilic balance, including representative drugs administered orally, are compared. A significant correlation can be derived between LogP and P_{app} for the tested molecules. Moreover, a sigmoidal model can be fitted to the plot of estimated oral bioavailability in humans *versus* the values of P_{app} for the tested molecules [6].

The good intestinal permeability of nimesulide depends on its favourable intrinsic hydro-lipophilic balance [7] and the low molecular weight (MW 308). Nimesulide solubility in water is rather modest and depends on the pH of the environment and on the temperature [7] (Tab. 2). Because of its pKa = 6.4, solubility at neutral or slightly basic pH results to be higher than in acidic media. By increasing the temperature from 25 °C to 37 °C, the solubility doubles. At pH 7.4 and 37 °C, nimesulide solubility is 82.87 mg/L.

The octanol/water partition coefficient of nimesulide, as expressed by the LogP value, has been evaluated in media with different pH [7]. The LogP values indicate marked hydrophobic properties for nimesulide (Tab. 3). The pH of the aqueous solution influences the LogP value: as expected on the basis of the pKa value, the LogP decreases with the pH increase.

Table 1 – Apparent permeability coefficients (P_{app}) of various compounds determined by Ussing chamber [6]

Compound	**MW**	**Log(P) octanol/water, pH 7.4**	**$P_{app} \times 10^{-6}$ cm/sec, 37 °C**
PEG-90	4000	–5.1	0.61
Mannitol	182	–3.1	2.41
Penicillin G	372	–1.8	3.22
Propranolol	259	1.5	9.37
Phenytoin	252	2.5	16.85
Naloxon	327	1.5	31.06
Diazepam	285	2.8	33.96
Nimesulide	308	1.8	48.04

Table 2 – Aqueous solubility of nimesulide [7]

Solvent	Temperature	Solubility (mg/L)
Water	25 °C	5.5
	37 °C	11.4
Saline (0.9% NaCl)	25 °C	5.5
	37 °C	12.2
Buffer, pH 1	25 °C	4.5
	37 °C	7.8
Buffer, pH 6.8	25 °C	12.0
	37 °C	27.6
Buffer, pH 7.4	25 °C	33.6
	37 °C	82.9

Table 3 – Octanol/water (1:10 v/v) partition coefficient of nimesulide [7]

Solvent	LogP, 25 °C
Water	2.5
Saline (0.9% NaCl)	2.5
Buffer, pH 1	2.6
Buffer, pH 6.8	2.2
Buffer, pH 7.4	1.8

The acid environment of the stomach seems therefore to be particularly favourable for the absorption of nimesulide. According to the Handerson-Hasselbach equation:

$$pH - pK_a = \log \frac{[A^-]}{[HA]}$$

in which $[A^-]$ and [HA] represent the molar concentrations of ionised and unionised nimesulide, respectively, we can easily calculate that at pH 3 (or lower) nimesulide is completely unionised (>99.9%). In this form, nimesulide passively crosses the lipidoidal mucosal membranes and is easily absorbed.

Similarly, the small bowel, characterised by a large absorption surface area and neutral properties of the lumen environment, appears to be favourable to nimesulide absorption. At pH 6 and 7, the percentage of unionised nimesulide decreases to 72% and 20%, respectively. In colon, nimesulide absorption is not favoured by a more limited absorption surface area and a slightly basic pH, which reduces the unionised form of nimesulide.

According to the biopharmaceutical classification system (BCS) [8], drugs may be divided in four groups:

Class 1: high solubility and high permeability
Class 2: low solubility and high permeability
Class 3: high solubility and low permeability
Class 4: low solubility and low permeability
On the basis of its properties, nimesulide can be included in Class 2.

It has been suggested [9] that one of the chemical features of nimesulide which accounts for its low gastrointestinal ulcerogenic activity is its high pKa (6.5). This contrasts with the lower pKa of other NSAIDs (e.g., carboxylates) which are more ulcerogenic than nimesulide. During the transit through the mucosal cells, carboxylates may dissociate intracellularly and release H^+ ions [10] which cause local acidification and cell necrosis. The high pKa of nimesulide may prevent a significant intracellular acidic dissociation and minimise the ulcerogenic potential. Reduced gastrointestinal side effects of nimesulide are also related to its mechanism of action: nimesulide exerts its anti-inflammatory activity by preferential inhibition of COX-2, with reduction of proinflammatory prostaglandins but not of cytoprotective molecules such as prostacyclin. In contrast, several NSAIDs inhibit COX-1, which causes reduction of the synthesis of cytoprotective compounds and unwanted gastrointestinal side effects.

Animal pharmacokinetics

The pharmacokinetic profile of nimesulide in animals was well described in male rats after administration of 2.5 mg/kg of the radiolabeled compound, [^{14}C] nimesulide, by intravenous (i.v.) and oral (p.o.) administration [11]. The plasma concentrations of total radioactivity, of the unchanged drug and of its main metabolite 4′-hydroxynimesulide (M1) were determined by liquid scintillation counting and by a validated HPLC/UV method.

After i.v. and oral administration to rats, the area under the curve (AUC) of unchanged nimesulide was similar to that of total radioactivity. This observation indicates that most of circulating radioactivity is represented by the unchanged drug and the presence of metabolites in the central compartment is limited (Fig. 1).

The main pharmacokinetic parameters for total radioactivity, unchanged nimesulide and the metabolite 4′-hydroxynimesulide (M1) are given in Table 4.

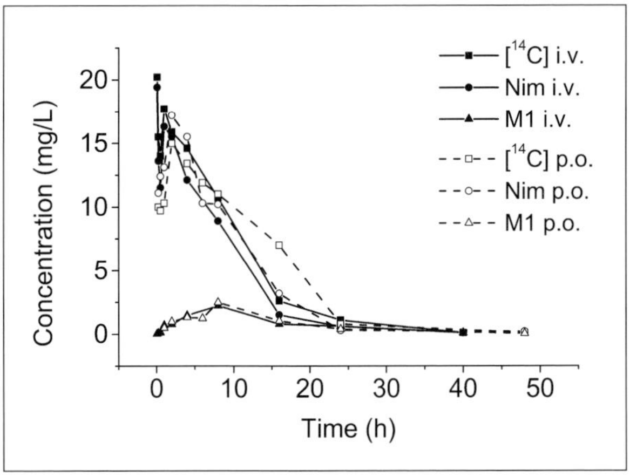

Figure 1
Plasma concentrations of total radioactivity ([14C]), nimesulide (Nim) and metabolite 4′-hydroxy-nimesulide (M1) in rats after i.v. and p.o. administration of 2.5 mg/kg [14C] nimesulide.

The volume of distribution (Vz) of nimesulide in the rat is low despite the good permeability properties of this drug. Vz represents only approximately 10% of the body volume (20% in humans [3]). This observation can be explained by a high binding of nimesulide to plasma proteins, which may retain the compound in the plasma compartment thus limiting nimesulide diffusion from plasma to the tissue interstitial space and cells. Protein binding studies have not been performed in animal plasma. However, in comparison with humans where 99% of the drug is bound to proteins [3], we can assume a high binding also in rats.

[^{14}C]Nimesulide was found in almost all the organs. The highest concentrations were attained in the fat tissue, the liver, kidneys, lungs, adrenals, gut, and heart between 1–4 h after the administration, whereas the brain showed low concentrations. The tissue-to-plasma concentration ratios for total radioactivity were generally lower than the unity during the entire observation interval (up to 48 h). This finding indicated a low affinity of the drug for tissue components and no accumulation in tissue compartments.

The systemic clearance (CL) evaluated in rats after i.v. administration is 16.1 mL/h/kg. Assuming that nimesulide oral bioavailability (F) in humans is close to the unity, the value of CL/F reported for humans after oral administration, that range from 31–106 mL/h/kg [3], appears to be higher than the clearance value observed in rats. Therefore, the rate of nimesulide elimination in rats appears to be from 2–7 times lower than in humans, probably due to a different rate and extent of drug metabolism. Indeed, the AUC ratio between M1 and nimesulide is 32% to 71% in humans (see page 87 of this chapter), and about 20% in rats.

Table 4 – Pharmacokinetic parameters in male rats after oral and i.v. administration of [^{14}C]nimesulide [11]

Route	**i.v. [^{14}C]**	**i.v. Nimesulide**	**i.v. M1**	**p.o. [^{14}C]**	**p.o. Nimesulide**	**p.o. M1**
Parameter	**(mg/L)**	**(mg/L)**	**(mg/L)**	**(mg/L)**	**(mg/L)**	**(mg/L)**
AUC (h.mg/L)	191.7	154.9	34.4	184.5	179.2	35.9
$t_{1/2,z}$ (h)	5.3	4.5	5.3	7.0	6.1	5.42
Vz (mL/kg)	100	104	–	–	–	–
CL (mL/h/kg)	13.0	16.1	–	–	–	–
MRT (h)	7.6	6.7	–	10.6	9.2	–
fe (faeces) (% dose)	68.0	–	–	61.3	–	–
fe (urine) (% dose)	28.9	–	–	26.5	–	–

AUC = area under the plasma concentration-time curve from time zero to infinity.
$t_{1/2,z}$ = apparent terminal half-life.
Vz = apparent volume of distribution in the post-distribution phase.
CL = systemic clearance.
MRT = mean residence time.
fe (faeces) = fraction of administered dose excreted in faeces 5 days after the administration.
fe (urine) = fraction of administered dose excreted in urine 5 days after the administration.

After oral administration, the AUC values for total radioactivity, nimesulide and M1 were similar to the corresponding values observed after intravenous administration. This indicates a complete absorption of the drug from the rat gastrointestinal tract and anticipates the excellent oral bioavailability of nimesulide found in humans.

Five days after the i.v. administration, the percentage of dose excreted was 28.86% in urine and 68.03% in faeces. Similar figures were found after oral administration (Tab. 4). Therefore, differently from humans (see page 71 of this chapter), the excretion of radioactive nimesulide and metabolites in the rat occurs mostly via the faeces, the renal excretion being a minor excretion route.

A large number of epidemiological studies have addressed the possible protective effect of anti-inflammatory drug use with regard to Alzheimer's disease (AD) [12]. Chronic use of NSAIDs in arthritis showed to have implications for prevention of progressive cognitive impairments and may decrease the risk of developing AD. Nimesulide concentration in the rat brain ranges between 400–700 ng/g within 16 h from administration. These concentrations are much greater than those that proved to exert neuroprotection in *in vitro* models. Treatment of rat neuronal B12 cells and mouse hippocampal HT22 cells with nimesulide *in vitro* was able to protect significantly from glutamate toxicity (10 mM, 24 h) at concentrations as low as 1×10^{-12} M (0.3 ng/L), as assessed by the lactate dehydrogenase assay [13]. Assuming that nimesulide unbound fraction in the rat brain tissue is 0.01 (like the unbound fraction in human plasma, *fu*), we may predict that after an oral administration of 2.5 mg/kg nimesulide, the unbound drug concentrations in the rat brain range from 4–7 ng/g, much larger than the concentrations showing neuroprotective activity on brain neurons *in vitro*. On this observation, clinical trials to investigate the efficacy of nimesulide in Alzheimer's disease may be envisaged.

In another study in rats nimesulide was given as a single 1 mg/kg i.v. bolus dose [14]. Multicompartmental pharmacokinetic analysis revealed values of systemic clearance, CL = 21.4 ± 1.10 mL/h/kg, volume of distribution, Vz = 187 ± 3.62 mL/kg, and apparent terminal half-life, $t_{1/2,z}$ = 3.94 ± 0.210 h, of nimesulide that are consistent with the estimates reported in Table 4 for unchanged nimesulide after a single i.v. administration of 2.5 mg/kg [^{14}C] nimesulide.

A parenteral formulation of nimesulide was administered i.m. to rats at doses of 1.5–25 mg/kg to determine the acute anti-inflammatory effects in the carrageenan paw oedema assay in relation to the pharmacokinetics of the drug at the highest dose [15]. The rate of absorption appeared to be slower than that observed following oral administration of the drug. Peak plasma concentrations of 23 mg/L were obtained at 115 min after injection then declined to half this value at 4–6 h. The plasma elimination half-life, $t_{1/2,z}$, was 4.2 h and the $AUC_{(0-6)}$ was 83.31 mg/L.h. The peak plasma concentration of nimesulide coincided with the maximal time for inhibition of the paw oedema which occurred at 2–3 h past injection.

Table 5 – Pharmacokinetic properties of nimesulide given 5 mg/kg by single intravenous, intramuscular and oral routes of administration to dogs [16]

Parameters (units)	**i.v.**	**i.m.**	**p.o.**
$t_{1/2,z}$ (h)	8.5	14	6.2
AUC (mg h/L)	351	228	173
t_{max} (h)		10.9	6.1
C_{max} (mg/mL)		6.1	10.1
CL (mL/kg/h)	15.3	–	–
Bioavailability %	–	69	47
V_{SS} (L/kg)	0.18	–	–

CL = plasma clearance.
$t_{1/2,z}$ = apparent terminal half-life.
AUC = area under the plasma concentration-tiome curve from 0 to infinity.
C_{max} = maximum plasma concentration.
t_{max} = time to C_{max}.
V_{SS} = steady state volume of distribution.

Toutain and co-workers [16, 17] undertook a detailed investigation of the pharmacokinetics of nimesulide in dogs in relation to its pharmacodynamic properties comparing COX-2 inhibition, anti-inflammatory and antipyretic properties. The authors employed a nominal dosage of 5 mg/kg which was given i.v., i.m. and p.o. (single and multiple doses). The nominal dosage, emerged from later pharmacodynamic studies on anti-inflammatory/analgesic and antipyretic effects, was the optimal dosage for these therapeutic properties (these aspects are discussed in detail in Chapter 5). The pharmacokinetic properties of nimesulide given by these three routes of administration are summarised in Table 5. These studies reveal that the plasma clearance of i.v. nimesulide is relatively slow and the plasma $t_{1/2,z}$ is long (8.5 h). The $t_{1/2,z}$ after oral administration of the drug was shorter (6.2 h) and that from i.m. injection longer (14 h). The later suggests that the lipophilic characteristics of nimesulide account for some retention of the drug in the aqueous environment of muscle tissue. The volume of distribution is also low and this would be expected to be related to the plasma protein binding and physicochemical properties (LogP, pKa) of the drug. The plasma concentrations at which the ED_{50} analgesic activity was achieved was 6.25 mg/L (at 1.34 mg/kg dose) and for antipyretic activity this was 2.72 mg/L (at a dose of 3.0 mg/kg).

Pharmacokinetics in humans

In most pharmacokinetic studies of nimesulide in healthy volunteers and different patient populations, the concentrations of the parent compound and of the main metabolite, i.e., the 4′-hydroxy derivative (M1), in plasma and urine were determined by HPLC [18–28]. Sample handling involves the extraction of nimesulide, M1 and the internal standard from acidified biological samples using organic solvents. After solvent evaporation, the extract residue is dissolved in the mobile phase and analysed by reverse phase HPLC with UV/VIS detection. Accuracy and precisions evaluated in plasma and urine samples for nimesulide and M1 are satisfactory for application of the methods to the analysis of biological samples in pharmacokinetic studies. The lower limit of quantitation (LLOQ) ranges from 25–50 ng/ml. A column-switching technique was also introduced [22]. This involves the direct injection of deproteinised plasma samples into an ODS extraction column, followed by chromatographic separation of nimesulide, M1 and the internal standard on a C_{18}-analytical column. UV detection is made at 330 nm.

Absorption

The favourable physical–chemical properties of nimesulide presented earlier in this chapter may explain the good oral bioavailability of this drug, evaluated in several studies in healthy individuals [23–26]. Nimesulide is rapidly absorbed from the gastrointestinal tract and the rate and the extent of nimesulide absorption are similar whether the drug is administered in tablet, suspension or granular form. Indeed, similar maximum concentration (C_{max}), time to C_{max} (t_{max}), and AUC values have been estimated after oral administration of different formulations to fasting healthy individuals (Tab. 6).

After oral administration of a 100 mg dose to healthy fasting subjects, a mean C_{max} of 2.86–6.50 mg/L was achieved within 1.22–2.75 h [19, 23–30]. Nimesulide concentrations of approximately 25–80% of the C_{max} appeared at the first sampling time, 30 min after administration. Pharmacological effectiveness appears to be exhibited earlier than time to C_{max}, from 30 to 60 min after administration [31, 32]. In 100 hospitalised children with acute upper respiratory tract infections and fever (body temperature 38.5 °C), the mean body temperature was decreased significantly 1 h after administration of a single dose of nimesulide suspension 5 mg/kg [31]. In the same study, the t_{max} in paediatric patients receiving nimesulide 50 mg (granules) was close to 2 h.

No studies of intravenously administered nimesulide were performed in this study and, therefore, the absolute bioavailability (F) of oral nimesulide has not been evaluated. However, the extent of oral nimesulide absorption may be deduced from mass balance studies (Tab. 7).

Table 6 – Pharmacokinetic parameters for nimesulide in healthy adult volunteers after single and multiple doses. Mean values [3]

No. subjects and gender	Study design	Dosage form	Dose (mg)	C_{max} (mg/L)	R_{max}	t_{max} (h)	C_{12} (mg/L)	R_{min}	AUC_{0-12} (mg/L.h)	R_{av}	AUC (mg/L.h)	$t_{1/2,z}$ (h)	CL/F (mL/h/kg)	Vz/F (L/kg)	Ref.
12M+12F	SD	Tablets	100	6.50		2.32	1.31				51.8	4.52	39.33	0.20	26
6M	SD	Granules	25	1.36		2.17					12.30	4.44	29.75	0.18	19
		Granules	50	2.30		3.00					22.57	4.63	34.40	0.20	
		Granules	100	4.80		2.50					54.09	4.73	31.02	0.18	
12M	SD	Tablets	50	1.98		2.51					7.87	1.64	104.59	0.22	27
		Tablets	100	3.42		1.67	0.13				14.65	1.80	106.16	0.26	
		Tablets	200	5.81		2.13	0.21				25.00	1.82	121.60	0.29	
18M	SD	Tablets	100 (fasted)	3.83		1.86	0.18				17.67	2.00	82.34	0.22	23
		Tablets	100 (fed)	3.02		1.75	0.21				15.89	2.21	90.88	0.27	
		Granules	100 (fasted)	4.11		1.34	0.19				18.30	2.27	81.30	0.24	
18M	SD	Suspension	100	4.58		1.22	0.15				18.37	2.00	81.81	0.22	24
		Suspension	100	4.18		1.89	0.15				17.50	2.06	86.18	0.25	
		Granules	100	4.26		1.78	0.12				17.32	1.96	86.41	0.23	
6M+6F	SD	Tablets	100	2.86		2.63	0.57		19.07		22.69	3.63	74.77	0.39	28
	MD[a]	Tablets	100 bid × 7 days	3.11	1.09	2.67	0.77	1,36	22.56	1.18		4.00[b]	75.23[b]	0.44[b]	
	SD	Tablets	100	2.93		2.75	0.60				23.84	3.84	70.96	0.39	
	SD	Suppository	200	2.32		4.17	1.08		18.36		27.26	5.75	130.98	1.25	
	MD[a]	Suppository	200 bid × 7 days	2.94	1.27	4.08	1.05	0,97	24.36	1.33		4.76[b]	152.55[b]	1.08[b]	
	SD	Suppository	200	2.14		4.58	0.98				25.11	5.17	139.89	1.15	

Table 6 – (continued)

No. subjects and gender	Study design	Dosage form	Dose (mg)	C_{max} (mg/L)	R_{max}	t_{max} (h)	C_{12} (mg/L)	R_{min}	AUC_{0-12} (mg/L.h)	R_{av}	AUC (mg/L.h)	$t_{1/2,z}$ (h)	CL/F (mL/h/kg)	Vz/F (L/kg)	Ref.
3M+3F	SD	Tablets	100	3.61		2.67	0.54				28.03	2.98	70.24	0.27	29
6M+6F	SD	Tablets	200	9.85		3.17	2.44				81.97	4.95	50.76	0.35	30
	MD[c]	Tablets	100 bid × 7 days	6.17		2.50	1.69		43.00		57.82	4.75	35.38	0.19	
	MD[a]	Tablets	100 bid × 7 days	8.37	1.36	2.50	2.70	1.60	66.13	1.54		4.81	31.49	0.19	
6M	SD	Tablets	200	4.72		2.31	1.03				39.57	3.61	76.18	0.33	25
	SD	Granules	200	5.60		2.08	1.29				46.14	3.41	73.33	0.31	

[a] *At day 7.*

[b] *Data not available in the reference [28], calculated by the author of this chapter using a model-independent approach.*

[c] *At day 1.*

Symbols and abbreviations: C_{max} = maximum plasma concentration; t_{max} = time to C_{max}; C_{12} drug concentration observed in plasma 12 h after administration; AUC_{0-12} and AUC = area under the plasma concentration-time curve from 0 to 12h and to infinity; $t_{1/2,z}$ = apparent terminal half-life; CL/F = total plasma clearance; Vz/F = volume of distribution in the postdistribution phase; R_{max} = ratio of C_{max} at steady state to C_{max} after the first dose; R_{min} = ratio of trough concentrations (C_{12}) at steady state and after the first dose; R_{av} = ratio of AUC_{0-12} values at steady state and after the first dose; SD and MD = single and multiple dose study; M = males; F = females; bid = twice daily.

Table 7 – Excretion pattern in healthy volunteers after a single oral dose administration of nimesulide [3]

No. of subjects and gender	Administered drug	Dose (mg)	Collection interval (days)	AUC_{nim} / $AUC_{[14C]}$ (%)	Excretion in urine (% dose)							Total in faeces (% dose)	Ref.
					nime-sulide	M1	M2	M3	M4	M5	Total		
6M	[^{14}C]nimesulide	100	5	55							60.2[a]	17.9[a]	33
6M	[^{14}C]nimesulide	200	10	46	<1	6.1	4.5	1.7	0.1	6.4	50.5[a]	29.2[a,b]	34
4M	[^{14}C]nimesulide	100	7	48	5.6	14.1	2.6		4.5	11.0	62.5[a]	36.2[a]	35
4M+4F	Nimesulide	200	3		<1	32.7	3.8	2.4	2.3	29.3	70.5	21.5[c]	37
3M+3F	Nimesulide	200	4		<1	17.6	0.7	2.5		19.0	39.8		38
6M+6F	Nimesulide	200	3		<1	16.9	2.4	4.2		8.4	31.9		39

[a] *Total radioactivity.*

[b] *Mainly unconjugated M2 (4.1%), M3 (3%), and M5 (2.3%).*

[c] *Unconjugated M3 (19.7%) and M5 (1.8%).*

AUC_{nim} = area under the plasma concentration-time curve for nimesulide; $AUC_{[14C]}$ = area under the plasma concentration-time curve for total radioactivity; M1 = 2-(4′-hydroxyphenoxy)-4-nitro-methansulfonanilide; M2 = 2-phenoxy-4-amino-methansulfonanilide; M3 = 2-(4′-hydroxyphenoxy)-4-amino-methansulfonanilide; M4 = 2-phenoxy-4-N-acetylamino-methansulfonanilide; M5 = 2-(4′-hydroxyphenoxy)-4-N-acetylamino-methansulfonanilide.

In different studies where [^{14}C] nimesulide was given orally [33–35], the radioactivity excreted in urine ranged from 50.5–62.5%, and in faeces from 17.9–36.2%. In another study with oral nimesulide 200 mg in tablet form, urinary recovery was 70.5% and faecal recovery 21.5% [37]. In urine the parent drug is almost absent and the recovery is attributable to metabolite excretion [33–39]. Only 6.3% [34] to 8.7% [36] of the parent drug was found in faeces after [^{14}C] nimesulide administration. From these studies we can deduce that about 50–70% of the administered dose is absorbed by the gastrointestinal tract and enters the systemic circulation before being excreted in urine. Given that excretion of unchanged nimesulide in faeces is less than 10% of the administered dose, and that nimesulide metabolites excreted in faeces reflect, at least in part, the proportion of drug that is absorbed and then passes, after biotransformation, into the bile or otherwise into the gut, nimesulide absorption after oral administration may be assumed to be complete. This conclusion is also supported by observations in rats [11]. After intravenous administration of radiolabeled nimesulide to male rats, 68% of the dose was recovered in faeces. This indicates that a large excretion of the parent compound and/or metabolites into the bile or the gut occurs in animals and possibly in man.

After single oral administration of 100 mg nimesulide in fasting volunteers, the area under the plasma concentration-time curve (AUC) values range from 14.65–54.09 mg/L.h (Tab. 6). Appreciable nimesulide concentrations of 0.12–1.31 mg/L were still measurable 12 h after administration, the time at which the successive dose is given in the recommended multiple dose regimen.

Regional absorption

For most oral drugs, the optimal absorption site is the small bowel, despite the rapid transit time in this region of the gastrointestinal (GI) tract, typically no more than 4–6 h [40]. The regional absorption of nimesulide in the GI tract was studied in nine healthy subjects using a special delivery system (InteliSite® capsules) combined with gamma-scintigraphy [41]. These capsules are radio-frequency activated, non-disintegrating devices capable of delivering therapeutic agents to specific regions of the GI tract in a non-invasive manner [42]. A ^{111}In (1 MBq) marker was incorporated into the radioactive tracer port of the capsule in order to assess the movement of the capsule through the gut. In order to establish that the drug was released from the device at the desired activation site, a radiolabeled marker (4 MBq ^{99m}Tc-DTPA) was added to the drug reservoir with the liquid drug formulation. Capsules were filled with 100 mg nimesulide dissolved in PEG 400 and activated either in the proximal small-bowel, the distal small bowel or the ascending colon. The control leg was a radiolabeled immediate release gelatin capsule filled with the same solution.

Nimesulide resulted to be well absorbed by the GI tract when released in the stomach and in the proximal small bowel. Stomach and proximal small bowel account for about 40% of nimesulide absorption. A clear-cut differentiation between the contributions given by each of these two absorption sites to the extent of drug absorption is problematic because the drug transit through the stomach into the proximal small bowel is rapid.

The distal small bowel appears also to have an important role in the absorption process. This region is responsible for about 50% of the whole nimesulide absorption. The colon contributes marginally to nimesulide absorption. The drug bioavailability in the ascending colon relative to that observed for the control leg was only 10%. Most of the GI tract appears therefore to be involved in nimesulide absorption, from the stomach to the distal small bowel, whereas the colon showed a poor nimesulide absorption capacity. Metabolite M1 measurements confirm the same trend observed for the parent drug and indicate that the metabolic pattern of nimesulide is not affected by the absorption site.

This study proved to be very informative regarding the actual sites of nimesulide absorption and to assess whether a modified release formulation for nimesulide could be envisioned. The drug is currently administered twice daily as fast release formulations (tablets, granules, suspension). In chronic treatment of inflammation and pain it is often preferable to administer NSAIDs as modified release preparations which minimise peak and trough concentration fluctuations in plasma, provide relatively steady plasma concentrations over the time interval between successive doses and improve patient compliance. A condition for the suc-

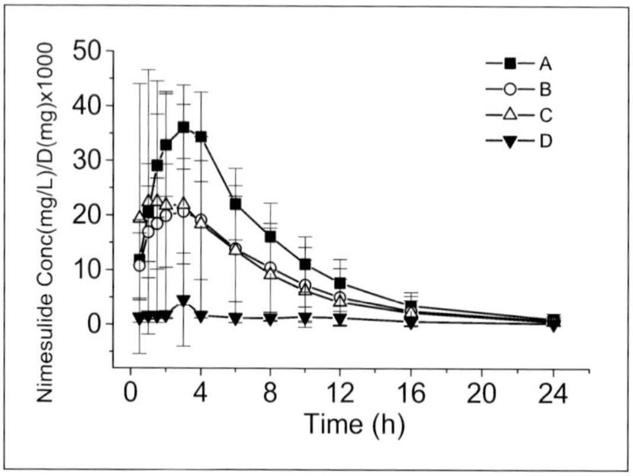

Figure 2
Mean (± SD) dose-normalised concentrations of nimesulide in plasma of healthy volunteers after administration of 100 mg nimesulide dissolved in PEG400 as a gelatine capsule (A, control leg), and as an InteliSite® capsule with drug release in the proximal small-bowel (B), in the distal small bowel (C) or in the ascending colon (D).

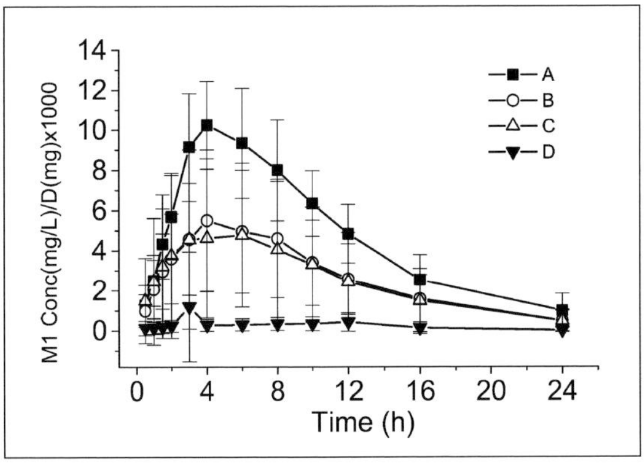

Figure 3
Mean (± SD) dose-normalised concentrations of M1 in plasma of healthy volunteers after administration of 100 mg nimesulide dissolved in PEG400 as a gelatine capsule (A, control leg), and as an InteliSite® capsule with drug release in the proximal small-bowel (B), in the distal small bowel (C) or in the ascending colon (D).

cessful development of a modified release formulation is that the drug is absorbed throughout the whole intestine, including the colon, thus ensuring a prolonged drug uptake into the systemic circulation and sustained plasma concentrations.

The results of this study did not support the possible development of a once-a-day modified release formulation for nimesulide. In fact, gastric empting time ranges in general from 0.5–1 h and small bowel transit time from pyloric sphincter to ascending colon is about 3–4 h post-dose. These transit times are rather reproducible, particularly in the fasted state. Therefore, a modified release oral formulation, intended for a 24 h drug delivery, is expected to spend about 20 h in the colon, i.e., 80% of total intestinal transit time, and to deliver most of the dose in that part of the GI tract. As a consequence, a good drug bioavailability from the colon is an essential factor in developing a once-a-day modified release formulation. In the case of nimesulide, due to the limited absorption properties of the colon for this drug, a modified release formulation is not expected to add a significant clinical value to the currently used formulations. On the contrary, the administration of a modified release formulation might significantly reduce the nimesulide bioavailability.

Effect of food on oral absorption

The presence of food has a limited influence on the rate and extent of nimesulide absorption. Oral administration of nimesulide 100 mg tablets to healthy males

Table 8 – Pharmacokinetic parameters for 4′-hydroxynimesulide (M1) after single and multiple oral doses of nimesulide to healthy volunteers. Mean values [3]

No. of subjects and gender	Dosage form	Dose (mg)	C_{max} (mg/L)	R_{max}	t_{max} (h)	C_{12} (mg/L)	R_{min}	AUC_{0-12} (mg/L.h)	R_{av}	AUC (mg/L.h)	$t_{1/2,z}$ (h)	Ref.
Single dose												
6M	Granules	25	0.32		4.33					6.10	6.79	19
	Granules	50	0.49		7.33					8.45	8.70	
	Granules	100	0.96		6.33					17.96	8.72	
12M	Tablets	50	0.85		3.08	0.17				6.52	3.93	27
	Tablets	100	1.43		2.93	0.27				10.90	3.58	
	Tablets	200	2.49		3.01	0.44				18.69	3.41	
18M	Tablets	100, fasted	1.36		3.25	0.29				11.06	3.16	23
	Tablets	100, fed	1.27		3.76	0.34				11.32	3.27	
	Granules	100, fasted	1.35		2.95	0.28				11.06	3.42	
18M	Susp.	100	1.57		2.61	0.28				12.43	3.28	24
	Susp.	100	1.53		3.28	0.31				12.68	3.06	
	Granules	100	1.54		2.81	0.25				11.76	2.89	
6M+6F	Tablets	200	3.03		5.33	1.45				35.49	4.78	30
Multiple doses												
	Tablets	100 bid, at day 1	1.60		4.67	0.95		13.91				30
	Tablets	100 bid, at day 7	2.34	1.46	4.25	1.17	1.23	20.29	1.46			

Symbols and abbreviations: C_{max} = *maximum plasma concentration;* t_{max} = *time to* C_{max}*;* C_{12} *drug concentration observed in plasma 12 h after administration;* AUC_{0-12} *and AUC = area under the plasma concentration-time curve from 0 to 12 h and to infinity;* $t_{1/2,z}$ = *apparent terminal half-life;* R_{max} = *ratio of* C_{max} *at steady state to* C_{max} *after the first dose;* R_{min} = *ratio of trough concentrations* (C_{12}) *at steady state and after the first dose;* R_{av} = *ratio of* AUC_{0-12} *values at steady state and after the first dose; M = males; F = females; bid = twice daily.*

after a standard American breakfast resulted in C_{max} of approximately 20% lower than that obtained under fasting conditions [23]. However, neither t_{max} nor AUC were significantly modified by food intake (Tab. 6). For the main nimesulide metabolite, M1, the C_{max}, t_{max}, and AUC values after a meal were similar to those under fasting conditions (Tab. 8).

Distribution

The plasma concentration-time profiles obtained after oral administration of nimesulide have mostly been analysed in accordance with a model-independent approach. In some studies, a bi-exponential modelling was proposed, in which the first exponential term represented the absorption process and the second the elimination process [28, 30]. A clear-cut distribution phase cannot be usually identified from the plasma concentration-time curve by use of a semi-logarithmic scale, indicating that the nimesulide distribution process is fast. Therefore, a one-compartment open model is generally appropriate to describe the pharmacokinetic profile of nimesulide after oral administration. In a few individuals the plasma kinetic profile was described by a tri-exponential equation [30]. In such cases, a definite distribution phase emerged, and a two-compartment open model was considered to be more appropriate for describing the data.

The extent of drug distribution can be evaluated by estimation of the volume of distribution in the post-distribution phase (Vz/F), which represents the actual volume of distribution, assuming that F is close to unity (see 'Absorption' section). After single oral 100 mg dose administration, Vz/F values range from 0.18–0.39 L/kg (Tab. 6), indicating that nimesulide is mainly distributed in the extracellular fluid compartment. Nearly all NSAIDs have a relatively small volume of distribution, Vz/F usually ranging from 0.1–0.2 L/kg [43–50]. In fact, with the exception of salicylates [46], NSAIDs are generally extensively bound to human serum albumin and less than 1% of the total plasma concentrations are in an unbound form, available to distribute to extravascular tissues [43]. On the basis of the low estimates of Vz/F, there is no evidence that nimesulide might accumulate in tissue compartments.

Specific distribution studies have been performed with oral nimesulide in female genital tissues [51] and in the synovial fluid of patients with rheumatoid arthritis [52]. Twelve women undergoing hysterectomy and salpingo-oophorectomy received a single oral dose of nimesulide 100 mg 1–6 h before surgery. The nimesulide concentrations in the cervix, fundus, oviduct and ovaries ranged from 0.3–0.55 µg/g at 1 h, 0.58–0.97 µg/g at 2 h, 1.11–1.79 µg/g at 4 h, and 0.37–0.76 µg/g at 6 h. Although cervical tissue comprises mainly collagen and fundal uterine tissue comprises smooth muscle, there was no significant difference in the distribution of nimesulide. At the fourth hour, when tissue concentrations were highest, the tissue-to-serum ratio was lower than the unity, as observed in the rat, ranging from 0.39–0.62 [51]. A good clinical response to nimesulide was seen patients with dysmenorrhoea and corresponded to the distribution of the drug in the genital tissues [32].

The penetration of nimesulide into the articular cavity was evaluated in six patients with rheumatoid arthritis treated with nimesulide 100 mg tablets twice daily for 7 days. Three and 12 h after the last dose, nimesulide concentrations in the synovial fluid were 2.39–1.38 mg/L, respectively. The synovial fluid-to-plasma concentration ratios were 0.44 (3 h) and 0.54 (12 h) [52].

Binding to blood components

The low tissue:plasma and synovial fluid:plasma ratios may be related to high plasma protein binding, as with other NSAIDs, which keeps the drug predominantly in the plasma compartment. The plasma protein binding of nimesulide has been studied *in vitro* using equilibrium dialysis. In a first study, at plasma concentration of 0.5–10 mg/L, the unbound fraction (*fu*) of nimesulide in human plasma varied from 0.7–4%, which is indicative of extensive plasma protein binding [53]. In a second study, the serum binding of nimesulide was constant (*fu* 1%) over the concentration range of 0.77–20 mg/L [54]. Using pure human serum albumin (735 µM), nimesulide binding was non-saturable and super-imposable to that observed using human serum. A weak binding to α_1-acid glycoprotein and lipoproteins was observed, whereas there was no binding to gamma-globulin. Erythrocyte-bound nimesulide was found only in the buffer rather than in plasma, indicating a strong affinity for plasma proteins [54].

After oral administration of 100 mg [^{14}C] nimesulide to healthy volunteers, the whole blood:plasma ratio of mean total radioactivity was approximately 0.6 at 4, 8 and 24 h. This suggests that nimesulide (and minor other radiolabeled components) are not associated *in vivo* with blood cells and do not significantly enter the erythrocytes [36].

Elimination

After single dose oral administration of nimesulide 100 mg, plasma concentrations of the parent drug declined mono-exponentially following the peak. The apparent mean elimination half-life ($t_{1/2,z}$) varied from 1.80–4.73 h (Tab. 6). The variation in $t_{1/2,z}$ values can, at least in part, be attributed to the different methods of data analysis used – non-compartmental analysis in some studies [19, 23–27], multi-exponential modelling (bi- or tri-exponential models) with the use of weighting factors, in others [28, 30].

The total plasma clearance (CL/F) of nimesulide varied from 31.02–106.16 mL/h/kg after oral administration of a 100 mg dose, and was almost exclusively attributable to metabolic clearance (Tab. 6). Nimesulide is a low-clearance drug: assuming that the liver is the only organ for metabolising this drug and that the absorption across the gut wall is complete (F = 1), the hepatic extraction ratio, calculated from the ratio of CL/F to hepatic plasma flow, is approximately 0.1. As a consequence of the low extraction ratio, the CL/F of nimesulide may, in principle, vary proportionally with any possible change in *fu* caused by physiopathological factors or drug–drug interactions.

Excretion

The excretion of the parent drug in urine and faeces resulted to be negligible in most of oral [18, 33–39] and rectal [55] administration studies, with only 6.3–8.7% of the parent drug found in faeces after [^{14}C] nimesulide administration [34, 36]. Indeed, nimesulide is mainly eliminated following metabolic transformation.

In dose balance studies involving oral administration of [^{14}C] nimesulide [34, 36], 78.1–98.7% of the radiolabeled dose was recovered, of which urinary excretion accounted for 50.5–62.5% and faecal excretion 17.9–36.2% of the administered dose (Tab. 7). These results indicate that nimesulide and its metabolites are mainly excreted by the renal route.

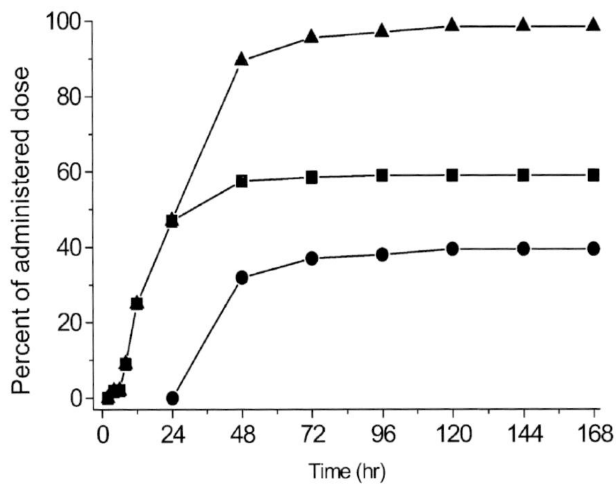

Figure 4
Mass balance of [^{14}C] nimesulide in healthy volunteers after oral administration of 100 mg radiolabeled drug. Percent of administered dose excreted in urine (■), faeces (●) and in total (▲).

In volunteers treated orally with unlabelled oral nimesulide 200 mg [37–39], urinary excretion of the known nimesulide metabolites accounted for 31.9–70.5% of the administered dose. Faecal excretion was 21.5% [37] (Tab. 7). The low urinary recovery of nimesulide in some studies with administration of unlabeled nimesulide is likely due to incomplete urine collection or incomplete mass balance (nimesulide metabolites identified later were not included in the mass balance estimation).

Metabolism

Nimesulide is extensively metabolised. A total of 16 metabolites of nimesulide were identified and the biotransformation of nimesulide in man was shown to proceed by three principle routes, cleavage of the molecule at the ether linkage, reduction of the NO_2 group to NH_2 and phenoxy ring hydroxylation. Other metabolites arise from the concomitant hydroxylation and reduction, acetylation of the amino group, conjugation with either glucuronic acid or sulphate of hydroxylated metabolites [34, 36, 56, 57].

A comprehensive determination of nimesulide metabolic pathway has been established in fasted male volunteers who received a single oral dose of 100 mg [^{14}C] nimesulide [36]. Radiolabeled metabolites were identified by LC-MS and LC-MS/MS with reference to synthesised metabolite standards. Recovery of the dose was essentially quantitative (>97%) for all subjects. The major proportion of the administered radioactivity was excreted via urine, accounting for 59–66% (mean value 62.5%). Radioactivity excreted in the faeces accounted for a further 33–39% (mean value 36.2%). Greater than 92.4% of the urinary (0–24) radioactivity was now accounted for by characterised metabolites. Methanol extraction of faeces recovered approximately 60% of the faecal radioactivity and greater than 40% of this radioactivity was identified as nimesulide.

Nimesulide was identified in extract of faeces and in plasma. Cleavage of the ether linkage gives metabolite 6 which is conjugated with glucuronic acid to give metabolite 17. Reduction of the NO_2 group to NH_2 is proposed to produce the intermediate metabolite 2, which is hydroxylated to produce metabolite 3. Metabolite 3 is acetylated to produce metabolite 5. An alternative route is possible for the production of metabolite 5. It is proposed that metabolite 2 is acetylated to the postulated intermediate metabolite 4 which is in turn hydroxylated to give metabolite 5. This second pathway is less likely because the acetylation of the NH_2 group is thought to occur in the kidney and therefore is a terminal metabolic reaction. Metabolite 5 is conjugated with sulphate and glucuronide to give metabolites 18 and 14, respectively. Ring hydroxylation of nimesulide gives metabolite 1; conjugation of this molecule with sulphate gives metabolite 9. Conjugation of metabolite 1 with glucuronic acid gives metabolite 10. It is proposed that the principle position of hydroxylation is consistent with the reference standard M1; however a

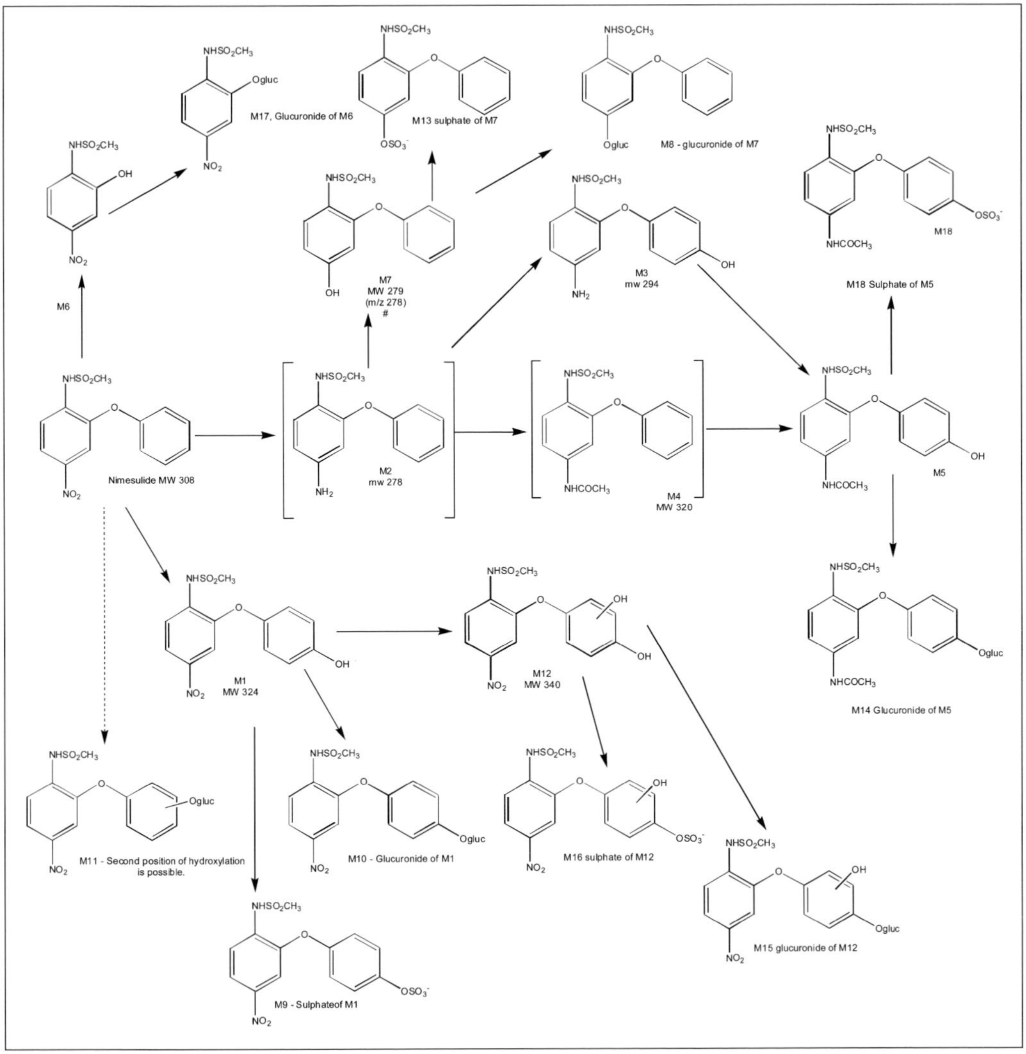

Figure 5
Metabolic pattern of nimesulide in humans (Based on data in Ref. 36)

second position of hydroxylation is proposed to give rise to a second glucuronide conjugate of molecular weight 500 (metabolite 11). Metabolite 1 is hydroxylated in a second position to give metabolite 12 which is conjugated with glucuronide and sulphate to give metabolite 15 and 16, respectively. Greater than 92.4% of the urinary (0–24) radioactivity is accounted for by characterised metabolites.

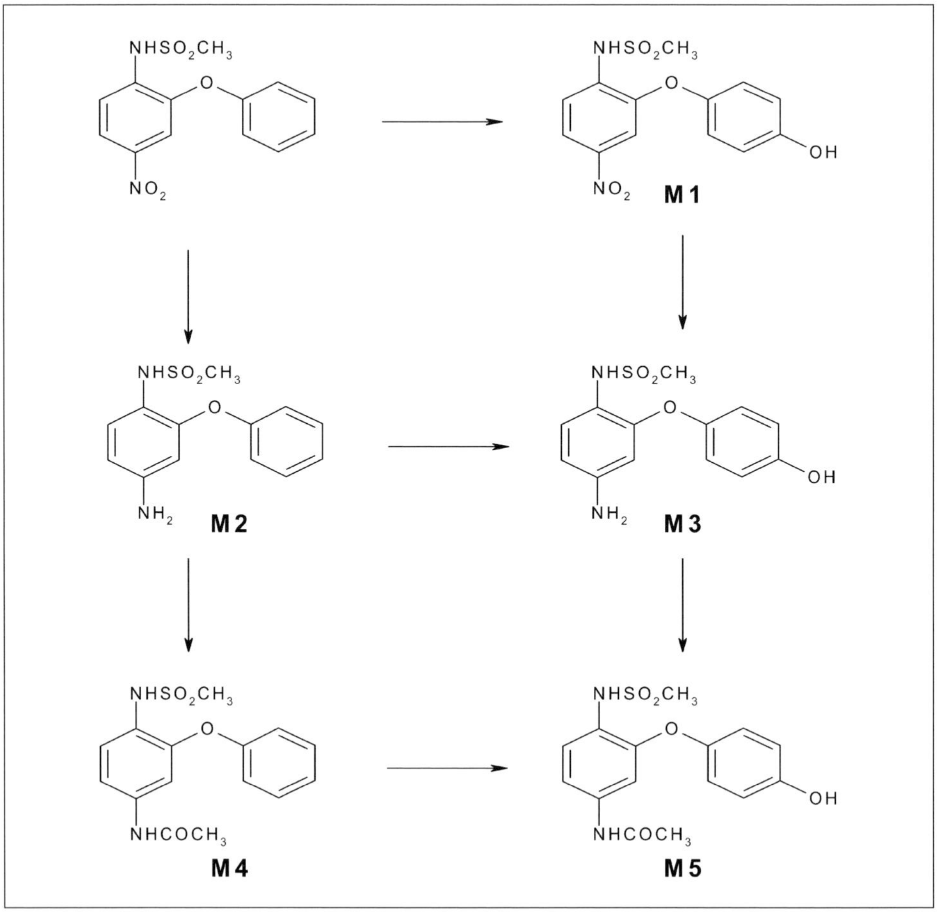

Figure 6
Simplified metabolic patterns of nimesulide in humans (reproduced from A. Bernareggi [3] with permission).

The metabolites identified during this investigation extended the observations from previous studies [34, 37, 38, 56, 57] in which only 5 metabolites of nimesulide were found in human urine (M1–M5).

Metabolites M1, M2 and M5 were confirmed during the more recent study [36] but M3 and M4 were not detected. These metabolites were previously reported as being present at low concentrations. As a consequence they are proposed as intermediates in the full biotransformation pathway. Additional phase 1 metabolites (M6, M7 and M12) have been identified which were not previously detected. The structural assignments of M6 and M7 and their glucuronide conjugates were confirmed with authentic reference standards. A large portion of the administered dose of nimesulide was excreted as glucuronide and sulphate conjugates.

The main metabolites are represented by M1, found in plasma and urine, and M5, found in urine and faeces. In urine, M1 and M5 are present almost completely in conjugated form. In faeces, M5 is mainly unconjugated. Tables 7 and 9 provide comprehensive quantitative data of the excretion of unchanged nimesulide and its metabolites according to the different authors.

No differences were observed in the metabolic profile between males and females [56].

The only important metabolite that can be followed in plasma is the 4′-hydroxy-derivative, M1. Earlier studies indicated that the isozyme of the cytochrome P450 family CYP1A2, may be responsible for the hydroxylation of nimesulide to M1 [58]. However, it has also been proposed that CYP2C9 and CYP2C19 may be implicated in nimesulide hydroxylation reactions [59]. Other important enzymes involved in nimesulide biotransformation are the nitroreductases that are flavoproteins responsible for the reduction of nitro-arenes to amino-arenes through the formation of reactive species such as the nitroso-group and the hydroxyl-

Table 9 – Nimesulide metabolite excretion in 0–24 h urine [36]

Metabolite	% administered dose
M1	0.4
M1 glucuronide (M10)	14.6
M1 isomer (a)	1.0
M1 isomer glucuronide (M11)	2.4
Total M1	**18.4**
M5 glucuronide (M14)	4.6
M5 sulphate (M18)	2.5
Total M5	**7.1**
M6 glucuronide (M17)	5.1
M7 glucuronide (M8)	6.4
M12 glucuronide (M15)	2.3
M12 sulphate (M16)	2.8
Total M12	**5.1**
Unknown	4.1
Unknown glucuronide conjugates	7.4
Total metabolites excreted in urine	**53.6**

(a) hydroxylation of the phenoxy-ring in a position different than 4′.

amine group with generation of the superoxide anion and ROS providing oxidative stress to cells. Other enzymes, involved in phase II reactions for nimesulide metabolism, are the N-acetyl-transferase (NAT), the uridine diphosphate glucuronosyl-transferase (UGT) and sulphotransferase (ST).

After administration of [^{14}C] nimesulide, the ratio between AUC values of the parent drug and total radioactivity in plasma ranged from 46–55% [33, 34, 36] (Tab. 7). Therefore, unchanged nimesulide in the plasma compartment represents approximately half of the circulating nimesulide-related species. Most of the remaining radioactivity AUC was attributable to M1; other metabolites in plasma, if any, are of minor importance.

The cumulative plasma concentration-time curve obtained by adding M1 concentrations to those of the parent drug was almost super-imposable to the total radioactivity profile, confirming that no other metabolic species are present in significant amount in the plasma compartment [34, 36].

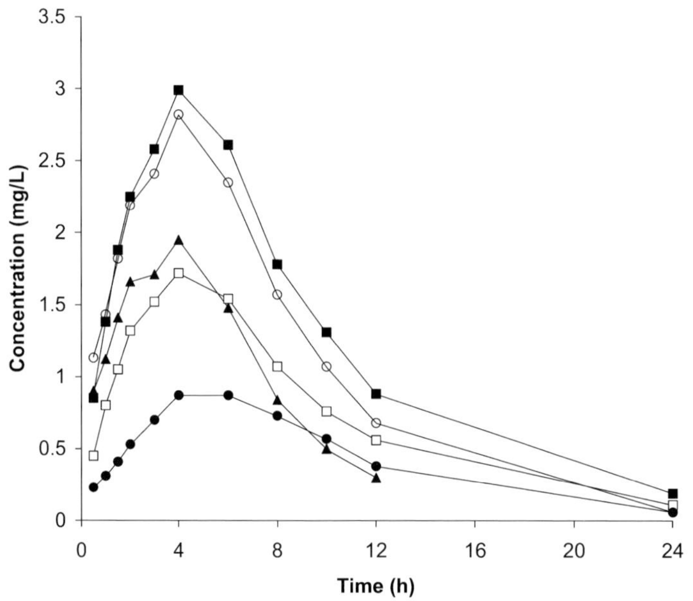

Figure 7
Temporal profiles of concentrations of total radioactivity in whole blood (□) and plasma (■), of unchanged nimesulide (▲) and its main metabolite 4′-hydroxynimesulide (●) in plasma, of combined nimesulide and M1 (○) after administration of 100 mg [^{14}C]nimesulide in healthy individuals.

Some activity and toxicity data have been generated for nimesulide metabolites. In *in vitro* models, that is, NADPH-dependent lipid peroxidation in rat liver microsomes and xanthine/xanthine oxidase iron-promoted depolymerisation of hyaluronic acid, nimesulide and its main metabolites M1 and M5 exhibited dose-dependent radical scavenging activity. In lipid peroxidation assays, M1 was more active (IC_{50} = 30 µM) than M2 (IC_{50} = 500 µM) and nimesulide (IC_{50} was 800 µM) [60]. Nimesulide, M1 and M2 can protect hyaluronic acid from oxidative stress; M1 and M2 are far less active than the parent drug in this assay [60].

Pharmacological tests *in vivo* showed that metabolites M1 to M5 are endowed with anti-inflammatory and analgesic properties, although their potency is lower than that of nimesulide [60, 61]. In the carrageenan oedema test in rats, different oral doses of nimesulide and its metabolites were administered 1 h before carrageenan challenge. The anti-inflammatory effect was observed 1, 3 and 5 h after carrageenan administration. At the third hour, a similar reduction of oedema *versus* the control group was generally achieved with metabolite doses at least 10-fold greater than the nimesulide dose. M1 proved to be more potent than M2, M3 and M4; M5 was almost inactive. ED_{50} values were 1.7 mg/kg (nimesulide), 40 mg/kg (M1), 55 mg/kg (M2), 62 mg/kg (M4). Similar findings were observed in the writhing test in the mouse. Nimesulide and its metabolites were administered orally 30 min before para-phenylquinone administration. The analgesic effect was evaluated for 30 min after administration of different doses of nimesulide and its metabolites. The parent drug proved to be 5- to 10-fold more potent than M1, M2, and M4. ED_{50} values were 5.5 mg/kg (nimesulide) and 54 mg/kg (M1). Metabolite toxicity (Irwin test) was evaluated for M2 and M3. Both metabolites did not induce gene mutations in strains of *Salmonella typhimurium* [61, 62].

Plasma pharmacokinetics of 4′-hydroxynimesulide (M1)

The pharmacokinetic profile of M1, the only nimesulide metabolite detected in plasma, has been studied after oral administration of the parent drug [19, 23, 24, 27, 30]. The main pharmacokinetic parameters are reported in Table 8.

After single dose administration of nimesulide 100 mg, the C_{max} of M1 ranged from 0.96–1.57 mg/L and was attained within 2.61–6.33 h (t_{max}), that is, 1–3 h later than that of the parent drug. The percentage ratio of M1 AUC to unchanged nimesulide AUC [corrected for the different molecular weights (MW) of the parent drug (MW 308) and its metabolite (MW 324)], ranged from 32–71%. This value indicates that the biotransformation of nimesulide to the hydroxylated metabolite represents a major elimination pathway for the drug.

The apparent terminal phase of the pharmacokinetic profile of M1 after oral administration of nimesulide is characterised by a $t_{1/2,z}$ value of 2.89–8.72 h, 1.5-

to 2-fold higher than that of nimesulide. This observation indicates that the elimination rate of M1 is not limited by the formation rate from the parent drug, but only by its own elimination characteristics. Therefore, the observed terminal half-life represents the actual elimination half-life of M1.

Linearity

Linearity is characterised by dose-proportionality for C_{max} and AUC, and by the absence of change, as a function of the administered dose, in those pharmacokinetic parameters that express the rate of drug absorption (e.g., t_{max}), the extent of drug distribution (e.g., Vz), and the efficiency of the eliminating organs at removing the drug from the body (e.g., CL). A linear pharmacokinetics suggests that in the range of doses used, no saturation of absorption, distribution and elimination mechanisms occur and no metabolic induction/inhibition mechanisms are present.

Nimesulide kinetics appears to be linear up to 100 mg, whereas indications of nonlinearity seem to emerge after administration of doses as high as 200 mg. This conclusion is supported by different studies. A crossover pharmacokinetic study in six males treated with oral doses of nimesulide 25, 50, and 100 mg (granules) [19], showed a proportional increase in the C_{max} and AUC of the parent drug with the administered dose (Tab. 6). After normalisation of C_{max} and AUC values for the respective doses, C_{max}/D (0.054, 0.046, and 0.048 L^{-1}) and AUC/D (0.49, 0.45, 0.54 h/L) were relatively constant over the tested dose range. No significant change in the t_{max}, CL/F, $t_{1/2,z}$, and Vz/F of nimesulide were found between doses (Tab. 6). The pharmacokinetic parameters for M1, indicate that the extent of drug metabolism does not vary significantly in the nimesulide dose range tested (Tab. 8). After administration of 25, 50 and 100 mg, C_{max}/D (0.013, 0.010, 0.010 L^{-1}) and AUC/D (0.24, 0.17, 0.18 h/L) values were relatively constant.

In another crossover pharmacokinetic study, 12 male subjects received oral doses of nimesulide 50, 100, and 200 mg in a tablet form [27]. In the tested dose range, increases in C_{max} and AUC were not proportional to the dose increase (Tab. 6). C_{max}/D values (0.040, 0.034, and 0.029 L^{-1}) decreased with increasing doses of nimesulide from 50 to 200 mg. The same trend was observed for AUC/D values (0.16, 0.15, 0.13 h/L). The values of t_{max} and $t_{1/2,z}$ were relatively insensitive to the dose escalation, however CL/F and Vz/F increased slightly with the dose increase (Tab. 6). Parameters for M1 are consistent with these findings (Tab. 8). A progressive decrease in C_{max}/D (0.017, 0.014, 0.012 L^{-1}) and AUC/D (0.13, 0.11, 0.09 h/L) was observed with increasing doses of nimesulide (50–200 mg). Similar conclusions can be drawn from the results of another investigation in which nimesulide was administered at doses of 100 and 200 mg [30]. Mean C_{max} and AUC values were less than proportional to the dose increase (Tab. 6). C_{max}/D (0.062

and 0.049 L^{-1}) and AUC/D (0.58 and 0.41 h/L) decreased slightly from 100 to 200 mg. The t_{max} and $t_{1/2,z}$ did not change significantly, however CL/F and Vz/F increased remarkably with the dose increase (Tab. 6). C_{max}/D for M1 was 0.015 and 0.016 after nimesulide 100 and 200 mg.

Similar observations of nonlinearity at higher dose levels have been reported for other NSAIDs, e.g., phenylbutazone [63], naproxen [64], diflunisal [65], as a consequence of non linear binding to plasma proteins. Since NSAIDs have low intrinsic clearance and are highly protein bound, small increases of *fu* that may occur at the higher dosages result in an increase in total clearance and a decrease in plasma concentrations. Nimesulide protein binding in human serum has been reported to be constant (99%) over a concentration range (0.77–20 mg/L) that covers nimesulide therapeutic levels after administration of doses up to 200 mg [54]. Therefore, apparent nonlinearity of nimesulide pharmacokinetics after administration of a 200 mg dose may also be attributable to other factors, for example a slightly reduced bioavailability. Increased drug metabolism at a higher dose can probably be excluded because M1 pharmacokinetic parameters C_{max}/D and AUC/D followed the same trend of the parent drug parameters as a function of the nimesulide dose size.

Rectal administration

Nimesulide is well absorbed when given rectally. In healthy volunteers, the extent of rectal bioavailability has been estimated as 54–64% of the bioavailability with an oral tablet formulation [28] (Tab. 6). By comparing the AUC values obtained in two different studies in paediatric patients, the rectal bioavailability was estimated to be 54% of the bioavailability after granule administration (Tab. 11).

Rectal administration of nimesulide 200 mg as a suppository resulted in a longer t_{max} and reduction in C_{max}/D when compared with values observed after oral administration. After administration of two different suppository formulations containing nimesulide 200 mg, a mean C_{max} of 2.32 and 2.14 mg/L were attained at 4.17 and 4.58 h; AUC values were 27.26 and 25.11 mg/L.h [28] (Tab. 6). Rectal administration provides a prolonged plasma concentration-time profile. Concentrations of nimesulide of 0.98 to 1.08 mg/L are still measurable in plasma 12 h after rectal administration and the $t_{1/2,z}$ observed after rectal administration is higher than that observed after oral treatment [28] (Tab. 6). These results suggest that after rectal administration the apparent terminal phase of nimesulide is affected by the prolonged absorption phase and that the estimated $t_{1/2,z}$ represents the absorption half-life of nimesulide through the intestinal mucosa, since absorption via this route is the rate-limiting step for the pharmacokinetics of nimesulide.

Because of the lower extent of nimesulide bioavailability in suppository form, CL/F and Vz/F estimates are higher than those obtained after oral administration [28] (Tab. 6).

Multiple dose administration

After multiple dose oral administration of nimesulide 100 mg in tablet form twice daily for 7 days, the pharmacokinetic profile of nimesulide appears to be time-independent; that is, the multiple dose regimen does not affect the pharmacokinetic properties of the drug as evaluated in single dose studies [28, 30] (Tab. 6). At steady-state, the t_{max}, $t_{1/2,z}$, CL/F and Vz/F values do not differ from those obtained after administration of the first dose. Accumulation ratios, R_{max}, R_{av} and R_{min}, indicate that only modest accumulation of nimesulide occurs in the body with multiple dose administration (Tab. 6). With repeated rectal administration of nimesulide 200 mg twice daily for 7 days, C_{max} and AUC_{0-12} again increase slightly at steady state relative to the first administration, whereas the C_{min} values are unchanged. The accumulation factors R_{max}, R_{min} and R_{av} were 1.27, 0.97 and 1.33, respectively [28]. Values for t_{max}, $t_{1/2,z}$, CL/F and Vz/F did not alter with multiple dose administration via the rectal route (Tab. 6).

On the basis of the pharmacokinetic parameter estimates obtained from single dose oral studies, steady state plasma concentrations are predicted to occur within a time corresponding to 5–7 half-lives, i.e., within 24–36 h (after 2–3 administrations). This prediction was confirmed by experimental findings. In a multiple dose study in which nimesulide 100 mg was administered twice daily in tablet form, AUC_{0-12} evaluated on day 7 (22.56 mg/L.h) overlapped AUC from time zero to infinity evaluated after a single dose treatment (22.69 mg/L.h). This clearly indicated that steady state was achieved 7 days after treatment was initiated [28]. The same conclusions can be drawn from the data of another study where AUC_{0-12} at steady-state was 66.13 mg/L.h and AUC on day 1 was 57.82 mg/L.h [30].

Considering that terminal half-life of nimesulide after rectal administration is a little longer than that observed after oral administration, steady state is expected to occur within 36–48 hours (3–4 administrations) [28]. Experimental data confirmed this prediction. After rectal administration of nimesulide 200 mg twice daily for 7 days, AUC_{0-12} at steady-state (24.36 mg/L.h) was almost superimposable to the AUC on day 1 (27.26 mg/L.h) [25] (Tab. 6).

As with the parent compound, no accumulation of M1 in the body is foreseen. Indeed, the accumulation of M1 in the body is modest during multiple dose administration of the parent drug. Thus following administration of nimesulide 100 mg in tablet form twice daily for 7 days, the C_{max} increased from 1.60 (day 1) to 2.34 mg/L (day 7), the trough levels C_{12} from 0.95 to 1.17 mg/L, and AUC_{0-12}

from 13.91 to 20.29 (Tab. 8). The accumulation ratios, R_{max}, R_{av} and R_{min} were 1.46, 1.46 and 1.23, respectively [30].

Topical administration

In recent years there has been considerable interest in the development of topical NSAIDs, particularly to deliver the drug to the site of action thus minimising the systemic exposure. Tolerability, efficacy and pharmacokinetic studies have been successfully performed with various gel formulations of nimesulide in animals [67], and in humans [68] (see also Chapter 5).

Following initial *in vitro* investigations [68] Helsinn identified a gel formulation (coded GEL 6TRC) which they went on to investigate its skin irritancy [69], pharmacokinetic properties in humans and clinical evaluation in acute tendonitis and ankle sprains (reviewed Chapter 5). The skin irritation and sensitisation potential of different topical formulations of nimesulide were evaluated in healthy volunteers using the repeated insult patch test [69]. The results showed that topical formulations did not produce either irritant or sensitisation reactions at the test sites.

In the first of the pharmacokinetic studies [70] 18 healthy male volunteers participated in a crossover study in which they applied a single dose of 3% nimesulide gel containing 200 mg of the drug to the back of the knees and after a 7 day washout period they ingested one oral nimesulide 100 mg tablet. The low nimesulide concentrations observed in plasma with the topical drug application indicated a limited systemic exposure to the drug. Nimesulide was detected in plasma of six out of 18 subjects between 1.5–24 h after gel application. The highest plasma concentration, 9.77 ng/mL, was observed in one individual 24 h after topical administration. M1 was not detected in the plasma after topical administration although it was found following oral administration.

In the second pharmacokinetic study a single day's dose followed by repeated daily doses of 3% nimesulide (90 mg) gel t.i.d. for the subsequent 7 days were applied to about 200 cm^2 of the outer part of the shaven right thigh for 8 days [71]. On day 8 before the last morning administration two microdialysis probes were inserted into the *vastus medialis* muscle for collection of interstitial fluid samples. Nimesulide and its principal 4′-hydroxy-metabolite were assayed in plasma, urine and intestinal fluids. The results of this study showed that nimesulide was rapidly absorbed from the gel and, after single dose, maximum plasma concentrations were 13.9 ng/mL while after repeated application for 8 days the minimum and maximum plasma concentrations were 26.5 and 37.3 ng/mL, respectively. The $AUC_{(0-24\,h)}$ values for plasma nimesulide after a single dose were 208.93 ng · h/mL and after repeated dose were about three times high being 725.5 ng · h/mL. The 4′-hydroxy metabolite was present in plasma to about one-third that of the parent drug after single dose ($AUC_{0\text{-}24\,h}$ 75.2 ng · h/mL) while at steady state the AUC values were about half that of nimesulide ($AUC_{0\text{–}24\,h}$ 371.4 ng · h/mL). It appeared, by comparison with the orally administered drug, that the fraction of skin absorption was about 0.4–1.4% of the oral formulation. Thus, the topical formulation has modest systemic impact and is likely to have very little impact on liver drug metabolism.

In the interstitial fluid, nimesulide concentrations >5 ng/mL (LLOQ) were detected in seven out of 72 samples. This finding may be related to the very high protein binding of nimesulide as only free drug would be measurable in the interstitial fluids.

Another indication of transdermal absorption is the urinary excretion data for the 4′-hydroxy metabolite. At steady state the urinary excretion in the 0–8 h period averaged 210 µg. By comparison with the urinary excretion following oral intake of the drug in the first pharmacokinetic study [70] it is estimated that 0.7–3.9% of the applied dose is absorbed transdermally.

This second pharmacokinetic study is interesting, in comparison with the first, where relatively little nimesulide was absorbed into the plasma. The differences could be related to site of application of the gel formulations. In the first study the gel was applied behind the knee while in the second it was on the thigh. The extent of vascularisation action in the skin behind the knee is relatively low where there is relatively little muscle. In contrast, the thigh has higher vascularisation and of course greater muscle mass. Thus, the permeation and extraction of drug would be expected to be greater when applied to the thigh than from the skin behind the knee.

Studies on the permeation of nimesulide through hairless rat skin under different conditions of skin preservation were contrasted with the relatively polar compound melatonin [72]. Skin stored at 4 °C for 2 days showed similar flux of nimesulide compared with that of fresh skin, and then was elevated progressively to be 3.5-fold after 14 days; melatonin had similar flux for up to 7 days and then progressively increased to be 2.4-fold after 14 days. Frozen skin (–22 °C) with or without glycerol preservative showed no difference in flux of nimesulide up to 4 days compared with that of fresh skin. Melatonin showed similar flux to normal skin when stored under the same conditions of freezing for up to 14 days. These studies have important practical implications for studying mechanisms of uptake of nimesulide but as no comparable data are available in human skin the results have limited relevance so far to the human situation. Of interest, however, are the values for the permeability coefficients for nimesulide which ranged from 4.5–5.3 $\times 10^{-2}$ cm/h over 7 days storage at 4 °C and were 5.3×10^{-2} and 4.7×10^{-2} cm/h after storage for 14 days at –22 °C without as with glycerol. Similar ranges of permeability coefficients were observed with melatonin showing that nimesulide in contrast to the more similar compound, melatonin, has excellent permeation characteristics.

Influence of gender

After single and multiple oral administrations of nimesulide tablets the pharmacokinetic parameters of nimesulide are similar in both males and females (Tab. 10).

In Figure 8, the ratios of the mean parameter values found in males and females in the various studies [26, 28–30] are plotted for C_{max}, t_{max}, C_{12}, AUC, AUC_{ss}, $t_{1/2,z}$, CL/F and Vz/F. In general, the pharmacokinetic parameter ratios were randomly scattered around the line (ratio = 1) to indicate that gender does not substantially affect the pharmacokinetics of nimesulide. The only exception is the C_{max} ratio, which was slightly less than unity in all studies and which may indicate faster absorption of nimesulide in females. This finding is consistent with the lower t_{max} values observed in females in six out of eight studies (Tab. 10).

Pharmacokinetic parameters for M1 in males and females can be derived only from one study in which nimesulide tablets 100 and 200 mg were administered [30] (Tab. 10, Fig. 9). The few data available for the metabolite support earlier conclusions that there are no major differences in the kinetics of nimesulide between males and females.

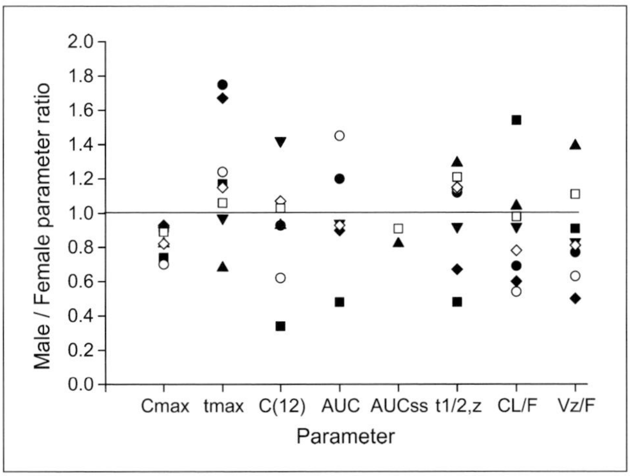

Figure 8
Correlation between nimesulide pharmacokinetic parameters and gender (reproduced from A. Bernareggi [3], with permission).

Table 10 – Influence of gender on the pharmacokinetics of nimesulide and 4′-hydroxynimesulide (M1) in healthy volunteers after single and multiple oral doses of nimesulide. Mean values and standard deviation in parentheses [3]

No. subjects and gender	Study design	Dose (mg)	C_{max} (mg/L)	R_{max}	t_{max} (h)	C_{12} (mg/L)	R_{min}	AUC_{0-12} (mg/L.h)	R_{av}	AUC (mg/L.h)	$t_{1/2,z}$ (h)	CL/F (mL/h/kg)	Vz/F (L/kg)	Ref.
Nimesulide														
12M	SD	100	5.52		2.50	0.67				33.57	2.94	47.73	0.19	26
12F	SD	100	7.48		2.13	1.95				70.07	6.11	30.93	0.21	
6M	SD	100	2.78		3.50	0.55		18.51		25.99	4.06	57.92	0.34	28
6F	SD	100	3.08		2.00	0.59		19.62		21.70	3.63	84.00	0.44	
6M	MD[a]	100 bid × 7 days	2.80	1.01	2.17	0.74	1.35	20.32	1.10		4.51	76.59	0.51	
6F	MD[a]	100 bid × 7 days	3.41	1.11	3.17	0.80	1.36	24.80	1.26		3.49	73.87	0.37	
6M	SD	100	2.71		2.58	0.71				21.96	3.48	71.82	0.35	
6F	SD	100	3.01		2.67	0.50				23.43	3.77	77.71	0.42	
3M	SD	100	3.48		3.33					26.54	2.38	52.68	0.18	29
3F	SD	100	3.73		2.00					29.52	3.57	87.80	0.36	
6M	SD	200	8.09		3.50	1.87				97.15	5.27	35.64	0.27	30
6F	SD	200	11.61		2.83	3.01				66.80	4.64	65.87	0.43	
6M	MD[b]	100 bid × 7 days	5.56		2.67	1.76		40.62		55.85	5.08	31.04	0.17	
6F	MD[b]	100 bid × 7 days	6.78		2.33	1.64		45.37		59.79	4.41	39.72	0.21	
6M	MD[a]	100 bid × 7 days	7.42	1.33	2.67	2.78	1.58	59.55	1.47		5.95	30.84	0.21	
6F	MD[a]	100 bid × 7 days	8.30	1.22	2.51	2.71	1.65	65.58	1.45		4.91	31.44	0.19	

Table 10 – (continued)

No. subjects and gender	Study design	Dose (mg)	C_{max} (mg/L)	R_{max}	t_{max} (h)	C_{12} (mg/L)	R_{min}	AUC_{0-12} (mg/L.h)	R_{av}	AUC (mg/L.h)	$t_{1/2,z}$ (h)	CL/F (mL/h/kg)	Vz/F (L/kg)	Ref.
4'-hydroxynimesulide (M1)														
6M	SD	200	2.56		5.33	1.33				32.31	4.99			30
6F	SD	200	3.49		5.33	1.57				38.67	4.57			
6M	MD[b]	100 bid × 7 days	1.36		4.50	0.82		11.77						
6F	MD[b]	100 bid × 7 days	1.84		4.83	1.06		15.70						
6M	MD[a]	100 bid × 7 days	1.85	1.36	4.50	1.07	1.30	17.43	1.48					
6F	MD[a]	100 bid × 7 days	2.82	1.53	4.00	1.27	1.20	23.16	1.48					

[a] *At day 7.*

[b] *At day 1.*

Symbols and abbreviations: C_{max} = maximum plasma concentration; t_{max} = time to C_{max}; C_{12} drug concentration observed in plasma 12 h after administration; AUC_{0-12} and AUC = area under the plasma concentration-time curve from 0 to 12h and to infinity; $t_{1/2,z}$ = apparent terminal half-life; CL/F = total plasma clearance; Vz/F = volume of distribution in the postdistribution phase; R_{max} = ratio of C_{max} at steady state to C_{max} after the first dose; R_{min} = ratio of trough concentrations (C_{12}) at steady state and after the first dose; R_{av} = ratio of AUC_{0-12} values at steady state and after the first dose; SD and MD = single and multiple dose study; M = males; F = females; bid = twice daily.

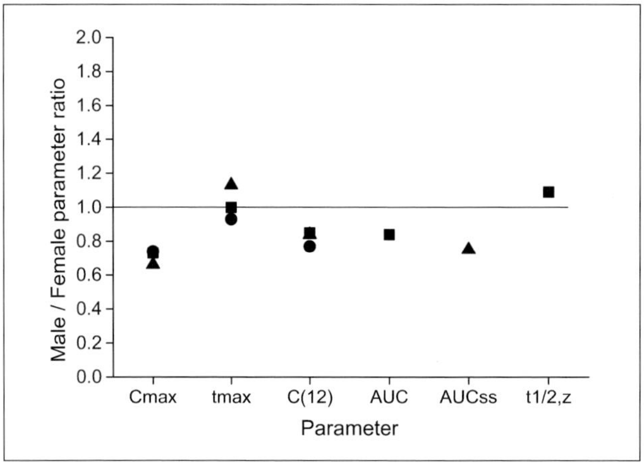

Figure 9
Correlation between M1 pharmacokinetic parameters and gender (reproduced from A. Bernareggi [3], with permission).

Effect of age

Age was found to have a minor effect on the pharmacokinetics of nimesulide in four studies, two in paediatric patients after single oral administration of nimesulide 50 mg (granules) [31] and single rectal administration of 100 mg (suppositories) [66] and two in the elderly after single and multiple dose oral administration of 100 mg (tablets) [73, 74]. The pharmacokinetic parameters of nimesulide and M1 are reported in Tables 11 and 12, respectively.

The oral doses of 50 mg in children and 100 mg in adults and elderly are rather different when expressed per bodyweight (BW) unit (mean values 1.9 [31], 1.3 [23], 1.2 [73] mg/kg, respectively), but similar when expressed per body surface area (BSA) unit (mean values 53.1 [31], 50.9 [23] and 50.3 [73] mg/m^2, respectively).

The individual body surface area (m^2) was calculated, when not available, according to the Mosteller formula [75]:

$$BSA = \sqrt{\frac{H \cdot BW}{3600}}$$

In which H is the height expressed in cm and BW is the bodyweight expressed in kg.

Children

The paediatric use of nimesulide is no longer recommended by the European Medicines Evaluation Agency (EMEA). Doses that have been employed in clinical studies (50 mg orally or 100 mg rectally) were half the dose proposed for adults and provide plasma concentrations similar to those observed in adults who receive the full dose (100 mg orally, 200 mg rectally).

Fourteen children with hypoglycaemia of either sex aged 7–9 years (mean 8.22 years) received a single oral dose of nimesulide 50 mg (granules) [31]. C_{max} (3.46 mg/L) and t_{max} (1.93 h) values were similar to the corresponding values observed in healthy adults after single dose oral administration of nimesulide 100 mg (C_{max} = 2.86–6.50 mg/L; t_{max} = 1.22–2.75 h) and the AUC (18.43 mg/L.h), was within the range of values reported for adults (14.65–54.09 mg/L.h). After normalisation for the body weight, C_{max}/D and AUC/D are significantly lower in children than in adults. This seems to be due to higher systemic clearance (138.59 mL/h/kg) and volume of distribution (0.41 L/kg) in children than in adults (assuming the same extent of oral bioavailability across the two populations).

The pharmacokinetic profile of nimesulide after a single dose of 100 mg in suppository form was studied in 38 children of either genders undergoing minor surgery and requiring anti-inflammatory treatment [66]. The age ranged from 4.1–15 years (mean 8.53 years). Only one blood sample was taken from each child. The pharmacokinetic parameters of nimesulide (Tab. 11) can be compared with those found after the administration of nimesulide 200 mg in suppository form in adults (Tab. 6). In children, the C_{max} (2.24 mg/L) was similar to that in healthy adults after administration of nimesulide 200 mg in suppository form (C_{max} = 2.14–2.32 mg/L). The AUC and t_{max} in children were 20.08 mg/L.h and 3 h, respectively, and lower than the values reported for adults (25.11 and 27.26 mg/L.h and 4.17 and 4.58 h, respectively). The terminal half-life of nimesulide was 3.15 h in children and 5.17 and 5.75 h in adults.

In summary, the pharmacokinetic profile of nimesulide after oral and rectal administration in children is similar to that in adults. However, some minor differences were seen in children after oral administration, including a rather larger distribution (Vz/F) and a more efficient elimination (CL/F) compared to adults.

The elderly

The pharmacokinetic profile of nimesulide was evaluated in the elderly after single and multiple doses of nimesulide 100 mg tablets [73, 74].

Ten elderly male patients aged 65–73 years (mean 68.6 years) with normal plasma creatinine concentrations (0.87–1.19 mg/dL), received a single dose of nimesulide on days 1 and 6. From days 2–5, they received nimesulide twice daily.

Table 11 – Pharmacokinetic parameters for nimesulide in special populations after single and

No. of subjects and gender	Mean age (and range) (years)	Mean plasma creatinine (and range) (mg/dL)	Study design	Dosage form	Dose (mg)	Day of treatm.	C_{max} (mg/L)	R_{max}
Paediatric								
8M+6F[a]	8.22 (7–9)	0.58 (0.4–0.8)	SD	Granules	50		3.46	
29M+16F[b]	8.57 (4.1–17)		SD	Supposi-tory	100		2.24[c]	
Elderly								
10M	68.6 (65–73)	1,02 (0.87–1.19)	SD	Tablets	100		3.73	
			MD		100 bid[e]	Day 6	4.24	1,14
3M+3F[f]	69.5 (65–79)	0.96 (<1.2) (0.74–1.11)	MD	Tablets	100 bid	Day 1	4.61	
					100 bid	Day 7	6.31	1,37
4M+2F[f]	72.8 (67–80)	1.40 (>1.2) (1.22–1.77)	MD	Tablets	100 bid	Day 1	5.70	
					100 bid	Day 7	5.74	1,01
Patients with moderate renal insufficiency								
8M+2F	61.2 (38–70)	41.7 (32–61)	SD	Tablets	100		4.76	
3M+2F,V	40.2 (22–50)	114.6 (100–127)	SD	Tablets	100		4.78	
9M+1F	61.6 (49–69)	47.5[h] (27–74.1)	SD	Tablets	100 bid	Day 1	3.70	
						Day 8	4.50	1,22
Patients with severe hepatic insufficiency								
5M+1F	57.3 (49–71)		SD	Tablets	100		8.33	
6M,V	31.3 (22–37)		SD	Tablets	100		4.66	

[a] *Paediatric patients with hypoglycaemia.*

[b] *Paediatric patients undergoing minor surgery. Three males and 4 females were excluded from the pharmacokinetic evaluation.*

[c] *Calculated by multiplying the maximum concentration (0.075 mg/L/kg) or the AUC (0.671 mg/L/kg.h) of the estimated best fit curve by the mean bodyweight (29.92 kg) of children considered for the data analysis (n = 38).*

[d] *Calculated from the biexponential fitting function.*

[e] *On days 1 and 6, one administration; from day 2 to 5, twice daily administration.*

[f] *First dose of day 1 of a multiple dose regimen (bid × 7 days).*

[g] *Mean creatinine clearance (and range) (mL/min).*

multiple doses. Mean values. Paediatric patients (P) and elderly (E) [3]

t_{max}	C_{12}	R_{min}	AUC_{0-12}	R_{av}	AUC	$t_{1/2,z}$	CL/F	Vz/F	Ref.
(h)	(mg/L)	(mg/L.h)		(mg/L.h)	(h)	(mL/h/kg)	(L/kg)		
1.93	0.26				18.43	2.36	138.59	0.41	31
3.00	1.09[d]				20.08[c]	3.15			66
1.65	0.41		19.63		22.43	3.24	63.60	0.26	73
1.36	0.47	1.15	21.97	1.12		3.24	65.40	0.26	
2.67	1.53		31.29		50.03	5.86	53.10	0.32	74
3.67	2.16	1.41	46.24	1.48		7.94	37.80	0.44	
3.33	2.38		42.89		75.12	8.72	31.30	0.40	
2.83	2.30	0.96	42.78	1.00		6.58	33.00	0.28	
1.90	1.02				36.73	4.50	40.25	0.25	76
2.30	0.44				24.60	3.02	70.14	0.28	
2.10	0.58		20.4		25.00	3.09	87.31	0.29	77
1.90	0.63	1.09	22.8	1.12		2.92	94.12	0.34	
2.33	4.44				234.84	28.68	7.58	0.28	78
2.58	1.20				43.83	5.43	33.70	0.22	

[h] *Values are expressed as mL/min/1.73 m^2.*

Symbols and abbreviations: C_{max} = maximum plasma concentration; t_{max} = time to C_{max}; C_{12} drug concentration observed in plasma 12h after administration; AUC_{0-12} and AUC = area under the plasma concentration-time curve from 0 to 12h and to infinity; $t_{1/2,z}$ = apparent terminal half-life; CL/F = total plasma clearance; Vz/F = volume of distribution in the postdistribution phase; R_{max} = ratio of C_{max} at steady state to C_{max} after the first dose; R_{min} = ratio of trough concentrations (C_{12}) at steady state and after the first dose; R_{av} = ratio of AUC_{0-12} values at steady state and after the first dose; V = control group of healthy volunteers; SD and MD = single and multiple dose study; M = males; F = females; bid = twice daily.

Table 12 – Pharmacokinetic parameters for 4′-hydroxynimesulide (M1) after single and repeated administration of nimesulide in special populations. Mean values. Paediatric patients (P), elderly (E), patients with moderate renal insufficiency (R) and severe hepatic impairment (H) [3]

No. of subjects and gender	Study design	Dosage form	Dose (mg)	C_{max} (mg/L)	R_{max}	t_{max} (h)	C_{12} (mg/L)	R_{min}	AUC_{0-12} (mg/L.h)	R_{av}	AUC (mg/L.h)	$t_{1/2,z}$ (h)	CLR[f] (mL/h/kg)	Ref.
Paediatric														
8M+6F[a]	SD	Granules	50	1.34		3.50	0.36				11.60	4.18		31
Elderly														
10M	SD	Tablets	100	1.10		4.10	0.48		9.46		14.59	6.04		73
	MD	Tablets	100 bid, day 6[b]	1.65	1.50	2.30	0.59	1.23	12.81	1.35		5.30		
3M+3F[c]	MD	Tablets	100 bid, day 1	1.67		5.00	0.77		12.31		21.36	7.45		74
			100 bid, day 7	2.67	1.60	6.50	1.26	1.63	20.27	1.65		8.35		
4M+2F[d]	MD	Tablets	100 bid, day 1	1.61		5.67	1.15		12.42		29,.51	9.74		
			100 bid, day 7	3.03	1.88	2.83	1.70	1.47	23.78	1.91		12.22		

Table 12 – (continued)

No. of subjects and gender	Study design	Dosage form	Dose (mg)	C_{max} (mg/L)	R_{max}	t_{max} (h)	C_{12} (mg/L)	R_{min}	AUC_{0-12} (mg/L.h)	R_{av}	AUC (mg/L.h)	$t_{1/2,z}$ (h)	CLR[f] (mL/h/kg)	Ref.
Patients with moderate renal insufficiency														
8M+2F	SD	Tablets	100	1.03		4.4	0.57				14.33	6.05	2.99	76
3M+2F,V	SD	Tablets	100	1.43		3.6	0.45				13.39	4.02	5.62	
9M+1F	MD	Tablets	100 bid, day 1	1.48		3.8	0.50		10.6		15.4	5.15	13.57	77
			100 bid, day 8	1.93	1.30	3.2	0.60	1.20	14.3	1.35		4.61		
Patients with severe hepatic insufficiency														
5M+1F	SD	Tablets	100	0.38		18.03	0.33				23.37[e]	38.78[e]		78
6M,V	SD	Tablets	100	1.39		5.71	0.71				17.95	5.95		

[a] *Paediatric patients (mean age 8.22 yrs) with hypoglycaemia.*
[b] *On days 1 and 6, one administration; from day 2 to 5, twice daily administration.*
[c] *Creatinine plasma concentration: <1.2 mg/dL.*
[d] *Creatinine plasma concentration: >1.2 mg/dL.*
[e] *n = 5.*
[f] *Conjugated form.*

Symbols and abbreviations: *C_{max} = maximum plasma concentration; t_{max} = time to C_{max}; C_{12} drug concentration observed in plasma 12 h after administration; AUC_{0-12} and AUC = area under the plasma concentration-time curve from 0 to 12h and to infinity; $t_{1/2,z}$ = apparent terminal half-life; R_{max} = ratio of C_{max} at steady state to C_{max} after the first dose; R_{min} = ratio of trough concentrations (C_{12}) at steady state and after the first dose; Rav = ratio of AUC_{0-12} values at steady state and after the first dose; CL_R = renal clearance. V = control group of healthy volunteers; SD and MD = single and multiple dose study; M = males; F = females; bid = twice daily.*

The pharmacokinetic profile of parent compound and M1 were assessed after the single dose administration and at steady state on day 6 [73]. The pharmacokinetic parameters for unchanged nimesulide and M1 are reported in Tables 11 and 12, respectively. All pharmacokinetic parameters for the elderly fell within the ranges of values found in young adults. In particular for nimesulide, CL/F was 63.60 mL/h/kg (adults 31.02–106.16 mL/h/kg), Vz/F was 0.26 L/kg (adults 0.18–0.39 L/kg) and $t_{1/2,z}$ was 3.24 h (adults 1.80–4.73 h). Therefore, we may conclude that the pharmacokinetic profile of nimesulide is similar in elderly and adults and that no dose adjustment is advisable in patients aged <80 years.

The AUC_{0-12} of nimesulide on day 6 of multiple dose administration overlapped with the AUC after single dose administration. This indicates that on day 6 of a twice daily regimen steady state was achieved. As with the young individuals, at steady state modest accumulation of nimesulide and M1 occurred in the elderly. Plasma concentration data showed R_{max}, R_{min} and R_{av} values of 1.14, 1.15 and 1.12, respectively, for nimesulide and 1.50, 1.23 and 1.35 for M1. The values of CL/F, Vz/F and $t_{1/2,z}$ at steady state were the same or similar to those observed after the first administration, indicating no time-dependency of the pharmacokinetics of nimesulide in the elderly.

In a second study, 12 elderly patients of either sex, aged 65–80 years (mean 71.5 years), were divided into two groups of six, according to their plasma creatinine concentration. In Group 1, the creatinine concentration was <1.2 (mean 0.96) mg/dL; in Group 2 (which included some patients with moderately impaired renal function) creatinine concentration was >1.2–2 (mean 1.40) mg/dL [68]. Each individual received nimesulide 100 mg twice daily on days 1–7 and a single dose on day 8. In Group 1, after the first administration (day 1), pharmacokinetic parameters for nimesulide and M1 were comparable to those observed in young healthy volunteers, with the exception of the half-life of nimesulide, and the C_{max} and AUC for M1, which were higher. After the first administration (day 1), Group 2 showed lower CL/F (31.30 mL/h/kg *versus* 53.10 mL/h/kg) and longer $t_{1/2,z}$ (8.72 h *versus* 5.86 h) for nimesulide than Group 1. Nimesulide volume of distribution for the two groups was similar (Vz/F 0.32 and 0.40 L/kg, respectively). Some differences were found also for M1. Group 2 showed higher AUC (29.51 mg.h/L *versus* 21.36 mg.h/L) and $t_{1/2,z}$ (9.74 h *versus* 7.45 h) values than Group 1. At steady state, accumulations of nimesulide and M1 in Group 1 were comparable to those observed in young adults (see the accumulation factors in Tabs 11 and 12). A greater accumulation was found for M1 in Group 2 (R_{av} 1.91). These results might suggest reduced elimination efficiency in the group that includes some patients with moderately impaired function. However, no correlations between plasma creatinine concentrations and $t_{1/2,z}$ and CL/F of nimesulide were found, indicating no relevant influence of mild renal impairment on nimesulide elimination. This is consistent with the fact that nimesulide is eliminated almost completely by metabolic biotransformation. Renal impairment should not

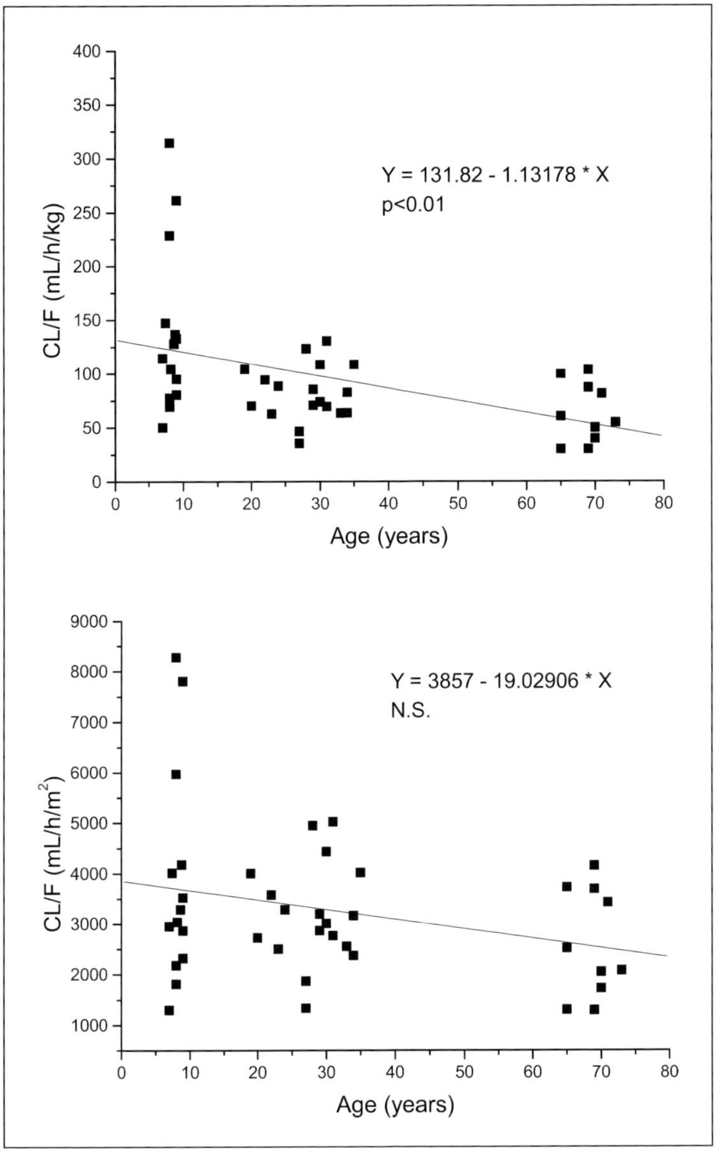

Figure 10
Correlation between systemic clearance of nimesulide and age.

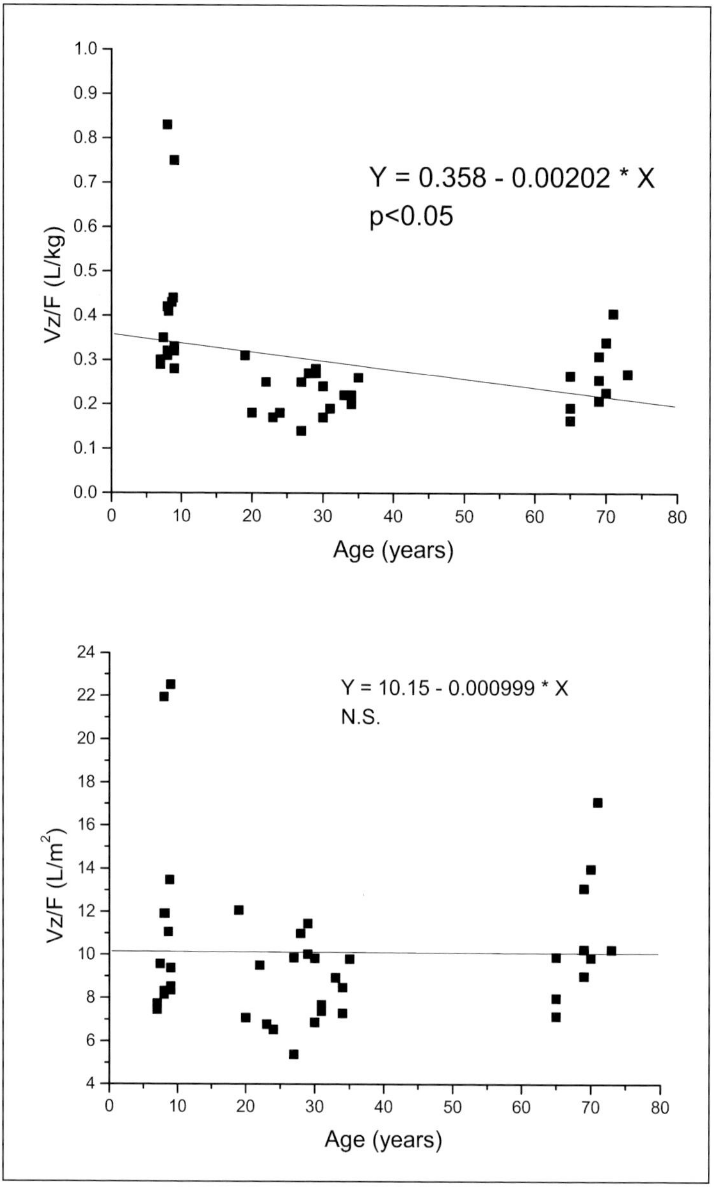

Figure 11
Correlation between volume of distribution of nimesulide and age.

even affect M1 elimination in that the metabolite is excreted almost entirely in conjugated form [34, 35, 37–39].

In both groups, the urinary excretion of nimesulide and unconjugated M1 in urine on days 1 and 7 was <1% of the administered dose. Conjugated M1 was not measured.

The reviewed studies show some differences in the pharmacokinetic profiles of nimesulide in children, adults and elderly, although they may be considered of minor importance from a clinical standpoint. It is interesting to note that significant linear correlations can be found between the individual values of CL/F, Vz/F and age only when the pharmacokinetic parameters are expressed per BW unit (kg). No significant correlations can be observed when these parameters are expressed per BSA unit (m^2) (Figs 10, 11).

The apparent decreasing systemic clearance and volume of distribution from children to elderly observed on BW basis could be the result of combined physiological and anatomical factors associated with age, including:

i) Different concentrations of binding protein in plasma. Total protein concentrations in plasma, including the concentration of albumin, the most relevant nimesulide binding protein, are usually lower in plasma of children compared to those in adult and elderly individuals. In the studies conducted with nimesulide, mean (SD) total plasma protein levels in children [31], adults [23] and elderly [73] were respectively 6.4 (0.4) g/dL, 7.3 (0.5) g/dL and 7.2 (0.4) g/dL. Therefore, we might expect a higher unbound fraction of nimesulide in plasma of children. Considering that nimesulide is a drug with low extraction ratio (ER = 0.1), its elimination efficiency (CL/F) is related to the extent of binding to plasma proteins and is expected to be greater in children than in adults. Similarly, a lower binding to plasma proteins in children may explain a higher volume of distribution.
ii) A progressive reduction with age of extracellular fluids. In adults, a nimesulide volume of distribution of 0.22 L/kg indicates that the drug is mainly distributed in the extracellular fluid compartment. As a consequence, nimesulide is expected to have a larger distribution in children than in adults.
iii) Different rates of liver metabolism with age. In general, enzyme activity increases with age and reaches adult levels by puberty. This factor should be in favour of a lower CL/F in children than in adults, the opposite of what we observed. Therefore, we may deduce that the effect of protein binding on CL/F is prevalent over that of age-related variability in liver enzyme activity between children and adults. Likely, nimesulide metabolism does not differ between the three patient populations as indicated by the mean AUC_{M1}/AUC_{Nim} ratio, which is 0.63 in children, 0.63 in adults and 0.65 in elderly.

Effect of moderate renal insufficiency

The pharmacokinetics of nimesulide in patients with moderate renal impairment was evaluated in two studies, after single [76] and multiple dose [77] oral administration of nimesulide 100 mg (tablets).

The results of these studies do not show unequivocally that the pharmacokinetic profiles of nimesulide and its hydroxylated metabolite are altered in patients with moderate renal failure. No dose adjustment in patients with $CL_{CR} > 1.6$ L/h can be advised.

In one study, 10 patients with moderate renal impairment (creatinine clearance CL_{CR} 1.92–3.66 L/h), aged between 38–70 years (mean 61.2 years), received a single oral dose of nimesulide 100 mg. In parallel, a group of five healthy volunteers, aged 22–50 years (mean 40.2 years) with a CL_{CR} of 6–7.62 L/h (mean 6.88 L/h), received the same treatment [76]. The C_{max}, t_{max} and Vz/F values for nimesulide were similar in the two groups, whereas the AUC and $t_{1/2,z}$ values were significantly higher in patients with renal impairment than in the healthy volunteers (Tab. 11). The mean CL/F in those with renal insufficiency (40.25 mL/h/kg) was significantly lower than the corresponding value (70.14 mL/h/kg) in the healthy volunteers. According to the results of this study, moderate renal insufficiency seems to affect the elimination of nimesulide. Metabolite and urinary excretion data support this conclusion. The AUC and $t_{1/2,z}$ of M1 were higher in patients with renal insufficiency than in healthy volunteers (Tab. 12). The mean cumulative urinary excretion of M1 (conjugated form) was 3.18 mg for patients with renal impairment and 5.17 mg in healthy volunteers. The renal clearance was 2.99 mL/h/kg in patients with renal impairment and 5.62 mL/h/kg in healthy volunteers (Tab. 11). However, all the pharmacokinetic parameter values for nimesulide and M1 observed in patients with renal impairment fell in the range observed for healthy volunteers (Tab. 6).

In a second study, 10 patients, aged between 49–69 years (mean 61.6 years), with moderate renal impairment (CL_{CR} 1.62–4.45 L/h) received a single dose of nimesulide 100 mg on days 1 and 8, and 100 mg twice daily on days 2–7 [77]. The pharmacokinetic parameters of nimesulide and M1 are detailed in Tables 11 and 12. After the first dose, all the pharmacokinetic parameters for nimesulide and M1 fell within the range of parameter values observed for healthy individuals. Following twice daily administration, steady state was achieved after the second administration. The accumulation of nimesulide and M1 in the plasma compartment was modest: the accumulation factors R_{max}, R_{min} and R_{av} for both species were slightly higher than the unity and were similar to the corresponding values found in healthy individuals. The $t_{1/2,z}$, CL/F, and Vz/F of nimesulide after the first dose and at steady state were similar (Tab. 11) and indicated that the pharmacokinetics of nimesulide in moderate renal failure is time-independent. After the first dose, the mean cumulative urinary excretion of M1 (free and conjugated, after enzymatic hydrolysis) in patients with renal impairment accounted for 12.8 mg of the dose. The renal clearance was 13.57 mL/h/kg in patients with renal impairment and was not correlated with the creatinine clearance.

The binding of nimesulide in serum samples obtained from patients with renal failure was lower than in serum collected from healthy volunteers. Indeed, the *fu* measured in six patients ranged from 1.53–2.33% (mean 1.98%), whereas the *fu* of control group averaged 1.14% [54]. Values of *fu* were inversely proportional to the albumin concentration.

Effect of severe hepatic failure

The pharmacokinetic profile of nimesulide and M1 was studied in six patients with severe hepatic failure and cirrhosis, after administration of a single oral dose of nimesulide 100 mg in a tablet form. A control group of six healthy subjects was treated in parallel with the same dose [78]. The severity of hepatic disease was assessed as grade B or C, as defined in Pugh's classification. The clinical and biological symptoms considered for the classification were the stage of hepatic ecephalopathy, presence of ascites, bilirubin concentration, albuminaemia, and Quick time.

The pharmacokinetic parameters of nimesulide and M1 are detailed in Table 11 and 12. In the control group, all parameters for nimesulide and M1 were within the range of values reported for healthy individuals, with the exception of $t_{1/2,z}$ of nimesulide which slightly exceeded the upper limit of the normal range. Hepatic insufficiency modified the pharmacokinetic profile of nimesulide and its hydroxy-metabolite to a significant extent. This was expected because nimesulide is almost exclusively eliminated by hepatic metabolism. The nimesulide parameters C_{max}, C_{12}, AUC, and $t_{1/2,z}$ were much higher than the corresponding values in healthy individuals, whereas CL/F was significantly lower. As for M1, the C_{max} was much lower than that in healthy subjects and was reached later. The AUC_{0-24} of M1 was lower in patients with hepatic insufficiency than in healthy subjects, whereas at infinity the AUC was more similar. The $t_{1/2,z}$ of M1 reached a mean value of 38.78 h in patients with hepatic impairment. The results of the aforementioned study clearly indicate that hepatic impairment reduces the rate of elimination of nimesulide and M1 substantially.

The binding of nimesulide in serum samples obtained from patients with hepatic insufficiency was lower than that found in serum collected from healthy volunteers. Indeed, *fu* measured in five patients with hepatic insufficiency ranged from 2.73–6.26% (mean 4.16%), whereas *fu* in the control group averaged 1.14% [54]. Values of *fu* were inversely proportional to the albumin concentration. An increase of *fu* may explain the higher Vz/F in patients with hepatic impairment in comparison with healthy volunteers. However, it is worth noting that the Vz/F in patients with hepatic impairment fell within the range of normal values.

Drug interaction studies

Pharmacokinetic interactions occur when the absorption, distribution and/or elimination processes of a drug are altered by the concomitant administration of another drug. Several studies have examined the effect of concomitant drug administration on the pharmacokinetic profile of nimesulide (Tab. 13).

In general, pharmacokinetic interactions between nimesulide and other drugs are absent or marginal, and are unlikely to be of clinical relevance [79].

Glibenclamide

The possible occurrence of a pharmacokinetic interaction between nimesulide and glibenclamide was studied in 12 healthy subjects, aged 25–39 (mean 31.5) years, in a single dose crossover study [80]. The participants received either nimesulide 100 mg (tablets) or glibenclamide 5 mg (tablets) or the two drugs together. Mean C_{max}, t_{max} and AUC values showed that the oral bioavailability of both drugs was unaffected by the concomitant administration. Therefore, the presence of a pharmacokinetic interaction between the two drugs can be ruled out (Tab. 13).

Cimetidine

Nimesulide 100 mg (tablets) was administered alone or in combination with cimetidine 400 mg (tablets) to 12 healthy subjects of both genders, aged 18–25 (mean 20.5) years, in a single dose crossover study [81]. The bioavailability of nimesulide was not influenced by the co-administration of cimetidine. Indeed, the C_{max}, t_{max}, AUC_{0-24}, AUC, and $t_{1/2,z}$ of nimesulide did not differ between the two treatment groups (nimesulide alone or with cimetidine) (Tab. 13). In addition, the model-dependent pharmacokinetic parameters, for example, lag-time and absorption rate constant, showed no statistical differences between the two treatment groups. The pharmacokinetic data of 4′-hydroxynimesulide (C_{max}, t_{max} and AUC_{0-24}) did not show significant differences after taking nimesulide alone or in combination with cimetidine, thus confirming that the administration of cimetidine does not alter the pharmacokinetic profile of nimesulide.

Antacids

The effect of co-administration of an antacid comprising magnesium hydroxide 3.65 g plus aluminium hydroxide 3.25 g in 100 g suspension (Maalox® suspension) on the pharmacokinetic profile of nimesulide, was evaluated in a single dose crossover study [82]. Nimesulide 100 mg in tablet form was administered alone

Table 13 – Pharmacokinetic interaction studies in healthy volunteers with single oral administration of nimesulide. Mean values of pharmacokinetic parameters [3]

No. subjects and gender	Study design	Drug administered[a]	Dose (mg)		C_{max} (mg/L)	t_{max} (h)	AUC_{0-24} (mg/L.h)	AUC (mg/L.h)	$t_{1/2,z}$ (h)	CL/F (mL/h/kg)	Vz/F (L/kg)	Ref.
6M+6F	SD	Nimesulide	100			Nimesulide parameters						
		Glibenclamide	5	Alone	4.80	2.83		37.83	3.62			80
				In combination	4.02	2.96		35.14	3.73			
						Glibenclamide parameters						
				Alone	123.5	2.92		637.0	4.94			
				In combination	122.1	3.33		654.4	4.69			
6M+6F	SD	Nimesulide	100			Nimesulide parameters						
		Cimetidine	400	Alone	4.73	2.58	34.19	35.01	3.77			81
				In combination	5.05	2.58	35.20	36.44	4.50			
						Hydroxy-nimesulide parameters						
				Alone	1.02	4.67	10.74					
				In combination	1.13	4.42	11.25					
6M+6F	SD	Nimesulide	100			Nimesulide parameters						
		Maalox[b]	15 mL	Alone	5.06	2.67	36.27	36.75	4.03			82
				In combination	5.07	2.83	36.75	37.56	3.91			
						Hydroxy-nimesulide parameters						
				Alone	0.96	5.00	11.03					
				In combination	0.98	5.25	11.79					

Table 13 – (continued)

No. subjects and gender	Study design	Drug administered[a]	Dose (mg)		C_{max} (mg/L)	t_{max} (h)	AUC_{0-24} (mg/L.h)	AUC (mg/L.h)	$t_{1/2,z}$ (h)	CL/F (mL/h/kg)	Vz/F (L/kg)	Ref.
8M	MD	Nimesulide	200 bid			Nimesulide parameters						
		Furosemide	40 bid	In combination, day 5				46.2	2.72	74.0	0.26	83
				In combination, day 10				47.6[d]	3.04	83.3	0.31	
						Furosemide parameters						
				Alone, day 4				2.89	2.14	207	0.68	
				In combination, day 5				2.36	2.00	266	0.77	
				In combination, day 10				2.18	1.64	297	0.72	
5M+5F	MD	Nimesulide	100 bid			Nimesulide parameters						
		Theophylline[c]	200 bid	In combination, day 8	4.8	2.9		41.8	5.3			84
						Hydroxy-nimesulide parameters						
				In combination, day 8	1.30	4.4		12.8	8.2			
						Theophylline parameters						
				Alone, day1	13.0	5.7		133.1	11.7			
				In combination, day 8	12.2	4.8		118.5	11.3			

Table 13 – (continued)

No. subjects and gender	Study design	Drug administered[a]	Dose (mg)		C_{max} (mg/L)	t_{max} (h)	AUC_{0-24} (mg/L.h)	AUC (mg/L.h)	$t_{1/2,z}$ (h)	CL/F (mL/h/kg)	Vz/F (L/kg)	Ref.
12M	MD	Nimesulide	100			Nimesulide parameters						
		Warfarin	5	Alone, day 1	4.06	1.92		18.26	2.16			85
				In combination, day 11	4.07	1.58		19.09[d]	2.27			
						Hydroxy-nimesulide parameters						
				Alone, day 1	1.58	3.25		11.14	3.62			
				In combination, day 11	1.59	2.60		10.47[d]	4.12			
						Warfarin parameters						
				Alone, day 14	1.45	2.88	24.38					
				In combination, day 11	1.46	2.21	23.32					

[a] *Formulation is tablet unless otherwise stated.*
[b] *Composition of the antacid Maalox® suspension: magnesium hydroxyde 3.65 g and aluminium hydroxyde 3.25 g in 100 g suspension.*
[c] *Sustained release tablet formulation.*
[d] *AUC_{0-12}.*

Symbols and abbreviations: C_{max} = maximum plasma concentration; t_{max} = time to C_{max}; AUC0-24 and AUC = area under the plasma concentration-time curve from 0 to 24h and to infinity; $t_{1/2,z}$ = apparent terminal half-life; CL/F = total plasma clearance; Vz/F = volume of distribution in the postdistribution phase; SD and MD = single and multiple dose study; M = males; F = females; bid = twice daily.

or in combination with 15 mL of antacid suspension to 12 healthy subjects of both genders, aged 18–25 (mean 20.8) years. The bioavailability of nimesulide was not influenced by the combined administration of the antacid. The C_{max}, t_{max}, AUC_{0-24}, AUC and $t_{1/2,z}$ of nimesulide, and C_{max}, t_{max} and AUC_{0-24} of the hydroxylated metabolite M1, did not differ significantly between the two treatment groups (Tab. 13). Thus, the administration of the magnesium hydroxide/aluminium hydroxide suspension does not alter the pharmacokinetic profile of nimesulide.

Furosemide

Oral furosemide 40 mg twice daily was administered for 10 days to eight healthy males, aged 20–34 (mean 25) years. Nimesulide 200 mg twice daily was administered in study days 5–10 [83].

A significant decrease (about 20%) in furosemide AUC was observed at days 5 and 10 compared with day 4 (Tab. 13). The cumulative excretion of furosemide was also significantly decreased on days 5 and 10, compared with day 4. The natriuretic and, to a lower extent, the kaliuretic effect of furosemide decreased after nimesulide administration. The diuretic response was reduced after multiple dose administration of nimesulide. However, the renal clearance of furosemide was unaffected by nimesulide. These results suggest a reduction in furosemide bioavailability induced by the concomitant administration of nimesulide. The interaction between the two drugs appears to involve other mechanisms in addition to reducing furosemide absorption through the gut. Indeed, the *fu* of furosemide increased slightly (from 2.54% to 2.88%) but significantly between days 4 and 5, and days 5–10, possibly as a consequence of displacement from plasma protein binding sites. Binding of nimesulide to plasma proteins remained stable (99.5%).

Theophylline

The potential pharmacokinetic and pharmacodynamic interactions between nimesulide and theophylline were studied in 10 patients aged 18–65 years, receiving NSAID and a maintenance therapy comprising slow-release theophylline for the treatment of chronic airflow obstruction [84]. On the first study day, patients received only theophylline, from the second day onward (to the end of study – day 8) they received theophylline (200 mg twice daily) and nimesulide (100 mg twice daily). The pharmacokinetic parameters of nimesulide and M1 observed on study day 8 (Tab. 13) were not substantially different in comparison with data reported for healthy individuals. The pharmacokinetic profile of theophylline on days 1 and 8 was essentially the same (Tab. 13), with the exception of a modest, but statistically significant, decrease in AUC on day 8. This result was interpreted as evidence of a nimesulide-induced increase in theophylline clearance, although theophylline $t_{1/2,z}$ was unaffected by nimesulide treatment. An alternative explanation, such as a decreased theophylline bioavailability, similarly to that reported with furosemide [83], should, therefore, be considered. However, the changes in AUC were small and clinically irrelevant, as confirmed by the lack of alteration in the parameters of respiratory function [84].

Warfarin

A possible drug–drug interaction between nimesulide and warfarin has been evaluated in pharmacodynamic terms by measuring the prothrombin time (Quick time) and by calculating the derived coumarin dose index (CDI), an indicator of warfarin effectiveness, and in pharmacokinetic terms by investigating the influence of nimesulide on the plasma profile of warfarin and the influence of the latter on the pharmacokinetic profile of nimesulide and M1 [85]. Twelve healthy males aged 23–39 (mean 30.2) years, received single oral doses of nimesulide 100 mg (tablets) on days 1 and 11, and twice daily doses on days 2–10. The concomitant treatment consisted of single or multiple doses of warfarin 5 mg in tablet form on days 2–14. The first warfarin dose was 20 mg and the second 10 mg; all other dosages depended on the individual's daily prothrombin time and ranged from 2.5–7.5 mg.

The pharmacokinetic profile of nimesulide and M1 observed after administration of nimesulide 100 mg alone (day 1) resembled that observed on day 11, when nimesulide was given in combination with warfarin (Tab. 13). Indeed, the C_{max}, t_{max}, AUC and $t_{1/2,z}$ remained substantially unchanged from day 1 to day 11. Similarly, the C_{max}, t_{max} and AUC_{0-24} of warfarin did not differ from day 11 (when warfarin was given in combination with nimesulide) to day 14 (warfarin alone) (Tab. 13). Although nimesulide and warfarin are highly bound to plasma proteins, co-administration did not alter the pharmacokinetic profile of either of these drugs.

Steady state values for prothrombin time and CDI resulting from combined treatment with warfarin and nimesulide at constant doses remained unchanged after cessation of nimesulide medication. Hence, nimesulide had no influence on the monitored pharmacodynamic parameters, whether direct or mediated by the action of warfarin, emerged from the aforementioned study. However, a few patients did show some increase in anticoagulant activity, suggesting that it would be prudent to monitor coagulation tests when nimesulide is given in combination with warfarin.

Digoxin

The potential interaction between nimesulide and digoxin was studied in nine patients (six males and three females), aged 57–70 (mean 67) years, with mild heart failure [86]. All patients, who were receiving maintenance therapy with digoxin

0.25 mg/day orally were treated with oral nimesulide 100 mg twice daily for 7 days. Serum digoxin concentrations, measured daily at 8 a.m. and 6 p.m. for 4 days before and throughout the nimesulide treatment period, remained within the normal therapeutic range in all patients despite large inter-individual variation. Mean digoxin concentrations in the afternoon (range 0.98–1.17 ng/mL) were higher than those observed in the morning (range 0.77–0.98 ng/mL) during the entire study period. The concomitant administration of nimesulide did not significantly change the morning and afternoon serum digoxin concentrations at steady state. Therefore, the results of this study indicate that short-term administration of therapeutic doses of nimesulide does not affect the pharmacokinetics of digoxin in patients with mild heart failure treated with a maintenance dose of this cardiac glycoside.

Alteration of protein binding

Nimesulide is extensively bound to plasma proteins; therefore, pharmacokinetic interactions at the protein binding level are expected. *In vitro* interaction studies with nimesulide and other drugs in human serum, showed that the free fraction of nimesulide was not altered significantly by the presence of therapeutic concentrations of furosemide, cefoperazone, glibenclamide, warfarin, tamoxifen, methotrexate or digoxin. Similarly, the *fu* of nimesulide was unaffected by the addition of fenofibrate or M1, but increased at higher concentrations of these two compounds [54]. To a minor extent nimesulide may be displaced from binding sites by tolbutamide, salicylic acid [53] and valproic acid [54]. Nimesulide has been shown to displace furosemide [83], methotrexate, valproic acid [54], and salicylic acid, but not warfarin [53] from plasma proteins. However, all these interactions appear to be of marginal or no clinical significance.

Nimesulide, like that of other NSAIDs (e.g., diclofenac, indomethacin, oxaprozin, salicylate) displaces tryptophan from its binding sites on albumin [87] and this may account for 5-hydroxytryptamine (serotonin) formation in the central nervous system and subsequent contribution to analgesia by serotoninergic activation of different pathways that lead to a gate control of pain stimuli at the level of the dorsal horn.

Conclusions

In male rats, after single intravenous administration, nimesulide is distributed throughout the body and the highest concentrations are attained in the fat tissue, the liver, kidneys, lungs, adrenals, gut, and heart between 1–4 h after the administration. Oral absorption is complete. Nimesulide is preferentially eliminated by metabolic biotransformation followed mainly by faecal excretion.

The pharmacokinetic data available from investigations in healthy volunteers provide useful information for the rational and safe use of nimesulide in the clinical setting. Such studies have shown that nimesulide is rapidly and completely absorbed by the stomach and the small bowel, is quickly distributed throughout the body and is principally eliminated by metabolic transformation. Metabolites are then preferentially excreted through the kidney. Tablet, granule and suspension formulations, provide the same rate and extent of nimesulide absorption and oral nimesulide can be administered with food without reducing the rate or extent of absorption. Twice-daily administration of nimesulide in oral or suppository formulations to a maximum dosage of 100 mg or 200 mg for suppositories twice daily in adults enables steady state to be achieved 24–48 hours after the first dose. The pharmacokinetic profiles of nimesulide and M1 are affected by severe hepatic insufficiency, marginally by age and moderate renal impairment, not by gender. The drug is contraindicated in patients with hepatic insufficiency and severe renal impairment. In the elderly (aged <80 years), and those with moderate renal insufficiency a dose adjustment is considered unnecessary. Caution should be employed when nimesulide is administered in combination with drugs that modify the coagulation process. In general, no recommendations can be given for use of nimesulide in combination with other drugs aside from those reported in the Summary of Product Characteristics.

References

1. Davis R, Brogden RN (1994) Nimesulide. An update of its pharmacodynamic and pharmacokinetic properties, and therapeutic efficacy. *Drugs* 48: 431–454
2. Bernareggi A (1993) The pharmacokinetic profile of nimesulide in healthy volunteers. *Drugs* 46 (Suppl. 1): 64–72
3. Bernareggi A (1998) Clinical pharmacokinetics of nimesulide. *Clinical Pharmacokinetics* 35: 247–274
4. Olive G, Rey E (1993) Effect of age and disease on the pharmacokinetics of nimesulide. *Drugs* 46 (Suppl. 1): 73–78
5. Bernareggi A (2001) Clinical pharmacokinetics and metabolism of nimesulide. *Inflammopharmacology* 9: 81–89
6. Regazzoni S (2001) Coefficienti di permeabilità apparente tramite la camera di Ussing e assorbimento di farmaci nell'uomo. Thesis, Milan: Faculty of Pharmacy, University of Milan, Italy
7. Bugatti C, Livi V (2000) Nimesulide: determination of aqueous solubility and logP. Helsinn Healthcare, data on file
8. Löbenberg R, Amidon GL (2000) Modern bioavailability, bioequivalence and biopharmaceutics classification system. New scientific approaches to international regulatory standards. *Eur J Pharmaceut Biopharm* 50: 3–12

9. Rainsford KD (1999) Profile and mechanism of gastrointestinal and other side effects of nonsteroidal anti-inflammatory drugs (NSAIDs). *Am J Med* 107: 27S–36S
10. McCormack K, Brune K (1987) Classical absorption theory and the development of gastric mucosal damage associated with the non-steroidal anti-inflammatory drugs. *Arch Toxicol* 60: 261–269
11. Casciarri I, Bernareggi A (1991) Disposition of total radioactivity and plasma levels of nimesulide and OH-nimesulide in rats after intravenous and oral administration. Helsinn Healthcare, data on file
12. McGeer PL, Schulzer M, McGeer EG (1996) Arthritis and anti-inflammatory agents as possible protective factors for Alzheimer's disease: a review of 17 epidemiological studies. *Neurology* 47: 425–432
13. Pasinetti GM (1999) US patent No. 5,985,930, Treatment of neurodegenerative conditions with nimesulide
14. Boje KMK, Jaworowicz Jr D, Raybon JJ (2003) Neuroinflammatory role of prostaglandins during experimental meningitis: Evidence suggestive of an *in vivo* relationship between nitric oxide and prostaglandins. *Pharmacol Exper Therap* 304: 319–325
15. Gupta SK, Bhardwaj RK, Tyagi P, Sengupta S, Velpandian T (1999) Anti-inflammatory activity and pharmacokinetic profile of a new parenteral formulation of nimesulide. *Pharmacol Res* 39: 138–141
16. Toutain PL, Cester CC, Haak T, Metge S (2001) Pharmacokinetic profile and *in vitro* selective cyclooxygenase-2 inhibition by nimesulide in the dog. *J Vet Pharmacol Therap* 24: 35–42
17. Toutain PL, Cester CC, Haake T, Laroute V (2001) A pharmacokinetic/pharmacodynamic approach vs. a dose titration for the determination of a dosage regimen: the case of nimesulide, a COX-2 selective nonsteroidal anti-inflammatory in the dog. *J Vet Pharmacol Therap* 24: 43–55
18. Castoldi D, Monzani V, Tofanetti O (1988) Simultaneous determination of nimesulide and hydroxynimesulide in human plasma and urine by high-performance liquid chromatography. *J Chromat B* 425: 413–418
19. Bernareggi A, Castoldi D, Nava ML, Ratti E (1998) Linear pharmacokinetics of oral nimesulide in healthy male volunteers treated with 25, 50 and 100 mg. Helsinn Healthcare, data on file
20. Remuzzi G, Gaspari F, Taiocchi L (1991) Development of micromethods to study the pharmacokinetic profile for brodimoprim and nimesulide in small volumes of biological fluids. Helsinn Healthcare, data on file
21. Giachetti C, Tenconi A (1998) Determination of nimesulide and hydroxynimesulide in human plasma by high performance liquid chromatography. *Biomed Chromatogr* 12: 50–56
22. Abbiati G, Rigoldi M, Parisi S, Arrigoni M (1996) Farmacocinetica e biodisponibilità di nimesulide beta-ciclodestrina granulato nel volontario sano. *Giorn Ital Ric Clin Terap* 17: 1–4

23. Lücker PW (1992) Study on the pharmacokinetics (food/drug interaction) and relative bioavailability of three different treatments with nimesulide in 18 healthy male volunteers. Helsinn Healthcare, data on file. TSD No. 5565E
24. Lücker PW (1991) Report on the pharmacokinetics and relative bioavailability/bioequivalence of three different nimesulide formulations in 18 healthy male volunteers. Helsinn Healthcare, data on file. TSD No. 5356E
25. Alessandrini A, Ballarin E, Bastianon A, Migliavacca C (1986) Confronto di biodisponibilità tra due diverse forme farmaceutiche orali equidosate di nimesulide in volontari sani. *Clinica Terapeutica* 118: 177–182
26. Scheen A (1997) Open, randomized, crossover, bioequivalence study of three different formulations of nimesulide, after single oral dose administration in 24 healthy subjects. Helsinn Healthcare, data on file. TSD No. 7215E
27. Lücker PW (1993) Report of the study on the pharmacokinetics of nimesulide and its main metabolite 4-OH-nimesulide after oral administration of three different doses in healthy male volunteers. Helsinn Healthcare, data on file. TSD No. 5781E
28. De Caro G (1989) Studio comparativo di farmacocinetica e biodisponibilità nell'uomo su formulazioni a base di nimesulide per uso orale (cpr 100 mg) e rettale (supposte da 200 mg) delle ditte LPB (Mesulid®) e BBR (Aulin®) in somministrazione singola e ripetuta. Helsinn Healthcare, data on file. TSD No. 5437E
29. Bernasconi PC (1989 Pharmacokinetics of nimesulide: relative bioavailability of the rectal *versus* the oral administration. A single and repeated dose study in healthy volunteers. Helsinn Healthcare, data on file. TSD No. 5252E
30. Gandini R, Montalto C, Castoldi D, Monzani V, Nava ML, Scaricabarozzi I, Vargiu G, Bartosek I (1991) First dose and steady-state pharmacokinetics of nimesulide and its 4-hydroxy metabolite in healthy volunteers. *Farmaco* 46: 1061–1079
31. Ugazio AG, Guarnaccia S, Berardi M, Renzetti I (1993) Clinical and pharmacokinetic study of nimesulide in children. *Drugs* 46 (Suppl. 1): 215–218
32. Pulkkinen M (1993) Nimesulide in dysmenorrhoea. *Drugs* 46 (Suppl. 1): 129–133
33. Diamond G (1976) Standard dossier – Volume 2 – pp 181–210. Single dose study using R805-^{14}C. Helsinn Healthcare, data on file. TSD No. 3007E
34. Allemon AM, Lebacq E (1993) Study of the excretion balance and metabolism of [^{14}C] nimesulide after single oral dose administration in 6 healthy male volunteers. Helsinn Healthcare, data on file. TSD No. 5755E
35. McCracken NW, Sanderson BJ, Young CG (1997) The excretion and plasma kinetics of [^{14}C] nimesulide in man following a single oral administration. Helsinn Healthcare, data on file. TSD No. 7277E
36. Macpherson D, Best SA, Gedik L, Hewson AT, Rainsford KD, Parisi F (2004) The biotransformation and pharmacokinetics in humans of single dose of ^{14}C-nimesulide. Helsinn Healthcare, data on file
37. Brambilla A, Maffei Facino R (1989) Metabolismo del farmaco antiinfiammatorio nimesulide nell'uomo. Thesis, Milan: Faculty of Pharmacy, University of Milan, Italy

38. Maffei Facino R, Carini M, Stefani R, Marinello C (1997) *In vitro* metabolism of the antiinflammatory drug nimesulide in man: simultaneous determination of the main urinary metabolites by HPLC with UV-DAD detection. Proceedings of the 7th meeting on Recent developments in Pharmaceutical Analysis (RDPA '97), Sept. 16–22, 1997, Isola d'Elba, p. 28
39. Castoldi D (1989) Escrezione urinaria della nimesulide dopo somministrazione unica al volontario sano. Helsinn Healthcare, data on file
40. Davis SS, Hardy JG, Fara JW (1986). Transit of pharmaceutical dosage forms through the small intestine. *Gut* 27: 886–892
41. Pharmaceutical Profiles (Nottingham, UK), Study Number PPL-322 (1999) Pharmacoscintigraphic evaluation of the regional absorption of nimesulide delivered using the InteliSite® capsule in healthy volunteers, Helsinn Healthcare, data on file
42. Gardner D, Casper R, Leith F, Wilding I (1997) Non-invasive methodology for assessing regional drug absorption from the gastrointestinal tract. *Pharm Tech Europe* 203
43. Lin JH, Cocchetto DM, Duggan DE (1987) Protein binding as a primary determinant of the clinical pharmacokinetic properties of non-steroidal anti-inflammatory drugs. *Clin Pharmacokinet* 12: 402–432
44. Verbeeck RK (1990) Pharmacokinetic drug interactions with nonsteroidal anti-inflammatory drugs. *Clin Pharmacokinet* 19: 44–66
45. Brouwers J, de Smet P (1994) Pharmacokinetic-pharmacodynamic drug interactions with nonsteroidal anti-inflammatory drugs. *Clin Pharmacokinet* 27: 462–485
46. Graham GG, Champion GD, Day RO, Paull PD (1977) Patterns of plasma concentrations and urinary excretion of salicylate in rheumatoid arthritis. *Clin Pharmacol Ther* 22: 410–420
47. Davies NM, Anderson KE (1977) Clinical pharmacokinetics of diclofenac. *Clin Pharmacokinet* 33: 184–213
48. Davies NM (1995) Clinical pharmacokinetics of flurbiprofen and its enantiomers. *Clin Pharmacokinet* 28: 100–114
49. Davies NM (1996) Clinical pharmacokinetics of tiaprofenic acid and its enantiomers. *Clin Pharmacokinet* 31: 331–347
50. Davies NM, Anderson KE (1997) Clinical pharmacokinetics of naproxen. *Clin Pharmacokinet* 32: 268–293
51. Pulkkinen MO, Vuento M, Macciocchi A., Monti T (1991). Distribution of oral nimesulide in female genital tissues. *Biopharm Drug Disp* 12: 113–117
52. Lignière GC, Tamborini U, Panarace G, Abbiati G, Montagnani G (1990) La nimesulide nel liquido sinoviale di paziente con artrite reumatoide. *Farmaci e Terapia* 7: 173–176
53. Vilageliu J (1983) Nimesulide evaluation of protein binding and the effects of possible displacers. Helsinn Healthcare, data on file. TSD No. 3545E
54. Bree F, Nguyen P, Urien S, Albengres E, Macciocchi A, Tillement JP (1993) Nimesulide binding to components within blood. *Drugs* 46 (Suppl. 1): 83–90
55. Milvio I (1984) Confronto di biodisponibilità tra due diversi lotti con massa witepsol. Helsinn Healthcare, data on file. TSD No. 4038

56. Maffei Facino R, Carini M, Brambilla A, Casciarri I, Scaricabarozzi I, Tofanetti O (1989) Metabolism of nimesulide in man and radical scavenging activity of its main metabolites. 3rd Interscience World Conference on Antirheumatics, Analgesics, Immunomodulators, Montecarlo, March 15–18, 1989. Abstract Book p. 244
57. Carini M, Aldini G, Stefani R, Marinello C and Facino RM (1998) Mass spectometric characterisation and HPLC determination of the main urinary metabolites of nimesulide in man. *J Pharm Biomed Anal* 18: 201–211
58. Ogaki J (1999) Studies on the regulational activities of NIM-03 on human liver microsomal CYP isozymes. Helsinn Healthcare, data on file
59. Rainsford KD (2004) Pharmacology and toxicology of COX-2 inhibitors. In: Pairet M, van Ryn J (eds): *COX-2 inhibition*. Birkhäuser Verlag, Basel, 67–131
60. Maffei Facino R, Carini M, Aldini G (1993) Antioxidant activity of nimesulide and its main metabolites. *Drugs* 46 (Suppl. 1): 15–21
61. Pirovano R (1988) Study of the capacity of the test article BBR 2335/7 to induce gene mutations in strains of *Salmonella typhimurium*. Helsinn Healthcare, data on file
62. Pirovano R (1989) Study of the capacity of the test article BBR 2395 to induce gene mutations in strains of *Salmonella typhimurium*. Helsinn Healthcare, data on file
63. Brodie B, Lowman E, Burns J, Lee P, Chenkin T et al. (1954) Observations on the antirheumatic and physiologic effects of phenylbutazone and some comparisons with cortisone. *Am J Med* 16: 181–190
64. Runkel R, Forchielli E, Sevelius H, Chaplin M, Segre E (1974) Nonlinear plasma level response to high doses of naproxen. *Clin Pharmacol Ther* 15: 261–266
65. Lin JH, Hooke KF, Yeh KC, Duggan DE (1985) Dose-dependent pharmacokinetics of diflunisal in rats: dual effects of protein binding and metabolism. *J Pharmacol Exp Ther* 235: 402–406
66. Schärli AF, Brülhart K, Monti T (1990) Pharmacokinetics and therapeutic study with nimesulide suppositories in children with post-operative pain and inflammation. *J Int Med Res* 18: 315–321
67. Gupta SK, Prakash J, Awor L, Joshi S, Velpandian T, Sengupta S (1996) Anti-inflammatory activity of topical nimesulide gel in various experimental models. *Inflamm Res* 45: 590–592
68. Menegatti E (1996) *In vitro* permeability test on different formulations. Helsinn Healthcare, Report No. TSD 7565E
69. Berardesca E (1998) Evaluation of skin irritation and sensitization potential of different nimesulide topical formulations by repeated insult patch test in healthy volunteers. Helsinn Healthcare, Report No. TSD 6103
70. Thompson CA, Dennis G (1997) A comparative kinetic study of nimesulide following administration of an oral formulation and a topical formulation to 18 healthy male volunteers. Helsinn Healthcare, Report No. TSD 7224 E
71. Eichler HG (1999) Pharmacokinetics profile and transdermal penetration of nimesulide in male, healthy volunteers after single and multiple epicutaneous administration of a new 3% gel formulation. Helsinn Healthcare, Report No. TSD 7781 2 E

72. Babu RJ, Kanikkanun N, Kikwai L, Ortega C, Andega S, Ball K, Yim S, Singh M (2003) The influence of various methods of cold storage of skin on the permeation of melatonin and nimesulide. *J Controlled Rel* 86: 49–57
73. Lücker PW (1992) Study on the pharmacokinetics (single and multiple dose) of nimesulide in 10 elderly subjects. Helsinn Healthcare, Report No. TSD 5619E
74. Pontiroli AE (1991) Report on the pharmacokinetics of nimesulide in elderly patients with normal and elevated creatinine plasma concentrations after single and repeated doses. Helsinn Healthcare, Report No. TSD 5520E
75. Mosteller RD (1987) Simplified calculation of body surface area. *New Engl J Med* 317 (17): 1098
76. Fillastre JP (1991) Pharmacokinetics of nimesulide after single oral dose in patients with moderate renal insufficiency. Helsinn Healthcare, Report No. TSD 4871E
77. Fouarge M (1992) Pharmacokinetic profile of nimesulide and its OH-metabolite in plasma and urine in 10 subjects with moderate renal failure during repeated oral administration. Helsinn Healthcare, Report No. TSD 5611E
78. Olive G (1993) Pharmacokinetics of nimesulide after single oral administration to healthy subjects and those with hepatic insufficiency. Helsinn Healthcare, Report No. TSD 5498E
79. Perucca E (1993) Drug interactions with nimesulide. *Drugs* 46 (Suppl. 1): 79–82
80. Fouarge M (1993) Pharmacokinetic interaction study between nimesulide and glibenclamide after single dose cross-over administration in 12 healthy subjects. Helsinn Healthcare, Report No. TSD 5687E
81. Ugazio G (1993) Pharmacokinetic interaction study between nimesulide tablets and cimetidine tablets after single dose administration to 12 healthy volunteers. Helsinn Healthcare, Report No. TSD 5691E
82. Ugazio G (1993) Pharmacokinetic interaction study between nimesulide tablets and Maalox suspension after single dose administration to 12 healthy volunteers. Helsinn Healthcare, Report No. TSD 5692E
83. Steinhauslin F, Munafo A, Buclin T, Macciocchi A (1993) Renal effects of nimesulide in furosemide-treated subjects. *Drugs* 46 (Suppl. 1): 257–262
84. Auteri A, Bruni F, Pasqui AL, Saletti M, Di Renzo M, Blardi P, Domini L, Verzuri S, Di Perri T (1991) Pharmacokinetics and pharmacodynamics of slow-release theophylline during treatment with nimesulide. *Int J Clin Pharmacol Res* 11: 211–218
85. Lücker PW (1993) Report of the study on the possible drug-drug interaction of warfarin and nimesulide in 12 healthy male volunteers. Helsinn Healthcare, Report No. TSD 5686E
86. Baggio E, Maraffi F, Montalto C, Nava ML, Torti L, Casciarri I (1993) A clinical assessment of the potential for pharmacological interaction between nimesulide and digoxin in patients with heart failure. *Drugs* 46 (Suppl. 1): 91–94
87. Rainsford KD, Omar H, Ashaf A, Hewson AT, Bunning RAD, Rishiraj R, Shepherd P, Seabrook RW (2002) Recent pharmacodynamic and pharmacokinetic findings on oxaprozin. *Inflammopharmacology* 10: 185–239

Pharmaceutical formulations of nimesulide

A. Maroni and A. Gazzaniga

Istituto di Chimica Farmaceutica e Tossicologica, Università degli Studi di Milano, Viale Abruzzi 42, 20131 Milano, Italy

Introduction

Nimesulide is widely used to accomplish both topical and systemic therapeutic goals. Topical multi-dose semi-solid formulations for cutaneous application as well as single-dose preparations for bioadhesion onto the buccal mucosa were proposed for local and regional treatments. On the other hand, many different oral, rectal and injectable dosage forms were devised in order to attain a systemic therapeutic effect through the administration of nimesulide as described in Chapter 1. However, only a very limited number of the above dosage forms were developed and finally put on the marketplace.

This chapter principally refers to the formulations that have been developed by Helsinn (the worldwide licensor of the original molecule).

Formulations for topical application

Nimesulide is successfully used for local and regional therapies. In this respect, a gel product was approved in many European and non-European countries for the symptomatic relief of pain and inflammation associated with sprain and tendonitis. The concentration of the active principle in the semi-solid preparation is of 3% w/w and was established on the basis of safety considerations. In fact, considering that a gel quantity of approximately 3 g has to be applied on the diseased site in order to achieve an adequate symptom control, the systemic availability of nimesulide could be assumed to be by far lower than the minimum effective concentration and, even in the only theoretically possible case of complete drug absorption through the skin, not to exceed the plasma levels related to a standard single oral dose (100 mg). A micronised form of nimesulide was selected for use in the topical formulation so as to facilitate the homogeneous dispersion of the active powder within the formulation itself. Moreover, due to the well-known poor wettability and water solubility characteristics which are to be faced when dealing with nimesulide, the relevant solubilisation was unlikely to occur, at least to a significant extent, in the hydrophilic gel base. Hence, the use of a micronised

Nimesulide – Actions and Uses, edited by K. D. Rainsford

form of the drug was also meant to improve the product *in vivo* performance as well as its reproducibility. Apart from the active ingredient, the typical excipients of gel preparations for cutaneous application were included in the formulation, i.e., vehicles, gelifying additives, penetration enhancers, skin sensitive promoters, cation chelating agents and antimicrobial preservatives.

As a consequence of the generally experienced efficacy of the topical proprietary formulation, further studies were undertaken on nimesulide gel formulations, in which the effect on the skin permeation profile of a different qualitative and quantitative composition in penetration enhancers or of the drug entrapment into special multiparticulate systems, such as niosomes, was evaluated [1, 2].

An alternative dosage form for topical application was proposed for the treatment of stomatological lesions. It consisted in a tablet provided with a mucoadhesive layer containing nimesulide, either as such or in the form of sodium salt, and a protective layer meant to prevent drug leaching into the oral cavity. Satisfactory bioadhesion behaviour and patient compliance were demonstrated *in vivo* on 10 volunteers over a period of 8 h [3, 4].

Formulations for systemic administration

Major interest is focused on nimesulide dosage forms indicated for systemic treatments, which can be pursued via different routes of administration. In particular, the intramuscular, oral and rectal routes have been related to the possibility of attaining plasma levels of nimesulide in the therapeutic range. Actually, the preparation of pharmaceutically acceptable parenteral formulations is hindered by the poor wettability and solubility characteristics exhibited by the drug molecule. Therefore, although different approaches ranging from the hydrotropic solubilisation to the encapsulation in biocompatible and biodegradable microparticles of nimesulide were attempted [5, 6], no injectable preparation is available on the marketplace for the present.

On the other hand, minor formulation problems were encountered in the development of the 200 mg nimesulide-containing product intended for rectal administration. In fact, the mixture of surfactants and glycerides employed for the preparation of suppositories definitely constituted a suitable base for nimesulide conveyance.

As generally observed in the case of most drugs, also for nimesulide the oral route is by far preferred. Oral nimesulide formulations include tablets and granules, commercialised in Italy for the first time in 1985. Both dosage forms have become quite popular among patients as well as physicians under Aulin® and Mesulid® trademarks, thanks to the advantageous efficacy and tolerability characteristics of the active principle. The formulation and manufacturing method of oral solid preparations able to give rise to the expected, reproducible nimesulide immediate release performances had to meet the challenge of the rather unfavourable physical–chemical properties of the drug substance. Particularly, the already mentioned low water solubility and poor wettability, which are worth being emphasised when dealing with the oral route, may pose noteworthy bioavailability problems in the case of tablets. Therefore, great effort was taken by the relevant pharmaceutical development, which required a special micronisation process of the drug raw material to increase dissolution rate, the use of the surfactant sodium docusate within a wet granulation phase to enhance formulation wettability, and the addition of the disintegrant sodium starch glycolate at two distinct mixing stages to improve tablet disintegration. Moreover, a suitable dissolution test had to be purposely devised for the formulation screening and control. In fact, relying on the Biopharmaceutics Classification System (BCS), which was introduced in recent years by the US Food and Drug Administration (FDA) with the aim of establishing when the *in vitro* dissolution test can be exploited to predict bioavailability, on account of its solubility and lipophilicity characteristics nimesulide might reasonably be considered as a Case 2 drug, including poorly soluble and highly permeable active principles. For these molecules, dissolution is supposed to represent the rate-controlling step in systemic absorption and, consequently, particular attention is focused on the *in vitro* dissolution test [7, 8].

Oral cyclodextrin formulations

The poor wettability and solubility of nimesulide later suggested the idea of undertaking the pharmaceutical development of a further 100 mg oral solid proprietary product, in which the drug was formulated with β-cyclodextrin. Cyclodextrins are cyclic oligosaccharides consisting of 6–8 glucopyranose moieties, delimiting an inner relatively hydrophobic cavity in which various lipophilic molecules can be lodged according to their size and physical–chemical properties. In principle, as compared to the drug, the resulting inclusion compounds exhibit improved hydrophilic characteristics and, therefore, increased solubility and dissolution rate in the aqueous environment [9]. The above formulation was shown to lead *in vivo* to a faster onset of nimesulide therapeutic plasma levels and more effective relief of pain, inflammation and fever [10].

Oral modified-release formulations

The use of β-cyclodextrin aimed at obviating the inherent solubility problems of nimesulide also turned out to be an interesting scientific cue, which was seized by several research groups. Hence, many different strategies were described in the literature to promote and support the formation of drug–cyclodextrin complexes with improved dissolution rate of nimesulide. In particular, spray and freeze drying processes, kneading and co-evaporation methods, co-grinding, supercritical fluid impregnation [8, 11–13] as well as the employment of cyclodextrin or nimesulide derivatives, namely hydroxypropyl β-cyclodextrin and nimesulide-L-lysine salt [14, 15], were pursued to achieve this goal. Moreover, the influence of α, β and γ-cyclodextrin on the dissolution behaviour of nimesulide or nimesulide-L-lysine was comparatively investigated [15, 16].

Besides those containing cyclodextrins, various further formulations were proposed within the efforts directed to the enhancement of dissolution rate and/or bioavailability of nimesulide from oral preparations. The main attempts yielded surface-activated powder mixtures [17], quaternary or ternary liquid systems based on water, alcohol, oil and surfactant components [18, 19], solid dispersions in opportunely selected pharmaceutical adjuvants [20–22] and fast-disintegrating mouth dissolve tablets [23]. Some authors even suggested that prolonged-release systems, either consisting in matrix or osmotic pump devices, would help to circumvent the drawbacks connected with the slow dissolution of nimesulide from most conventional dosage forms by exerting a programmed control on its release rate [24, 25].

Generic formulations

In the 90s, the production and commercialisation of a number of further oral nimesulide-containing formulations have been triggered by the remarkable scientific and commercial success reached by the proprietary products. Therefore, a wide variety of nimesulide preparations are, at least in theory, presently available in pharmacies worldwide. From the regulatory standpoint, most of them are handled as generics, i.e., “Interchangeable multi-source pharmaceutical products” according to the World Health Organization (WHO) official definition. For the purpose of interchangeability, the assessment of bioequivalence to an already marketed reference product is mandatory for generics, since it is accepted as a proof of therapeutic equivalence [26, 27]. The above-mentioned spread of nimesulide tablet preparations and their challenging bioavailability, which might impair bioequivalence to the innovator product, have lately aroused the interest of many research groups in comparative investigations into the *in vitro* as well as *in vivo* performances of the most representative nimesulide generics available on the marketplace *versus* Aulin® and the co-marketed preparation Mesulid® [28–30]. The relevant results, published not only by the scientific but also by the mass-circulation press, have drawn a general attention concerning the therapeutic reliability of generics. Hence, it seems helpful and interesting to briefly review in scientific terms all the information provided about different nimesulide proprietary as well as generic products and relevant *in vitro/in vivo* performances.

The first article published in this respect was focused on an evaluation of the *in vitro* behaviour of Sulidamor® (Farmaceutici Damor S.p.a., Italy) and Nimesulide Dorom (Dorom S.r.l., Italy), which were at that time the best-selling nimesulide copy and, respectively, generic preparations on the Italian market, *versus* Aulin® (Roche S.p.a., Italy, under licence of Helsinn Healthcare SA, Switzerland) and Mesulid® (Novartis Farma S.p.a., Italy, under licence of Helsinn Healthcare SA, Switzerland), all in their 100 mg tablets formulation [28]. This comparative study was mainly based on the *in vitro* dissolution test, considered by the authors as an important investigation tool due to the poor hydrosolubility and high lipophilicity characteristics of nimesulide, for which the dissolution step is therefore reasonably supposed to control the kinetic aspect of bioavailability. A series of preliminary experiments were carried out with the aim of setting up the operating conditions of the test, since neither pharmacopoeial nor compendial reference monographies were available at that time on nimesulide preparations. A USP 24 paddle dissolution apparatus was employed on account of its general use when tablet units are dealt with. Owing to the physical–chemical characteristics of the molecule, the main difficulties were met in the selection of the appropriate disso-

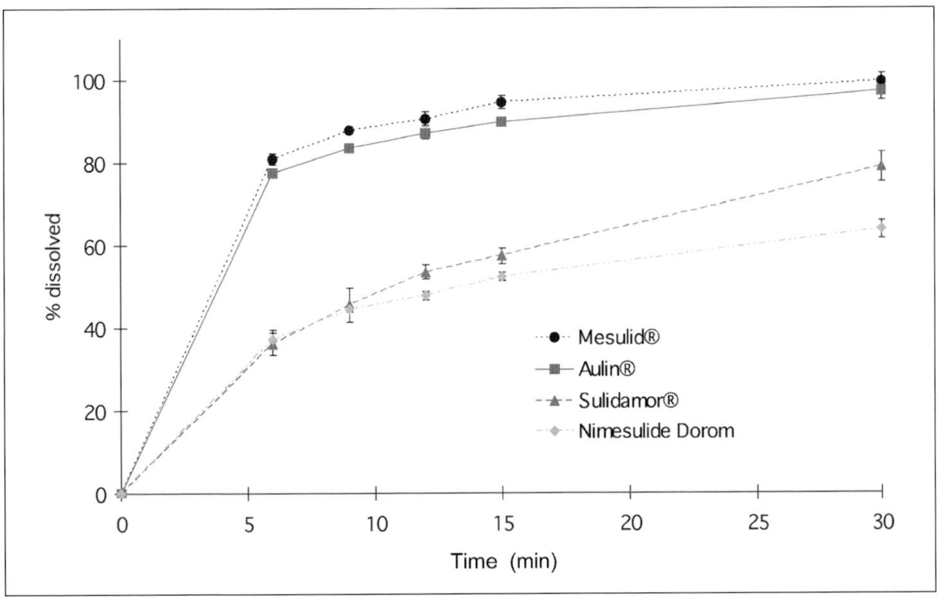

Figure 1
Dissolution profiles of nimesulide from Aulin®, Mesulid®, Sulidamor® and Nimesulide Dorom 100 mg tablets (USP 24 paddle apparatus, 1,000 mL of simulated intestinal fluid without enzymes + Tween® 80 2.5% w/v, 100 rpm, 37 ± 0.5°C; n = 6, arithmetic means ± standard deviations). Adapted from [28].

lution medium, which was expected to enable discrimination among the products in exam on one hand, and to allow sink conditions to be maintained throughout the whole test on the other. The study pointed out marked differences in the *in vitro* behaviour of the generic and copy as compared to the original products (Fig. 1). Thirty minutes after the test start, practically the entire drug labelled content was dissolved from Aulin® and Mesulid® tablets, whereas Sulidamor® and Nimesulide Dorom did not exceed the average release of about 80% and 65%, respectively. The differences exhibited by both copy and generic *versus* the innovator tablets were demonstrated to be statistically significant for all the considered experimental points ($P \leq 0.05$). The authors of this investigation, however, did not fail to underline that the obtained data could not be considered as predictive of the *in vivo* behaviour of the examined formulations, unless a suitable *in vitro/in vivo* correlation was previously established.

Subsequently, the relative bioavailability of Nimesulide Dorom and Sulidamor® *versus* Aulin®, all 100 mg tablets, was explored within two different *in vivo* studies [29, 30]. Both investigations were carried out in Germany by an internationally renowned contract research organisation (CRO) in agreement with the procedure adopted worldwide for bioequivalence assessment and with the recognised GXPs (Good "X" Practices) in force. The investigations were performed according to a single-dose, randomised, two-way crossover design on 18 healthy male Caucasian

Table 1 – Nimesulide pharmacokinetic parameters obtained after single oral administration of Aulin® and Nimesulide Dorom 100 mg tablets (n = 18, arithmetic means and standard deviations)

	Aulin®		**Nimesulide Dorom**	
	mean	**SD**	**mean**	**SD**
AUC_{0-z} (mg·h/L)	19.608	8.219	8.977	4.164
$AUC_{0-\infty}$ (mg·h/L)	19.926	8.299	9.319	4.252
C_{max} (mg/L)	4.668	1.142	1.601	0.481
t_{max} (h)	1.57	0.75	3.25	1.48
λ_z (h^{-1})	0.365	0.107	0.342	0.108
$t_{1/2,z}$ (h)	2.084	0.697	2.258	0.817

AUC_{0-z}: area under the plasma concentration-time curve from time of administration (t_0) to the last sample with quantifiable concentration; $AUC_{0-\infty}$: area under the plasma concentration-time curve from time of administration (t_0) to infinity; C_{max}: maximum plasma concentration; t_{max}: time to C_{max}; λ_z: terminal elimination rate constant; $t_{1/2,z}$: terminal elimination half-life. Adapted from [29].

volunteers, who were enrolled after providing written informed consent. Following an overnight fast, each subject received a single dose of the test preparation with 200 mL of water. Prior to intake and in the subsequent 24 h, blood samples were collected at predetermined time intervals. The concentration of nimesulide and its main metabolite 4′-hydroxy-nimesulide was determined in plasma specimens through a validated analytical method based on high performance liquid chromatography (HPLC) combined with UV detection technique. From the relevant values, plasma concentration-time curves were plotted for each preparation in exam, and the parameters describing the respective bioavailability were calculated.

Within the first *in vivo* study, the relative bioavailability of the generic Nimesulide Dorom and innovator product Aulin® was comparatively investigated [29]. The obtained results, reported in Table 1 and Figure 2, showed that an almost two-fold drug absorbed amount, expressed by the area under the plasma concentration-time curve extrapolated to infinity ($AUC_{0-\infty}$), was reached following ad-

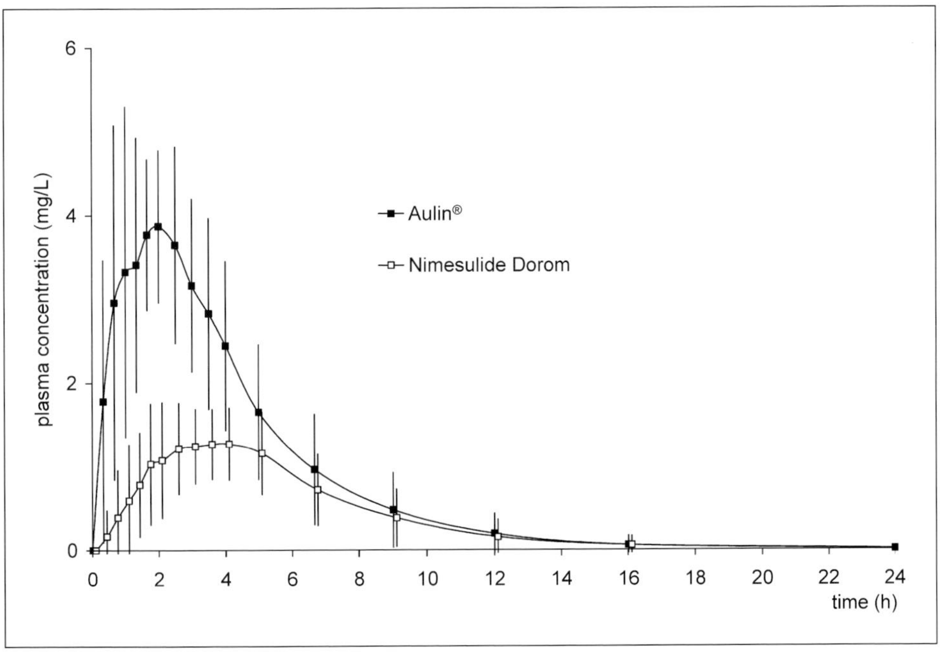

Figure 2
Plasma concentration profiles of nimesulide after single oral administration of Aulin® and Nimesulide Dorom 100 mg tablets to healthy volunteers (n = 18, arithmetic means ± standard deviations). Adapted from [29].

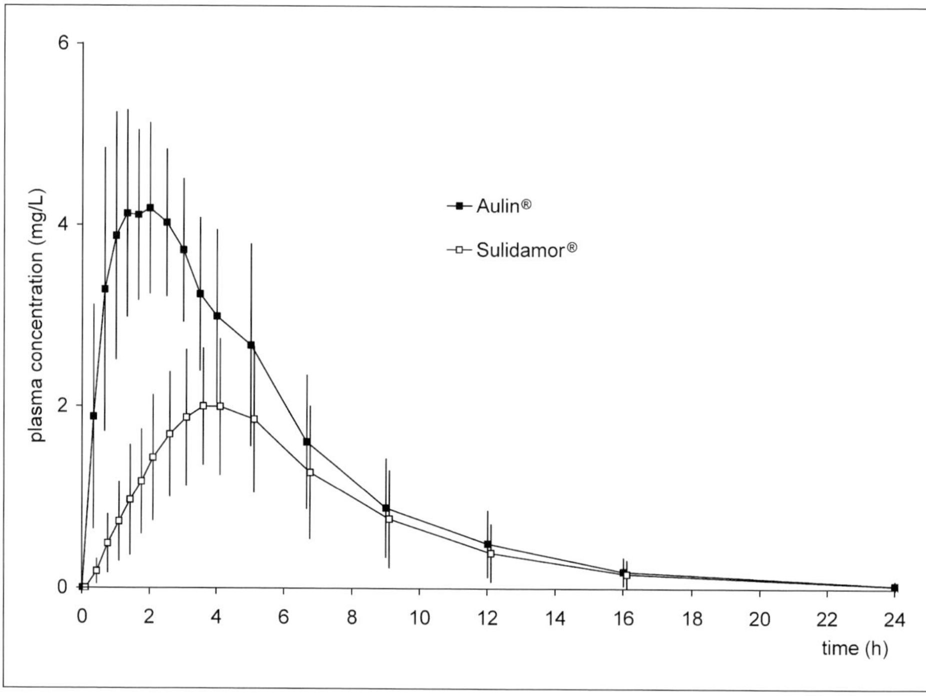

Figure 3
Plasma concentration profiles of nimesulide after single oral administration of Aulin® and Sulidamor® 100 mg tablets to healthy volunteers (n = 18, arithmetic means ± standard deviations). Adapted from [30].

ministration of Aulin® as compared to Nimesulide Dorom tablets. Furthermore, the absorption of nimesulide was much faster in the case of the innovator product, as pointed out by the higher maximum plasma concentration (C_{max}) and $C_{max}/AUC_{0-\infty}$ ratio, as well as by the lower time to C_{max} (t_{max}). Analogous considerations could be addressed to the metabolite 4′-hydroxy-nimesulide pharmacokinetics. The differences observed between Nimesulide Dorom and Aulin® with respect to the bioavailability parameters $AUC_{0-\infty}$ and $C_{max}/AUC_{0-\infty}$ turned out to be significant through statistical analysis, thus pointing out bioinequivalence of the two products.

The latter study, performed according to the same experimental plan, was aimed at exploring the bioequivalence of Sulidamor® *versus* Aulin® tablets [30]. Figure 3 and Table 2 show that the $AUC_{0-\infty}$ obtained after intake of the original product exceeds 175% of that pertaining to Sulidamor®. In addition, C_{max} and t_{max} values of 4.723 mg/L and 1.63 h, and 2.343 mg/L and 4.07 h were observed

Table 2 – Nimesulide pharmacokinetic parameters obtained after single oral administration of Aulin® and Sulidamor® 100 mg tablets (n = 18, arithmetic means and standard deviations)

	Aulin®		**Sulidamor®**	
	mean	**SD**	**mean**	**SD**
AUC_{0-z} (mg·h/L)	27.219	8.786	15.478	7.010
$AUC_{0-\infty}$ (mg·h/L)	27.530	8.951	15.728	7.146
C_{max} (mg/L)	4.723	0.811	2.3428	0.704
t_{max} (h)	1.63	0.98	4.07	1.18
λ_z (h^{-1})	0.267	0.106	0.268	0.092
$t_{1/2,z}$ (h)	2.856	0.765	2.840	0.836

AUC_{0-z}: area under the plasma concentration-time curve from time of administration (t_0) to the last sample with quantifiable concentration; $AUC_{0-\infty}$: area under the plasma concentration-time curve from time of administration (t_0) to infinity; C_{max}: maximum plasma concentration; t_{max}: time to C_{max}; λ_z: terminal elimination rate constant; $t_{1/2,z}$: terminal elimination half-life. Adapted from [30].

for Aulin® and Sulidamor®, respectively, which indicate a lower absorption rate in the case of Sulidamor®. Again, the differences found out in the investigated pharmacokinetic parameters relevant to the two compared preparations were proven statistically significant.

Relying on the evidence of bioinequivalence highlighted in the above-reviewed studies, it would ensue that neither the generic Nimesulide Dorom nor the copy Sulidamor® could be regarded as therapeutically equivalent to the reference preparation Aulin®. Moreover, analogous conclusions of bioinequivalence might be drawn for the same generic and copy products *versus* Mesulid®, which is co-marketed with Aulin®. These findings appear particularly critical in view of the major role played by the onset and intensity of action in pharmacotherapies based on nimesulide, which is especially used in the symptomatic treatment of acute phlogosis and pain conditions. Therefore, the interchangeability principle seems to apply neither to Nimesulide Dorom nor to Sulidamor®: the possible substitution of Aulin® and Mesulid® with such generic and copy preparations might have given rise to a therapeutic effect far from meeting the prescriber's expectations and the needs related to the pathology. It is noteworthy that, some time after publication of the quoted studies, both Nimesulide Dorom and Sulidamor® have been withdrawn from sale on own initiative of the respective manufacturing companies [31, 32]. Hence, no questionable interchange involving such products may any longer occur to the detriment of the patients. Based on the critical biopharmaceutical features of the molecule, however, a question spontaneously arises: can every nimesulide formulation really be relied on when an acute pain condition is being experienced?

References

1. Güngör S, Bergişadi N (2004) Effect of penetration enhancers on *in vitro* percutaneous penetration of nimesulide through rat skin. *Pharmazie* 59(1): 39–41
2. Shahiwala A, Misra A (2002) Studies in topical application of niosomally entrapped nimesulide. *J Pharm Pharmaceut Sci* 5(3): 220–225
3. Ceschel GC, Maffei P, Lombardi Borgia S (2001) Design and evaluation of a new mucoadhesive bi-layered tablet containing nimesulide for buccal administration. *S.T.P. Pharma Sci* 11(2): 151–156
4. Maffei P, Lombardi Borgia S, Sforzini A, Bergamante V, Ceschel GC, Fini A, Ronchi C (2004) Mucoadhesive tablets for buccal administration containing sodium nimesulide. *Drug Del* 11(4): 225–230
5. Agrawal S, Pancholi SS, Jain NK, Agrawal GP (2004) Hydrotropic solubilization of nimesulide for parenteral administration. *Int J Pharm* 274: 149–155
6. Vandelli MA, Ruozi B, Forni F (1999) PLA microparticles for the prolonged release of nimesulide: effect of preparative variables. *S.T.P. Pharma Sci* 9(61): 567–572
7. Amidon GL, Lennernas H, Shah VP, Crison JR (1995) A theoretical basis for a biopharmaceutic drug classification: The correlation of *in vitro* drug product dissolution and *in vivo* bioavailability. *Pharm Res* 12(3): 413–420
8. Miro A, Quaglia F, Calignano A, Barbato F, Cappello B, La Rotonda MI (2000) Physicochemical and pharmacological properties of nimesulide/beta-cyclodextrin formulations. *S.T.P. Pharma Sci* 10(2): 157–164
9. Thompson DO (1997) Cyclodextrins – Enabling excipients: Their present and future use in pharmaceuticals. *Crit Rev Ther Drug Carrier Syst* 14(1): 1–104
10. Vizzardi M, Sagarriga Visconti C, Pedrotti L, Marzano N, Berruto M, Scotti A (1998) Nimesulide beta cyclodextrin (nimesulide-betadex) *versus* nimesulide in the treatment of pain after arthroscopic surgery. *Curr Ther Res Clin Exp* 59(3): 162–171
11. Chowdary KPR, Nalluri BN (2000) Nimesulide and beta-cyclodextrin inclusion complexes: Physicochemical characterization and dissolution rate studies. *Drug Dev Ind Pharm* 26(11): 1217–1220
12. Adhage NA, Vavia PR (2000) Beta cyclodextrin inclusion complexation by milling. *Pharm Pharmacol Commun* 6(1): 13–17
13. Moneghini M, Kikic I, Perissutti B, Franceschinis E, Cortesi A (2004) Characterisation of nimesulide-betacyclodextrins systems prepared by supercritical fluid impregnation. *Eur J Pharm Biopharm* 58(3): 637–644
14. Vavia PR, Adhage NA (1999) Inclusion complexation of nimesulide with beta-cyclodextrins. *Drug Dev Ind Pharm* 25(4): 543–545

15. Piel G, Pirotte B, Delneuville I, Neven P, Llabres G, Delarge J, Delattre L (1997) Study of the influence of both cyclodextrins and L-lysine on the aqueous solubility of nimesulide; Isolation and characterization of nimesulide-L-lysine-cyclodextrin complexes. *J Pharm Sci* 86(4): 475–480
16. Nalluri BN, Chowdary KR, Murthy KR, Hayman AR, Becket G (2003) Physicochemical characterization and dissolution properties of nimesulide-cyclodextrin binary systems. *AAPS PharmSciTech* 4(1): article 2
17. Gohel MC, Patel LD (2000) Improvement of nimesulide dissolution by a co-grinding method using surfactants. *Pharm Pharmacol Commun* 6(10): 433–440
18. Meriani F, Coceani N, Sirotti C, Voinovich D, Grassi M (2003) Characterization of a quaternary liquid system improving the bioavailability of poorly water soluble drugs. *J Colloid Interface Sci* 263(2): 590–596
19. Meriani F, Coceani N, Sirotti C, Voinovich D, Grassi M (2004) *In vitro* nimesulide absorption from different formulations. *J Pharm Sci* 93(3): 540–552
20. Gohel MC, Patel LD (2002) Improvement of nimesulide dissolution from solid dispersions containing croscaramellose sodium and Aerosil 200. *Acta Pharm* 52(4): 227–241
21. Gohel MC, Patel LD (2003) Processing of nimesulide-PEG400-PG-PVP solid dispersions: Preparation, characterization, and *in vitro* dissolution. *Drug Dev Ind Pharm* 29(3): 299–310
22. Murali Mohan Babu GV, Ravi Kumar N, Himasankar K, Seshasayana A, Ramana Murthy KV (2003) Nimesulide-modified gum karaya solid mixtures: Preparation, characterization, and formulation development. *Drug Dev Ind Pharm* 29(8): 855–864
23. Gohel MC, Patel M, Amin A, Agrawal R, Dave R, Bariya N (2004) Formulation design and optimization of mouth dissolve tablets of nimesulide using vacuum drying technique. *AAPS PharmSciTech* 5(3): article 36
24. Madhuri K, Prasnthi E, Manasa, Durvasa Rao B, Goureenadh N, Chowdary YA, Murthy TEGK (2004) Design and *in vitro* evaluation of nimesulide controlled release tablets. *Pharma Rev* 2(9): 107–108
25. Verma RK, Mishra B (1999) Studies on formulation and evaluation of oral osmotic pumps of nimesulide. *Pharmazie* 54(1): 74–75
26. WHO (World Health Organization) (1993) Interchangeable multi-source pharmaceutical products. WHO draft guideline on marketing authorization requirements
27. A.F.I. Working Party (1996) Generici, aspetti regolatori e tecnici. *Acta Technol Legis Med* VII(1): 1–29
28. Butler D, Bonadeo D, Maroni A, Foppoli A, Zema L, Giordano F (2000) Comparative *in vitro* evaluation of nimesulide-containing preparations on the Italian market. *Boll Chim Farm* 139(6): 237–241
29. Hutt V, Waitzinger J, Macchi F (2001) Comparative bioavailability study of two different nimesulide-containing preparations available on the Italian market. *Clin Drug Invest* 21(5): 361–369
30. Hutt V, Waitzinger J (2001) Generics, copies and original drugs: Are they really interchangeable? Investigations on nimesulide-containing preparations. *J Clin Res* 4: 77–89

31. Ministero della Salute (2003) Revoca, su rinuncia, dell'autorizzazione all'immissione in commercio della specialità medicinale per uso umano "Nimesulide". *Gazzetta Ufficiale della Repubblica Italiana* 95: 90
32. Ministero della Salute (2003) Revoca, su rinuncia, dell'autorizzazione all'immissione in commercio della specialità medicinale per uso umano "Sulidamor". *Gazzetta Ufficiale della Repubblica Italiana* 59: 52

Pharmacological properties of nimesulide

K. D. Rainsford[1], M. Bevilacqua[2], F. Dallegri[3], F. Gago[4], L. Ottonello[3], G. Sandrini[5], C. Tassorelli[5] and I. G. Tavares[6]

[1]Biomedical Research Centre, Sheffield Hallam University, Howard Street, Sheffield, S1 1WB, UK; [2]U O Endocrinologia e Diabetologia, Ospedale L Sacco-Polo Universitario, I-20157, Milano, Italy; [3]First Clinic of Internal Medicine, Department of Internal Medicine, University of Genova Medical School, I-16132, Genova, Italy; [4]Departamento de Farmacologia, Universidad de Alcalá, E-28871, Alcalá de Henares, Madrid, Spain; [5]IRCCS Fondazione "Istituto Neurologico C. Mondino", Dipartimento di Scienze Neurologiche, Università di Pavia, Via Mondino 2, 27100 Pavia, Italy; [6]Academic Department of Surgery, Guy's, King's and St Thomas' School of Medicine, The Rayne Institute, London, SE5 9NU, UK

Introduction

The pharmacological and toxicological properties of nimesulide have been previously reviewed (see [1–8]). The major part of these reviews has been concerned with the preclinical actions of the drug. A key issue concerning the interpretation of the *in vitro* effects of nimesulide has been the relationship of these to the plasma or synovial fluid concentrations of the drug, which are found during therapy. The review by Bennett and Villa [4] is noteworthy for having discriminated the *in vitro* effects, which are known to occur at therapeutic drug concentrations with those which are above this range. Thus, generally speaking although nimesulide has preferential COX-2 selectivity it is also an inhibitor of histamine release and actions, leukotriene B_4 and C_4, and platelet activating factor (PAF) production, the adherence and activation of neutrophils, collagenase and other metalloproteinases, glucocorticoid receptor phosphorylation, interleukin-6 production, calcium channel activation and is an antioxidant within the range of drug concentrations encountered therapeutically [4]. Thus, nimesulide can be regarded as having multifactorial actions in relation to its anti-inflammatory activity.

In vivo pharmacological actions

Models of acute inflammation

Swingle, Moore and co-workers in their preclinical pharmacological investigations of nimesulide (then coded R-805) at Riker Laboratories showed that the drug had relatively potent acute and chronic anti-inflammatory, analgesic and antipyretic

Nimesulide – Actions and Uses, edited by K. D. Rainsford

effects, in conventional animal models [1, 9]. A summary of the acute anti-oedemic potencies of nimesulide in rats compared with that of some standard reference non-steroidal anti-inflammatories (NSAIDs) shows it has anti-inflammatory effects in the rat paw carrageenan assay some 2–3 times more so than that of indomethacin and naproxen and is also more potent than other NSAIDs (Tab. 1 [9]).

Data on the acute therapeutic index (TI) is shown in Table 1. The TI in this case is the measure of the ED_{50} mg/kg values from the carrageenan assay compared with that of the lethal toxicity in a 14 day assay following oral administration of a single oral dose of the drugs and which was then used to calculate the LD_{50} (mg/kg) using the standard Litchfield and Wilcoxon method. This showed that nimesulide was the least toxic among the NSAIDs tested and consequently the drug has a high therapeutic index [9] (Tab. 1).

Adrenalectomy had no effect on the acute anti-inflammatory effects of nimesulide in rats although the anti-inflammatory effects of phenylbutazone (15 mg/kg, p.o.) were reduced in adrenalectomised rats. This indicates that with the latter drug the anti-inflammatory effects are mediated, in part, by adreno-cortical stimulation, a phenomena not apparent with nimesulide.

The oral ED_{50} for nimesulide for inhibition of the ultraviolet (UV)-induced erythema (assayed using the Winter method) in guinea pigs was 2.3 (confidence intervals, CI = 1.9–2.9) and 1.4 (CI = 2.3–3.1) mg/kg, respectively, in two separate assays [9]. This was about five times that of phenylbutazone [9]. In comparison with published data from other NSAIDs, nimesulide is slightly more potent

Table 1 – Acute oral therapeutic indices for non-steroidal anti-inflammatory drugs in rats. (From [9])

Drug	ED_{50} mg/kg[1] in carrageenan assay	LD_{50}, mg/kg[2] (95% confidence limits)	Therapeutic Index (LD_{50}/ED_{50})
Nimesulide (R-805)	1.25	324 (295–356)	260
Naproxen	2.10	395 (281–557)	190
Ibuprofen	13.5	923 (833–1020)	68
Diflumidone	38.0	750 (694–811)	20
Flufenamic acid	14.7	249 (221–280)	17
Phenylbutazone	29.5	406 (375–440)	14
Acetylsalicylic acid	135	1520 (1360–1710)	11
Indomethacin	2.95	21.0 (19.0–23.0)	7

[1] *Calculated from the 3 point regression of response on dose.*
[2] *Calculated by the method of Litchfield and Wilcoxon.*

than indomethacin and ibuprofen and of greater potency than aspirin and most other NSAIDs in the UV erythema assay [10].

Swingle and co-workers determined the chronic anti-inflammatory effects of nimesulide in the rat mycobacterial adjuvant-induced arthritis assay [9]. The hind paw swellings in this assay were prevented in the established disease by nimesulide 0.2 mg/kg/d p.o. when given over the period of 14–30 days after induction of the disease [9]. Higher doses of 0.6 and 1.8 mg/kg/d of nimesulide led to almost complete suppression of the disease [9] (Fig. 1).

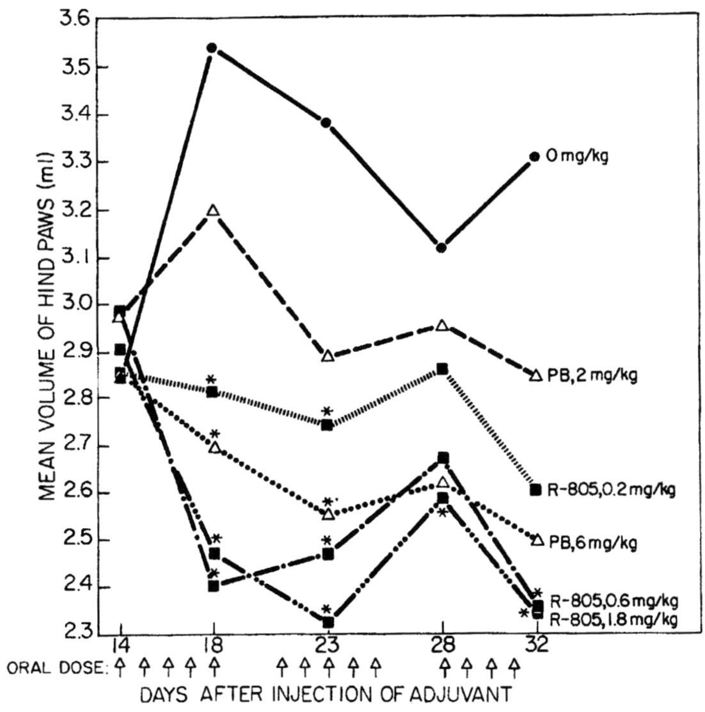

Figure 1

Therapeutic effect of nimesulide (R-805) compared with phenylbutazone (PB) as established adjuvant-induced arthritis in rats. The animals were dosed orally with the drugs from day 14 after injection of the adjuvant on days as shown by the arrows. The mean values shown are from data of readings from 6 rats. The two highest doses of nimesulide (0.6 and 1.8 mg/kg/d) produced almost complete inhibition of adjuvant disease from day 18.) Data significantly different from control P > 0.05. From [9]. Reproduced with permission of the former Editor of Archiv Int Pharmacodyn (no longer in publication).*

In comparison, phenylbutazone was about 10 times less potent. At a dose of 1.8 mg/kg/d nimesulide significantly reversed the body weight loss that occurs in adjuvant disease and improved the general 'state of health' of the animals [9].

These acute and chronic anti-inflammatory effects were largely confirmed in later studies by Tanaka et al. [11]. These authors were using nimesulide as one of a number of comparator NSAIDs to determine the relative anti-inflammatory, analgesic and antipyretic effects of related novel sulphonanilide drug, T-614 (3-formylamino-7-methylsulphonylamino-6-phenoxy-4*H*-1-benzopyran-4-one) [1]. A summary of the data of these authors are shown in Table 2.

Similar oral dose ranges for inhibition of carrageenan paw oedema in rats to those reported by Swingle et al. [9] and Tanaka et al. [11] have been reported by other workers [12–14].

The combined oral administration of nimesulide with aspirin did not show added anti-oedemic effects over that of the drugs alone in the carrageenan paw oedema assay in rats [9]. This was notable even when relatively low doses of nimesulide were given (0.7 or 2.0 mg/kg) with a fixed low dose of aspirin (60 mg/kg; a dose which causes only ~30% inhibition of oedema) [9]. These data suggest that there is unlikely to be a limitation of effects due to reaching the upper range of the dose-response curve, but rather due to some pharmacological antagonism between the two drugs.

Intramuscular nimesulide 1.5–25 mg/kg has been found to produce a dose-related inhibition in carrageenan paw oedema in rats at 2 h (which is the time for peak plasma levels of nimesulide) as well as at 3 and 4 h post-treatment [15]. The anti-oedemic effects of nimesulide were slightly greater than those from the same dose-range of diclofenac [15].

Topical nimesulide 50 mg in a 1% gel (of unknown pharmaceutical composition) applied to the top part of one of the paws of rats 1 h prior to sub-plantar injection of carrageenan produced a reduction of 71.2% in the paw swelling compared with the same dose of diclofenac which produced a reduction of 64.4% in the paw oedema [16]. While there were no data on dose-response or pharmacokinetics of these drugs for comparison the results suggest that nimesulide gel has good anti-inflammatory effects.

In the carrageenan pleural oedema model in rats the ED_{50} values for inhibition of leucocyte infiltration and the ED_{30} value for reduction in pleural fluid are comparable for both nimesulide and indomethacin (Tab. 3) [17]. In the carrageenan air pouch model both drugs appeared less potent [18]; this may be a reflection of the drug accumulation in the air pouch being less than that in the inflamed pleural cavity. The differences may account in part for the claim by Wallace and co-workers [18] that COX-2 inhibition may lead to limited anti-inflammatory effects.

The vascular permeability induced by i.p. acetic acid in rats and in the writhing test in mice was found to be inhibited to about the same extent as the anti-oedemic effects in the air pouch oedema in rats [11]. The ED_{50} for the Evans blue pleural

Table 2 – Summary of acute oral anti-inflammatory effects of nimesulide in animal models (from [11])

MODEL	Acute Paw Oedema in Rats Induced by:					UV Erythema	
	Carrageenan ED_{50} mg/kg (CI)	Dose mg/kg	Kaolin % Inhibition ± s.e.m @ 5 h	Dextran % Inhibition (± s.e.m) @ 1 h	Bromelain % Inhibition (± s.e.m) @ 1 h	In Guinea Pigs ED_{50} mg/kg (CI) after	
DRUG						**Single dose**	**Two doses**
Nimesulide	2.4 (0.95–6.3)	30	55.5 ± 5.1*	28.8 ± 2.1*	31.3 ± 6.3*	4.5 (1.7–12)	2.3 (0.99–5.2)
Ibuprofen	25 (6.5–98)		N/D	N/D	N/D	7.7 (2.7–22)	7.8 (3.4–18.0)
Indomethacin	1.9 (0.75–4.9)	10	45.8 ± 7.0*	7.4 ± 1.2*	2.1 ± 3.6	2.8 (0.63–13)	1.3 (0.68–2.5)
Ratio-Nimesulide/ Indomethacin	1.3		1.2	3.9	14.9	1.6	1.8
Nimesulide/ Ibuprofen	0.096		N/D	N/D	N/D	0.6	0.3

CI = 95% confidence interval estimates.

* *Statistically significant differences compared with control ($p < 0.05$).*

N/D = not determined.

Table 3 – Components of the oral anti-inflammatory effects of nimesulide in rats (from [11, 18])

MODEL	Percent of Inhibition of Paw Swelling in:		Reduction in Carrageenan Air Pouch [18]		Inhibition of Acetic Acid-induced Capillary Permeability in Rats [11]
DRUG	**Adjuvant Arthritis [11] ED_{40} (CI) mg/kg/d**	**Paper Disk granuloma [11] ED_{30} (CI) mg/kg**	**Exudate Volume ED_{30} mg/kg**	**Leucocytes ED_{50} (CI) mg/kg**	
Nimesulide	1.6 (0.07–38)	0.65 (0.18–2.3)	~1.0	~1.5	11 (3.9–28)
Ibuprofen	75 (11–500)	45 (10–198)	N/D	N/D	16 (6.1–40)
Indomethacin	0.8 (0.03–23)	1.2 (0.31–5.9)	~0.3	~0.2	1.1 (0.4–3.0)
Ratio-Nimesulide/ Indomethacin	2.0	0.54	0.3	7.5	10.0
Nimesulide/ Ibuprofen	0.021	0.014	N/D	N/D	0.69

CI = 95% confidence interval.
N/D = not determined.

effusion for nimesulide was found to be 21 mg/kg in mice and 11 mg/kg in rats, which is higher than the doses required for acute anti-inflammatory effects. Indomethacin was more potent in these models with ED_{50} values of 0.57 mg/kg in mice and 1.1 mg/kg in rats [11]. Likewise, ibuprofen was also more potent than nimesulide having ED_{50} values of 7.5 mg/kg (mice) and 16.0 mg/kg (rats) [11]. These data suggest that the effects of nimesulide on vascular permeability while not potent may, like other NSAIDs, also contribute to its acute anti-inflammatory activity.

Relationship of acute anti-inflammatory effects to prostaglandin production

The relationship between inhibition of prostaglandin (PG) production *in vivo* and anti-inflammatory effects of nimesulide has been investigated in a number of studies in rats [17–20]. Nakatsugi and co-workers [17] observed reduction by nimesulide of PGE_2 concentrations in the pleural cavity following intrapleural injection of carrageenan in rats with an ED_{50} of 0.75 mg/kg. In contrast the pleural exudate volume was significantly inhibited at higher doses of 3 and 10 mg/kg of this drug suggesting that inhibition of PGE_2 production by nimesulide is both within the range of dosage required for inhibition of paw oedema and lower than that required for the post-venule vascular changes that are responsible for fluid accumulation. Similarly lower doses were required for inhibition of PGE_2 production in the pleural cavity by indomethacin (ED_{50} 0.25 mg/kg) and ibuprofen (ED_{50} 6.9 mg/kg) than required for statistically significant reduction in exudate volume by indomethacin (≥3 mg/kg) or ibuprofen (30 mg/kg) respectively. These authors also showed the presence of COX-1 as well as COX-2 protein by Western blotting in the pleural cell extracts and that COX-2 expression was unaffected by any of the NSAIDs (including nimesulide). This suggests that any effect of these drugs on COX-2 mediated PGE_2 production is due to direct effects on this enzyme and not its synthesis. As discussed later in the section on "Inhibition of the synthesis of COX-2" (page 160), there is evidence from *in vitro* studies to suggest that nimesulide may inhibit the synthesis of the COX-2 enzyme. Whether this effect is cell or tissue specific is as yet unresolved so it is not possible to conclude if COX-2 synthesis is affected by nimesulide in various models of inflammation *in vivo*.

In the carrageenan air pouch model Wallace et al. [18] found that PGE_2 concentrations were reduced in the pouches at much lower doses of nimesulide or indomethacin than those at which there was reduction in exudates volume or leucocyte numbers; the difference being in the order of about ten-fold.

Tanaka et al. [19] showed that PGE_2 levels in carrageenan-impregnated sponges were significantly reduced to about 24–25% of control values following 3 h treatment with 0.3–20 mg/kg. p.o. nimesulide. The same degree of inhibition of PGE_2 levels was observed with 1.0 mg/kg indomethacin and 100 mg/kg ibuprofen p.o. Since statistically significant inhibition of PGE_2 levels was observed with 0.3 mg/kg nimesulide p.o. (25% of control values) this effect on PGE_2 production is within the dose range for inhibition of paw oedema [11].

In the carrageenan-soaked sponge model in rats, Tofanetti et al. [20] observed reduction in PGE_2 and thromboxane B_2 (TxB_2) concentrations by nimesulide with IC_{50} values of 1.2 (95% CI = 0.88–180) mg/kg and 1.56 (95% CI = 0.82–2.92) mg/kg respectively. Again, these values are within the range of those at which inhibition of carrageenan paw oedema has been observed (Tab. 2) [11]. In comparison, indomethacin reduced PGE_2 and TxB_2 with IC_{50} values of 0.92 (95% CI = 0.7–1.15) mg/kg and 0.94 (95% CI = 0.66–1.34) mg/kg respectively, which are also within the values for inhibition of carrageenan paw oedema (Tab. 2) [11]. The conclusion from these studies is that nimesulide exhibits acute anti-inflammatory effects by inhibition of PGE_2 production coincident with reduction in leucocyte accumulation, although the latter may occur at higher doses than those required for inhibition of prostaglandin production.

In other models of acute paw inflammation in rats nimesulide is appreciably less potent in its effects than in the carrageenan paw oedema and UV erythema in guinea pigs (Tab. 2). Similar less potent inhibition of rat paw oedema than that from carrageenan has been observed with several other NSAIDs and may be related to their differential effects on production of inflammatory mediators produced during the acute phases of inflammation [10].

Harada and co-workers [21] observed that nimesulide, like that of other COX-2 selective inhibitors, reduced intrapleural concentrations of the COX-2-derived prostacyclin metabolite, 6-keto-$PGF_{1\alpha}$ and of PGE_2, but not the COX-1 derived TxB_2, in the carrageenan-induced pleurisy in rats. Since there was induction of PGHS-2 protein in the pleural exudates cells [12] this suggests that nimesulide induced reduction of PGE_2 and 6-keto-$PGF_{1\alpha}$ was a consequence of the selective inhibition of COX-2 activity.

In a study in which the long-term acute effects of nimesulide 3 mg/kg were investigated in pleural effusion of rats 14 h after injection of carrageenan (in the so-called late phase of this pleurisy model) Hatanaka et al. [22] found reduction in 6-keto-$PGF_{1\alpha}$, PGE_2, but not TxB_2 in the pleural exudates when the drug was given orally 9 h after the induction of pleurisy. In these studies COX-2 was induced at 3–20 h in the pleural mesothelial cells, so the effects of nimesulide are probably due to the direct inhibitory effects of the drug on COX-2 activity.

Models of chronic inflammation

The mycobacterial adjuvant-induced arthritis in rats is a well-established model for determining chronic anti-inflammatory activity with some NSAIDs [10], and close parallels in aetio-pathology with that of rheumatoid arthritis (RA) in humans [23]. In their initial screening for anti-inflammatory activity of nimesulide

(R-805) Swingle and co-workers [9] had observed almost complete suppression in the therapy of this disease by relatively low doses of 1.8 mg/kg/d of nimesulide (Fig. 1). When compared with the acute effects in the carrageenan paw oedema model [9, 11] (Tabs 1 and 2) there is striking overlap in the doses required for both chronic as well as acute effects in this drug. Aside from indomethacin and a few other NSAIDs of like potency this correspondence of acute with chronic anti-inflammatory effects is not often seen with NSAIDs [10]. These differences may be due to differences in the mode of action of the various NSAIDs. With nimesulide it is possible that its multifactorial activities are evident in both acute and chronic inflammation. Also, it is possible that other NSAIDs may have more pronounced effects on COX-2-derived prostaglandin production and lesser effects on leucocyte or other components of the inflammatory response.

Tanaka and co-workers [11] and Qui et al. [12] have also observed effects of nimesulide in the therapeutic mode of treatment in adjuvant arthritis within the same dose range as observed by Swingle et al. [9].

Topical formulations of nimesulide have been found to have acute and chronic anti-inflammatory activities in experimental animal models [16]. The application to the upper surface of the right paws of rats of a 1% gel formulation of nimesulide 50 mg on day 1 followed by 25 mg thereafter caused a reduction in paw swelling of the same paw given a sub-plantar injection of Freund's adjuvant [16]. By 18 h there was a reduction of 36.5% and by 18 days it was 52.8%. In comparison, the same doses of diclofenac and piroxicam gels administered in the same way reduced paw swelling by about a third of that of controls. The amounts of the drugs applied in these studies are relatively high in relation to body weight. The 25 mg nimesulide dose (as a 1% gel) yielded peak plasma concentrations of 23 µg/mL at about 2 h after application and was still present in the plasma by 6 h. The peak plasma concentration was about three times that observed in humans after 100 or 200 mg oral doses of nimesulide (Chapter 2; by A. Bernareggi and K.D. Rainsford; Tab. 6) The $AUC_{0\text{–}6\,h}$ was 83 µg/mL h [15] which is about three times that obtained in humans after 100 mg oral dose of nimesulide (Chapter 2; by A. Bernareggi and K.D. Rainsford; Tab. 6).

Collagen II arthritis (with Freund's complete adjuvant as immuno-stimulant) given in mice was found to be inhibited from 6 weeks post-induction by 1.0 and 3.0 mg/kg/d nimesulide p.o. given three times per week for up to 10 weeks [24]. Collagen II antibody levels were also reduced by the lower dose of nimesulide but not by 3 mg/kg indomethacin given under the same dosage regime and even though the joint inflammation was reduced by this drug [24].

Gilroy and co-workers [25] undertook a study in which Freund's adjuvant was injected into the air pouch of rats and after 3 days the animals were given nimesulide 0.5 or 5.0 mg/kg/d, aspirin 10 or 200 mg/kg/d or the selective COX-2 inhibitor, NS-398, 0.1–10 mg/kg/d were given daily up to 28 days post-induction of the granulomas. The weights of the granulomas, vascularity and PGE_2 concentrations in the granulomatous tissues were variably affected by nimesulide. There was no indication in the methods if the mycobacterial adjuvant had been delipidated prior to injection so there may be a possibility of endotoxin contamination in the preparation that was injected. With the higher dose of the drug the weights of the granulomas and vascularity were, paradoxically, increased at day 7 compared with controls but not at later times. The higher dose of 5 mg/kg/d of nimesulide also resulted in an increase in PGE_2 concentrations in the granulomas at days 5 and 21 but not at other times compared with controls. PGE_2 was significantly reduced by both dose levels of aspirin, but granuloma weights were unaffected except by the higher dose on day 14 only and vascularity was unaffected by this drug. NS-398 had no effects on any of the parameters.

There are several puzzling aspects about these studies. The results do not agree with the potent anti-inflammatory effects of nimesulide in the carrageenan pouch granuloma and disc granuloma models in rats observed by Tanaka et al. [11]. Majima and co-workers [26] observed that nimesulide and NS-398 reduced PGE_2 levels in rat sponge granulomas and neovascularisation. While neither of these studies were in animals given Freund's adjuvant in the air pouch as used by Gilroy et al. [25] it is difficult to see how the anti-inflammatory effects of nimesulide and NS-398 could not have been fully expressed in their studies especially since nimesulide has potent anti-inflammatory effects in Freund's adjuvant arthritis in rats [9, 11, 12]. There is also the possibility of endotoxin contamination as noted above and this could lead to production of a wide range of proinflammatory cytokines and other inflammatory mediators.

In order to explain the effects of nimesulide in increasing granuloma weight and on the PGE_2 concentration Gilroy et al. [25] suggested that there might be an increase in mRNA coding for PGHS-2 by nimesulide so leading to increased COX-2 activity but no evidence was provided from experiments in their pouch model in support of this suggestion. In other studies in carrageenan-induced pleural inflammation there is no evidence for increased COX-2 protein [26]. Overall, the lack of clear time-, or dose-dependent effects of the drugs along with the high variability in the results and lack of correspondence with other studies makes these data very difficult to explain.

Analgesic activities

The oral analgesic activities of nimesulide in rodents compared with that of some other NSAIDs given orally is shown in Table 4 [11]. In contrast with the anti-oedemic effects, nimesulide is less potent in rat and mouse writhing models (see also [12]), but in the pain threshold in the Randall-Selitto test and adjuvant induced hyperalgesic in rats nimesulide is more potent and the dose-effects in these models overlaps that for anti-oedemic effects (c.f. Tab. 2). Ibuprofen and indo-

Table 4 – Oral analgesic activity of nimesulide in rodents

MODEL	Writhing response ED_{50} (CI) mg/kg induced in Mice by:			Writhing response in Rats induced by Acetic Acid	Pain threshold in Randall-Selitto test in Rats	Adjuvant-induced Hyperalgesia in Rats
DRUG	Acetic Acid	Acetylcholine	Phenylquinone	ED_{50} (CI) mg/kg	ED_{50} (CI) mg/kg	ED_{50} mg/kg
Nimesulide	40 (6–240)	10 (5.8–18)	18 (8.9–35)	21 (6.6–64)	3.5 (1.3–9.4)	2.4
Ibuprofen	45 (14–147)	8.5 (4.9–15)	3.1 (0.86–11)	7.5 (1.5–37)	26 (7.6–87)	29
Indomethacin	4.1 (1.5–11)	0.45 (0.2–1.0)	1.2 (0.66–2.1)	0.57 (0.15–2.2)	1.9 (0.5–7.0)	3.1
Ratio – Nimesulide/ Indomethacin	9.8	22.2	10.0	36.8	1.8	0.8
Nimesulide/ Ibuprofen	0.9	1.2	5.8	2.8	0.13	0.08

From [11].

methacin show similar differences in relative potencies in these assays [10] implying that these differences among the NSAIDs may be a more common feature of the drugs in these models.

In the acetic acid writhing model in rats in which lipopolysaccharide was given to enhance the production of PGHS-2 protein, Matsumoto and co-workers [27] observed reduction in the elevated levels of 6-keto-$PGF_{1\alpha}$ by nimesulide as well as by some other COX-2 selective drugs in the peritoneal exudates.

Intraperitoneal administration of nimesulide leads to more pronounced inhibition of the acetic acid writhing test in mice with an ED_{50} of 7.6 mg/kg [28]. Intrathecal nimesulide produces even greater inhibition of writhing in this model and the relative potency of nimesulide in this model is much higher compared with other NSAIDs [28, 29] suggesting there is a strong component of spinal analgesia exhibited by nimesulide in this model.

Using models of lameness induced in the hind limb of dogs given intra-articular injection of Freund's complete adjuvant or sodium urate crystals, Toutain and co-workers [30] observed that concurrent treatment with a single oral dose of nimesulide 5 mg/kg in the former and 3–9 mg/kg in the latter significantly reduced the lameness scores. The time course of the relief of lameness in both models showed this peaked at about 2–4 h and extended to about 12 h in the urate crystal model. In the Freund's adjuvant model this period was longer and appeared to extend to about 72 h. The authors modelled these changes in relation to the pharmacokinetics of nimesulide in dogs. With the maximum plasma concentrations being at 5.3 h (8.5 µg/mL) and the half-life of plasma elimination (in the β-phase) being 7.2 h it is apparent that the therapeutic action of the drug extends beyond the time of drug levels in the plasma and this is born out by the curve fitting models obtained by the authors [30]. Using the same approach to modelling there was nearly complete inhibition of COX-2 activity *in vitro* at therapeutic concentrations of nimesulide in the dog [31]. Further aspects of this are discussed on page 149 in the section on "Effects of nimesulide on arachidonic acid metabolism *in vitro, ex vivo* and *in vivo*".

Other aspects about the mode of actions of nimesulide in animal models of analgesia are discussed elsewhere in this chapter, and a comparison with the coxib sub-class of NSAIDs is also discussed in Chapter 5.

Antipyretic effects

Nimesulide is a relatively potent antipyretic drug compared with that of other NSAIDs. In the yeast-induced fever model in rats nimesulide has an ED_{50} of 0.21 (95% CI = 0.85–0.52) mg/kg and is appreciably more potent than indomethacin (ED_{50} = 1.8 [95% CI = 0.28–12] mg/kg), ibuprofen (ED_{50} = 3.7 [95% CI = 0.6–23.0] mg/kg) and aspirin (ED_{50} = 25 [95% CI = 9.9–64] mg/kg) [11]. Nimesulide 2 mg/kg reduced the elevated body temperature following injection of peptone i.v. in rabbits [12]. Both orally and rectally administered nimesulide were found to be more effective in lowering the febrile response in rats injected with brewer's yeast than with paracetamol [32].

Using conscious guinea pigs that several days previously had been fitted with individually cannulas, Steiner and co-workers [33] found that nimesulide 0.3–3.0 mg/kg given by i.p. or intracerebroventricular (i.c.v) injection 30 min prior to i.v. lipopolysaccharide (LPS) caused a dose-related reduction in the second of the biphasic elevations in core body temperature (at 1.5–3.0 h). The LPS-induced febrile response was almost completely abolished by the highest dose of nimesulide. Nimesulide 0.3 and 3.0 mg/kg i.p. also reduced both plasma and brain concentrations of PGE_2 induced by i.v. or i.c.v. LPS. Both i.p. and i.c.v. nimesulide 3.0 mg/kg reduced the entire febrile response to i.c.v. LPS coincided with reduction in brain but not plasma PGE_2. These results are interesting for showing that nimesulide exerts antipyretic effects by crossing the blood–brain barrier. Furthermore, this drug can produce antipyretic effects over both phases of LPS induced fever, although when given i.p. it tends to have greater effects on the long second phase. Finally, there is a clear relationship between antipyretic effects of nimesulide and the reduction in brain PGE_2 which is produced by LPS. The COX-1 inhibitor, SC-560 5 mg/kg i.p., did not affect the febrile response to LPS in this model, but indomethacin 10 mg/kg i.p. did reduce both the first and second phases of the LPS induced febrile reactions. These results imply that only COX-2, and not COX-1, underlies the development of fever; the fact that indomethacin i.p. reduced both phases might suggest that there may be some component of COX-1 derived PG's that contributed with the PGE_2 from COX-2 inducible by LPS that affects both phases of fever. The case of a COX-2 selective inhibitor with the COX-1 inhibitor, SC-560, might have established if a form of cross-talk between COX-1 and COX-2 exists as postulated by others for control of inflammation [34].

Mechanisms of action of nimesulide on pathways of inflammation

Concepts of the actions of nimesulide on the pathways of inflammation that are considered to be involved in arthritic diseases, especially osteoarthritis and related musculoskeletal conditions are shown in Tables 5A and 5B. When these are considered in relation to the present concepts of the actions of nimesulide (Tab. 5A and 5B) it is seen that there are a considerable number of inflammation pathways that are affected by this drug.

The major pathways of significance in the actions of nimesulide for control of acute and chronic inflammation, pain and fever [4, 6–8, 35] (Tab. 5A and 5B) are:

- Arachidonic acid metabolism, especially production of COX-2 derived prostaglandins and leukotrienes
- Proinflammatory cytokine production and actions, especially of tumour necrosis factor-α (TNFα) and interleukins (IL), 1, 6 and 8

- Complement activation
- Leucocyte recruitment and activation at inflamed sites
- Superoxide production, hydroxyl-radical scavenging, lipid peroxidation reactions and effects on nitric oxide production
- Intracellular signalling and expression of cell surface adhesion molecules
- Histamine and other basophils/mast cell mediators, including platelet activating factor (PAF)
- Metalloproteinase production and chondrocyte apoptosis in inflamed/arthritic joints
- Plasminogen activator inhibitor production
- Peroxisomal proliferator-activator receptors (PPAR) transcription pathways
- Glucocorticoid receptor activation

Table 5A – Some pharmacological effects of nimesulide relevant to its anti-inflammatory activity

Actions of nimesulide at normal doses or concentrations, or at therapeutically relevant plasma concentrations *in vitro* (up to 0.06 µg ml^{-1}; approximately 0.2 µ/mol/L in the absence of albumin). All the effects listed are inhibitory (Bennett [6]).

Pathway	Man	Lab Animals	Cells *in vitro*	Dose/conc.
COX-2 activity	leucocytes *ex vivo*			100 mg b.d
			human leucocytes	Therapeutic range 200 mg
Superoxide formation	leucocytes *ex vivo*			
Histamine action	skin *in vivo*			200 mg
Histamine release		guinea pigs		1.6 µmol/kg
Histamine release		guinea pigs		0.1–1 mg/kg i.v
Cytokine action		Rats		7 mg/kg
COX-2 formation			human synoviocytes	0.03 µg/mL
Metalloprotease formation			human synoviocytes	0.03 µg/mL
Collagenase	Synovial fluid			2 µmol/L
Chondrocyte apoptosis		Rat	chondrocytes	1 pmol/L–10 nmol/L

Table 5B – In vitro *effects of nimesulide (in presence or absence of serum or albumin) at high therapeutic or supratherapeutic plasma/ blood concentrations*

Maximum plasma concentration of nimesulide during therapy is approximately 20 µmol/L (6 µg/mL) FCS = Foetal Calf Serum; Alb = albumin. Addition of these proteins would be expected to reduce the free concentration of the drug by about 1/2–1/3 the total drug added (Bennett and Villa [4]).

Pathway	**Cells/Tissue**	**Nimesulide**	**Protein**	**Effect (%)**
Urokinase synthesis	Synovial fibroblasts	0.3 µg/mL	1% FCS	↓ 50%
Plasminogen activator inhibitor	Synovial fibroblasts	0.3 µg/mL	1% FCS	↑ 50%
Interleukin-6 synthesis	Synovial fibroblasts	0.3 µg/mL	1% FCS	↓ 50%
Histamine action	Bronchial muscle	0.3 µg/mL	0	↓ 20%
Cyclic AMP	PMN leucocytes	1 µmol/L	0	↑ 40%
Phosphodiesterase Type-4	PMN leucocytes	10 µmol/L	0	↓ 20%
Collagenase	Bacterial	1.9 µmol/L	0	↓ 50%
Platelet activating factor synthesis	Eosinophils	3 µmol/L	0.5% Alb	↓ 40%
Leukotriene C_4	Eosinophils	3 µmol/L	0.5% Alb	↓ 40%
Anti-oxidant activity	Liposomes	5 µmol/L	0	↓ 50%
Histamine release	Basophils	10 µmol/L	0.05% Alb	↓ 20%
Histamine release	Mast cells	10 µmol/L	10% FCS	↓ 20%
Myeloperoxidase/hypochlorous acid	Neutrophils	20 µmol/L	0	↓ 50%
α1-antitrypsin inactivation	Neutrophils	20 µmol/L	0	↓ 80%
Cell adherence	Neutrophils	20 µmol/L	?	↓ 20%
Cell migration	Neutrophils	50 µmol/L	?	↓ 20%
Glucocorticoid receptor phosphorylation	Synovial fibroblasts	0.3–30 µmol/L	0.5% FCS	↑ 55%
Calcium channels	Myometrial cells	100 µmol/L	0	block 35%

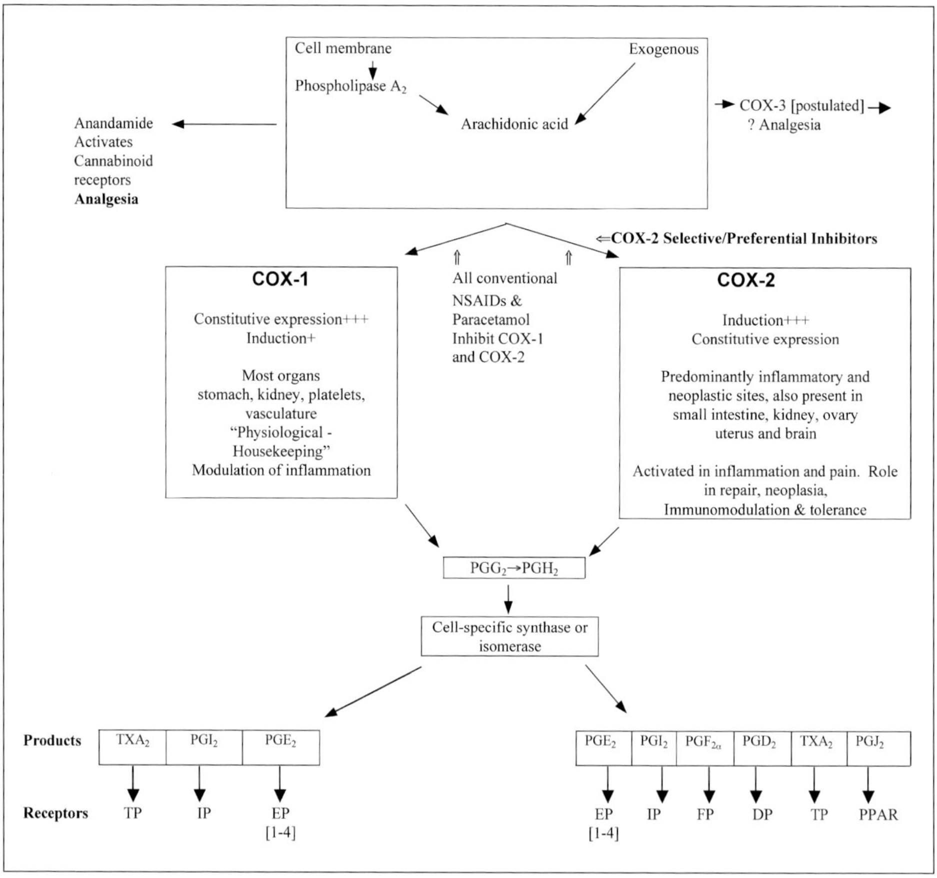

Figure 2
Pathways of cyclooxygenase (COX)-1 and 2 activities showing the main involvement of products of COX-1 in controlling physiological functions and COX-2 in inflammation and pain. The actions of individual prostanoids on receptors leads to their specific actions on cells. From Rainsford KD (2004) [40]. Reproduced with permission of the publishers, Wiley, Chichester.

The concept has emerged following extensive studies on the actions of nimesulide in different pathways of inflammation that this drug has multiple modes of action [4, 6–8, 35]. The multifactorial actions of nimesulide may have particular advantage in enabling its potent actions in relief of pain, diverse inflammatory reactions and fever. The fact that it is not simply a COX-2 inhibitor may separate nimesulide from the coxib class of inhibitors, some of whom can in some conditions be less effective therapeutically compared with nimesulide, e.g., in pain relief in humans and animals (see Chapter 5 and later section).

Table 5A summarises the effects of nimesulide that are considered to be therapeutically relevant at plasma concentrations up to 60 ng/mL or 200 nmol/mL in the absence of albumin, or at concentrations that might be slightly higher than these values. The latter group of effects might be pharmacologically significant in relation to control of inflammation by nimesulide.

Effects of nimesulide on arachidonic acid metabolism *in vitro, ex vivo* and *in vivo*

The fatty acid arachidonic acid, released from phospholipids by the enzyme phospholipase A_2, is a substrate for cyclooxygenase enzyme (Fig. 2 and 3) [36].

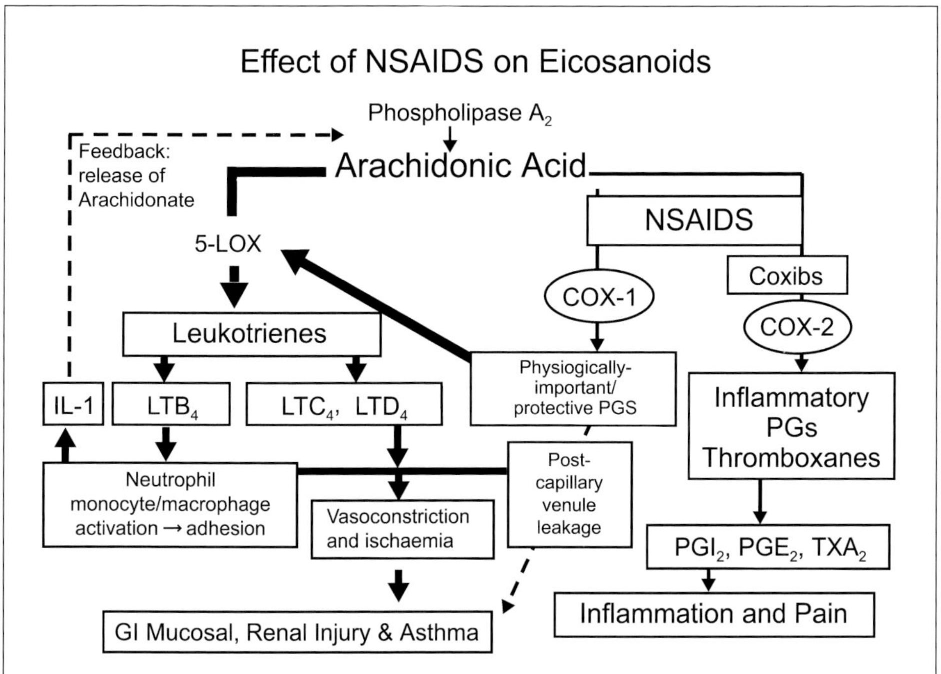

Figure 3
Inter-relationships between proinflammatory cytokines (e.g., interleukin (IL)-1 and the induction of phospholipases (e.g., PLA_2) leading to arachidonic acid release and induction of cyclooxygenase-2 (COX-2). These amplifying processes lead to increase in production of prostanoids and leukotrienes, whose end products can repress (e.g., PGE_2) or stimulate (e.g., LTB_4) PLA_2 and COX-2. From Rainsford KD (2004) [40]. Reproduced with permission of the publishers, Wiley, Chichester.

Arachidonic acid is oxidised by cyclooxygenase first to prostaglandin G_2 followed by reduction to prostaglandin H_2, thus initiating the cascade where various synthases then act to produce prostaglandins or thromboxane, collectively termed prostanoids. Arachidonic acid is also the substrate for lipoxygenase enzymes that convert it to hydroxyl-fatty acids and leukotrienes.

In 1990, it was discovered that there are at least two forms of cyclooxygenase – a constitutive enzyme, COX-1 and an inducible enzyme, COX-2 [36–38]. COX-1 is mainly involved in normal physiological functions while COX-2 is usually involved in the inflammatory response or some physiological functions that require prostaglandins transiently such as in gastric ulcer healing, some renal functions, atherothrombosis, bone metabolism, insulin secretion, vascular functions, regulation of immune functions and ovulation (Fig. 3) [41–44]. In the stomach, COX-1 forms the prostaglandins that help maintain gastric mucosal integrity that include reduction in acid output, increase in mucus discharge, increase in bicarbonate secretion, enhancement of blood flow and protection against mast cell degranulation [42]. In addition, prostanoids formed by COX-1 have diverse physiological roles that include facilitation of renal function [43] as well as effects on blood clotting and pressure [44] which are affected by NSAIDs. Prostanoids are also important mediators of inflammation, pain and fever, where they act by sensitising pathways to bradykinin, histamine and serotonin. For the most part, prostaglandins that contribute to the complex inflammatory process are formed by the inducible COX-2 found in leucocytes and other cells at inflammatory sites [45]. Inhibition of prostaglandins formed by COX-1 generally lead to unwanted side effects such as gastric bleeding, while inhibition of prostaglandins formed by COX-2 generally relieve pain and inflammation [27, 38]. Leukotrienes also play important roles as mediators of the vascular and some of the immunological components of inflammation [36, 40].

At the early stage of discovery and development of the sulphonanilides at Riker there was no indication or little evidence that there could be effects on PG production, since Vane's discovery of the effects of aspirin and other analgesics in 1971 post-dated the period when the medicinal chemistry concepts for the development of these drugs were being formulated. Moreover, the later discovery of the different isoforms of cyclooxygenase about two decades after Vane's discovery had not led to the identification of the COX-2 selective effects of nimesulide [39, 40]. Between these two periods work on the actions of NSAIDs, including nimesulide, on prostaglandin production in various cellular systems was undertaken without knowledge of the existence of the COX-isoforms or many of the details concerning the mechanisms of action of these drugs on components of arachidonic acid metabolism. After the discovery of the leukotrienes and the lipoxygenases (LOX) involved in their production in the 1980s that interest focussed on the potential for some NSAIDs to affect the second pathway (LOX) of arachidonic acid metabolism [36, 40].

Initially, nimesulide was found to be a weaker inhibitor of cyclooxygenase compared with other NSAIDs. In 1977 Vigdahl and Tukey [46] showed that nimesulide inhibited prostaglandin biosynthesis by bovine seminal vesicle microsomes and aggregation of human platelets in a concentration-dependent manner but to a lesser extent than indomethacin (Tab. 6). These results showed that nimesulide has relatively weak effects on prostaglandin synthesis in an *in vitro* system, which with hindsight is probably a COX-1 preparation. Rufer et al. (1982) [47] using ram seminal vesicles in the presence and absence of cofactors, confirmed that indomethacin was greater than one order of magnitude more effective in inhibiting PG synthesis than nimesulide.

These results show that COX-1 inhibition by nimesulide in these microsomal preparations shows poor relationship to inhibition of carrageenan paw oedema in rats and platelet aggregation inhibition (Tab. 6). With current knowledge of the mode of action of nimesulide being predominantly on COX-2 it is not surprising that there is a poor relationship between microsomal COX-1 inhibition and *in vivo* inhibition of acute inflammation, which is primarily due to COX-2 effects.

Rufer et al. [47] also investigated the effects of a number of sulphonanilides with varying pKa's on the peroxidase (PEROX) reactions of prostaglandin synthase compared with the COX-1 activity (Fig. 4). Those with higher pKa values (that ranged from 6.5 to 9.36) had increasing trend to greater PEROX than COX-1 activity (Fig. 4). Nimesulide had both COX-1 and PEROX activity. The

Table 6 – Effects of nimesulide on inhibition of microsomal prostaglandin synthesis compared with effects on carrageenan paw oedema in rats and platelet aggregation

Drug	Prostaglandin synthetase Inhibition* IC_{50} (μmol/L)	Carrageenan Bioassy ED_{50} (mg/kg)	Platelet aggregation Inhibition** IC_{50} (μmol/L)
Nimesulide	25.0	1.25	8.5
Indomethacin	0.5	2.95	0.7
Flufenamic Acid	4.0	14.7	30.0
Phenylbutazone	9.8	29.5	65.0

* *Drug concentration inhibiting 50% arachidonic acid conversion to PGE_2 and $PGE_{2\alpha}$ in bovine seminal vesicles probably a COX-1 preparation.*

** *Drug concentration inhibiting 50% platelet aggregation in citrated platelet-rich human plasma.*

From Vigdahl and Tukey [46].

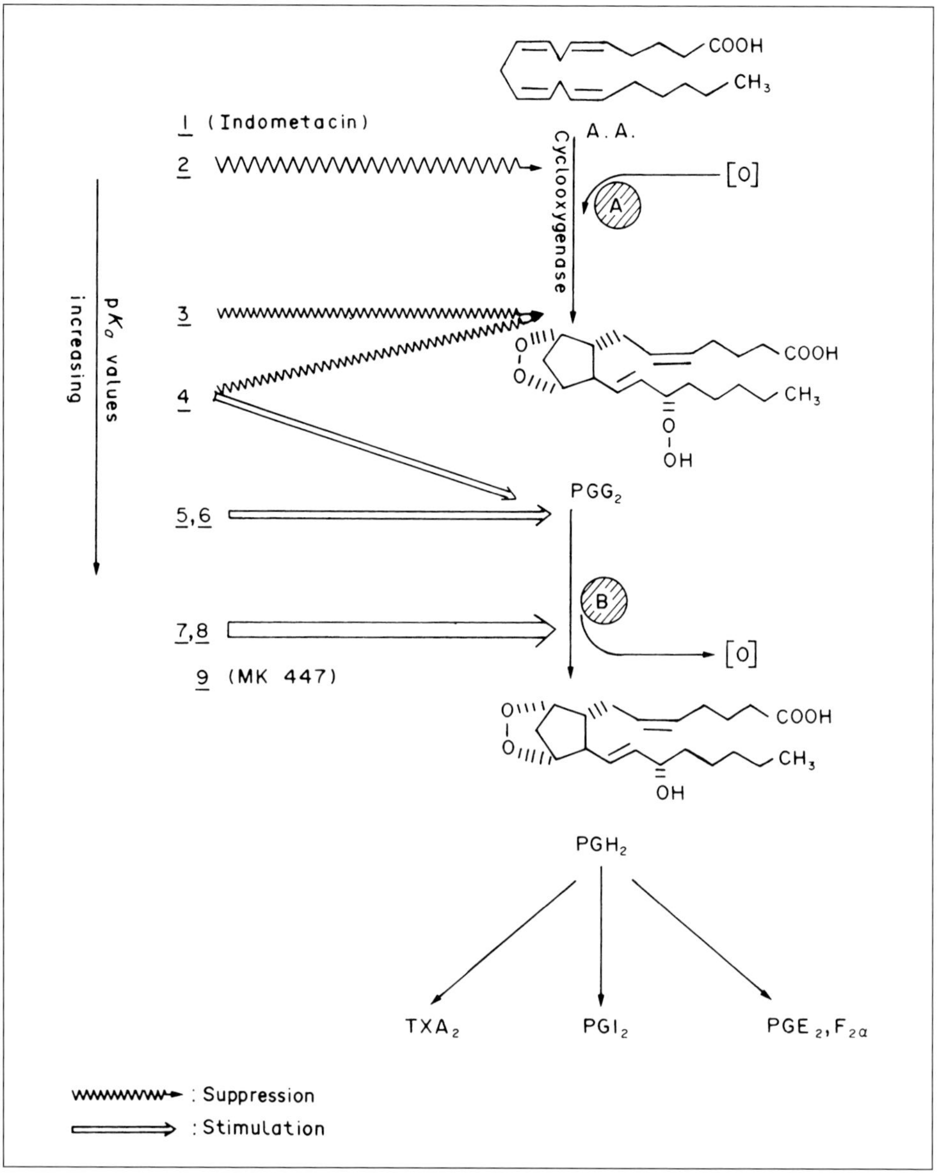

Figure 4
Actions of nimesulide and other methane sulphonanilides (compounds #2-#8 whose sites of action are shown on left hand side), indomethacin (compound #1) and MK-447 (compound #9) that have varying pKa on the cyclooxygenase ("A") and peroxidase ("B") activities of prostaglandin synthetase. From Rufer et al. (1982) [47]. Reproduced with permission of the publishers of Biochemical Pharmacology.

significance of the effects on PEROX activity is that this related to scavenging by phenolic compounds (e.g., MK-447; 2-aminomethyl-4-*tert*-butyl-6-iodo-phenol) and similar agents of the peroxy-radical cleaved from the 15-hydroperoxy group of PGG_2 by PEROX [47]. This oxyradical cleaved by PEROX assists in initiating the COX activity and thus scavenging of this indirectly by phenolic compounds and those with antioxidant activity like nimesulide (see section on Nimesulide and neutrophil functional responses, page 173) effectively blocks the initiating reactions (Fig. 4). Thus, an explanation of the relatively weak COX-1 inhibition by nimesulide is that this is only one part of the enzymatic system in PGHS's affected by the drug; the other is the antioxidant effect in the PEROX reaction.

In 1987 Böttcher et al. [48] showed that in rats nimesulide was as effective as indomethacin in carrageenan oedema, adjuvant arthritis, acetic acid writhing and yeast fever but nimesulide caused substantially less ulcers and blood loss compared to indomethacin. Their study showed that inhibition by nimesulide was weaker compared to indomethacin for prostaglandin synthesis by bovine seminal vesicles but both were more effective in zymosan-stimulated murine macrophages (Tab. 7). Nimesulide did not influence 5-HETE production in murine macrophages, so implying there is no effect of the drug on 5-LOX activity.

Nimesulide effectively inhibited prostaglandin formation at sites of inflammation in the rat, but only poorly in rat gastric tissue. Tofanetti et al. [20] demonstrated that a single oral administration of nimesulide in rats decreased PGE_2 and TXB_2 synthesis ten times more potently in inflammatory exudates than in gastric mucosa, while indomethacin was potent at both sites (Tab. 8). Previously, in 1986, Carr et al. [49] had shown that the threshold dose for gastrointestinal (GI) blood loss in the rat for nimesulide and indomethacin following 10 days once daily administration were 100 and 4 mg/kg p.o. respectively, while the ED_{90} from the serum of clotted blood was >10 and 5 mg/kg p.o., respectively.

Ceserani and co-workers [50] showed that oral administration of nimesulide 1.0–9.0 mg/kg to rats caused no significant effect on renal excretion of PGE_2,

Table 7 – Effect of nimesulide and other compounds on arachidonic acid metabolism

Drug	Microsomal Cyclooxygenase IC_{50} (μmol/L)	Murine macrophages PGE_2 IC_{50} (μmol/L)
Nimesulide	300	2.8
Indomethacin	2	0.03
Benoxaprofen	60	10

Adapted from Böttcher et al. [48].

Table 8 – Inhibition by nimesulide and indomethacin of prostaglandin synthesis in rat inflammatory exudate and gastric mucosal tissues

	Inhibition ED_{50} (mg/kg p.o.)			
	Nimesulide		Indomethacin	
	Inflammatory exudate	Gastric Mucosal tissue	Inflammatory exudate	Gastric Mucosal tissue
PGE_2	1.26	15.7	0.92	1.76
TXB_2	1.56	17.9	0.94	<0.4

Adapted from Tofanetti et al. [20].

Table 9 – Urinary PGE_2 concentrations after oral nimesulide and indomethacin administration in rats

Drug/dose (mg/kg orally)		PGE_2 (ng/24 h)
Control		44.3 ± 13.9
Nimesulide	1	44.2 ± 19.6
	3	27.0 ± 16.6
	9	22.1 ± 10.6
Indomethacin	1	21.3 ± 7.7
	3	15.0 ± 8.1*
	9	3.2 ± 1.1*

* $p < 0.0001$ versus *control.*
Adapted from Ceserani et al. [50].

whereas indomethacin 3 and 9 mg/kg, but not 1.0 mg/kg, caused dose-related reduction in urinary PGE_2 (Tab. 9). The pharmacological and toxicological significance of these effects are discussed in Chapter 6 (see page 357).

COX-2 selectivity

In studies with fresh human gastric mucosa pieces from gastrectomy operation specimens, Tavares et al. in 1995 [51] showed that indomethacin was 6 to 22-

Table 10 – Effect of nimesulide and indomethacin on basal eicosanoid accumulation in incubates of human gastric mucosal pieces and in stimulated human leucocytes

	Gastric Mucosa IC_{30} (µmol/L)		**Stimulated Leucocytes IC_{50} (µmol/L)**	
	Nimesulide	**Indomethacin**	**Nimesulide**	**Indomethacin**
PGE_2	14.8	2.5	0.22	0.15
6-Keto-$PGF_{1\alpha}$	12.8	1.0	0.93	0.18
TxB_2	31.1	1.4	0.42	0.15

Adapted from Tavares et al. [51].

fold more potent than nimesulide in causing inhibition of PGE_2, 6-keto-$PGF_{1\alpha}$ and TxB_2 accumulation (IC_{50}: 2.5 *versus* 14.8 µmol/L; 1.0 *versus* 12.8 µmol/L; 14 *versus* 31.1 µmol/L respectively, $p < 0.05–0.02$) (Tab. 10). In the same study in LPS-stimulated human leucocytes indomethacin was only 1.5 to 5-fold more potent than nimesulide (approximate IC_{50} for PGE_2, 6-keto-$PGF_{1\alpha}$ and TxB_2 0.15 *versus* 0.22 µmol/L, 0.18 *versus* 0.93 µmol/L; 0.15 *versus* 0.42 µmol/L) (Tab. 10). With COX-1 from ram seminal vesicles, nimesulide did not inhibit PGE_2 production from arachidonic acid while indomethacin caused a concentration-related inhibition (IC_{50} 0.6 µmol/L). PGE_2 production from arachidonic acid with COX-2 from sheep placenta was inhibited by both nimesulide and indomethacin using a 5 min pre-incubation of enzyme with drug (IC_{50} 90.3 and 4.1 µmol/L, respectively). However, in 1995, Vago et al. [52] using a 2, 5, 10 and 15 min pre-incubation of enzyme with drug prior to adding arachidonic acid, achieved an IC_{50} of 70, 5, 0.05 and 0.07 µmol/L for nimesulide demonstrating the important time-dependent mechanism between NSAIDs and COX-2. Taniguchi et al. [53] with a 10 min pre-incubation with drugs observed an IC_{50} of 7.1 µmol/L for COX-2 inhibition by nimesulide. The time-dependent changes in inhibition by nimesulide of COX-2 activity are features common to COX-2 selective inhibitors and reflect slow interactions with the active site [39, 40].

To account for the binding of NSAIDs to plasma proteins, Patrignani et al. [54] developed an assay in whole human blood in which COX-1 activity was measured by assay of TxB_2 in clotting blood at 1 hr, and LPS-stimulated monocyte COX-2 activity by assay of PGE_2 after 24 hr incubation. The ratio of the IC_{50} values for COX-2 and COX-1 was 0.1 for nimesulide while ibuprofen, naproxen and indomethacin had ratios of 2.0, 1.8 and 0.5, respectively. Using this method, Cryer and Feldman [55] found that nimesulide had a ratio of 0.017 while ibuprofen, naproxen and indomethacin had ratios of 1.69, 0.88 and 1.78, respectively (Tab.

Table 11 – Drug concentration (IC_{50}) for 50% inhibition of cyclooxygenase activity in blood and in gastric mucosa

Drug	COX-1 In Blood (μMol/L)	COX-2 In Blood (μMol/L)	COX-2/ COX-1 Ratio	Gastric Mucosa (μMol/L)
Ketoprofen	0.11	0.88	8.16	0.08
Indomethacin	0.21	0.37	1.78	0.85
Diclofenac	0.26	0,01	0.05	0.23
Ketorolac	0.27	0.18	0.68	0.33
Flurbiprofen	0.41	4.23	10.27	0.23
Tolmetin	1.08	2.25	2.09	3.50
Mefenamic acid	1.94	0.16	0.08	0.70
Piroxicam	2.68	2.11	0.79	0.87
Fenoprofen	2.73	14.03	5.14	0.17
Aspirin	4.45	13.88	3.12	0.03
Ibuprofen	5.90	9.90	1.69	0.70
Nimesulide	10.48	0.18	0.017	1.49
Oxaprozin	14.58	36.67	2.52	2.62
Etodolac	19.58	2.47	0.12	3.20
NS-398	21.93	0.92	0.042	100.00
6-MNA	31.01	19.84	0.64	0.48
Naproxen	32.01	28.19	0.88	0.52
Valeryl salicylate	32.64	0.04	0.001	>100.00
Nabumetone	33.57	20.83	0.62	20.09
Sulindac	41.26	24.94	0.61	>100.00
Paracetamol	42.23	10.69	0.25	>100.00
Dexamethasone	59.95	0.13	0.002	>100.00
Bismuth subsalicylate	75.24	37.50	0.50	>100.00
Salicylic acid	>100.0	14.08	0.13	>100.00
Salsalate	>100.0	39.90	0.29	>100.00

6-NMA = 6-methoxy naphthalene acetic acid (metabolite of nabumetone)
Adapted from Cryer and Feldman [55].

Table 12 – Inhibition of PGE_2 production in CHO cells stably transfected with human COX-1 and COX-2

	IC_{50} values (nMol/L)		
Drug	**COX-1**	**COX-2**	**COX-2/COX-1**
Flurbiprofen	1.8	4	2.2
Diclofenac	4	1.3	0.33
Ketoprofen	6.1	119	19.5
Indomethacin	18	26	1.4
Sulindac sulphide	28	4	0.14
Dup 697	59	2.1	0.036
Naproxen	62	26	0.41
Ibuprofen	470	670	0.14
Nimesulide	780	9	0.012
Meloxicam	1810	6	0.003
NS-398	1900	6	0.002
Piroxicam	3460	35	0.01
6-MNA	2290	~5000	~2.18
SC-57666	6000	3.2	0.0005
CGP 28238	8100	8	0.001
SC-58125	12000	10	0.001
L-745,337	~5000	60	~0.001
Etodolac	~5000	41	~0.001
DFU	>5000	41	>0.001

Adapted from Riendeau et al. [56].

11). In minced gastric mucosal biopsy samples from healthy volunteers the IC_{50} values for nimesulide were 1.49 for nimesulide and 0.70, 0.52 and 0.85 for ibuprofen, naproxen and indomethacin, respectively (Tab. 11).

Stably transfected Chinese Hamster Ovary (CHO) cells expressing either human COX-1 or human COX-2 that were assayed for the production of PGE_2 offered a system where COX-1 and COX-2 could be monitored under identical conditions of a 15 min pre-incubation with drug followed by challenge with 10 µmol/L arachidonic acid then a further 15 min incubation [56]. However, the incubation medium did not contain serum albumin so resulting in low IC_{50} values. Nimesulide had an IC_{50} COX-2/COX-1 ratio of 0.012 while ibuprofen, naproxen and indomethacin had ratios of 0.14, 0.41 and 1.4, respectively (Tab. 12).

Miralpeix and co-workers [57] investigated the kinetics of COX-2 expression in IL-1β compared with phorbol-12-myristate-13-acetate (PMA) stimulated human umbilical vein cell line (HUV-EC-C; which is of normal human origin) and found that nimesulide, like some other COX-2 selective drugs showed greater inhibitory potency in PMA stimulated cells. In PMA treated cells the COX-2/COX-1 ratio for nimesulide was 0.03 and for NS-398 and SC-58125 were 0.001 and 0.006 respectively which is in the order of selectivity of the coxibs [39, 40, 44, 51–56]. The authors claimed that since this data is from a stably developed normal human cell system the results probably more closely related to normal conditions.

Warner et al. [58, 59] developed the human whole blood assay in which COX-1 activity was determined following incubation with calcium ionophore stimulation for 30 min and COX-2 following addition of LPS and incubation for 18 h (Whole Blood Assay or "WBA" method). In addition, a modified whole human blood assay was used, with interleukin-1β pre-stimulated human A549 cells as a source of COX-2, that were further stimulated with A23187 and incubated for 30 min (William Harvey Modified Assay or "WHMA" method). The two methods for COX-2 gave different results. Hence for nimesulide the WBA assay gave an IC_{50} COX-2/COX-1 ratio of 0.19 while in the WHMA assay the ratio of the activities was 0.038. Ibuprofen, naproxen and indomethacin had ratios of 0.9, 3.0 and 80 with the WBA method and 2.6, 3.8 and 10 with the WHMA method (Tab. 13). The authors suggested that it would be more appropriate to use IC_{80} than IC_{50} values since the steady-state plasma concentrations of these drugs on average caused an inhibition of 80% in their system. This suggestion does not, however, appear to have been taken up by other researchers. Indeed it could be that kinetic conditions at high concentration-response curves (i.e., at 80% inhibition values) where there is non-linearity and high error could lead to aberrations in the results. Also, peak-trough plasma concentrations are probably more valid for making comparisons with *in vitro* data [60]. There may be just as valid comparisons at the low end of the plasma concentrations where there may be different kinetic responses with NSAIDs.

The relationship of plasma concentrations of NSAIDs to their expected COX-1 and COX-2 inhibition based on *in vitro* data has been explored for both relevance to the clinical outcomes (pain, anti-inflammatory activities) as well as *in vivo* situations [60, 61]. In a comparison of pharmacokinetics of nimesulide after its administration by various routes to dogs with effects on COX isoforms *in vitro* using the whole blood assay, Toutain et al. [31] observed that the IC_{50} for inhibition of COX-2 and COX-1 was 1.6 and 20.3 μmol/L, respectively. They established that the ratio of the IC_{50} values for COX-2/COX-1 was 13, which is a similar degree of COX-2 selectivity as in other species. Selectivity for COX-2 was found at concentrations within those observed in plasma (8–10 μg/mL; 26–32 μmol/L) after a dosage of 5 mg/kg p.o. At this dosage anti-inflammatory and analgesic activity in dogs was achieved as noted earlier [30].

Ex vivo determination of COX-1 and COX-2 activities using the whole blood assay was applied by Cullen et al. [62] to a study comparing the effects of nime-

Table 13 – Inhibition of COX-1 and COX-2 in human whole blood

	COX-1 IC_{50} µMol/L	**WBA-COX-2 IC_{50} µMol/L**	**WHMA-COX-2 IC_{50} µMol/L**	**IC_{50} ratios COX-2/**	
				WBA COX-1	**WHMA COX-1**
Aspirin	1.7	>100	7.5	>100	4.4
Carprofen	0.087	4.3	n.d.	50	n.d.
Diclofenac	0.075	0.038	0.020	0.5	0.3
Fenoprofen	3.4	41	5.9	12	1.7
Flufenamate	3.0	9.3	n.d.	3.1	n.d.
Flurbiprofen	0.075	5.5	0.77	73	10
Ibuprofen	7.6	7.2	20	0.9	2.6
Indomethacin	0.013	1.0	0.13	80	10
Ketoprofen	0.047	1.0	22	61	5.1
Ketorolac	0.00019	0.086	0.075	453	395
Meclofenamate	0.22	0.7	0.2	3.2	0.91
Mefenamic acid	25	2.9	1.3	0.11	0.049
Naproxen	9.3	28	35	3.0	3.8
Niflumic acid	25	5.4	11	0.22	0.43
Piroxicam	2.4	7.9	0.17	3.3	0.1
Sulindac sulphide	1.9	55	1.21	29	0.64
Suprofen	1.1	8.7	8.3	7.7	7.3
Tenidap	0.081	2.9	n.d.	35.2	n.d.
Tolmetin	0.35	0.82	1.3	2.3	3.8
Tomoxiprol	7.6	20	0.32	2.7	0.042
Zomepirac	0.43	0.81	0.096	1.9	0.22
Celecoxib	1.2	0.83	0.34	0.7	0.3
Etodolac	12	2.2	9.4	0.2	0.1
Meloxicam	5.7	2.1	0.23	0.37	0.040
Nimesulide	10	1.9	0.39	0.19	0.038
L745,337	>100	8.6	1.3	<0.01	<0.01
6MNA	42	146	n.d.	3.5	n.d.
NS398	6.9	0.35	0.042	0.051	0.0061
Rofecoxib	6.3	0.84	0.31	0.013	0.0049

Adapted from Warner et al. [58].

sulide 100 mg/d bid with that of aspirin 300 mg/d tid both taken for 14 days. Production of PGE_2 from LPS-treated whole blood, a marker of COX-2, was uniformly reduced at 2, 5, 10 or 14 days by nimesulide to the extent of ≤10% of controls. By 2–5 days washout there was recovery of PGE_2 production. Aspirin in contrast did not result in any significant reduction of PGE_2 production in this assay system. Serum concentrations of TxB_2 (a marker of COX-1 activity) were unaffected by nimesulide treatment, whereas they were markedly reduced in subjects that took aspirin, and in the washout period were significantly reduced at days 2 and 5 following the last period of aspirin intake.

Cullen et al. also measured urinary output on day 14 of TxB_2, and the 11-dehydro-metabolites of TxB_2, as an indication of COX-1 inhibition *in vivo* and urinary 6-keto-$PGF_{1\alpha}$ and its 2,3 dinor metabolite, as a marker of COX-2 inhibition *in vivo* [62]. Both TxB_2, and the 11-dehydro-metabolites of TxB_2 were unaffected by intake nimesulide, but these were reduced by about one-half by aspirin compared with control values. Nimesulide and aspirin both reduced urinary output of 6-keto-$PGF_{1\alpha}$ and its 2,3 dinor metabolite by about one-half compared with controls. The levels of TxB_2 were markedly reduced in the serum of subjects that took aspirin but not those that had nimesulide. Plasma levels of PGE_2 were reduced to about 5% with nimesulide but were unaffected by aspirin treatment.

Thus, the evidence reviewed here shows that in both human and animal models there is unequivocal evidence that nimesulide exhibits COX-2 selectivity *in vitro* in relation to the pharmacokinetics of the drug as well as *ex vivo*.

Inhibition of the synthesis of COX-2

Another site of action of nimesulide on the systems involved in the production of prostaglandins produced in inflammatory reactions (e.g., PGE_2, PGI_2), aside from the direct inhibitory effects on the enzymatic activity of COX-2, is in the synthesis of the COX-2 (or more precisely the PGHS-2) enzyme protein. Fahmi et al. [63] and in similar studies also from the Pelletiers' laboratory by Di Battista et al. [64] showed that nimesulide 30 or 300 ng/mL inhibited the production of IL-1β-induced production of COX-2 protein as well as mRNA coding for this protein. Naproxen did not affect COX-2 expression.

In contrast with these results, Taniguchi and co-workers [65] found that nimesulide did not affect the synthesis of mRNA coding for COX-2 or the COX-2 enzyme protein in rat peritoneal macrophages stimulated with opsonised zymosan although the drug reduced the production of PGE_2. The differences between these results might be due to the human synovial tissues used to prepare the fibroblasts may have already been sensitised by the chronic inflammatory disease of the osteoarthritis (OA) and rheumatoid arthritis (RA) patients in the studies by Fahmi, Di Battista and their co-workers [63, 64], as distinct from the acute inflammatory response induced *in vitro* in the rat peritoneal macrophages used by Taniguchi et al. [65].

Leukotriene production and lipoxygenase activity

Tool and Verhoeven [66] found that nimesulide 1.0–100 µmol/L produced a concentration-related inhibition of the production of leukotriene B_4 (LTB_4) in serum-treated zymosan (STZ) – and formyl-methionyl-leucyl-phenylalanine (fMLP) – stimulated polymorphonuclear (PMN) neutrophil leucocytes. The IC_{50} values for inhibition by nimesulide were approximately 10 and 50 µmol/L respectively in the presence of these two stimuli. Similar effects of nimesulide were observed on the production of platelet activating factor although the IC_{50} values were slightly lower being 20 µmol/L with STZ but less so with fMLP as a stimulus where the IC_{50} was 30 µmol/L. The effects of nimesulide were ascribed to increase intracellular cyclic-adenosine monophosphate which activated protein kinase A. In contrast with these results, nimesulide, like that of aspirin and indomethacin, had no effects on the production of the peptidoleukotrienes, LTC_4, LTD_4 and LTE_4 in the calcium ionophore (A23187)-stimulated blood from aspirin-sensitive patients [67]. This suggests that nimesulide may selectively inhibit production of the chemoattractant, LTB_4, while not affecting production of the peptide-leukotrienes and that the former effect might contribute to the anti-inflammatory effects of nimesulide.

Nimesulide does not appear to affect breakdown of leukotrienes or prostanoids whereas diclofenac and indomethacin inhibit the activities of enzymes involved in their breakdown [68].

Anandamide production

An alternative fate of arachidonic acid is the pathway leading to the formation of the endogenous cannabinoid, anandamide (*N*-arachidonyl-ethanolamine) [69]. The interaction of anandamide with CB1 receptors in the nervous system is important in control of pain, while activation by this endogenous ligand of CB2 receptors is important in modulating the immune system [69]. The synthesis of anandamide occurs from phospholipids precursors while the breakdown is catalysed by fatty acid amide hydrolase to yield arachidonic acid and ethanolamine [69]. This enzyme is the site of inhibition by a number of acidic NSAIDs [70]. Nimesulide has no effect on this enzyme [70], possibly as a consequence of its higher pKa. However, inhibition of COX-2 by nimesulide has been shown to reduce CB1-receptor mediated GABA-ergic transmission by a process known as depolarisation-induced suppression of inhibition (DSI) [71]. Nimesulide prolongs the DSI suggesting that this represents a protraction of the effects of the cannabinoid [71]. Further studies examining the effects of nimesulide on the turnover of anandamide and its effects on CB receptors would seem essential.

Structural aspects of cyclooxygenase (COX) activity and COX-2 inhibition by nimesulide

Introduction

Prostaglandin-endoperoxide synthases 1 (PGHS-1 = COX-1) and 2 (PGHS-2 = COX-2)[1] are bifunctional enzymes that catalyse two sequential reactions in spatially distinct, but mechanistically coupled active sites. The first reaction, at the cyclooxygenase (COX) site, converts the achiral arachidonic acid (AA) to prostaglandin G_2 (PGG_2), which has five chiral centres, by addition of two molecules of oxygen. PGG_2 then undergoes a two-electron reduction to PGH_2 at the peroxidase site, and this short-lived intermediate is in turn converted by tissue-specific isomerases to other prostanoids. The close coupling of the two active sites arises because not only is the product of the cyclooxygenase reaction the substrate of the peroxidase reaction but also this latter reaction is required to initiate the former. The necessary translocation of PGG_2 from one active site to another is efficiently accomplished because the enzyme is tightly associated with one monolayer of the membrane on the luminal surfaces of both the endoplasmic reticulum and the nuclear envelope of different cell types [72].

Despite the similar subcellular location and the overall sequence similarity (Fig. 5), biochemical and pharmacological differences exist between the two isoforms. For example, COX-2 accepts a wider range of fatty acids as substrates than does PGHS-1, and when acetylated by aspirin on Ser-530, COX-1 does not oxidise AA whereas similarly acetylated PGHS-2 will still function as a 15-lipoxygenase, oxidising AA to 15(*R*)-hydroxy-eicosatetraenoic acid (15-HETE) [73]. There is considerable variation in inhibitory effects of NSAIDs on both isoforms [39–40, 51–59, 73–75]. The slow conversion between an initial reversible complex and a functionally irreversible one is thought to be responsible for the selectivity of inhibition of COX-2 over COX-1 [75–77].

Since inhibition of COX-2 alone has been considered sufficient to obtain an anti-inflammatory effect and most mechanism-based side effects result from blockade of COX-1 activity in normal tissues, targeting COX-2 stimulated development of new agents (coxibs) with an improved safety profile [78]. Several classes already

[1] *The terms PGHS-1 and PGHS-2 refer to the proteins that have cyclooygenase-1 (COX-1), COX-2 as well as peroxidase activities, respectively.*

```
                                                                                                                             119
COX1_SHEEP   MSRQSISLRFPLLLLLLSPSPVFSADPGAPAPVNPCCYYPCQHQGICVRFGLDRYQCDCTRTGYSGPNCTIPEIWTWLRTTLRPSPSFIHFMLTHGRWLWDFVNAT-FIRDTLMRLVLTV
COX1_HUMAN   -MSRSLLLRFLLFLLLLPPLPVLLADPGAPTPVNPCCYYPCQHQGICVRFGLDRYQCDCTRTGYSGPNCTIPGLWTWLRNSLRPSPSFTHFLLTHGRWFWEFVNAT-FIREMLMRLVLTV
COX2_HUMAN   ---------------MLARALLLCAVLALSHTANPCCSHPCQNRGVCMSVGFDQYKCDCTRTGFYGENCSTPEFLTRIKLFLKPTPNTVHYILTHFKGFWNVVNNIPFLRNAIMSYVLTS
COX2_MOUSE   ---------------MLFRAVLLCAALGLSQAANPCCSNPCQNRGECMSTGFDQYKCDCTRTGFYGENCTTPEFLTRIKLLLKPTPNTVHYILTHFKGVWNIVNNIPFLRSLIMKYVLTS
                            :*   :: *  . . ..****  ***::* *:  *:*:*:*******: * **: * : * ::  *:*:*.  *::*** : .*:.**   *:*. :*  ***

             120                                                                                                         239
COX1_SHEEP   RSNLIPSPPTYNIAHDYISWESFSNVSYYTRILPSVPRDCPTPMGTKGKKQLPDAEFLSRRFLLRRKFIPDPQGTNLMFAFFAQHFTHQFFKTSGKMGPGFTKALGHGVDLGHIYGDNLE
COX1_HUMAN   RSNLIPSPPTYNSAHDYISWESFSNVSYYTRILPSVPKDCPTPMGTKGKKQLPDAQLLARRFLLRRKFIPDPQGTNLMFAFFAQHFTHQFFKTSGKMGPGFTKALGHGVDLGHIYGDNLE
COX2_HUMAN   RSHLIDSPPTYNADYGYKSWEAFSNLSYYTRALPPVPDDCPTPLGVKGKKQLPDSNEIVEKLLLRRKFIPDPQGSNMMFAFFAQHFTHQFFKTDHKRGPAFTNGLGHGVDLNHIYGETLA
COX2_MOUSE   RSYLIDSPPTYNVHYGYKSWEAFSNLSYYTRALPPVADDCPTPMGVKGNKELPDSKEVLEKVLLRREFIPDPQGSNMMFAFFAQHFTHQFFKTDHKRGPGFTRGLGHGVDLNHIYGETLD
             ** ** ******  :.* ***:***:***** **.*. *****:*.**:*:***:: : .:.****:*******:*:****************. * **.**..*******.****:.*

             240                                                                                                         359
COX1_SHEEP   RQYQLRLFKDGKLKYQMLNGEVYPPSVEEAPVLMHYPRGIPPQSQMAVGQEVFGLLPGLMLYATIWLREHNRVCDLLKAEHPTWGDEQLFQTARLILIGETIKIVIEEYVQQLSGYFLQL
COX1_HUMAN   RQYQLRLFKDGKLKYQVLDGEMYPPSVEEAPVLMHYPRGIPPQSQMAVGQEVFGLLPGLMLYATLWLREHNRVCDLLKAEHPTWGDEQLFQTTRLILIGETIKIVIEEYVQQLSGYFLQL
COX2_HUMAN   RQRKLRLFKDGKMKYQIIDGEMYPPTVKDTQAEMIYPPQVPEHLRFAVGQEVFGLVPGLMMYATIWLREHNRVCDVLKQEHPEWGDEQLFQTSRLILIGETIKIVIEDYVQHLSGYHFKL
COX2_MOUSE   RQHKLRLFKDGKLKYQVIGGEVYPPTVKDTQVEMIYPPHIPENLQFAVGQEVFGLVPGLMMYATIWLREHNRVCDILKQEHPEWGDEQLFQTSRLILIGETIKIVIEDYVQHLSGYHFKL
             ** :********:***::.**:***:*::: . * **  :* : ::*********:****:***:**********:** *** *********:**************:***:****.::*

             360                                                                                                         479
COX1_SHEEP   KFDPELLFGAQFQYRNRIAMEFNQLYHWHPLMPDSFRVGPQDYSYEQFLFNTSMLVDYGVEALVDAFSRQPAGRIGGGRNIDHHILHVAVDVIKESRVLRLQPFNEYRKRFGMKPYTSFQ
COX1_HUMAN   KFDPELLFGVQFQYRNRIAMEFNHLYHWHPLMPDSFKVGSQEYSYEQFLFNTSMLVDYGVEALVDAFSRQIAGRIGGGRNMDHHILHVAVDVIRESREMRLQPFNEYRKRFGMKPYTSFQ
COX2_HUMAN   KFDPELLFNKQFQYQNRIAAEFNTLYHWHPLLPDTFQIHDQKYNYQQFIYNNSILLEHGITQFVESFTRQIAGRVAGGRNVPPAVQKVSQASIDQSRQMKYQSFNEYRKRFMLKPYESFE
COX2_MOUSE   KFDPELLFNQQFQYQNRIASEFNTLYHWHPLLPDTFNIEDQEYSFKQFLYNNSILLEHGLTQFVESFTRQIAGRVAGGRNVPIAVQAVAKASIDQSREMKYQSLNEYRKRFSLKPYTSFE
             ********. ****:**** *** *******:**:*.:  *.*.::**::*.*:*:::*:  :*::*:** ***:.****:    :  *:   * :** :: *.:******* :*** **:

             480                              513       523                                                              594
COX1_SHEEP   ELTGEKEMAAELEELYGDIDALEFYPGLLLEKCHPNSIFGESMIEMGAPFSLKGLLGNPICSPEYWKASTFGGEVGFNLVKTATLKKLVCLNTKTCPYVSFHVPDPRQEDRPGV------
COX1_HUMAN   ELVGEKEMAAELEELYGDIDALEFYPGLLLEKCHPNSIFGESMIEIGAPFSLKGLLGNPICSPEYWKPSTFGGEVGFNIVKTATLKKLVCLNTKTCPYVSFRVPDASQDDGPAV------
COX2_HUMAN   ELTGEKEMSAELEALYGDIDAVELYPALLVEKPRPDAIFGETMVEVGAPFSLKGLMGNVICSPAYWKPSTFGGEVGFQIINTASIQSLICNNVKGCPFTSFSVPDPELIKTVTINASSSR
COX2_MOUSE   ELTGEKEMAAELKALYSDIDVMELYPALLVEKPRPDAIFGETMVELGAPFSLKGLMGNPICSPQYWKPSTFGGEVGFKIINTASIQSLICNNVKGCPFTSFNVQDPQPTKTATINASASH
             **.*****:***: **.***.:*:**.**:** :*::****:*:*:*********:** **** ***.*********:::**:::.*:* *.* **:.** * *.    .   :

                         595  601
COX1_SHEEP   ------------ERPPTEL
COX1_HUMAN   ------------ERPSTEL
COX2_HUMAN   SGLDDINPTVLLKERSTEL
COX2_MOUSE   SRLDDINPTVLIKRRSTEL
                         :. .***
```

Figure 5

Alignment [126] of mouse and human PGHS 1 and 2. Residue numbering follows the convention for the ram PGHS-1 structure (PDB entry 1prh). An asterisk under a given amino acid means identity in that position for mouse and human PGHS-2 enzymes; the = sign stands for a conserved residue in both PGHS-1 and PGHS-2 isozymes from the three species. The valine that occupies the same position as Ile-523 in PGHS-1 and the arginine that replaces His-513 are labelled and highlighted in bold type.

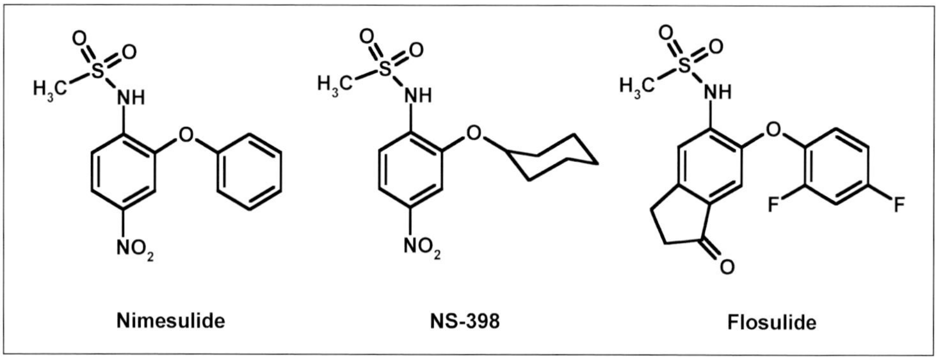

Figure 6
Chemical structures of COX-2-selective methanesulphonanilides.

exist of compounds that display such selectivity, and many of them have in common the presence of two appropriately substituted aromatic rings on adjacent positions about a central, usually heterocyclic ring [79]. Remarkably, nimesulide shares some of these characteristics (Fig. 6) [46], but was already on the market as an anti-inflammatory agent without the gastrointestinal (GI) side effects of classical NSAIDs prior to the discovery of the second COX isoform [40]. Indeed, a selectivity of this drug towards the PGHS involved in inflammation was suggested more than 25 years ago when a lack of correlation was found between its potency to inhibit PGHS in preparations from bovine seminal vesicles and its anti-inflammatory potency *in vivo*, which was comparable to that of indomethacin [46]. More recently, nimesulide has been demonstrated to be a potent time-dependent inhibitor of COX-2 [39, 40, 77], like that of other COX-2 inhibitors, and has been shown to competitively inhibit binding of [^{3}H]-valdecoxib to the His207→Ala mutant COX-2 enzyme with a K_i value of 174 ± 47 nM [80].

Structural overview of PGHS

Difficulties associated with the crystallisation of membrane proteins delayed the acquisition of detailed atomic information about these important enzymes for many years but a number of methodological advances have made it possible to obtain crystals of both isoforms suitable for X-ray diffraction studies. Initial work with PGHS-1 from ovine seminal vesicles [81] was rapidly extended to PGHS-2 from human cells [82] and mouse skin fibroblasts [83]. The freely available Protein Data Bank (PDB) [84] contains over 20 three-dimensional structures of both PGHS-1 and PGHS-2 complexed with several substrates, products and inhibitors [85]. These protein crystallography studies, together with site-directed mutagenesis experiments, spectroscopic measurements and kinetic characterisations, have helped enormously in our understanding of these two pharmacologically important enzymes. In particular, they have revealed how the intricate arrangement of active site atoms can restrain the flexible AA substrate and cause it to adopt a conformation that will yield a product with exact stereochemistry at five nascent chiral centres. They have also provided important insights into the molecular basis of selectivity even though for certain NSAIDs there appear to be no direct ligand-protein interactions with amino acid residues that are unique to PGHS-2 [82, 86]. For nimesulide, no complex with PGHS has been reported yet but some structural knowledge has been gained by use of homology modelling, automated docking techniques, and molecular dynamics simulations, as discussed below.

The alignment of the primary sequences of PGHS-1 and PGHS-2 from different sources demonstrates that the majority of the changes between the two isoforms occur at the N-terminal region and at the C-terminal tail (Fig. 5). The COX active sites of PGHS-1 and PGHS-2, on the other hand, are very similar, the only differences in the first shell of residues lining the cavity being a His→Arg and an Ile→ Val substitution at positions 513 and 523, respectively.

All available X-ray crystal structures of PGHS-1 and PGHS-2 enzymes reveal homodimers showing simple two-fold symmetry and an overall ellipsoidal shape (Fig. 7). In each monomer three distinct folding units can be discerned: (i) a short N-terminal region that gives rise to a compact domain similar to that of epidermal growth factor, (ii) a right-handed spiral of four amphipathic α-helices which make up the membrane-insertion domain (the protein is monotopic rather than transmembrane), and (iii) a C-terminal catalytic domain. The COX active site is located at the end of a hydrophobic channel that extends from the membrane-binding region towards Tyr-385 (standard PGHS-1 numbering), the catalytically essential amino acid that is strategically located between the haem cofactor and the bound substrate. When the enzyme reacts with hydroperoxides, a ferryl oxo porphyrin radical cation forms, which evolves to oxidise the side chain of Tyr-385. The resulting tyrosyl radical is then capable of abstracting the *pro-S* hydrogen from C13 of AA thereby initiating COX catalysis. In PGHS-1, AA adopts the proper orientation for attack by making multiple hydrophobic interactions with the residues lining the channel, meandering around the side chain of Ser-530, and positioning its carboxylate group to form both a salt bridge with the guanidinium group of Arg-120 and a good hydrogen bond with the phenolic oxygen of Tyr-355 [87, 88]. These two enzyme residues are similarly used to fix the prototypical carboxylate present in most of the non-selective NSAIDs [82, 89] which project their aromatic functionality into the COX active site toward Tyr-385. Intriguingly, when AA was co-crystallised with a mutant (His207→Ala), inactive form of PGHS-2 (deficient in peroxidase activity as it cannot bind the haem iron), the orientation that was observed for this substrate was opposite that found for PGHS-1, with the carboxylate group forming strong hydrogen bonds to the side chains of

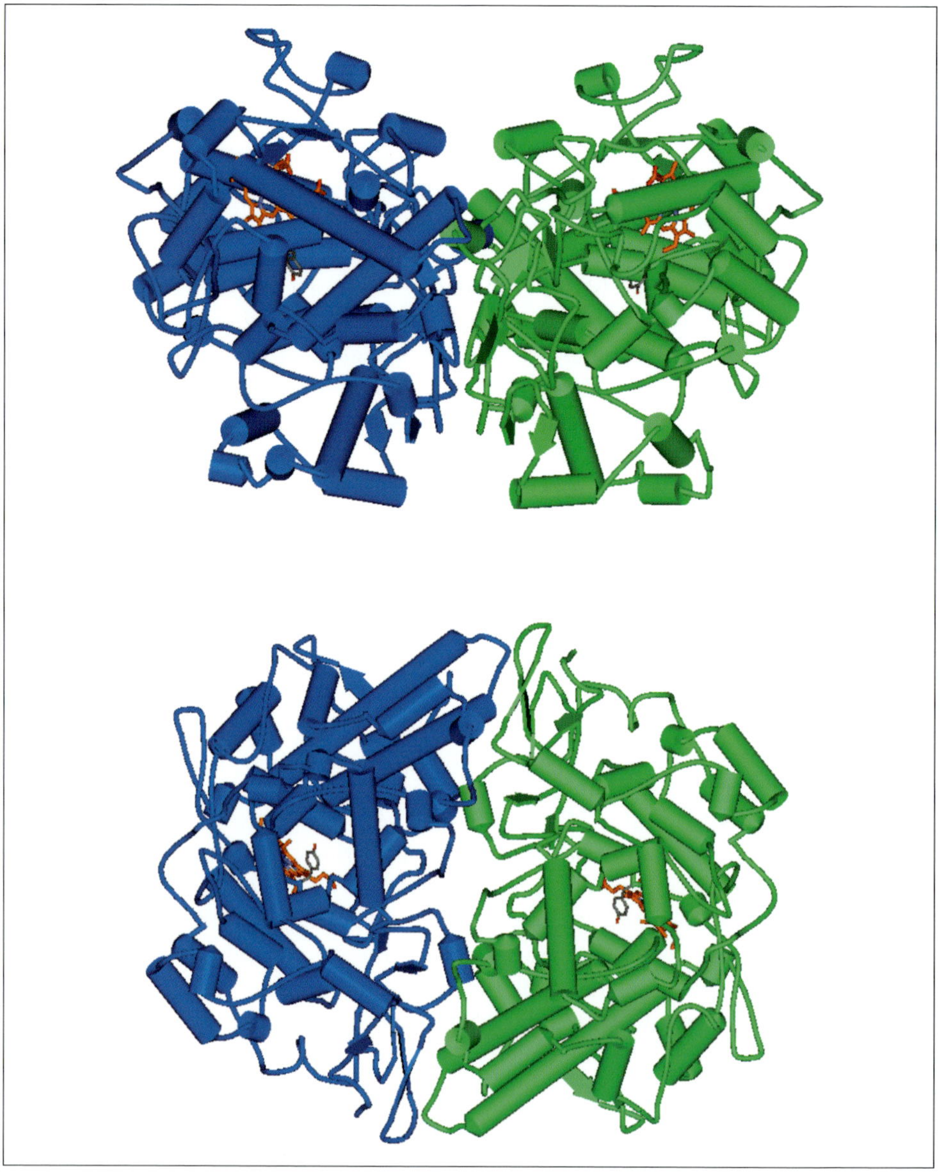

Figure 7
Schematic representation of the PGHS-2 homodimer (PDB entry 1PXX): α-helices and β-strands are depicted as cylinders and flat arrows, respectively. The bottom drawing is related to the top one by a 90° rotation about the X axis and allows visualisation of the substrate channel at the end of which protrudes the side chain of Tyr-385 (carbon atoms in grey and oxygen atom in red). The haem carbon atoms have been coloured orange.

Tyr-385 and Ser-530 [90]. Although this binding mode has to be considered non-productive, as it is not viable for catalysis, it is in consonance with the finding from site-directed mutagenesis experiments that an arginine at position 120 is not critical for substrate binding in PGHS-2 [91] unlike PGHS-1 for which this positively charged amino acid is a major determinant for binding of both AA [88] and many NSAIDs [92]. Moreover, this unexpected positioning of the carboxylate has also been found for diclofenac which, in its complex with PGHS-2, shows its carboxylic acid similarly coordinated to both Tyr-385 and Ser-530 [93].

The interaction between the side chains of these two protein residues has been recently characterised as a critical determinant of the selectivity of ASA for covalent modification of Ser-530 since acetylation of this serine is reduced by over 90% in a PGHS enzyme containing the Tyr385→Phe site-directed mutation [94].

Other more COX-2-selective NSAIDs, however, make use of what is probably the single most important difference between PGHS-1 and PGHS-2 [95], namely the replacement of isoleucine at position 523 with a valine (Fig. 5). The shorter side chain of this latter amino acid (Val-509 in PGHS-2 numbering) allows the binding site to be extended into a neighbouring side pocket in which an arginine (Arg-499) occupies the position of His-513 in PGHS-1. Therefore, the shape of the COX active site, the molecular electrostatic potential within this cavity [96], and the volume accessible to both substrates and inhibitors are different in PGHS-2 compared with PGHS-1 [78, 83, 95].

Structural studies on nimesulide

No experimentally determined structure of a complex between nimesulide and PGHS-2 has been disclosed in the literature as yet. The only structural details of the interaction of this molecule with COX enzymes have been provided by several molecular modelling studies involving wild-type and mutant ovine PGHS-1 [97, 98], as well as a homology-based model of human PGHS-2 [96, 99].

The 3-dimensional structure of nimesulide (Fig. 8) has been solved by X-ray crystallography [100] and is deposited in the Cambridge Structural Database [101] (ref. WINWUL).

Molecular dynamics simulations of this molecule in the absence of crystal lattice constraints showed that it can populate a limited repertoire of conformations, all of which were considered in subsequent docking calculations aimed at positioning nimesulide into the COX active site of a homology-based model of human PGHS-2 [96]. A binding mode was found that makes use of the side pocket adjacent to the substrate channel (access to which is made possible by the Ile-523→Val substitution), but two alternate binding orientations were obtained: the first one placed the methane sulphonamide moiety in the side pocket, leaving the nitro group close to Arg-106 (Arg-120 in PGHS-1) in the substrate channel (Fig. 9), whereas the sec-

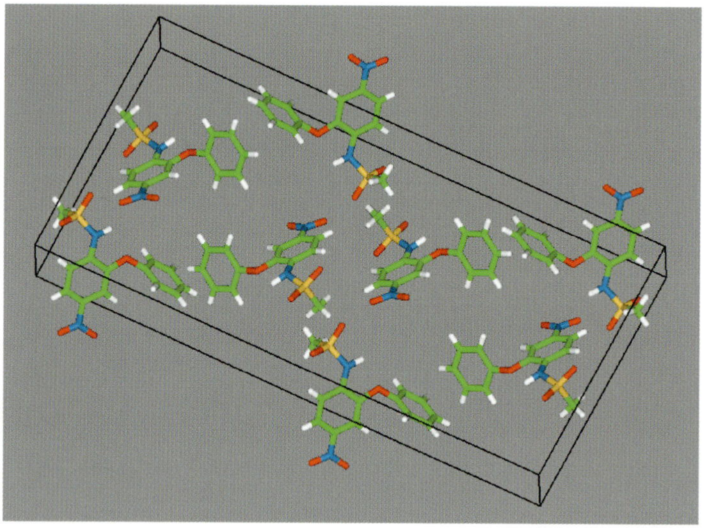

Figure 8
Three dimensional structure and packing interactions of nimesulide, as found in the unit cell of the crystal lattice solved by X-ray diffraction [100].

ond orientation placed the sulpho group in the vicinity of Arg-106 and the nitro group in the side pocket close to Arg-499 (not shown). In both cases, the phenoxy ring lies close and perpendicular to the aromatic ring of the catalytic tyrosine (Tyr-371, equivalent to Tyr-385 in PGHS-1) and in van der Waals contact with Leu-338 (Leu-352 in PGHS-1). Remarkably, nimesulide binding has been shown to alter the site of radical formation from Tyr-371 to Tyr-490, suggesting a change in the relative redox potentials of these two residues [102].

Differences in calculated interaction energies between these two complexes were found to arise mainly from electrostatic and van der Waals contributions emanating from the nitro and methyl groups in the two different enzyme environments. Nimesulide itself was found to be in a comparable low-energy conformation in both orientations. Ensuing molecular dynamics simulations of the complexes, carried out to take into account the reported flexible nature of the human COX-2 binding site [82], demonstrated the stability of the two proposed binding modes and revealed that the major contributors to the binding energy were Val-509 and Leu-338, and that the relative importance of Arg-499 and Arg-106 depended on the orientation considered.

Both binding orientations look chemically reasonable, yield very similar interaction energies with the enzyme, and are in agreement with the fundamental role played by the side chain of Val-509. Attempts to discriminate between the two

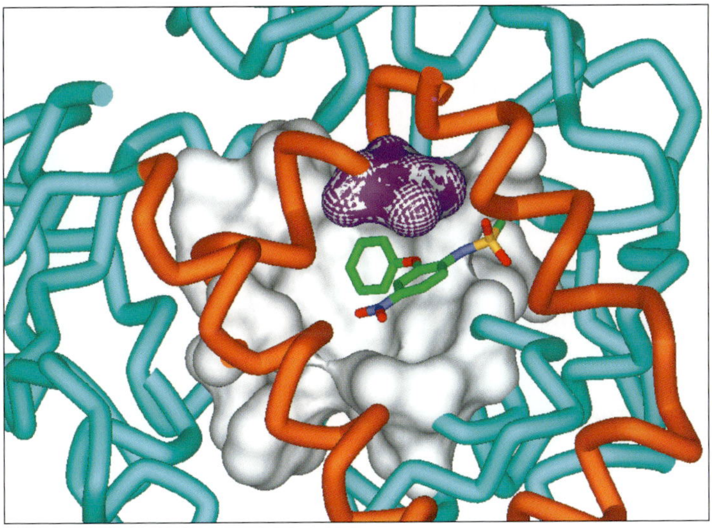

Figure 9
Nimesulide (carbon atoms coloured in green) bound in the cyclooxygenase active site of human PGHS-2 (C-α trace in orange for the membrane-insertion domain and hydrophobic "lobby" region, and cyan for the rest) in one of the proposed orientations. A white surface is used to highlight the volume available for NSAIDs that are able to occupy the side pocket adjacent to the substrate channel. Access to this cavity is provided by the side chain of Val-509 (pink surface) which is one carbon atom shorter than that of Ile-523 in PGHS-1.

models are hampered by the rather limited structure-activity data for nimesulide but both complexes can be examined in light of the experimental evidence available for this drug and other structurally and pharmacologically related compounds, such as NS-398 [103, 104] and flosulide [105] (Fig. 6). For these three methane sulphonanilides, inhibition of COX-1 activity is competitive and rapidly reversible but inhibition of COX-2 is characterised by being time-dependent [52, 74, 106–108]. In this respect, the structural similarities found in the crystallographic complexes of COX-1 with chemically related inhibitors that are either reversible competitive inhibitors or slow tight-binding inhibitors led to the suggestion that time-dependent and time-independent NSAIDs may differ only in the speed and efficiency with which they can enter into the substrate channel and the enzyme active site [109]. This claim supported previous theoretical studies [110] pointing to a possible mechanism based on differences in the ability to perturb the hydrogen bonding network around Arg-120, Tyr-355, and Glu-524.Nonetheless, comparison of kinetic data obtained during steady-state and time-dependent inhibition of PGHS-1 and PGHS-2 has provided evidence for a three-step reversible kinetic model for COX inhibition: 1) binding of the inhibitor to the enzyme near the solvent-accessible opening of the hydrophobic channel ("lobby" region); 2) translocation of the inhibitor along the length of this channel and subsequent association within the COX active site, and; 3) formation of the tightly bound enzyme-inhibitor complex, which involves the optimisation of inhibitor and protein conformational changes in the active site and the side pocket [77]. The first two steps have been postulated to be common to PGHS-1 and PGHS-2 during inhibition by the vast majority of NSAIDs. The third kinetic process, which appears to be irreversible, is only observed during inhibition of PGHS-2 by diarylheterocycles that contain a phenyl sulphonamide or a phenylsulphone moiety.

Experimental support for the proposed binding mode

The molecular basis of COX-2 inhibition by isoform-selective agents has been extensively probed by site-directed mutagenesis experiments on both PGHS-1 and PGHS-2. Two residues that are not conserved between the two isoforms and impinge on selectivity are Arg-499 and Val-509 of PGHS-2, which are equivalent, respectively, to His-513 and Ile-523 in PGHS-1 (Fig. 5). Thus, the single amino acid change of valine at position 509 of PGHS-2 to isoleucine results in a loss of sensitivity to inhibition by nimesulide and NS-398, among other COX-2 selective inhibitors, while inhibition by non-selective NSAIDs such as indomethacin remains unaffected [111]. The hydrophobic side chain of valine appears to be more important for nimesulide in order to form a tight complex with PGHS-2 than it is for other related inhibitors as the Val-509→Ala PGHS-2 mutant is inhibited in a time-dependent fashion by NS-398 but not by nimesulide [106, 107]. Similarly, Val-509→Lys and Val-509→Glu PGHS-2 mutants, like recombinant human PGHS-1, also show reversible but not time-dependent inhibition with nimesulide [107].

The role of Val-509 as an essential determinant in the differential interaction of PGHS-2 with selective and non-selective inhibitors is also patent from experiments with the Ile-523→Val PGHS-1 mutant, which displays increased sensitivity to various COX-2 selective inhibitors including NS-398 [108]. Interestingly, this sensitivity is not altered in a His-513→Arg PGHS-1 mutant but the simultaneous occurrence of both mutations translates into increased inhibition by NS-398 relative to the single Ile-523→Val mutant, and also in time-dependent inhibition. Nevertheless, although both mutations appear to be necessary to change the rapidly reversible mechanism of PGHS-1 inhibition to the time-dependent mechanism characteristic of PGHS-2 inactivation, not all the properties of the active site of PGHS-2 are restored by these two mutations. In fact, the double mutant does not synthesise any appreciable amount of 15-HETE when treated with 100 μM ASA, in contrast with ASA-inhibited PGHS-2, which can be an indication that additional amino acid changes may be involved. One example would be another Ile→

Val substitution at position 434 (PGHS-1 numbering), which has been proposed to facilitate access to the side pocket [83]. Incidentally, acetylation of Ser-516 (the equivalent of Ser-530 in PGHS-1) by ASA, or mutation of this residue to methionine, does not greatly affect the binding and inhibitory properties of NS-398, as opposed to meclofenamic acid or diclofenac, which are potent inhibitors of PGHS-2 but inhibit neither ASA-PGHS-2 nor the Ser-516→Met [112] and Ser-516→Ala PGHS-2 mutant enzymes. Interestingly, this latter mutation was also shown to eliminate time-dependent inhibition by nimesulide but not competitive inhibition [93], possibly indicating a role for the side chain of this serine in inhibition by this compound. If this is the case, the orientation that places the nitro group inside the substrate channel would be preferable although a bridging water molecule would probably be necessary to mediate a hydrogen bonding interaction between the nitro moiety of the drug and the hydroxyl group of Ser-516. In this respect, it is known that this nitro group cannot be replaced with a cyano group or a tetrazole ring [113], and that replacement by a hydroxyl (as in the main metabolite of nimesulide) is accompanied by a 20-fold loss of activity in whole blood assays *in vitro* [114].

The models presented above highlight the importance in PGHS-2 of a positively charged residue, in addition to Arg-106, in the pocket adjacent to the substrate binding channel that non-selective inhibitors do not occupy [83] for ionic interactions with nimesulide-like molecules. Remarkably, recent site-directed mutagenesis studies have demonstrated the importance of Arg-499 in the selective oxygenation of endocannabinoids by PGHS-2 [115], which pinpoints a possible physiological function for the side pocket that is targeted by most of the COX-2 selective inhibitors.

Biochemical [92] and structural evidence [81–83] attests to the importance of Arg-106 in PGHS-2 (or Arg-120 in PGHS-1) for interacting both with the carboxylic acid group of AA and with the free carboxylic acid moiety of several NSAIDs. Nevertheless, in line with the crystallographic evidence presented above, in PGHS-2 this positively charged amino acid contributes to ligand binding less than the equivalent residue in PGHS-1 and this difference has been effectively exploited as a new strategy for converting certain non-selective NSAIDs to COX-2-selective inhibitors [116]. When the arginine is replaced by a negatively charged glutamic acid, the Michaelis constant (K_M) increases ~30-fold in the case of PGHS-2 [117] but ~100-fold or more [118] in the case of PGHS-1. The effect of this charge reversal mutation results in a decrease in the inhibitory potency of flosulide and NS-398 of 600- and 1,000-fold, respectively, against human PGHS-2 (Arg106→Glu). This loss of effect is due to a difference in the kinetics of inhibition, with these two drugs displaying time-independent inhibition of this mutant enzyme but time-dependent inhibition of the wild-type human PGHS-2 [117]. In agreement with the involvement of this arginine in the tight binding of this class of inhibitors, the models we propose for nimesulide give rise to rather strong electrostatic interactions with both arginine residues in any of the two orientations considered.

The crystallographic studies have consistently shown Arg-120 in PGHS-1 to be engaged in a salt bridge with the carboxylic group of Glu-524 [81], and these two residues, together with Tyr-355, to participate in a hydrogen bonding network at the bottom of the COX active site in which ligand atoms are also involved. Tyr-355 is, in fact, a key determinant of the stereospecificity of PGHS-1 toward inhibitors of the 2-phenylpropionic acid class [118]. Glu-524, on the other hand, does not appear to be importantly involved in catalysis or substrate binding, as suggested by results obtained with Glu524→Asp, Glu524→Gln and Glu524→Lys PGHS-1 mutant enzymes [118]. The corresponding residues in PGHS-2 also participate in a similar hydrogen bonding network in the free enzyme and in the complexes with non-selective inhibitors [83] but, in some of the complexes with COX-2 selective inhibitors, the salt bridge between equivalent Arg-106 and Glu-510 is disrupted because the side chain of this latter residue is reoriented so as to form a salt bridge with Arg-499 on the other side of the extended binding site. This is observed, for example, in the complexes of mouse PGHS-2 with the celecoxib analogue, SC-558, and human PGHS-2 with RS-57067, an analogue of the non-selective NSAID zomepirac in which replacement of the carboxylic group with a pyridazinone ring leads to preferential inhibition of PGHS-2 [82, 119]. By contrast, this reorientation is not observed in the complex of human PGHS-2 with the related analog RS-104897, and the acylsulphonamide group of this drug interacts, in a manner similar to the carboxylic group of flurbiprofen or indomethacin [83, 120], not only with Arg-106 but also with Tyr-341 [82], a residue that is known to be involved in the molecular mechanism of time-dependent inhibition of PGHS-2 [119]. In our models with nimesulide, dual interactions with Arg-106 and Tyr-341 are also observed but, depending on the orientation of the drug in the binding site, it is either the nitro or the sulphonamide group that interacts with the side chain of either Arg-106 or Arg-499 (Fig. 9). Interestingly, during the molecular dynamics simulations of the complexes of human PGHS-2 with nimesulide, in addition to the reported hydrogen bonding network in the COX active site, a dynamic network of alternating salt bridges was observed [96] involving a number of residue pairs: Arg106–Glu510, Glu510–Arg499, Arg499–Glu506, Glu506–Arg453, and Arg453–Glu496. This dynamic picture complements the static X-ray data and deserves further study.

Finally, the sulphonanilide amino group of nimesulide (pKa = 6.5 [121]) does not appear to make any direct contacts with the protein. Its major role appears to be in limiting the conformational flexibility of the phenoxy moiety and in enforcing the co-planarity of the sulphonamide group with respect to the nitrophenyl ring. *N*-methylation in both nimesulide [122] and the related flosulide [105] (Fig. 6) has been shown to result in complete loss of *in vitro* COX-2 inhibitory activity. In the light of the present docking experiments, this is not surprising given that this chemical modification brings about a conformational change in the ligand [123] that is incompatible with the strict geometric requirements of this binding site (Fig. 9).

Inspection of ligand-receptor complexes deposited in the PDB with ReLiBase tools [124] reveals a similar protein environment for nitro and sulpho groups of ligands and a similar tendency of these moieties to interact with the guanidinium group of arginine residues. In accordance with this, both binding modes remain feasible. In this respect, it is of interest to note that multiple modes of binding have been suggested for the interaction between diarylheterocyclic compounds and PGHS-2 [83] and also that reversal of the functionalities in some substituted 1,5-diphenylpyrazoles brings about striking changes in potency and selectivity [125].

Conclusions

Any of the two possible orientations for nimesulide in the COX active site of human PGHS-2 suggested by the automated docking programs can account for the pharmacological profile of this agent as a COX-2 selective inhibitor. Nimesulide is proposed to bind PGHS-2 at the bottom of the substrate channel where it gains access to an adjacent pocket, the entrance to which is more restricted in PGHS-1 as a result of the presence of an isoleucine at position 523 in place of a valine (Figs 5 and 9). The two possible orientations that are found have in common the sandwiching of the ring bearing the nitrophenyl and sulphonamide groups in the hydrophobic environment between the side chain of this valine (Val-499) and the Cα and Cβ atoms of Ser-339. In both cases, the unsubstituted phenoxy ring lies in close proximity to Leu-338 and the catalytic tyrosine residue (Tyr-371) thus blocking the approach of the AA substrate. In one orientation the side chain of Arg-106 interacts with the nitro group whereas in the alternate one it is the sulphonamide group that interacts with this positively charged residue. Conversely, Arg-499 and His-75 can establish hydrogen bonding interactions with either the sulphonamide or the nitro group of nimesulide in the side pocket of this enlarged binding site.

Discrimination between these two binding modes was not possible on the basis of molecular dynamics simulations and energy analysis of the two complexes, so the possibility that nimesulide binds in the COX active site of human PGHS-2 in both orientations cannot be ruled out. Clearly, greater insights about the details of the interaction will be gained when the crystal structure of the complex is solved.

Nimesulide and neutrophil functional responses

Introduction

While the inhibition of prostaglandin synthesis through the blockade of cyclooxygenase is widely accepted as a mode of action of NSAIDs [39, 127], during the last 3–4 decades, a variety of non-prostaglandin mediated effects of non-steroidal anti-inflammatory drugs have been reported [128–133], suggesting that inhibition of cyclooxygenase does not represent the only explanation for the activity of these drugs. In this regard, neutrophils are considered a major potential target for NSAIDs because of the relevant role of these cells in natural and immune-driven inflammatory responses [128–130, 133–135]. Here we consider the role of neutrophils in inflammatory reactions and review *in vitro* and *in vivo* effects of nimesulide on activities of this cell population.

Hallmarks of neutrophil-mediated inflammation

Neutrophils represent the major population of circulating inflammatory cells naturally capable of responding to infectious and non-infectious tissue danger signals. They contain more than 40 hydrolytic enzymes and can generate various oxygen-derived oxidants. The number of toxic molecules is redundant; this being presumably related to the task of these cells, i.e., their microbiocidal activity. Unfortunately, these toxic agents cannot discriminate between exogenous microorganisms and tissue structures, implying potential histiotoxic activities directed to cells and tissues in the body [131, 135]. At sites of inflammation, activation of venular endothelium provides a pro-adhesive vascular surface for the local adherence of circulating neutrophils [136]. Then, adherent neutrophils, stimulated by platelet activating factor (PAF) and interleukin-8 (IL-8) are exposed on the surface of endothelial cells, migrate across the endothelial monolayer [136]. Migration of neutrophils through subendothelial tissues is also thought to involve a limited digestion of both the venular basement membrane and the components of the tissue matrix by serine proteases such as cathepsin G, elastase and proteinase 3 expressed on the surface of migrating cells [131]. Migration is directed by gradients of chemotaxins generated locally by complement activation (C5a), local cells such as macrophages, fibroblasts, endothelial and epithelial cells (IL-8) and, when present, microorganisms (formyl-peptides). Under the influence of local cytokines, mainly tumour necrosis factor-α (TNFα), IL-1 and granulocyte-macrophage colony stimulating factor (GM-CSF) initially released by local macrophages, recruited neutrophils undergo full activation [131]. These cytokines also promote the development of a cytokine-rich microenvironment prone to induce modifications of expression and activity of neutrophil adhesion molecules and chemokine receptors leading to the switch from a migratory to stationary phenotypes of recruited cells. Other ligands, such as for instance immune-complexes or antibody-coated surfaces, can also activate neutrophils. This results in the triggering of both respiratory burst and degranulation, with exocytosis of cytoplasmatic granules [137]. Respiratory burst is characterised by the rapid consumption of oxygen which is transformed into superoxide anion in turn dismutated to hydrogen peroxide. Hydrogen peroxide is then transformed by neutrophil myeloperoxidase into potent oxidants including hypochlorous acid and chloramines [131, 137]. These oxidants together with proteolytic enzymes, such as elastase and metalloproteases, contribute substantially to inflammation-dependent damage of local parenchymal cells and interstitial components of the inflamed tissue by mechanisms outlined in Figure 10. In spite of its misleading name, elastase is particularly toxic as it is capable of digesting several key elements of the extracellular tissue matrix, i.e., elastin, collagen type III and IV, laminin, fibronectin and core proteins of proteoglycans [137]. Neutrophil-mediated damage can be amplified by various pathways, two of them are presently considered of major relevance. First, locally recruited neutrophils are capable of undergoing activation and expression of genes coding for proinflammatory cytokines, such as TNFα and IL-1, and chemokines, such as IL-8. Synthesis of these mediators promotes waves of re-

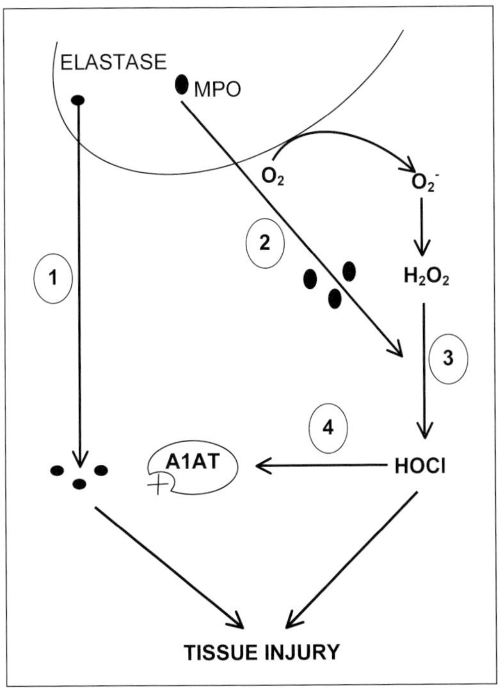

Figure 10
Pathways of neutrophil-mediated tissue injury: extracellular release of elastase (1); chlorinated oxidant production by hydrogen peroxide/MPO pathway (2,3); inactivation of alpha-1-antitrypsin (A1AT) by HOCl (4). These pathways converge to proteolytic and oxidative tissue injury (lower pathways).

cruitment of circulating neutrophils, thereby augmenting the pool of extravasated potentially dangerous cells [138]. Second, at sites of inflammation, neutrophils inactivate the neutral anti-proteases, both by oxidative and proteolytic mechanisms. In particular, as shown in Figure 10, neutrophils inactivates the physiologic inhibitor of their elastase, i.e., alpha-1-antitrypsin, in turn favouring the unrestrained digesting and tissue-damaging activity of elastase [131, 137].

In vitro effects of nimesulide on neutrophil functions

As shown in Table 14 and Figure 11, nimesulide inhibits various *in vitro* activities of normal human neutrophils, ranging from migration and oxidative respiratory burst to degranulation and production of proinflammatory mediators. In this re-

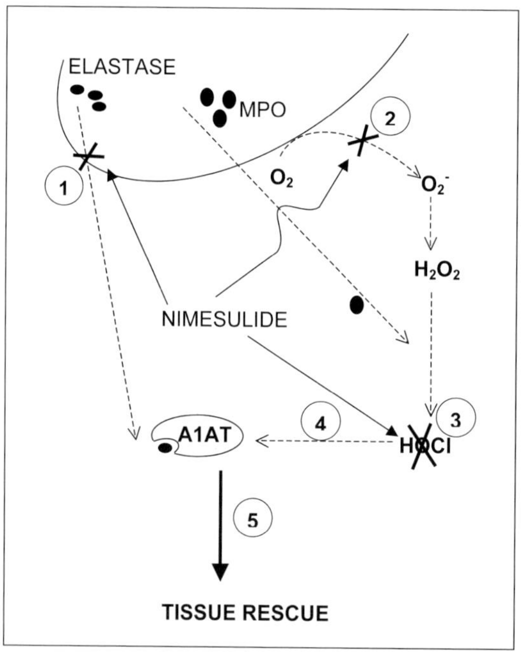

Figure 11
Inhibitory effect of nimesulide of histotoxic pathways of neutrophils. The drug inhibits the release of elastase (1); reduces the bioavailability of HOCl by impairing the production of the oxidant precursor O_2^- (2), reduces the bioavailability of HOCl (3); restores the anti-elastase activity of alpha-1-antitrypsin (A1AT) by neutrophil-mediated inactivation (4). Taken together, these activities results in nimesulide mediated tissue rescue from neutrophil histiotoxicity (5).

Table 14 – In vitro effects of nimesulide on neutrophil activities

NEUTROPHIL ACTIVITIES	DRUG EFFECT	DRUG CONCENTRATIONS (μmol/L)	REF.
Chemotaxis	inhibition	1–10	139
Superoxide production	inhibition	1–10	139, 142
Chemiluminescence production	inhibition	1–50	143, 144
Hypochlorous acid production	inhibition	10–20	145, 146
Chloramine production	inhibition	20	147
Elastase release	inhibition	10	148
β-glucuronidase release	inhibition	10	149
Transcobalamin-I release	inhibition	10	149
Oxidative inactivation α_1-AT	prevention	20–50	150, 151
Proteolytic inactivation α_1-AT	prevention	20	152
IL-8, IL-1 and IL6 production	inhibition	20–50 [IC_{50}]	153
PAF production	inhibition	20–50 [IC_{50}]	66
LTB_4 production	inhibition	10–30	66
Adherence	inhibition	20	140
Transendothelial migration	inhibition	50	140
L-selectin shedding	inhibition	50	141

gard, active concentrations of the drug are listed in Table 14. In particular, nimesulide inhibits the cell ability to migrate in response to chemotaxins, as measured in standard polycarbonate filter assays [139]. Moreover, the drug reduces the adherence of neutrophils to monolayers of cytokine-activated human endothelial cells, grown to confluence on filter surfaces and exposed to TNFα to stimulate venular walls at sites of inflammation [140]. This probably involves drug-mediated interferences with the expression and/or the activity of adhesion molecules on the neutrophil surface such as, e.g., L-selectins [141]. Owing to this effect on the cell adherence, the drug inhibits neutrophil migration across monolayers of activated endothelial cells, without interfering with the cytokine ability to convert resting endothelium to a pro-adhesive and pro-locomotory cell layer [140]. Together, these findings strongly support the concept of nimesulide as an anti-inflammatory drug endowed with the potential to reduce the recruitment of circulating neutrophils at tissue sites of inflammation. Once recruited at sites of inflammation, neutrophils undergo to full functional activation with consequent respiratory burst and degranulation [131]. These cell activities are susceptible to inhibition by nimesulide as well. Indeed, the drug is able to inhibit neutrophil respiratory burst in a concentration-dependent manner, as detected by measuring both the production of superoxide anions [139, 142] and cellular chemiluminescence [143, 144, 223]. The generation of oxidative derivatives of superoxide anions are also prevented by nimesulide [145, 224]. Nimesulide inhibits the activity of the myeloperoxidase system involved in the transformation of superoxide-derived hydrogen peroxide into hypochlorous acid [146] and chloramines [147].

On the other hand, nimesulide inhibits the release of elastase [148, 226] and β-glucuronidase [149] by activated neutrophils, suggesting that the drug interfere with the exocytosis of neutrophil primary granules. Finally, nimesulide reduces the release of transcobalamin-I [149], consistent with its ability to inhibit exocytosis of neutrophil secondary granules as well. As neutrophil-derived oxidants, particularly hypochlorous acid and its derivatives and primary granules constituents, such as elastase, are well-known to mediate tissue damage at inflamed tissue sites, the ob-

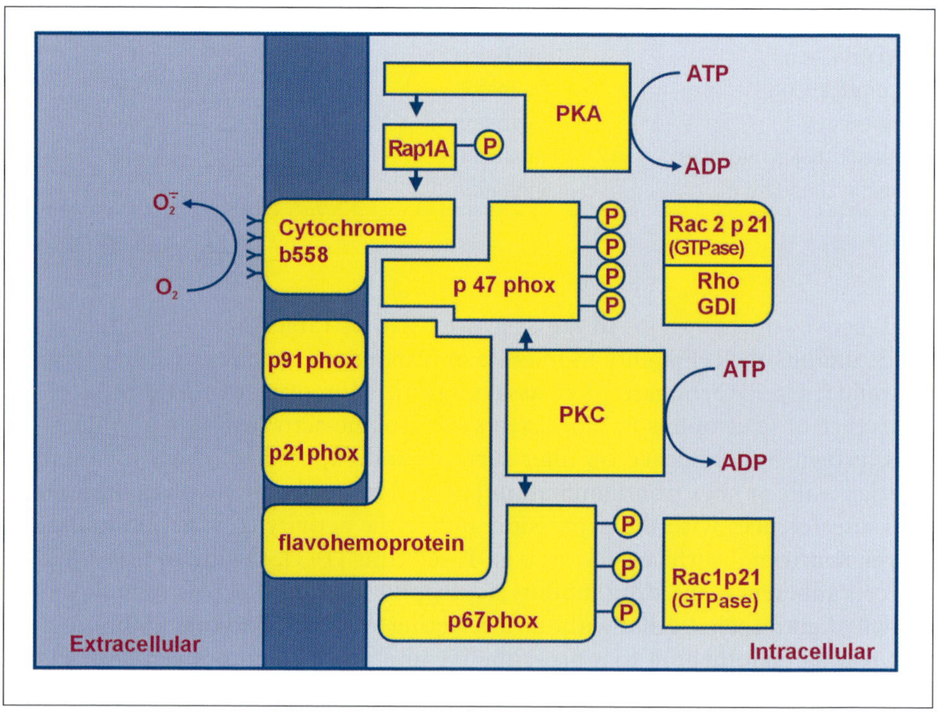

Figure 12
Protein kinase C activation (PKC) of neutrophils and subsequent activation of NADPH oxidase. This involves sequential phosphorylation of a variety of proteins by PKC. Increase in cAMP terminates this process.

served inhibitory activities of nimesulide suggest that the drug has potential histio-protective properties other than anti-inflammatory activity. Moreover, nimesulide can prevent both the oxidative [150, 151] and the proteolytic [152] inactivation of anti-proteases, such as alpha-1-antitrypsin, i.e., the well-known specific inhibitor of neutrophil elastase [131]. This is a particularly interesting action of the drug. It is indeed known that neutrophils, recruited at inflamed sites, promote the damage of the tissue by increasing the local burden of oxidants and proteases, such as elastase, and by reducing tissue defensive systems such as the anti-elastase alpha-1-antitrypsin screen [131]. Therefore, by reducing the burden of neutrophil-derived oxidants and proteolytic enzymes, such as elastase, and by rescuing alpha-1-antitrypsin from neutrophil-mediated inactivation, nimesulide is a candidate for the pharmacologic correction of oxidant–antioxidant and protease–antiprotease imbalances present at sites of neutrophilic inflammation and involved in the genesis of inflammation-related tissue damage. Finally, it is noteworthy that nimesulide is able to reduce neutrophil production of proinflammatory cytokines, such as IL-1 [153], and also the production of chemotaxins such as IL-8 [153], PAF [66] and leukotriene (LT) B_4 [66]. These inhibitory activities raise the possibility for the drug to interfere with proinflammatory feedback loops involved in the amplification of inflammatory responses including these involved in protein kinase C activation in leucocytes of the recruitment of circulating neutrophils. In conclusion, nimesulide appears to inhibit various steps of inflammatory reactions (Fig. 12) and may, therefore, be suitable for developing pharmacologic strategies to control neutrophil-mediated tissue injury.

Relevance of in vitro *findings and* ex vivo *studies*

It is known that the highest mean blood concentration of nimesulide after the oral administration of a standard dose of 100 mg is about 6 µg ml^{-1} (≅20 µmol/L) [6]. Nevertheless, it is known the nimesulide in blood is ~99% bound to albumin and, therefore, only ≅1% of the total amount is free and active [6]. Consequently, the concentration of free and active nimesulide after oral administration of 100 mg could be calculated to be about 0.06 µg/mL (=0.2 µmol/L). However, closer examination of the mechanisms of uptake of nimesulide reveals that the intracellular concentrations may be much higher than calculated on the basis of plasma concentrations. Thus, Bevilacqua and co-workers studied the uptake of nimesulide within neutrophils by the use of ^{14}C-labelled nimesulide (gift from Helsinn Healthcare, unpublished data) and discovered that intracellular concentrations of nimesulide are about 50–150 µmol/L. Hence it appears that the intracellular levels of nimesulide are well in the range of the data obtained in many laboratories regarding the effects of this drug on the respiratory burst.

The mechanism of intracellular accumulation of nimesulide has been studied recently [154]. In neutrophils there are two vacuoles whose pH is tightly regulated, the lysosomes and the phagosome. In lysosomes the pH is acidic and the addition to the medium of weak bases that can be uptaken by lysosomes may increase the pH of lysosomes and may ultimately affect in some so far unknown way also the respiratory burst (for reviews [155–160]). The pH of phagosome is also acidic (some pathogens escape to death just by rising slightly the pH of phagosome) [161] and becomes more alkaline [162] during phagocytosis. The intravacuolar pH of neutrophil phagosomes is less acidic that pH of other organelles and this has been attributed to the consumption of protons during the dismutation of superoxide. In these studies no significant change was observed in intracellular pH by nimesulide.

Ivanov and Tzaneva [163] have evaluated the ability of many NSAIDs to enter into cells in an acidic environment. This is relevant to the fact that commonly the pH of inflammatory foci is slightly acidic due do the accumulation of anaerobic metabolites, such as lactic acid. They found that in an acidic medium the accumulation of NSAIDs within erythrocytes ranked as aspirin < paracetamol < nimesulide < diclofenac < piroxicam < meloxicam < ibuprofen < naproxen < indomethacin. Moreover, there did not appear to be any relationship of intracellular accumulation of NSAIDs and their effects on respiratory burst in neutrophils. Some of those drugs that were found to poor accumulators in neutrophils (e.g., paracetamol, aspirin) were unable toaffect the burst whereas, paradoxically, others that were high accumulating drugs (e.g., ibuprofen, naproxen, indomethacin) were also ineffective. Nimesulide is potent inhibitor of respiratory burst and accumulated to a relatively considerable degree in neutrophils. The uptake of nimesulide into neutrophils does not occur by a simple chemical mechanism.

The concentration of free nimesulide at sites of inflammation may be higher than blood concentrations because of the slight acidic microenvironmental conditions in inflamed tissues [164]. It is known that various cells possess elaborate mechanisms to internalise albumin as a source of amino acids, making human albumin a suitable protein as potential drug delivery system [164]. For instance, methotrexate-albumin complexes are taken up by tumour cells and then methotrexate is released as an active compound into the cytosol to exert its action [165]. As neutrophils exert endocytosis of macromolecules in fluid phase easily [166], they can take up drug–albumin complexes that are also prone to diffuse into inflamed tissues because of the local enhanced microvascular permeability. These events might explain why higher concentrations of nimesulide may sometimes be required to reproduce *in vitro* events occurring *in vivo*.

In the light of these estimations of free and intracellular drug concentrations in leucocytes *in vitro* experiments carried out to test the effects of nimesulide on neutrophil function (Tab. 14) have generally been performed with plasma concentrations achievable *in vivo* after the oral administration of the drug. As *in vitro* assays have been carried out using cell-culture media without or with low levels of albumin, the concentrations found to be effective *in vitro* as far as chemotaxis and superoxide production are concerned (Tab. 14) appear to be 5-fold higher than the concentration of free nimesulide expected to be reached *in vivo*. Although nimesulide inhibits neutrophil transendothelial migration and L-selectin shedding at relatively high concentrations (50 µmol/L), the majority of the other *in vitro* inhibitory effects of nimesulide have been observed at concentrations ranging from 1–10 µmol/L. Similarly, the drug concentrations able to inhibit hypochlorous acid production, elastase release and LTB_4 synthesis (Tab. 14) appear to be 50-fold higher than those expected in plasma *in vivo*, but these are within the range observed in cellular uptake studies by Bevilacqua.

Two major observations concerning the relevance of *in vitro* effects to what may occur *in vitro* should be taken into consideration. First, consistent with the ability of nimesulide to inhibit *in vitro* neutrophil respiratory burst [139, 142–144], it was found that the oral administration of 200 mg nimesulide taken by healthy volunteers results in a reduced capacity of circulating neutrophils to generate superoxide anions in response to a soluble stimulus, such as formyl-peptides, as well as in response to phagocytosis of opsonised targets [167]. Percent reductions of superoxide generation were 67.62% ± 7.57 (mean ± 1SEM, n = 8) and 36.75% ± 7.92 (mean ± 1SEM, n = 8) in response to formyl peptides and particle phagocytosis, respectively [167]. This drug activity in isolated leucocytes after *in vivo* administration has been recently confirmed by other authors [168], by evaluating neutrophil chemiluminescence in response to phorbol myristate acetate or Ca^{++} ionophore or phagocytosable targets.

Second, consistent with *in vitro* observations showing that nimesulide inhibits neutrophil migration (Tab. 14), *in vivo* administration of nimesulide resulted in reduced ability of circulating neutrophils to migrate in response to casein as a standard chemotactic stimulus [168]. These data show that nimesulide inhibits neutrophil migration and oxidative burst after *in vivo* administration of the drug.

Apoptosis and superoxide release

During phagocytosis of microbes (mycobacteria, viruses), immune complexes and foreign bodies, the production by neutrophil superoxide anions can escape from the phagosome and thus are possibly harmful to the host tissue (for reviews see [169–186]). The death by necrosis of neutrophils may also induce the liberation into the medium of toxic substances. Apoptosis of neutrophils is possibly a more controlled mechanism to remove activated neutrophils. The resolution of neutrophil-mediated inflammation is based upon the activation of a cyclic AMP biochemical pathway that interrupts the production of superoxide anions and by an apoptosis differentiation programme. Through this mechanism the final stage of transcriptionally regulated neutrophil maturation is significantly accelerated and neutrophils undergo apoptosis and are phagocytosed by mononuclear cells [187–196].

Neutrophil apoptosis is strictly controlled by cAMP [188–190, 192–194, 197] though the effects of cAMP on apoptosis (inhibition of apoptosis) seems to be independent from protein kinase A activation [193, 195, 197]. Recent findings suggest a pivotal role of interleukin 10 in the acceleration of apoptosis in neutrophils [174].

Nimesulide has been shown to reduce apoptosis in a monocyte cell line [155]. Since the inhibition of monocyte apoptosis is associated *in vivo* with the accumulation of neutrophil destruction [188–192, 198–200] nimesulide may enhance the neutrophil death *in vivo*. However, it has been shown that oxidants generated from the oxidation of plasma membrane phosphatidylserine facilitate the recognition of neutrophils by macrophages [190, 201–203]. The inhibitory effect of nimesulide on the oxidant production by neutrophils might affect their apoptosis. It has also been shown that nimesulide stimulates apoptosis in some tumour cell lines [204], so there might be a common mechanism of this phenomenon in neutrophils and monocytes/macrophages as seen in tumour cells.

Regulation of NADPH oxidase

During phagocytosis there is activation of the respiratory burst NADPH oxidase [171, 181–185]. The first product of NADPH-oxidase is the O_2^- (superoxide anion) which is produced by univalent reduction of oxygen. Superoxide anion has a very poor antibacterial activity and undergoes dismutation via superoxide dismutase to produce H_2O_2. This in turn reacts with Cl^- ions by the actions of myeloperoxidase released into the vacuole from the cytoplasmic granules to produce hypochlorous acid (HOCl), a potent antimicrobial oxidant. Recently, the role of HOCl as the final common pathway for the microbicidal activity has been questioned [171]. On the basis of quantitative analysis of the ratio among the various chemical species of oxygen it has been postulated that the function of the neutrophil oxidative pathway is to provide optimal conditions for bacterial killing by neutrophil proteases stored in granules rather than by direct oxidative destruction.

The stimulation of NADPH oxidase by specific and/or non-specific physical reaction of neutrophils with foreign material occurs through the activation of protein kinase C pathway [180–185] and the sequential phosphorylation of various proteins whose embedding into the plasma membrane triggers the final activation of NADPH oxidase (see Fig. 12). The deactivation of the respiratory burst is obtained by the classical mechanisms of receptor internalisation [184] or by the "auto-termination" effect of the elevation of cAMP which in turn by activating protein kinase A terminates the burst. Increased cAMP in neutrophils can be achieved by activation of external receptors that are coupled to adenylate cyclase [205], such as beta adrenergic agents [206], PGE_2, adenosine, histamine or by the manipulation of the biochemical pathway to control intracellular production of cAMP. Subsequently, cAMP is rapidly degraded by phosphodiesterase type IV [207] in active metabolites. Interestingly, after addition of formylated peptides the increase of the respiratory burst is auto-terminated by an increase of endogenous cAMP induced by the same peptides [139].

The inhibitory effect of nimesulide on superoxide anion production has been shown to be linked with the inhibition of a specific phosphodiesterase of neutrophils (Type IV) [139]: the effect was observed just at 1 µmol/L and the IC_{50} of nimesulide on the enzyme was 49 µmol/L, a concentration readily attainable within the neutrophils (see above). At 1 µmol/L nimesulide also increased cyclic AMP in neutrophils [139], an effect that was confirmed by others who showed that nimesulide at 30 µmol/L decreased PAF and LTB_4 production and increased cAMP [66, 208].

In at least two laboratories [66, 139, 208, 209] it has been found that the nimesulide effects on superoxide anions, PAF, leukotrienes, neutrophil adhesion were blocked by H-89, a specific inhibitor of protein kinase A [66, 207, 208], the ultimate effector of cAMP. Interestingly, H-89 increases apoptosis in neutrophils [197]. Nimesulide inhibited also the eosinophil chemotaxis and synthesis of lipid mediators, again with an effect linked to the inhibition of cAMP degradation [208]. Recently, it has been shown that nimesulide is competitive to rolipram, a prototype phosphodiesterase IV inhibitor [210], an effect that is linked to the anti-inflammatory activities *in vivo* of nimesulide in animals, but not to its analgesic properties. Thus, inhibition of phosphodiesterase type IV by nimesulide (by analogy to rolipram) seems to be a mechanism of control of inflammation, whereas the analgesic properties of the drug are perhaps related to cyclooxygenase type II inhibition [51, 211].

Time-dependent effects

Capsoni et al. [142] were the first to demonstrate that near therapeutically relevant concentrations of nimesulide (around 10 mg/mL or about 30 µmol/L) were active *in vitro* against superoxide anion production in neutrophils. To replicate this experiment the neutrophil must be incubated with nimesulide for approximately 10 min before the addition of the secretagogues: this is important for the entry of the drug into neutrophils, a process that requires at least 10 min. Nimesulide, when added simultaneously to stimulants, was unable to affect the respiratory burst. Pre-incubation of neutrophils with some other NSAIDs has been found to enhance release of superoxide [211]. This effect of pre-incubating neutrophils with nimesulide may, therefore, not be seen with all NSAIDs. It suggests that the drug must enter within the neutrophil phagosome in order to inhibit the burst. The timing of the addition of nimesulide and the evaluation of the biochemical effects is also relevant for the study of the interaction of nimesulide with COX-2 [52]. In particular, the pre-incubation of the COX-2 with nimesulide from 1 to 10 min increased the inhibitory activity from about 70 to 0.07 µmol/L (in analogy to what was observed by other investigators for the *p*-nitro-methanesulphonanilide analogue, NS-398) [212, 213]. So the timing of the addition of the drug to the experimental system is fundamental to obtain reproducible results. Finally, recent biochemical investigation has failed to support the scavenging effect of nimesulide on HClO [214, 215], whereas it has confirmed the ability of the drug and of its metabolite 4-OH-nimesulide to scavenge other chemical species of oxygen, including hydroxyl and superoxide anions [216].

Phagosome and lysosome accumulation and protease inhibition

Recently, it has been shown that a group of methanesulphonanilide anti-inflammatory drugs (most of which act as COX-2 inhibitors) act as lysosomal protease inhibitors after being concentrated into the vacuoles of neutrophils where they block proteases without affecting the acidic pH of lysosomes [155].

The mechanisms by which methanesulphonanilides interact with lysosomal protease is unknown. However, it is known that there are strict conformational analogies between cyclooxygenase and metalloproteinase inhibitors. As evaluated by molecular modelling (docking) [217] the sulphonanilide group of nimesulide (and celecoxib) is a prerequisite for the inhibition of COX-2 but also for the inhibition of metalloproteinase [218]. Furthermore, a sulphonanilide moiety is also important for the inhibition of Tumour necrosis factor Alpha Converting Enzyme (TACE) [219, 220]. Further studies are necessary to evaluate if the intracellular accumulation of nimesulide into neutrophils is due to selective uptake by lysosomes or also involve other organelles including the phagosome, whose inside is also acidic. Nevertheless, it is possible that upon fusion of lysosomes with phagosome [221] the concentration of nimesulide within the phagosomes may reach effectively rather elevated concentrations.

Nimesulide has not been found to affect superoxide anion release in isolated plasma membranes of neutrophils [139] (but actually prevented the embedding of NADPH oxidase into plasma membranes) thus suggesting that nimesulide may act in some way within the phagosome. Interestingly, the biochemical machinery for the control of cAMP, i.e., adenylate cyclase, phosphodiesterase type IV (all the isoforms 4A, 4B, 4D) and the cAMP-dependent kinase (PKA) are localised at the phagosome during its formation suggesting that cAMP levels are focally regulated by PDE-4 at the nascent phagosome, and that PKA may phosphorylate protein associated with pseudopodia formation and phagosome internalisation [207]. The hypothesis is that nimesulide, by affecting phosphodiesterase type IV in neutrophils and increases endogenous levels of cAMP, might lead to prevention of the formation of the nascent phagosome: this could explain the reduced plasma membrane localisation of NADPH-oxidase, as well as the decreased production of superoxide anion, release of azurophylic granules and of many granule-related substances. Furthermore, this could explain why nimesulide affects the production of superoxide anion by all the stimulants so far tested, including phorbol diesters. In fact the formation of the nascent phagosome is a *sine qua non* for the production of superoxide anion by all stimulants in the whole neutrophil.

Other biochemical effects on leucocytes

Nimesulide also has a range of other biochemical effects on neutrophils, as well as in eosinophils and mast cells. The drug inhibits the release of elastase in neutrophils in normal conditions and after TNFα "priming" [148, 225]. Capecchi and co-workers [143] also showed that nimesulide reduced cytosolic calcium that is increased by formylated peptides or by ionomycin, an ionophore. Interestingly, in two independent studies the effects of nimesulide were reversed by employing theophylline (an adenosine antagonist) [143] or by adding the adenosine catabolising enzyme adenosine deaminase to the incubation mixture [148]. Therefore one of the possible mechanisms of action of nimesulide could be related to the adenosine enhancing activity [227–229] that was also shown for low-dose methotrexate and sulfasalazine. It is possible that nimesulide in analogy with the effects seen with sulfasalazine [230] (to which nimesulide might be considered to be chemically related) is the possibility of the inhibition of phosphoribosylaminoimidazolecarboxamide formyltransferase (AICAR transformylase, EC 2.1.2.3) and the related enzyme dihydrofolate reductase (EC 1.5.1.3). This has been shown to occur with many NSAIDs (sulindac, indomethacin, naproxen, salicylic acid, ibuprofen, piroxicam and mefenamic acid) [231]. By inhibiting AICAR transformylase (analogous to that observed with methotrexate) nimesulide might increase the tissue levels of 5-aminoimidazole-4-carboxamide ribonucleoside, AICAR (also known as acadesine) that in turn is a potent adenosine releaser [227–229, 231].

Nimesulide inhibits eosinophil chemotaxis and synthesis of lipid mediators, again with an effect linked to the inhibition of cAMP degradation [208]. Nimesulide (10 µmol/L) also inhibits the release of IgE-stimulated histamine from basophils and various other mediators and potentiates the effect of adenylate cyclase agonists such as forskolin and PGE_1 [143].

Complement activation

Complement activation is another mechanism involved in the chemotactic responses of phagocytic cells in inflammation [232]. In studies of the classical pathway, Auteri and co-workers [233] observed that complement activation was inhibited by 10 µmol/L nimesulide and progressed linearly to 100% inhibition with 100 µmol/L of the drug. Total haemolytic activity that was inhibited by the latter concentration of nimesulide was restored to normal when fresh serum containing complement components treated with anti-β1E or anti-C1q, but not anti-β1C globulins were added [233]. The activation of the alternate pathway was also inhibited in a linear fashion by approximately the same concentrations of nimesulide.

Endothelial reactions and angiogenesis

As outlined previously (see section on "Hallmarks of neutrophil-mediated inflammation" page 174) migration of leucocytes through endothelial cells are prone to expression of adhesion molecules and chemokine receptors. The expression of vascular adhesion molecules and angiogenesis also participate in the inflammatory reactions in a time-dependent process [232]. Angiogenesis is in part controlled by COX-2 derived PGE_2 and this in turn by growth factors [234, 235]. Thus, inhibition by nimesulide of basic fibroblast growth factor (bFGF)-induced angiogenesis in sponge implants in rats was considered to be related to COX-2 inhibition by this drug as well as some experimental COX-2 inhibitors [235]. The inhibition of angiogenesis by these drugs was related to reduction in the expression of vascular endothelial growth factor (VEGF) [235]. In hepatic stellate cells stimulated to produce COX-2 by exposure to hypoxic conditions increased expression of VEGF was reduced by prior treatment with nimesulide [236]. In a model of angiotensin-2 angiogenesis in mice, it was found that nimesulide 13 mg/L impaired the pro-angiogenic effect of angiotensin-2. The coincidental increase in VEGF was considered to be a possible target for the effects of nimesulide [237]. Using the *in vitro* model of angiogenesis in the chick chorioallantoic membrane (CAM) it was found that nimesulide, as well as celecoxib, had an anti-proliferative effect [238]. It was also found in these studies that nimesulide and celecoxib had anti-proliferative effects in a time- and concentration-dependent manner in a variety of human NPC cell lines at drug concentrations in the range of 8–200 µmol/L; most cell lines were affected at ≤50 µmol/L [238].

Summary and conclusions

Polymorphonuclear leucocytes (neutrophils) are endowed with potent biochemical machinery to kill bacteria and to destroy foreign bodies as well as to cause tissue injury when activated. This machinery is characterised by the formation of phagosomes in which NADPH oxidase produce superoxide anions that, in turn, after the action of myeloperoxidase, give origin to the important killing agent HOCl. The activation of neutrophils may be injurious to the host so, therefore, it must be terminated either by biochemical mechanisms (increase of endogenous production of cyclic AMP and activation of the protein kinase A, that blocks any further activation of neutrophils) and, perhaps more importantly, by apoptosis of the cells and their elimination by mononuclear cells. In the course of phagocytosis and in the case of activation of neutrophils in the context of various rheumatic and immune-mediated diseases, oxidants escaping the phagosome may be harmful to the host tissues, and this requires appropriate medications.

Nimesulide has been shown to accumulate into neutrophils, to inhibit the main important catabolizing enzyme of cAMP (phosphodiesterase type IV) and to increase cAMP in neutrophils and eosinophils. The compartmentalisation of the cAMP-related enzymes is typically found in the nascent phagosome and this suggests that nimesulide acts by decreasing the amount of nascent phagosome (this might also explain why nimesulide decreases the release of all the products by neutrophils, not only superoxide anions). Nimesulide does not decrease superoxide anion production in isolated plasma membranes suggesting that the effect is not directed at the NADPH oxidase.

The experiments carried out after oral administration of the drug strongly support the conclusion of nimesulide is a compound endowed with the ability to inhibit neutrophil functional responses relevant to inflammatory reactions.

The inhibitory effect of nimesulide on neutrophil activation [139, 140, 142, 143, 145–153, 167, 224–226] has been confirmed in a wide variety of experimental systems and in various laboratories.

Further investigations are required to clearly understand why relatively high concentrations of nimesulide are needed to reproduce *in vitro* effects detectable *ex vivo*. In this regard, recent investigations on albumin as potential drug delivery systems coupled with the particular ability of neutrophils to internalise macromolecules raise the possibility that presently uncovered mechanisms accounts for the mentioned *in vitro versus ex vivo* experimental discrepancies.

Analgesic actions of nimesulide in animals and humans

Molecular biology and neural mechanisms of pain

As shown by basic research advances in the mechanisms of neuronal plasticity, e.g., central sensitisation, and in the molecular neurobiology of pain [239], the discovery of neurotransmitters and neuromodulators involved in pain processing – for example, nitric oxide (NO), CGRP, substance P (SP), neuropeptide Y (NPY), vasoactive intestinal polypeptide (VIP) – has furthered our understanding of pain mechanisms and of the mode of action of analgesic compounds.

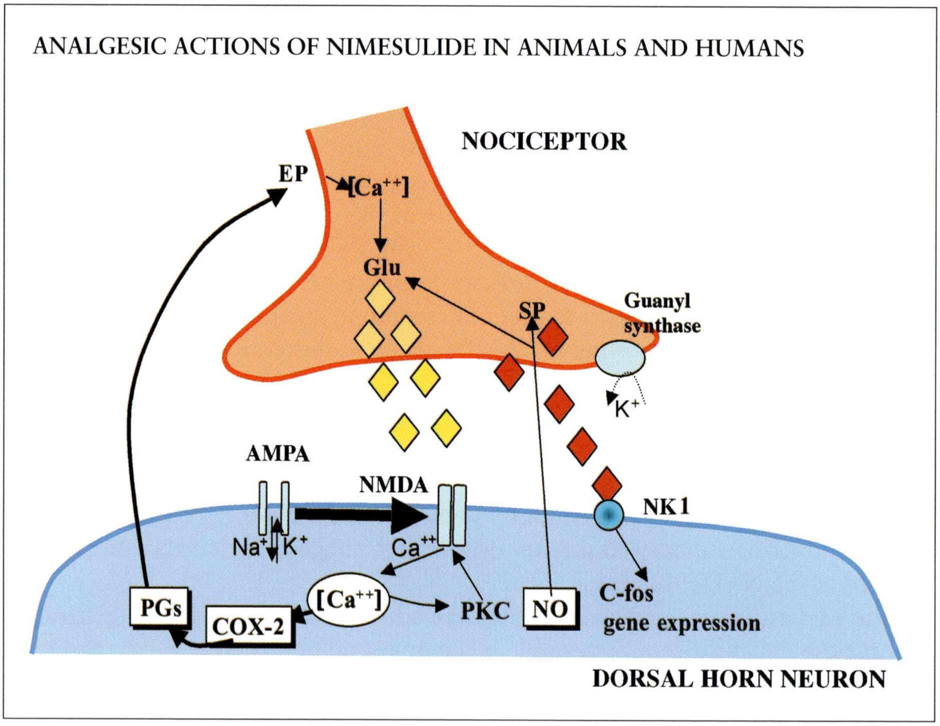

Figure 13

Putative mechanisms of hyperalgesia in the spinal cord: the role of cyclooxygenase, nitric oxide and NMDA-receptors. Incoming pain signals trigger the release of glutamate (Glu) into the synaptic cleft between nociceptors and dorsal horn neurons. Pain activates AMPA receptors on Na^+/K^+ channels. Prolonged activation alters the polarisation of the membrane: the magnesium plug in the Ca^{++} channels is removed and the NMDA receptors are primed for Glu activation. Ca^{++} flowing into the cell activates cyclooxygenase-2 (COX-2), protein kinase C (PKC) and nitric oxide (NO) synthase. COX-2 activation leads to the synthesis of prostaglandins (PGs), which can stimulate the prostaglandin receptor (EP) located pre-synaptically to increase Ca^{++} concentration and glutamate release in the pre-synaptic neurons. Newly synthesised NO diffuses to the nociceptor, where it stimulates guanyl synthase-induced closure of K+ channels, therefore inducing opiate resistance, and further Glu release. NO also stimulates release of substance P (SP), which binds to neurokinin 1 (NK1) receptors in the post-synaptic neuron and triggers gene expression (neuronal plasticity).

Indeed, recent data suggest that relief of pain by NSAIDs may occur via mechanisms other than inhibition of PG synthesis, including anti-nociceptive effects at peripheral and at central nervous system (CNS) levels.

Both the mRNA and the proteins for COX-1 and COX-2 are expressed in the brain and spinal cord. Whereas normal expression of COX-2 is induced by basal synaptic activity, its overexpression occurs as a nervous system response to a somatic or neural injury (primary inflammation) [240]. The fact that both isoforms of COX are constitutively expressed in the CNS, and that basal levels of PGs seem to exist normally in spinal cord perfusates, means that these eicosanoids serve some physiological functions in the spinal cord.

In inflammation PGs can produce sensitisation of pain receptors. Following damage or during inflammatory conditions, these neurons can become spontaneously active, present lowered thresholds to various stimuli, and show enhanced responses resulting in the clinical phenomenon of hyperalgesia and – in some instances – allodynia.

Sensitisation of primary nociceptive afferent neurons (hyperexcitability of A-delta and C-polymodal nociceptive afferent neurons), and concomitant expanded receptive fields of dorsal horn neurons constitute the basis of primary and secondary hyperalgesia. A substantial component of the hyperalgesia and allodynia that characterise post-injury hypersensitivity occurs in the CNS, at both spinal and supraspinal levels. It appears likely that COX-2 expression is increased via N-methyl D-aspartate (NMDA) receptor activation and a calcium-dependent mechanism in the CNS.

A study by Dolan and Nolan [241] demonstrated that NMDA-induced mechanical allodynia is blocked by both nitric oxide synthase (NOS) and COX-2 inhibitors. It is known that upregulation of spinal COX-2 and NOS expression [240, 242–245] occurs in response to peripheral inflammatory stimuli and that a complex dynamic interaction seems to exist between these two pathways, and these aspects account for an important part of the analgesia observed with NSAID use [246–249] (Fig. 13).

It has been shown that a close relationship exists between noxious stimulation and PG release in the spinal cord [240]. The PGE_2 increase in the early phase of the response is accompanied by enhanced release of the excitatory amino acids glutamate and aspartate, the inhibitory amino acids glycine and taurine, and the NOS product citrulline.

The complex array of multiple system responses explains how peripheral inflammation can result in a state of both peripheral and central hyperexcitability, i.e., wind-up and central sensitisation, mediated by PGs and other endogenous products such as oxygen free radicals, all conditioning and/or predisposing to persistent/chronic pain states.

Central sensitisation, the wind-up phenomenon and the role of nitric oxide

Experimental states of central sensitisation, presenting changes similar to those associated with clinical chronic pain, can be obtained through repetitive electrical stimulation, at critical/high frequency (greater than 3 Hz), of nociceptive C-fibres of dorsal horn neurons, which induces a slow temporal summation of evoked responses, progressively increasing in frequency, magnitude and duration (wind-up). Evidence has accumulated in recent years to indicate that prolonged after-responses and slow temporal summation are mediated by the co-release of glutamate and substance P and their respective activation of NMDA and neurokinin 1 and 2 receptors, leading to NO generation and prolonged depolarisations. Many of the effects of NMDA receptor activation are mediated by production of NO [242].

The free radical, NO, is highly reactive and unstable, and is a messenger molecule involved in various biological functions, including nociception, synthesised by a complex family of NOS enzymes. NO contributes to the development and maintenance of central sensitisation at spinal level, i.e., that sensitisation of pain pathways can be caused by or associated with activation of NOS and the generation of NO, and that sustained elevation of NO is critical in maintaining central sensitisation, whereas inhibition of NOS reduces central sensitisation in pain models [238]. Furthermore, during central sensitisation it has been demonstrated that administration of NO-donor nitroglycerin (NTG) induces a significant increase in NOS- and c-fos-immunoreactive neurons, which exert pro-nociceptive effects possibly through further production of other substances in the CNS [244, 245].

Experimental studies in laboratory animal models

Studies in laboratory animal models have provided evidence for both central as well as peripheral actions of NSAIDs in mediating pain responses [246, 247]. Thus C-fibre activity in the thalamus is blocked by NSAIDs [246]. Most of the data that further our understanding of the mode of action of NSAIDs is derived from animal models of hyperalgesia, which demonstrate that high anti-inflammatory doses of other NSAIDs do not affect physiological nociception in animals [247]. However, nimesulide has recently proved able to modulate nociceptive (physiological) pain [248]. Bianchi et al. [249] recently demonstrated that nimesulide completely prevents the development of thermal hyperalgesia induced by injection of formalin in the tail, producing an inhibitory effect that was more marked and complete than that of diclofenac and/or celecoxib. In addition, nimesulide was also capable of reducing the mechanical hind paw hyperalgesia induced by the intraplantar injection of Freund's complete adjuvant (FCA), with an effect that was significantly greater than that observed following administration of celecoxib and rofecoxib.

Thermal hyperalgesia induced by formalin injection in the tail is considered a model of centrally-mediated hyperalgesia [250, 251]. Prostanoids modulate sensory processing via an alteration of spinal excitability and PGE_2 has been involved in spinal nociceptive processing [252–258]. It has been shown that injection of a diluted formalin solution into the rat tail is associated with increased spinal PGE_2 release, which correlates with hyperalgesic behaviour [186]. From these data, we can infer that the anti-hyperalgesic activity of nimesulide is related to inhibition of PG formation in the spinal cord. However, it has been suggested that NSAIDs exert centrally-mediated analgesia by mechanisms independent of PG synthesis inhibition [258–260].

In addition, the intraplantar injection of FCA is associated with the development of mechanical hyperalgesia within a few days, which is confined to the ipsilateral paw [261, 262]. Prostaglandins can sensitise peripheral nociceptors [263–265] and PGE_2 increases locally following FCA injections into a rat's hind paw. A selective COX-2 inhibitor proved able to block FCA-induced increase in peripheral PGE_2 [266]. Therefore, the anti-hyperalgesic effects observed by Bianchi et al. [249, 257] in the FCA-induced inflammatory hyperalgesia may be related to the inhibition of peripheral PG production.

The role of NO as a modulator of nociceptive information processing in the CNS has already been described above. Previous studies by Tassorelli et al. showed that the NO-donor NTG may activate specific nociceptive nuclei in the rat [267–270] and induce a condition of hyperalgesia [271], as well as an increase in the discharge rate of spinal nociceptive neurons [272] and activation of NF-kB, a transcriptional factor involved in the mediation of pain and inflammation [273].

In a recent report [247], the effect of nimesulide was investigated on NTG-induced hyperalgesic state and the results showed that the drug proved effective in counteracting NTG-induced hyperalgesia both in the formalin (Fig. 14) and in the tail flick test. In addition, the brain mapping of nuclei activated by NTG administration showed that nimesulide pretreatment significantly inhibited neuronal activation in several areas of the CNS, namely the supraoptic nucleus, ventrolateral column of the periaqueductal grey, locus coeruleus, nucleus tractus solitarius, and area postrema.

NTG-induced hyperalgesia is detected 2 and 4 h after the drug administration. Pharmacokinetic studies show that NTG rapidly disappears from the blood compartment and peripheral tissues, while it accumulates in the brain, where it reaches maximal concentrations 2 h after its administration [274]. Together with the demonstration that intradermal NTG does not alter thermal pain threshold in humans [275], this suggests that NTG-induced hyperalgesia is mediated by an increased availability of NO at central sites, rather than in the periphery. Therefore, the findings regarding the effect of nimesulide on NTG-induced hyperalgesia strongly suggest that the mechanism of action of this NSAID is, at least partly, related to central, NO-mediated mechanisms. This is further supported by the data

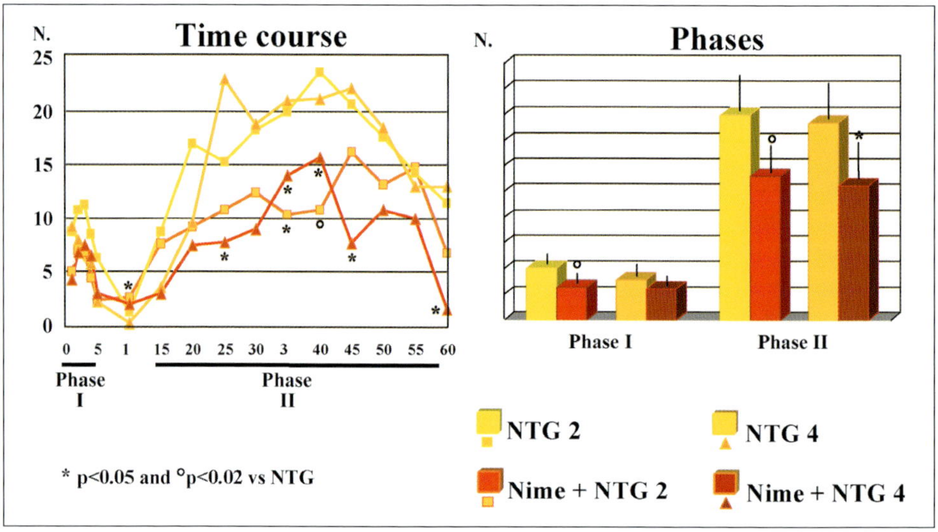

Figure 14
Pre-treatment with nimesulide induces a significant decrease in formalin-evoked nociceptive behaviour at 2 and 4 h after NTG administration. In this study, rats were treated with nimesulide 30 min before being injected subcutaneously with NTG. The formalin test was performed 2 or 4 h after NTG administration. Formalin-related nociceptive behaviour was quantified for 1 h by counting spontaneous flinches and shakes of the injected paw: over 60 s periods for the first 5 min (min 1, 2, 3, 4 and 5) and thereafter following 4 min pauses, for 1 min periods up to the hour. Phase I was defined as the period from 1–5 min, phase 2 was defined as the period from 10–60 min inclusive. Phase I is generally considered to reflect the chemical activation of the nociceptors, whereas phase II reflects the inflammatory reaction and central processing. Reproduced from [271] with permission.

we obtained in the tail flick test, where nimesulide showed an anti-hyperalgesic action even when it was administered 2 h after NTG (i.e., when the increased NO availability at peripheral level had disappeared).

The findings obtained with the formalin test further support a role of central mechanisms in the action of nimesulide. The analgesic and anti-hyperalgesic effect of nimesulide extended over both phases of the test, but was more marked during phase II. This phase corresponds to a prolonged tonic response in which inflammatory processes are involved and neurons in the dorsal horns of spinal cord are activated [276]. This is in agreement with human data obtained by Sandrini's group (see next section) by examining the actions of nimesulide on the spinal RIII reflex before and after NTG administration to healthy volunteers (Fig. 15). A central effect of nimesulide is also supported by its physical–chemical characteristics

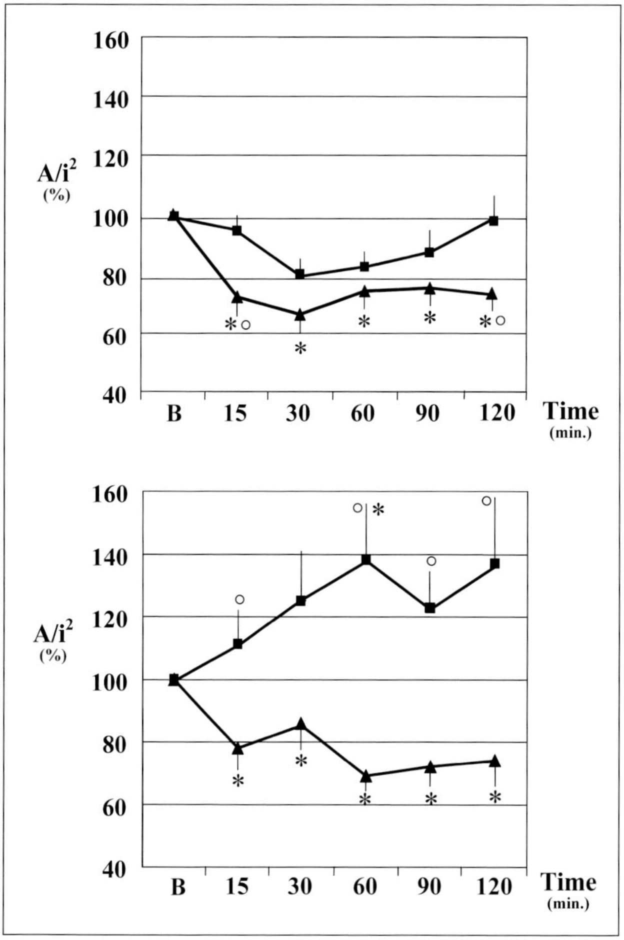

Figure 15

*Upper panel: Changes in RIII reflex (expressed as percent changes from baseline in A/i² ratio; see text for further details) after nimesulide/placebo. The RIII reflex was performed before nimesulide (▲) or placebo (■) administration (0) and 15, 30, 60, 90 and 120 min afterwards. Data are represented as means ± standard error. ANOVA for repeated measures: nimesulide, F = 3.89, p = 0.005; placebo, F = 1.91, p = 0.11. Post-hoc Duncan test *p < 0.05* versus base*line values, °p < 0.05 nimesulide versus placebo. Lower panel: Effect of NTG administration on RIII reflex (expressed as percent changes from baseline in A/i² ratio) following nimesulide (▲) or placebo (■) treatment. Baseline (0) RIII reflex was performed 2 h after nimesulide/placebo administration and 15, 30, 60, 90 and 120 min after NTG administration. Data are represented as means ± standard error. ANOVA for repeated measures: nimesulide, F = 2.95, p = 0.03; placebo, F = 4.16, p = 0.003; Post-hoc Duncan test *p < 0.05* versus *baseline values, °p < 0.05 nimesulide versus placebo. From Sandrini et al. (2002) [289].*

– a relatively high pKa (approximately 6.5) and a moderate lipophilicity – which suggest ready diffusion to the brain.

Orally administered nimesulide results in a brain level of approximately 1 µg equiv/g at 3 h after administration [277]. This tissue level of nimesulide corresponds to levels that induce inhibition of COX-2 activity [52, 53].

These findings widen the spectrum of mediators potentially implicated in the analgesic effect of nimesulide by including excitatory amino acids and peptides, such as glutamate and substance P, which are released from primary afferents and dorsal horn neurons [278–281]. As recently shown, formalin injection [282] into the paw of rats causes an increase of nociception, which seems to be mediated by the release of NO and PGE_2 in the spinal cord. Both COX isoforms are constitutively expressed in the spinal cord – COX-1 in dorsal horn glial cells and COX-2 in motoneurons of ventral horns [283, 284] – and they contribute to the nociception-induced PGE_2 increase associated with nociceptive behaviour. NO and PGs increase glutamate release [285, 286], which heightens the sensitivity of dorsal horn neurons. Spinal hyperalgesia induced by NO-donors and PG_2 may be blocked by NMDA receptor antagonists. In addition, the administration in mice and rats of NOS inhibitors reduces the pain-related behaviour induced by formalin test [287].

The data also seem to suggest that the anatomic circuitry involved is more widespread than previously suggested. The brain mapping of nuclei activated by NTG and inhibited by the pretreatment with nimesulide suggests that this NSAID acts, directly or indirectly, on several structures located in the CNS. The ventrolateral column of the periaqueductal grey plays an important role in the control of nociception and in the coupling of pain perception with autonomic response. The nucleus tractus solitarius, area postrema and supraoptic nucleus are deeply involved in the control of autonomic function. The locus coeruleus plays a pivotal role in the integration of autonomic and nociceptive function. Surprisingly, no significant inhibition of NTG-induced Fos expression was observed in the nucleus trigeminalis caudalis, a nucleus with a primary nociceptive function found, in previous experiments [269], to be inhibited by another NSAID, indomethacin. Taken together, these data on the effect of nimesulide on NTG-induced *Fos*-activation in the CNS support a role of supraspinal mechanisms in the analgesic effect of nimesulide. With the exception of the supraoptic nucleus, all the brain nuclei inhibited by nimesulide pretreatment receive a rich serotonergic innervation, which suggests that nimesulide may, at least partly, owe its analgesic effect to the interaction with the central serotonergic system, in line with what has been demonstrated for other simple analgesics [288, 289].

Experimental studies in humans

The nociceptive flexion reflex (RIII reflex) is a polysynaptic spinal reflex that can be elicited through electrical stimulation of the sural nerve and recorded via the flexor biceps femoris muscle. Several studies have shown a close correlation between the subjective pain sensation and the threshold and amplitude of the responses recorded, making the RIII reflex a useful model both in the neurophysiological investigation of pain and in evaluation of the effects on pain transmission of several compounds capable of modulating nociceptive activity [290, 291]. Although the nociceptive reflex response depends on the excitability of the spinal reflex arc, it is strongly modulated by multiple and remote supraspinal and midbrain areas.

From the perspective of objective electrophysiological testing, the RIII reflex is the most interesting parameter given that it provides not only a means of measuring subjective pain, but also information on the functional organisation of the pain control system and on the neuronal state of the spinal and supraspinal structures, which are the possible sites of pharmacological analgesic actions.

A large number of experimental studies have investigated the pharmacological modulation of the RIII reflex, documenting opiatergic control at spinal and supraspinal levels, and aminergic control at supraspinal level [292–294].

Inhibition of the RIII reflex occurs with some NSAIDs (ketoprofen, ibuprofen, and indomethacin) and analgesics such as acetaminophen, tramadol, nefopam and ketamine, documenting a central analgesic activity of these compounds. More recently studies were undertaken of the RIII reflex as an electrophysiological method for assessing the analgesic effects of nimesulide, as a preferential COX-2 inhibitor, in an attempt to elucidate further the possible central mechanisms of this drug during NO-induced hyperalgesia [295, 296]. The study was a double-blind, placebo-controlled, crossover trial in which each subject randomly underwent treatment with nimesulide 100 mg *per os* or placebo in two different sessions, separated by an interval of at least 4 days. In each session an NO-donor, NTG (0.9 mg sublingual), was administered 140 min after nimesulide or placebo. The RIII reflex responses were obtained at 1.5-fold the RIII thresholds, assessed before and at 15, 30, 60, 90, 120 min after nimesulide/placebo administration. The responses recorded were amplified, digitised and full-wave rectified and integrated, after which the area of each reflex response was calculated as percentage change from baseline. Given the linear correlation between intensity of the stimulus (i) and the area of the RIII reflex (A), we took the ratio between the area and the square of the stimulus intensity (A/i^2) as the index for monitoring the neurophysiological effects of the active drug and placebo.

The results (Fig. 15) showed that the A/i^2 ratio decreased in both the groups, but whereas the change was quick and marked in the nimesulide group, reaching statistical significance *versus* basal value at as early as 15 min and persisting for up to 2 h (ANOVA for repeated measures $p = 0.0052$), in the placebo group the

A/i^2 ratio reduction never reached a significant level and had disappeared by 120 min post-administration (Fig. 15, upper panel). Comparison of the groups revealed statistically significant differences at 15 and 120 min.

Following NTG administration, the *A/i*^{2} ratio reduction was persistent and statistically significant at each of the time points in the nimesulide group, whereas in the placebo group the *A/i*^{2} ratio showed a progressive increase, statistically significant 60 min after NTG administration. Comparison of the groups showed statistically significant differences at 15, 60, 90 and 120 min after NTG administration (Fig. 15, lower panel). These data revealed a significant inhibitory effect of nimesulide on the RIII reflex, expressed in terms of reduced RIII reflex area and/or increased RIII reflex threshold. In line with what has been demonstrated in relation to other NSAIDs in human and animals studies [246, 293, 297], we can speculate that the analgesic effect of nimesulide depends upon central (spinal/supraspinal) mechanisms.

Since significant COX-2 expression is found in CNS, particularly at spinal cord level, where it seems to play an important role in nociceptive transmission, it would appear that, as suggested by previous animal data, nimesulide exerts its analgesic activity, partially at least, via central inhibition of COX-2 , probably at spinal cord level.

In the present model, the progressive increase of the *A/i*^{2} ratio after the administration of NTG in the placebo group confirms previous studies showing a hyperalgesic action of NTG and suggests a sustained sensitisation phenomenon induced by NTG-derived NO, in accordance with the observation that NO is involved in several potential pro-nociceptive mechanisms. The effectiveness of nimesulide in counteracting NO-mediated hyperalgesia seems to suggest that COX-2 inhibition is a step that restricts NO-mediated hyperalgesic mechanisms. The interactions between NO and COX-2 in the CNS are not fully known. NMDA receptor activation in inflammation-induced mechanical allodynia has been shown to interfere with both NO and COX pathways. The intracellular cascade of molecular events initiated by glutamate release from nociceptive afferents and NMDA receptor activation includes the release of a number of intracellular second messengers such as NO and PGs and the overexpression of COX-2 by a calcium-dependent mechanism (Fig. 13).

Conclusions

The mechanisms involved in nimesulide-mediated analgesia involve both central and peripheral events and are not restricted to the inhibition of cyclooxygenase activity. Other mediators are likely to be involved in this analgesia, e.g., NMDA glutamate receptor activity modulation and NOS inhibition, although further investigation is needed in order to quantify the relative importance and the exact site and characteristics of their role in its dynamics.

Actions on joint destruction in arthritis

Joint destruction and effects of NSAIDs

Much interest has been shown in the past 2–3 decades on the actions of NSAIDs on joint destructive processes in arthritic diseases, especially in OA and RA [298–311]. The issue has been debated whether or not NSAIDs should be used in the treatment of OA because of claims that some of these drugs may accelerate cartilage or bone destruction in this condition [300, 302, 308]. Some have proposed that it is only pain relief that is needed in the treatment of osteoarthritis and not control of inflammation that paracetamol and other analgesics (including weak opioids) should be employed at least as first-line agents since these give adequate pain relief without having the risk of joint destructive changes thought to occur with some NSAIDs [300, 302–306, 308, 311] (see also Chapter 5). The situation is complicated because (a) analgesic activity which is provided by all these drugs, distinct from anti-inflammatory effects *per se* of the NSAIDs, may contribute to "over-use" of already degrading joints in OA [300, 304–306, 308, 312, 313], (b) that in contrast to the potential for over use exercise and physical activity are considered to promote musculoskeletal strengthening and enhanced vascular perfusion of joints that can override local destructive changes and have benefit in joint strengthening [312], (c) only a few NSAIDs have been shown in long-term well controlled trials to have adequate radiological evidence of either joint destructive changes or reduction in progression of destruction [302, 303, 314], (d) even so there are issues concerning the need for more sensitive radiological or magnetic resonance imaging (MRI) techniques to determine the anatomic locations where joint destruction is occurring in OA and the responses to therapy [309, 315–317], and (e) biochemical analysis has yet to reveal whether these drugs change the progression of joint changes in OA [316, 317]. So overall there are serious questions whether there is sufficient evidence to say whether the actions of all these drugs in promoting or protecting against joint changes in OA.

The idea that NSAIDs may be harmful to the joints of patients with OA originated from some key observations, (a) that there may be a condition which was originally described as the "analgesic hip" by the eminent radiologist, Ronald Murray in 1971, based on radiological observations of degeneration of patients with OA of the hip who had received long-term treatment with indomethacin (originally it was thought that corticosteroids may have contributed to this condition but later review of the cases highlighted indomethacin) [318, 319] and this was confirmed by Coke and others [320–322], (b) a condition described by Serup and Ovesen in 1981 [323] as "salicylate-arthropathy" which arose from observations in a case report of an 87-year old women who had been on long-term aspirin and dextropropoxyphene as well as having taken indomethacin (so highlighting a misnomer where a drug associated condition can be attributed to more than one causative agent!), and (c) in a long-term (≥1 year) study where patients with OA of the hip who were to undergo hip arthroplasty received either indomethacin (a potent prostaglandin synthesis [PG] inhibitor), or azapropazone (a weak PG synthesis inhibitor), the results of which showed that indomethacin produced significantly greater joint destruction observed radiologically and with evidence of greater bone destruction at operation [302, 303, 324]. Coinciding with these observations, the synovial tissues from patients that received indomethacin had lower levels of PGs and cartilage with lower proteoglycan (PrGn) concentrations than in those patients that received azapropazone, so implying that reduction in joint PGs may be somehow related to reduced levels of PrGns [302, 303].

Regulation by eicosanoids of cartilage–synovial–leucocyte interactions

The suggestion has been made recently that since COX-2-derived PGE_2 may underlie the inflammatory changes in joints that contribute to the destructive changes in cartilage in OA [314, 325–327]. COX-2 selective drugs (e.g., celecoxib, SC-236) appear to reverse cytokine-induced cartilage proteoglycan (PrGn) degradation and inhibition of PrGn synthesis [325–327]. Celecoxib has also been reported to counteract the depletion in hyaluronan concentration and to increase its synthesis; effects which are not observed with diclofenac [327].

In view of these observations concerning the roles of COX-2 derived PGE_2 in PrGn metabolism it is useful to review the roles that eicosanoids have in the regulation of matrix metabolism in cartilage and in bone functions.

Chondrocytes produce COX-2-derived PGE_2 and have EP_1, EP_2 and EP_4 receptors for PGE_2 [328–330]. PGE_2 production is increased by IL-1 in chondrocytes by induction of COX-2 along with induction of nitric oxide synthase (NOS-II or iNOS) and increase in NO production [331, 332]. NO can amplify the production of PGE_2 in chondrocytes of PGE_2 thus indicating that there is NO–COX-2 "cross-talk" in regulation of inflammatory mediator production [333]. COX-2 is also a regulator of IL-6 production by chondrocytes [334], this effect on IL-6 production could influence acute phase protein production in the liver. IL-15 may prime TNFα production, especially in RA, that in turn drives production of IL-1, and both IL-1 and TNFα increase production of metalloproteinases [335]. The involvement of mast cells in these events is seen to be central especially in RA, but also this may be significant in OA [335].

While IL-1 and other cytokines increase chondrocyte production of LTB_4 and LTC_4 [336, 337] via increased activity of PLA_2 [336], these cytokines reduce the synthesis of 5-lipoxygenase (5-LOX) [337]. The increased production of LTs may be a consequence of release of arachidonic acid via the transcellular movement of precursors from granulocytes in contact with chondrocytes [338] thus providing more substrate for the 5-lipoxygenase enzyme; the increased activity of this enzyme via intracellular translocation mechanisms being mediated by calcium [339], with production of anti-inflammatory lipoxins [340].

Against these proinflammatory changes are regulatory responses to the inflammatory reactions which results from (a) negative feedback by PGE_2 produced from IL-1 stimulation of the PLA_2 and COX-2 enzymes (which is probably a response mediated by increased cyclic AMP) [341–343], (b) the co-induction of COX-2 and NOS-II/iNOS [344–346], the stimulation of NO production by PGE_2 and LTB_4 [347], and the reduction of COX-2 activity by nitric oxide produced following induction of NOS-II/iNOS [344], (c) negative control of the production of IL-1 and TNFα by PGE_2 against which there is enhanced production of these cytokines by LTB_4, and negative controls of the T-cell mediated reactions by anti-inflammatory cytokines (IL-4, IL-10, IL-15, etc.) [347–352]. The extent of involvement of T-cells in mediating osteoarthritis is probably restricted to localised inflammatory reactions in inflamed joints probably "primed" largely by fragments or decomposition products of cartilage and bone (apatite) components that initiate localised inflammation [353]. This is, in contrast, to the more extensive systemic immuno-inflammatory reactions in rheumatoid arthritis, systemic lupus erythematosus, ankylosing spondylitis and other related conditions [353–355].

Localised joint destruction in OA involve (a) inflammatory reactions in synovial tissues that may be initiated by cartilage and bone fragments and decomposition products constituting neoantigens [356], (b) regional vascular ischaemia leading to production of tissue destructive oxyradicals and promoted by local arteriosclerosis in which there is restricted blood flow to both synovial tissues and sub-chondral bone and vascular inflammation [357, 358] which may also be initiated by "foam" cells or activated macrophages in the vascular wall, (c) infiltration and activation of neutrophils (see section on "Nimesulide and neutrophil functional responses", p. 173) with accompanying complement activation [233], (d) destruction and degradation of cartilage and involvement of associated subchondrial cells and bone driven by synovial–leucocyte interactions, the production of metalloproteinases and other proteases stimulated by proinflammatory cytokines [358, 359], (e) inhibition of the synthesis of proteoglycan and collagen components of cartilage by IL-1 and TNFα by cytokines [359, 360], (f) regional T- and B-cell activation leading to further promotion of the immune-based reaction in the OA synarthrodal joints [361–363], (g) changes in the osteoblasts and osteoclasts of subchondral bone leading to bone lysis [364] and (h) alteration in the production of growth factors some of which may be important in cartilage and bone repair [365].

Similar increases in eicosanoids initiated by IL-4, TNFα, γ-IFN and other proinflammatory cytokines occur with synovial cells, the synovial A-cell and infiltrated and activated macrophages [351] as well as from polymorphonuclear neutrophil leucocytes (see also early section on "Nimesulide and neutrophil functional responses", p. 173). The proportion of the principal COX-2 and 5-, 12- and 15-

LOX products (PGE_2, LTB_4, LTC_4 and lipoxins) that mediate the major part of synovial–cartilage inflammatory changes that are produced in the articular cartilage and synovial capsule will depend on the extent of the cytokine- and other inflammogenic stimuli, and the type and extent of T-cell immune stimulus in the inflamed synovium, the production of $sPLA_2$, COX-2 and 5-LOX enzymes in different cells in the synarthrodal region.

Diagrammatic representations of the inflammatory events in the synarthrodal joints in OA can be seen in Figures 16–18 showing:

- The anatomic changes in the synovial capsule featuring synovitis and hyperplasia, ischaemia and changes in subchondral bone and cartilage in synarthrodal joints in OA (Fig. 16).
- The cellular events in the cartilage–synovial–leucocyte interactions involving production of inflammatory mediators and oxyradicals, destruction of matrix macromolecules by activated or cytokine-mediated synthesis of metalloproteinases and the underplay of cytokine-eicosanoid interactions that mediate these destructive events (Figs 17 and 18). These figures also show the principal sites of action of nimesulide (NIM), the details of which will follow later.

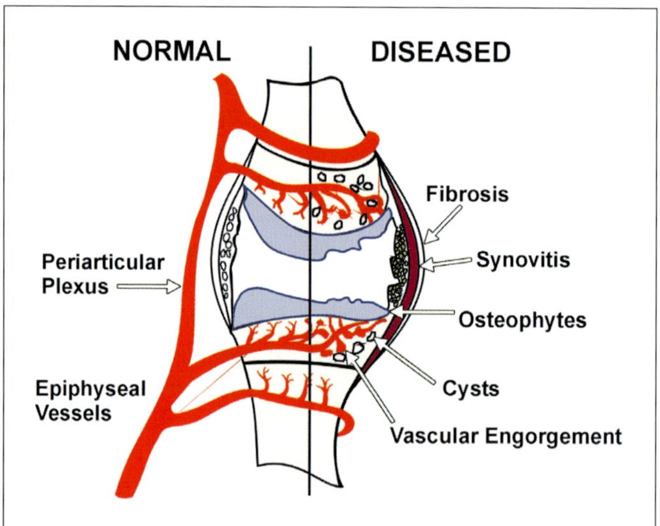

Figure 16
Synarthrodal joint showing areas of cartilage and bone destruction in arthritis and the associated involvement of synovitis and vascular changes. From Burkhardt and Ghosh (1987) [299]. Redrawn and reproduced with permission of the publishers of Seminars in Arthritis and Rheumatism.

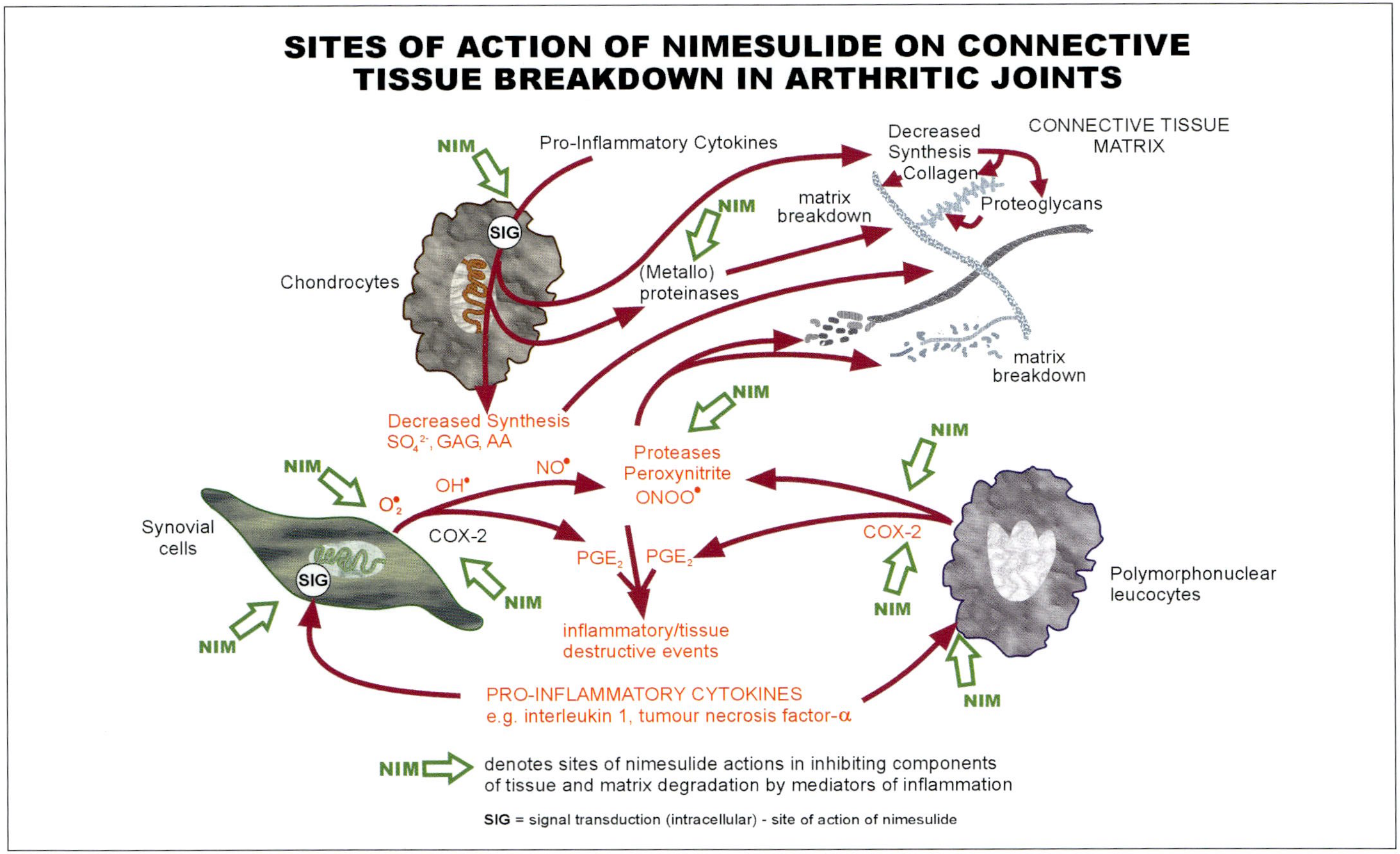

Figure 17
Cellular destructive changes and sites of action of nimesulide in connective tissue breakdown in arthritic joints. K. D. Rainsford©.

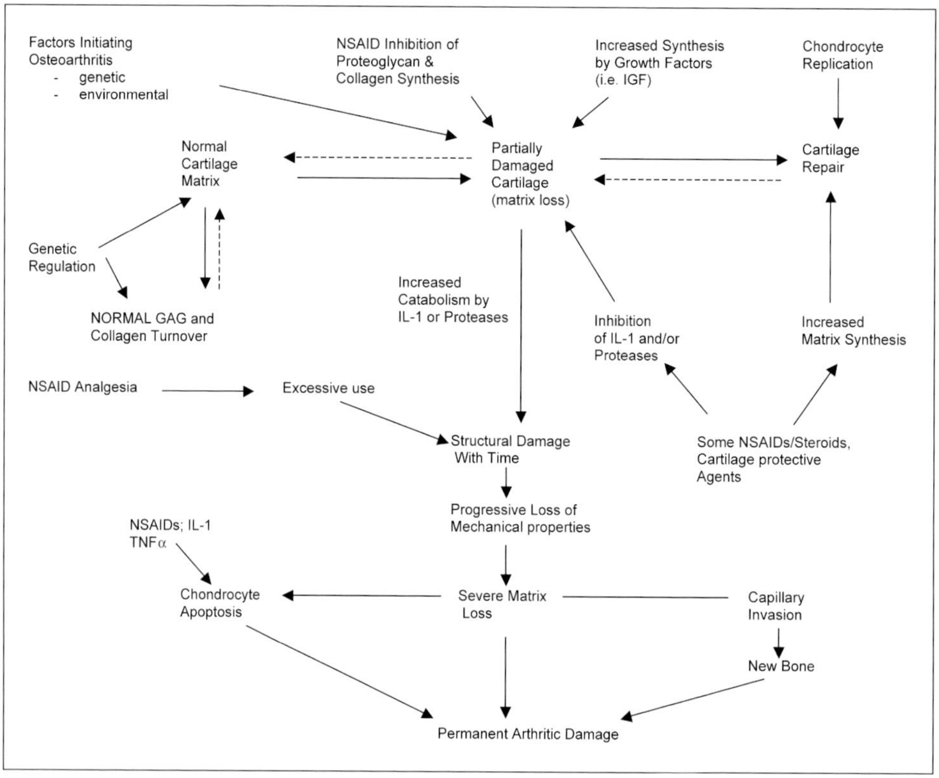

Figure 18
Effects of NSAIDs on cartilage matrix degradation in osteoarthritis.

- The molecular events involved in the destruction of cartilage proteoglycan and collagen, the influences of cytokines, environmental factors and NSAIDs on these processes (Fig. 18). It is important to note that not all NSAIDs act in the same way on these molecular components of cartilage degradation in OA; some may promote degradative changes (indomethacin), others have little or no effects (azapropazone, naproxen) and others have even claimed to protect against cartilage destruction in OA [302, 303, 366–390].

In vivo effects of nimesulide on cartilage and bone in experimental model systems

The injection of cell wall particles of heat-killed *Mycobacterium tuberculosis* (in the form of Freund's complete adjuvant), into the stifle joints of dogs or in rats has to be found to be a potent elicitor of joint inflammatory reactions and consequent cartilage and bone erosions. Botrel and co-workers [391] investigated the reactions to Freund's complete adjuvant injected into the stifle joints of dogs and established that nimesulide had protective effects against bone and cartilage destruction in this model of periarthritis.

In a shorter-term model using the same adjuvant treatment in rats, Gilroy and co-workers [392] injected 200 µg of mycobacterial adjuvant intra-articularly into the left stifle joint. This is a relatively large dose of this inflammogen and although given in sterile saline there is no indication from the description of the methods if they removed endotoxin and other lipid contaminants of the mycobacteria or sterilised the suspension before injection. The combined effects of the high dose (given to young rats) and potential for endotoxin and other lipids eliciting inflammatory reactions raises the possibility that the joint destruction may be very severe in the rats injected with the inflammogen.

These authors found that 0.5 mg/kg nimesulide given for 4 days did not cause any loss of GAGs from the dissected patella [392]. In contrast, piroxicam 10 mg/kg/d exacerbated the cartilage GAG-loss from the patella to the extent of about 25% above that in control animals. A high dose of 5 mg/kg/d nimesulide given for the same period produced a statistically significant increase in GAG loss but this was slight (about 8%) in comparison with the overall effects of the control group and the effects observed with piroxicam. Both drugs also caused reduction in joint swelling and local leucocyte inflammatory reactions. Neither nimesulide nor piroxicam had any effects on patella bone loss. The experimental COX-2 inhibitor, NS398 1 and 10 mg/kg/d did not cause any effects on cartilage or bone even though there was reduction in joint swelling.

These results present rather inconclusive information about the influence of nimesulide on processes underlying cartilage erosion and bone loss. The technical issue about the possibility of severe local inflammation from the combination of the high dose of mycobacterial adjuvant and the possibility of endotoxins and other lipids contributing to the joint inflammation is one aspect. Likewise, factors such as the timing of the induction of the joint disease and the lack of adequate dose-response data for effects of nimesulide and comparator NSAIDs in this model are important considerations.

Actions of nimesulide on cartilage degradation *in vitro*

When analysed in relation to the events involved in cartilage degradation in OA (Figs 17–19) nimesulide has the potential to act on a considerable number of these with the potential to at least have no effects on the promotion of cartilage destruction and possibly to even prevent such changes. In the absence of any definitive evidence of preventative events of nimesulide in animal models of OA or

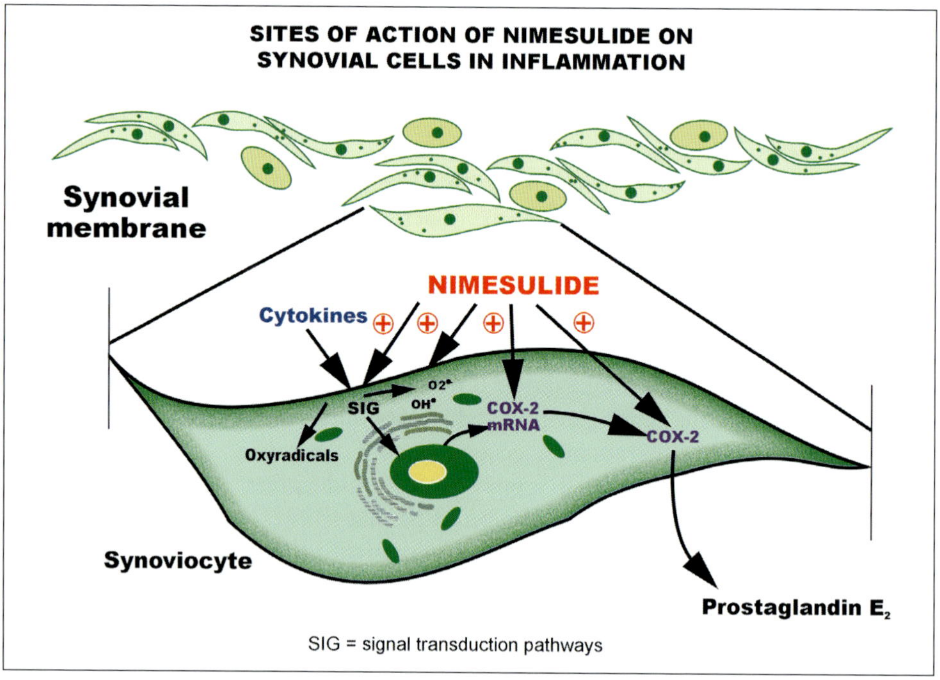

Figure 19
Multifactorial actions of nimesulide on the pathways leading to oxyradical production, intracellular signalling and the expression of cyclooxygenase-2 mRNA expression, protein synthesis and activities in human synovial fibroblasts. Cytokines (e.g., interleukin-1) increase production of oxyradicals (OH•; $O_2^{\bullet}$), activation of signalling pathways (e.g., NKκB-IκB) leading to stimulation of the production of COX-2 and metalloproteinases. Modified from Figure 1 of Di Battista et al. (2001) [404]. Permission obtained from the publishers of Clinical and Experimental Rheumatology *for use of the original figure.*

in the principal joints of patients with OA, the data reviewed and analysed below are indicative or suggestive evidence for the potential of nimesulide to control degenerative events at the molecular and cellular level of cartilage degradation in OA.

Issues, which are important to consider regarding the actions of nimesulide in controlling the molecular and cellular changes in the cartilage of patients with OA, are:

- The concentration ranges at which the drug is shown to act are within those encountered in therapy, at least the plasma concentrations (circa 20 μmol/L [6]), preferably those in synovial fluid or tissues or those in cartilage [366].

- The variable influence of pro- and anti-inflammatory cytokines and other mediators (e.g., lipoxins) which may act to limit drug effectiveness in preventing cartilage destructive changes (e.g., from high levels of IL-1, TNFα and IL-6) or variations in the effect of nimesulide to promote protection (e.g., from inhibition of the production of proinflammatory cytokines, metalloproteases or oxyradicals).
- Physical effects (joint loading, exercise, excessive, abnormal or unwarranted physical activities) may contribute to an over-use syndrome. The painful effects are masked by the analgesic effects of nimesulide or for that matter by any other analgesic or NSAID.
- Physiological or environmental effects that may influence joint function and use by OA patients.

Considering the molecular and cellular changes in OA cartilage that may be affected in cartilage, synovium and infiltrating leucocytes by nimesulide it is possible that this drug or its major 4′-hydroxy-metabolite can have multiple effects in controlling the joint destructive process in OA and other arthritic conditions. Among these effects [302, 366, 394–417] are the inhibition of:

- Production of IL-1 and TNFα
- Generation and actions of oxyradicals
- Ingress and activation of neutrophils and monocytes/macrophages into inflamed synovial tissues, synovial fluids or partially degraded cartilage, and subsequent activation to produce inflammatory mediators.
- COX-2-derived PGE_2, NOS-II (iNOS)-derived NO, LTB_4, complement products and platelet activation products (PAF) that are involved in the expression of inflammatory reactions.
- Metalloprotease and leucocyte-derived proteases and oxyradicals
- Initiation of apoptosis (programmed-cell-death) of chondrocytes and possible changes in their activation and morphology during OA.

Uptake of nimesulide into synovial tissues, synovial tissues and cartilage

Before considering any effects of nimesulide on cartilage destruction in OA it is important to consider at what concentrations the drug and the 4′-OH metabolite are present in these cellular compartments. As reviewed in Chapter 2 plasma concentrations of nimesulide and its 4′-hydroxy-metabolite observed during therapy with the standard dose of 100 mg of the drug are in the range of 6 µg/mL (20 µmo/L) and 1.5 µg/mL (4 µmol/L), respectively [6]. Synovial fluid concentrations in patients with arthritic conditions (principally RA) amount to about one-half to one-third of these values. Free concentrations of nimesulide in plasma during therapy are only likely to be about 1–4% those in the plasma and so are too low to be meaningful except in the more sensitive ranges for pharmacological effects (e.g., COX-2 inhibition) (Chapter 2). It is possible that the drug concentrations in inflamed sites are more likely to exceed the free plasma concentrations, and so to be pharmacologically-relevant in relation to joint destructive effects, as observed with other NSAIDs [384]. Cartilage uptake of nimesulide is about 1.4% of that in the media in the presence or absence of IL-1 and TNFα [366].

Production of PGE_2, cytokines and proteoglycans *in vitro*

Early studies by Pelletier and Martel-Pelletier [393] showed that nimesulide, like naproxen, reduced IL-1 induced proteoglycan degradation and the production of the cartilage-destructive, stromelysin (MMP-3) enzyme in human OA cartilage. Collagen synthesis was, however, affected by nimesulide. The exact meaning of the latter observation was not clear from these studies.

Henrotin et al. [378] observed that production by isolated human articular chondrocytes of proteoglycans (PrGns) was unaffected by 3 µg/mL (100 µmol/L) of nimesulide, a concentration which was within that in the synovial fluid during therapy with 100 mg/d of the drug. At 6 µg/mL (200 µmol/L) nimesulide there was inhibition of PrGn production in some, but not all the chondrocytes from RA patients. It is important to note that the chondrocytes used in this study were derived from patients with RA and not OA. The heavily diseased cartilage samples used in these studies may have an important bearing on the outcome in these studies.

Sanchez et al. [394] observed that nimesulide, like that of some other NSAIDs (aceclofenac, celecoxib, diclofenac sodium, ibuprofen, indomethacin, piroxicam and rofecoxib), added to chondrocytes of isolated human osteoarthritic cartilage cultured in alginate beeds all inhibited PGE_2 production but they had variable effects on basal and IL-1β stimulated IL-6 and IL-8, matrix metalloproteinase III (stromelysin) and aggrecan production. Nimesulide, like aceclofenac, diclofenac and indomethacin reduced basal and IL-1β stimulated IL-6 production, while celecoxib inhibited IL-1β-induced IL-6, and piroxicam and rofecoxib were without effects on the production of IL-6. Aside from aceclofenac and indomethacin, the other drugs had no effects on aggrecan production.

The responses to NSAIDs and cytokines in isolated chondrocytes differ from those in cartilage explants in organ culture [395]. The latter are to some extent a more realistic representation of the *in situ* responses to NSAIDs given to patients with inflamed joints. There are, however, some technical difficulties in employing cartilage explants for measuring production of inflammatory mediators (especially NO) because of the matrix in the cartilage explants limiting diffusion of these small molecules into the culture media. Hence, chondrocytes are preferable for these types of studies [366]. It should also be noted, however, that there may be loss of expression of surface receptors and changes in phenotype to dedifferentiated states (e.g., formation of "fibroblast-like" cells) with chondrocytes in alginate beads or other matrices are considered to retain to an extent most of the characteristics of their original phenotype.

Rainsford et al. [366] employed porcine bovine articular cartilage to examine the effects of nimesulide 1.0–100 µmol/L on cartilage proteoglycan (PrGn) destruction induced by IL-1, TNFα and the combination of these two cytokines. At low concentrations of nimesulide there were no changes but at high concentrations (50–100 µmol/l) of nimesulide, there was inhibition of PrGn degradation induced by the cytokines. Using isolated bovine chondrocytes these authors observed that there was inhibition by nimesulide over a wide concentration range of NO production induced by the cytokines alone or as a mixture. Since NO is thought to be a stimulus to inflammatory events leading to metalloprotease or other enzyme-induced PrGn destruction [396–398] it appears that the inhibitory effects of nimesulide on NO production may, in part, account for the reduction in PrGn degradation observed in these studies.

Pelletier and Martel-Pelletier [393] showed that nimesulide can inhibit stromelysin (metalloprotease-3, or MMP-3). As noted in the section on "Nimesulide and neutrophil functional responses" (p. 173) neutrophil-dependent inflammation enzymes are involved in tissue destruction and neutrophils may also be affected by nimesulide (although the ultimate effects on PrGn integrity have not been established). Studies by Barracchini et al. [399] found that nimesulide along with some other NSAIDs (meclofenamate sodium, meloxicam, piroxicam, sulindac and tolmelin) inhibited the isolated collagenase enzyme with IC_{50} values of 1.9–28 mol/l and values of apparent inhibition constants, K_i, of 0.83–21.8 mol/l. The effects were reversible as shown by restoration of enzyme activity upon dialysis.

Ex vivo studies on regulation of metalloproteinases in patients with OA

Preliminary studies by Bevilacqua compared the effects of administration for 28 days of nimesulide 200 mg/d with ibuprofen 1,200 mg/d on serum levels of metalloproteinase-3 (stromelysin-1) in patients with OA [400]. They also measured levels of the tissue inhibitor of MMPs, (TIMP), hyaluronan (HA) and YKL-40 (Chondrex), a biomarker of joint disease [400]. In 22 patients that received nimesulide there were statistically significant changes in serum concentrations of MMP-3 and HA but not TIMP-1 or YKL-40 before and after treatment. Of the 27 patients that had ibuprofen, statistically significant changes were observed in serum MMP-3 but not in any of the other parameters. At the end of 28 days treatment with the two drugs there were no significant differences in the four parameters between the drugs. There was high variability in the data and further studies are indicated with larger numbers of patients in each of the treatment groups and a placebo group should be included.

In another pilot study, Kullich and co-workers [401] determined serum levels of MMP-1, MMP-3, MMP-8 and cartilage oligomeric protein (COMP) in 20 patients with OA that received 100 mg b.i.d. nimesulide for 3 weeks and compared these results with a control group of 22 healthy subjects that were without pain. No placebo group was included in this study. Compared with baseline values (before drug treatment) there were statistically significant changes in the serum levels of MMP-3, MMP-8 and COMP but not of MMP-1 after 3 weeks treatment with nimesulide.

Taken together these preliminary studies in OA patients [400, 401] suggest that MMP-3 levels may be reduced by nimesulide. Thus, in addition to direct inhibition of MMP-3 enzyme activity nimesulide may inhibit the expression of this key cartilage destructive enzyme. The possibility that nimesulide influences the cytokine mediated induction of MMP-3 in an analogous way to that involving expression of COX-2 via inhibition of intracellular signalling pathways (e.g., NFκB/IκB) (Fig. 19) is worthy of future investigation in chondrocytes derived from patients with OA.

Glucocorticoid receptor activation and other signalling pathways

Nimesulide has been shown to reduce the synthesis of urokinase (uPA), an endogenous tissue plasminogen activator inhibitor (PAI), as well as PAI-1, and IL-6 in human synovial fibroblasts isolated from OA patients [402]. Based on these observations and the inhibitory effects of nimesulide on MMP synthesis, Pelletier and Di Battista et al. [403, 404] investigated the possibility that nimesulide may affect components of the glucocorticoid receptor system and that this might contribute to its anti-inflammatory activity. Using fibroblasts from synovial membranes derived at necropsy from donors without arthritic disease, the effects were investigated of nimesulide 0.3–30 µg/mL compared with that of naproxen 30 µg/mL and dexamethasone 0.01–1.0 µg/mL on the number of glucocorticoid binding sites on fibroblasts. These results showed that nimesulide did not affect the number of the glucocorticoid receptors, whereas naproxen and dexamethasone did. Internal redistribution of the glucocorticoid receptor levels was unaffected by nimesulide or naproxen whereas dexamethasone reduced the levels of immunoreactive glucocorticoid receptor and its mRNA. To further extend the possibility of glucocorticoid receptor effects of the NSAIDs, the effects were explored of these drugs on the phosphorylation reactions of the receptor. It was found that nimesulide 30 µg/mL increased the p44/42 mitogen activated protein kinase (MAPK) phosphorylation, although there was an initial slight decrease in phosphorylation. Interestingly, nimesulide caused hyper-phosphorylation of glucocorticoid receptors in a concentration-dependent manner whereas naproxen was without effects on this and MAPK phosphorylation [404, 405].

Using electrophoretic gel shift assays it was found that nimesulide 0.3–30 µg/mL caused a concentration- and time-dependent binding of nuclear protein extracts to a ^{32}P-labelled glucocorticoid receptor element (GRE) consensus sequence. In contrast, naproxen 90 µg/mL had no effects. The possibility that nimesulide affected the transcription factors, NF-1 and OCT-1, interacting with the glucocorticoid receptor promoter was ruled out in control experiments. Likewise, GRE associated effects on induction of COX-2 was ruled out as a site of action of nimesulide. Using a transfected promoter, it was found that there was an increased induction of glucocorticoid promoter activity. These and other studies suggest that the effects of nimesulide on the glucocorticoid receptor system may be unique to this NSAID. It may explain the effects of nimesulide on glucocorticoid target genes, e.g., MMPs.

A diagrammatic representation of the postulated effects of nimesulide on the glucocorticoid system is shown in Figure 20 (from [403]). This shows that the intracellular phosphorylation of the glucocorticoid receptor leads to its dissociation from heat shock protein binding. Subsequent translocation of the receptor leads to its binding to nuclear GRE components that are responsible for the corticosteroid like inhibition of transcription of MMP, COX-2 and iNOS proteins.

The importance of human synovial mast cell reactions that underlie the onset and promotion of synoviocytes and inflammatory reactions in arthritides as loci for the actions of nimesulide were investigated by de Paulis and colleagues [406]. Histamine release from basophils and mast cells has been found to be inhibited by therapeutic concentrations of nimesulide in response to IgE or protein kinase stimulation but not calcium ionophore [407]. 4-hydroxynimesulide also cause the same inhibitory effects as the parent drug. Likewise, de Paulis et al. [406] found that Anti-IgE-mediated histamine release from human synovial fibroblasts was observed with 1.0–100 µmol/L nimesulide. Less potent inhibition was observed with the same concentration of diclofenac; neither piroxicam nor aspirin had any effects. Prostaglandin D_2 (PGD_2) released by IgE from the human synoviocytes was inhibited by 0.1–10 µmol/L nimesulide, as well as by piroxicam 0.1–10 µmol/L, diclofenac 0.01–10 µmol/L, and 10–100 µmol/L aspirin. Nimesulide 1.0–10 µmol/L also inhibited the release of LTC_4 from human synovial fibroblasts treated with Anti-IgE but the other drugs were without effects. Tryptase release was also inhibited by nimesulide but at a lower concentration (10.0–100 µmol/L) than required for release of PGD_2, LTC_4 or histamine [407]. These results suggest a major role for mast cells in mediator release and this may have significance in the synovial inflammation in OA and other arthritides.

Chondrocyte programmed cell death, or apoptosis, is a major event in cartilage degradation in OA as well as age-related changes including reduction of in tissue cellularity and matrix decline [408–413]. The production of nitric oxide, PGE_2 and interleukin-1 has been linked especially to this process [408, 411, 412]. Ageing also plays a role in increasing susceptibility of chondrocyte apoptosis [413]. Using a rat chondrogenic cell line, RCJ3.IC5.18, stimulated with stau-

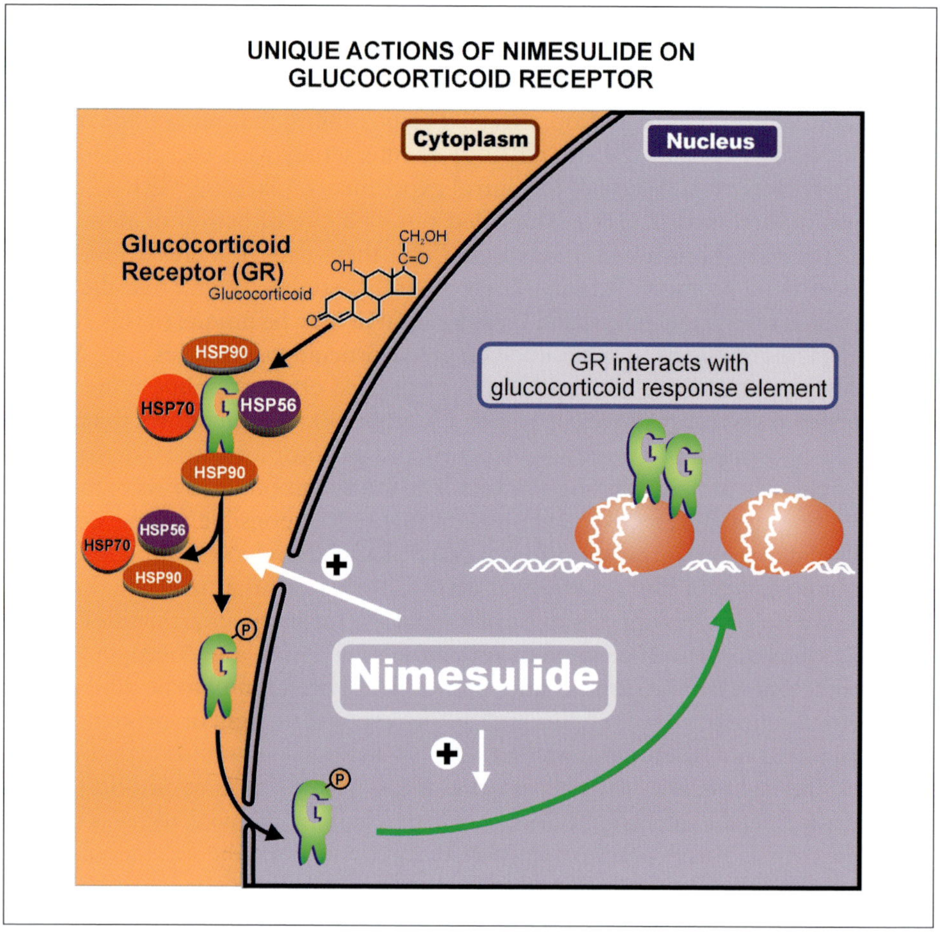

Figure 20
Stimulation by nimesulide of the intracellular phosphorylation pathways leading to activation of the glucocorticoid receptor and binding to Glucocorticoid Response Element (GRE) with subsequent blockade of the activation of genes controlling production of COX-2, metalloproteinases and cytokines that initiate or control inflammatory reactions. From Pelletier et al. (1999) [403]. Reproduced with permission of the publishers of Rheumatology.

rosporine to induce cell death, it was found that nimesulide 1 pmol/L–10 nmol/L and the same concentrations of ibuprofen reduced the loss of cell viability [414] which appeared to be about 20% of controls. The selective COX-2 inhibitor, NS-398, was without effect. Staurosporine-mediated caspase-3 activation was significantly reduced by nimesulide 1–10 nmol/L and ibuprofen 1–10 nmol/L coincident with improved morphology of the cells (Fig. 21) [414].

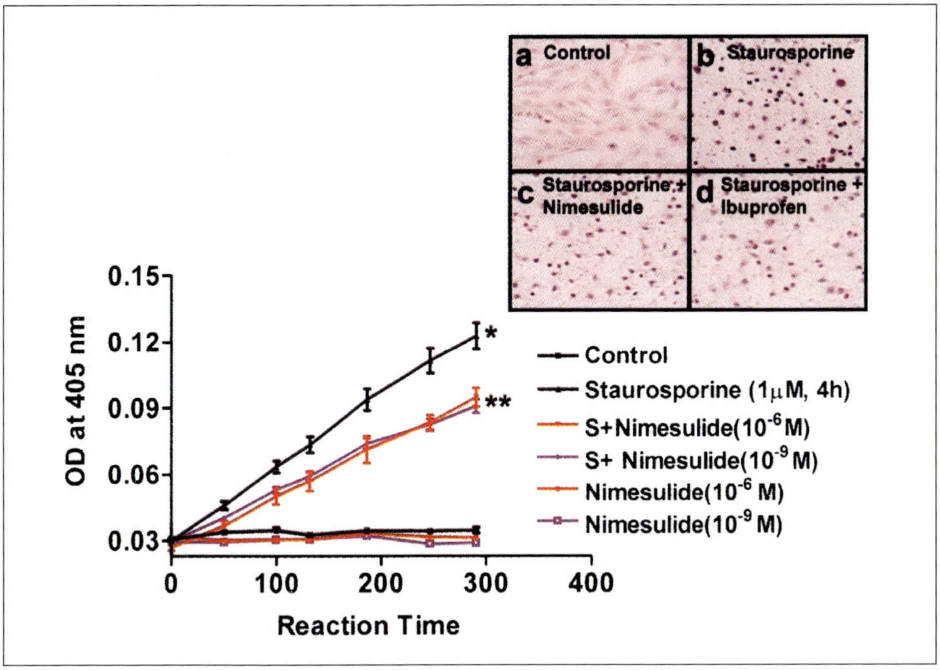

Figure 21
Action of nimesulide in controlling staurosporine (1.0 µmol/L)-induced caspase-3 induction with consequent reduction in morphological features of chondrocyte apoptosis (as observed in haematoxylin and eosin stained chondrocyte cultures shown in the insert). Caspase-3 activity was measured at 3 h after treatment with the drugs from the optical density (OD) changes (at 405 nm) occurring following hydrolysis by the caspase enzyme of the p-nitroaniline from the substrate. From [414]. Reproduced with permission of the publishers of Clinical and Experimental Rheumatology.

Human TC28a chondrocytes were employed by Mukherjee and Pasinetti [415] to investigate staurosporine-mediated cell death in a human chondrocyte cell system in which cDNA microarrays was used to screen gene expression, and responses to nimesulide. In these studies 12 genes were identified to be altered in expression in response to staurosporine-mediated cell death. Most of the genes involved coding for S15a, S27 and other ribosomal proteins and those controlling cell proliferating activity. In addition to reducing staurosporine cell toxicity, nimesulide 1 µmol/L prevented induction of these candidate genes [415]. Further studies are required to explore the concentration and time dependence of these effects and to establish if they have any significance in cartilage explants in culture or in cartilage biopsies or tissue taken at operation from patients on long-term therapy with nimesulide and comparator drugs.

Oxidant stress injury, peroxynitrite, cell injury and lipid peroxidation

The effects of nimesulide in protecting against apoptosis may have several underlying mechanisms. It is also possible that the apparent cell death that is seen in chondrocytes of patients with osteoarthritis and other rheumatic conditions may not necessarily involve programmed cell death as is seen in apoptosis at actual necrotic changes. Whatever the mechanism it is clear that oxidative stress injury and the production of peroxynitrite from reaction of hydroxyl radicals with nitric oxide could play a major part in this process of either necrosis or apoptosis.

As mentioned in Chapter 1 relating to the chemical properties of nimesulide and its principle metabolites for 4-hydroxy-nimesulide, these both have important effects in preventing the cellular injury from oxidant species and accelerating the composition of peroxynitrite. Moreover, nimesulide inhibits lipid peroxidation [416–418].

As indicated earlier, in the section on the effects of nimesulide on neutrophil mediated inflammation, production of superoxide by fMLP or PMA stimulated neutrophils, is blocked by nimesulide, probably by the inhibition by phosphodiesterase-IV. Evidence for antioxidant activity of nimesulide and its 4-hydroxy metabolite on the production of active oxygen species (ROS) in human chondrocytes was provided by the work of Zheng and co-workers [419]. Using electron-spin resonance (ESR) and 5,5-dymethylpyrroline-N-oxyde-DMPO as the spin trap-agent, it was found that both nimesulide and its metabolite 10–100 µmol/L was potent scavengers of hydroxyl and superoxide radicals respectively. Chemiluminescence generated by hypochlorous acid (HOCl) was inhibited in a concentration dependent manner by both nimesulide and its metabolite but the inhibitory effect of the metabolite process was more marked especially at high concentrations, 100 µmol/L. Thus, in addition to affecting oxyradical production in leucocytes it is clear from these studies that both nimesulide and its hydroxy metabolite can interfere with the production of these oxidant species in chondrocytes.

Regulation of other cytokine or cellular reactions that might be significant in controlling inflammation

A considerable number of other cytokine or cell-regulated processes may be affected by nimesulide that might have significance in the control of localised joint inflammation, cellular changes involving the regulation of phenotypic expression or other events by the drug in osteoarthritis and other joint inflammatory conditions. Among these are the observations by Stanford et al. [420] that inhibition of COX-2 may regulate the production of granulocyte-macrophage colony stimulating factor (GMC-SF) which has been demonstrated in human vascular cells and shown to be dependent on the effects of the drug on cyclic AMP formation. In these studies the authors were unable to show any effect of nimesulide on the production of IL-8.

Other metalloproteinases that may be affected by nimesulide include MMP-1 and MMP-9 that may be stimulated by platelet-activating factor [421].

The regulation by TNFα of interleukin-6 secretion may be another target for nimesulide. Using fresh villis fragments of term human placental tissue, Turner and co-workers [422] found that stimulation of IL-6 production by 1 µmol/L TNFα was markedly inhibited by 150 µmol/L nimesulide or indomethacin. These effects were not overcome by the addition of prostanoids to the cultures, indicating that the effects were independent of prostaglandins and possibly COX-2. These studies may represent a basis for further investigations of the effects of nimesulide on IL-6 production stimulated by TNFα using chondrocytes.

Another target of action of nimesulide involving the synthesis of fatty acid oxidation products may be the receptors for peroxisomal proliferation activated receptor (PPAR) signals. These PPAR's exist in three principle isoforms α, β and γ and their activation leads to nuclear translating factors for number of hormones as well as for leukotrienes and prostaglandins of the D and J series. The prostaglandin J series ligands are of particular significance in eliciting proinflammatory cytokine expression in macrophages as well as nitric oxide and matrix metalloproteinase-13 (MMP-13) in human chondrocytes [423–425]. Kalajdzic and co-workers [426] investigated the possibility that nimesulide and the sulphono-analogue, NS-398, might influence the ligand activation of PPARα and PPARγ using a cell transfected system derived of cells derived from human synovial fibroblasts. They found that both the activation of PPARα and PPARγ were inhibited by nimesulide in a concentration dependent manner with IC_{50} values of 0.602 and 0.8 µmol/L, respectively. Nimesulide also inhibited by nimesulide the PPAR-dependent transcription activation although the effects of the drug were not as potent as other PPAR activators including Wy-14,643 and ciglitazone. However, the combination of these agents with nimesulide led to a marked increased activation of PPARα and PPARγ. In the human synovial fibroblast cells that were transfected with a COX-2 promoter and the reporter luciferase measured it was found that the PPARα and PPARγ induced increase in COX-2 expression were inhibited by the addition of nimesulide 1 µmol/L; the effects being more pronounced than with ciglitazone and nimesulide [426].

The implication of these studies is somewhat difficult to determine at this stage. They do suggest that nimesulide has influences on another promoter effect that is influenced during COX-2 expression other than that previously mentioned concerning the glucocorticoid receptor system. Whether this has any significance or not in relationship to joint or other inflammatory reactions is at this stage indeterminate.

Smooth muscle and related pharmacological properties

NSAIDs have effects on the contractile properties of smooth muscle and seems in part related to their effects in affecting the actions of $PGF_{2\alpha}$ and PGE_2 [427, 428], possibly by affecting calcium channel activities [429] as well as production of NO and COX or LOX metabolites [430]. Smooth muscle tissues vary in their contractile responses to NSAIDs [428]. For example, low concentrations of aspirin or indomethacin may potentiate adrenaline or angiotensin-induced contractions in aortic muscle but reverse the effects of these agonists in portal venous muscle [428]. Guinea pig ileum also has varying sensitivity to the actions of NSAIDs according to the agonist employed [427].

In isolated rat uterine horns that had been obtained during the oestrogenic phase of the oestrous cycle, it was found that nimesulide inhibited the spontaneous contractility in a concentration related manner in the range of 0.1–20 μmol/L. The effects were comparable with those of indomethacin, while drugs such as mefenamic acid and naproxen were much less potent [431]. These effects suggest there may be smooth muscle effects of nimesulide possibly mediated through prostaglandins or nitric oxide, but as yet the mechanism is unknown.

Nimesulide has been reported to have smooth muscle relaxant effects in a number of systems including the endothelium denuded rat aorta, myometrial smooth muscle from pregnant or hormone sensitised women or rats and in the rat ductus arteriosus [432–440]. Clearly with much interest in the areas of nitric oxide and prostanoid involvement in smooth muscle contraction and the significance of this in a wide variety of physiological and physio-pathological conditions, further studies are warranted to investigate the mechanisms of smooth muscle relaxant effects of nimesulide like that of other NSAIDs.

The possibility that in addition to other inflammatory mediators (e.g. PGE_2 and NO) there might be influences on endothelin (ET) receptor mediated contraction was investigated by White et al. [440]. Induction of the endothelin receptor ETB by interleukin-1 in human temporal artery segments in organ culture was found to be inhibited by nimesulide as well as by rofecoxib, aspirin and indomethacin in a concentration related fashion. The effects of interleukin-1 resembled that of the endothelin receptor agonist, sarafotoxicon S6c. These results suggest that there may be a prostaglandin-related influence on endothelin related receptor mechanisms in initiation of contractile events and that the relaxant effects of nimesulide like that of other NSAIDs may be in part influenced through this mechanism. These results suggest that there may be disequilibrium between the inhibition of COX-2-derived PGE_2 by nimesulide and subsequent regulation *in vivo* of nitric oxide production [427].

These smooth muscle effects of nimesulide may have broad pharmacological significance but at this stage require more detailed investigation to establish their role in vascular, gastrointestinal or myometrial functions.

Conclusions

The studies reviewed in this chapter have shown the potent and diverse effects of nimesulide in controlling inflammatory reactions and particularly those that are important in chronic inflammation. It is clear that the inhibition of COX-2 forms a significant event in the actions of nimesulide in the systems but is not alone. The influence on a number of tissue destructive events that may be mediated by oxy-radicals and peroxynitrite, metalloproteinases and the influence on the production and action of proinflammatory cytokines such as tumour necrosis factor TNFα and interleukin-1 are among the multiple actions of this drug which are of considerable importance in mediating chronic as well as acute inflammation and the subsequent effects on pain production.

References

1. Swingle KF, Moore GGI (1984) Preclinical pharmacological studies with nimesulide. *Drugs Exptl Clin Res* 10: 587–597
2. Davis R, Brogden RN (1994) Nimesulide. An update of its pharmacodynamic and pharmacokinetic properties, and therapeutic efficacy. *Drugs* 48: 431–454
3. Famaey JP (1997) Review. *In vitro* and *in vivo* pharmacological evidence of selective cyclooxygenase-2 inhibition by nimesulide: An overview. *Inflamm Res* 46: 437–446
4. Bennett A, Villa G (2000) Nimesulide: an NSAID that preferentially inhibits COX-2, and has various unique pharmacological actions. *Exp Opin Pharmacother* 1: 277–286
5. Singla AG, Chawia AM, Sing A (2000) Review. Nimesulide: some pharmaceutical and pharmacological aspects – an update. *J Pharm Pharmacol* 52: 467–486
6. Bennett A (2001) Nimesulide: a well-established cyclooxygenase-2 inhibitor with many other pharmacological properties relevant to inflammatory diseases. In: Vane JR, Botting RM (Eds): *Therapeutic Role of Selective COX-2 inhibitors.* William Harvey Press, London. 521–540
7. Bennett A (2001) Clinical importance of the multifactorial actions of nimesulide. *Drugs of Today (Suppl B)* 37: 9–14
8. Rainsford KD (2004) Pharmacology and toxicology of COX-2 inhibitors. In: Pairet M, van Ryn J (Eds): *COX-2 Inhibitors.* 66–131, Birkhäuser, Basel
9. Swingle KF, Moore GGI, Grant TJ (1976) 4-Nitro-2-phenoxymethanesulfnanilide (R-805): a chemically novel anti-inflammatory agent. Ar*chiv Int Pharmacodyn* 221: 132–139
10. Rainsford KD (1999) Pharmacology and toxicology of ibuprofen. In: KD Rainsford (Ed): *Ibuprofen. A Critical Bibliographic review.* Taylor & Francis, London
11. Tanaka K, Shimotori T, Makino S, Aikawa Y, Inaba T, Yoshida C, Tanako S (1992) Pharmacological studies of the new anti-inflammatory agent 3-formylamino-7-methyl-

sulfonylamino-6-phenoxy-4H-1-benzopyran-4-one. 1st Communication: Antiinflammatory, analgesic and other related properties. *Arzneim-Forsch*, 42: 935–944

12. Qui J, Chen B-J, Zhang J-F (1993) Studies on pharmacodynamics of domestic nimesulide. *Chinese Pharmacol Bull* 9: 468–471 (Chinese)
13. Omata Y, Itokazu Y, Tsuzuike N, Inoue N, Segawa H, Tamaki H (1997) Zaltoprofen, a nonsteroidal anti-inflammatory drug, selectively inhibits prostaglandin G/H synthase/cyclooxygenase-2 (COX-2) activity *in vitro*. *Yakuri to Chiryo* 15: 2131–2136 (Japanese)
14. Scaglione F, Rossoni G (1998) Comparative anti-inflammatory effects of roxithromycin, azithromycin and clarithromycin. *J Antimicrob Chemotherap* 41 (Suppl B): 47–50
15. Gupta SK, Bhardwaj RK, Tyagi P, Sengupta S, Velpandian T (1999) Anti-inflammatory activity and pharmacokinetic profile of a new parenteral formulation of nimesulide. *Pharmacol Res* 39: 137–141
16. Gupta SK, Prakash J, Awaor L, Joshi S, Velpandian T, Sengupta S (1996) Anti-inflammatory activity of topical nimesulide gel in various experimental models. *Inflamm Res* 45: 590–592
17. Nakatsugi S, Terada N, Yoshimura T, Horie Y, Furukawa M (1996) Effects of nimesulide, a preferential cyclooxygenase-2 inhibitor, an carrageenan-induced pleurisy and stress-induced gastric lesions in rats. *Prost Leuk Essential Fatty Acids* 55: 395–402
18. Wallace JL, Chapman K, McKnight W (1999) Limited anti-inflammatory efficacy of cyclo-oxygenase-2 inhibition in carrageenan-air pouch inflammation. *Br J Pharmacol* 126: 1200–1204
19. Tanaka K, Makino S, Shimotori T, Aikawa Y, Inaba T, Yoshida C (1992) Pharmacological studies of the new anti-inflammatory agent 3-formylamino-7-methylsulfonylamino-6-phenoxy-4H-1-benzopyran-4-one. 2nd Communication: Effect on the arachidonic acid cascade. *Arzneim Forsch* 42: 945–950
20. Tofanetti O, Casciarri I, Cipolla PV, Cazzulani P, Omini C (1989) Effect of nimesulide on cyclooxygenase activity in rat gastric mucosa and inflammatory exudates. *Med Sci Res* 17: 745–746
21. Harada Y, Hatanaka K, Kawamura M, Saito M, Ogino M, Majima M, Ohno T, Ogino K, Yamamoto K, Taketani Y et al. (1996) Role of prostaglandin H synthase-2 in prostaglandin E_2 formation in rat carrageenin-induced pleurisy. *Prostaglandins* 51: 19–33
22. Hatanaka K, Kawamura M, Ogino M, Saito M, Ogino K, Sumitaka M, Harada Y (1999) Expression and function of cyclooxygenase-2 in mesothelial cells during the late phase of rat carrageenin-induced pleurisy. *Life Sci* 65: 161–166
23. Rainsford KD (1982) Adjuvant polyarthritis in rats: Is this a satisfactory model for screening anti-arthritic drugs? *Agents and Actions* 12: 452–458
24. Furukawa H, Kancoka H, Hoshi K, Kikukawa T, Abe C, Mizushima Y, Sakane T (1994) Effect of nimesulide on murine collagen-induced arthritis. *Jpn J Inflamm* 14: 31–34
25. Gilroy DW, Tomlinson A, Willoughby DA (1998) Differential effects of inhibition of isoforms of cyclooxygenase (COX-1, COX-2) in chronic inflammation. *Inflamm Res* 47: 79–85

26. Majima M, Hayashi I, Muramatsu M, Katada J, Yamashina S, Katori M (2000) Cyclooxygenase-2 enhances basic fibroblast growth factor-induced angiogenesis through induction of vascular endothelial growth factor in sponge implants. *Br J Pharmacol* 130: 641–649
27. Matsumoto H, Naraba H, Ueno A, Fujiyoshi T, Murakami M, Kudo I, Oh-ishi S (1998) Induction of cyclooxygenase-2 causes an enhancement of writhing response in mice. *Eur J Pharmacol* 352: 47–52
28. Miranda HF, Pincardi G (2001) Interactions of prazosin with non-steroidal anti-inflammatory drugs. *Pharmacol Res Commun* 11: 253–262
29. Miranda HF, Lopez J, Sierralta F, Correa A, Pinardi G (2001) NSAID antinociception measured in a chemical and a thermal assay in mice. *Pain Res Manag* 6:190–196
30. Toutain PL, Cester CC, Haak T, Laroute V (2001) A pharmacokinetic/pharmacodynamic approach *versus* a dose titration for the determination of a dosage regimen: the case of nimesulide, a Cox-2 selective nonsteroidal anti-inflammatory drug in the dog. *J Vet Pharmacol Therap* 24: 43–55
31. Toutain PL, Cester CC, Haak T, Metge S (2001) Pharmacokinetic profile and *in vitro* selective cyclooxygenase-2 inhibition by nimesulide in the dog. *J Vet Pharmacol Therap* 24: 35–42
32. Ceserani R, Carboni L, Germini M, Mainardi P, Passoni A (1993) Antipyretic and platelet antiaggregating effects of nimesulide. *Drugs* 46 (Suppl 1): 48–51
33. Steiner AA, Li S, Llanos-Q, Blatteis CM (2001) Differential inhibition by nimesulide of the early and late phases of intravenous- and intracerebroventricular-LPS-induced fever in guinea pigs. *Neuroimmunomodulation* 9: 263–275
34. Parnham MJ (1998) Is there a COX-fight during inflammation? *Inflamm Res* 47: 43
35. Bennett A, Berti F, Ferreira SH (1993) Nimesulide: A multifactorial therapeutic approach to the inflammatory process? A 7-year clinical experience. *Drugs* (Supp 1): 1–283
36. Murphy RC, Bowers RC, Dickinson J, Berry KZ (2004) Perspectives on the biosynthesis and metabolism of eicosanoids. In: Curtis Prior P (Ed): *The Eicosanoids*. John Wiley, Chichester. 3–16
37. Fu JY, Masferrrer IL, Seihert K, Raz A, Needleman P (1991) The induction and suppression of prostaglandin H_2 synthase (cyclooxygenase) in human monocytes. *J Biol Chem* 265: 16737–16740
38. Xie W, Chipman JG, Robertson DL, Erikson RL, Simmons DL (1991) Expression of mitogen responsive gene encoding prostaglandin synthase is regulates by mRNA splicing. *Proc Natl Acad Sci USA* 88: 2692–2696
39. Vane JR, Botting RM (2001) Formation and actions of prostaglandins and inhibition of their synthesis. In: Vane JR, Botting RM (Eds): *Therapeutic Roles of Selective COX-2 Inhibitors*. William Harvey Press, London. 1–47
40. Rainsford KD (2004) Inhibition of Eicosanoids. In: Curtis Prior P (Ed): *The Eicosanoids*. John Wiley, Chichester. 198–210
41. Mizuno H, Sakamoto C, Matsuda K, Wada K, Uchida T, Noguchi H, Akamatsu T, Kasuga M (1997) Induction of cyclooxygenase-2 in gastric mucosal lesions and its

inhibition by the specific antagonist delays healing in mice. *Gastroenterology* 112: 387–397

42. Droy-Lefais MT (1988) Prostanoids and stomach physiology. In: Curtis-Prior PB (Ed): *Prostaglandins: Biology and Chemistry of Prostaglandins and Related Eicosanoids*. Churchill Livingstone, London. 345–360
43. Kleinknecht D (1993) Diseases of the kidney caused by non-steroidal anti-inflammatory drugs. In: Stewart JH (Ed): *Analgesic and NSAID-induced kidney diseases*. Oxford University Press, Oxford. 160–179
44. Henzl MR (2004) Perspectives and clinical significance of eicosanoids in immunology, endocrinology and metabolic regulation. In: Curtis-Prior PB (Ed): *The Eicosanoids*. Wiley, Chichester. 229–236
45. Willoughby DA, Tomlinson A, Gilroy A and Willis D (1996) Inducible enzymes with special reference to COX-2 in inflammation and apoptosis. In: Vanel J, Botting J, Botting RM (Eds): *Improved non-steroid anti-inflammatory drugs – COX-2 – enzyme inhibitors*. Kluwer Academic Publishers & William Harvey Press, London. 67–83
46. Vigdahl RL, Tukey RH (1979) Mechanism of action of novel anti-inflammatory drugs diflumidone and R-805. *Biochem Pharmacol* 26: 307–311
47. Rufer C, Schillinger E, Böttcher I, Repenthin W, Herman C (1982) Nonsteroidal anti-inflammatories – XII: mode of action of anti-inflammatory methane sulfonanilides. *Biochem Pharmacol* 31: 3591–3596
48. Böttcher I, Schweizer A, Glatt M, Werner H (1987) A sulphonamidoinadanone COP 28237 (ZK34228), a novel non-steroidal anti-inflammatory agent without gastrointestinal ulcerogenicity in rats. *Drugs Under Exper Clin Res* 13: 237–245
49. Carr DP, Henn R, Green JR (1986) Comparison of the systemic inhibition of thromboxane synthesis, anti-inflammatory activity and gastrointestinal toxicity of non-steroidal anti-inflammatory drugs in the rat. *Agents Actions* 19: 374–375
50. Ceserani R, Casciarri I, Cavaletti E, Cazzulani P (1991) Action of nimesulide on rat gastric prostaglandins and renal function. *Drug Invest* 3 (Suppl 2): 14–21
51. Tavares IA, Bishai PM, Bennett A (1995) Activity of nimesulide on constitutive and inducible cyclooxygenases. *Arzneim Forsch* 45: 1093–1095
52. Vago T, Bevilacqua M, Norbiato G (1995) Effect of nimesulide action time dependence on selectivity towards prostagandin G/H synthase/cyclooxygenase activity. *Arzneim Forsch* 45: 1096–1098
53. Taniguchi Y, Jkesue A, Yokoyama K, Noda K, Debucchi H, Nakamura T et al. (1995) Selective inhibition by nimesulide, a novel non-steroidal anti-inflammatory drug with prostaglandin endoperoxide synthase-2 activity *in vitro*. *Pharm Sci* 1: 173–175
54. Patrignani P, Panara MR, Sciulli MG, Santini G, Renda G, Patrono C (1997) Differential inhibition of human prostaglandin endoperoxide synthase-1 and -2 by nonsteroidal anti-inflammatory drugs. *J Physiol Pharmacol* 48: 623–631
55. Cryer B, Feldman M (1998) Cyclooxygenase-1 and cyclooxygenase-2 selectivity of widely used nonsteroidal anti-inflammatory drugs. *Am J Med* 104: 413–421

56. Riendeau D, Charleson S, Cromlish W, Mancini JA, Wong E, Guay J (1997) Comparison of the cyclooxygenase-1 inhibitory properties of nonsteroidal anti-inflammatory drugs (NSAIDs) and selective COX-2 inhibitor, using sensitive microsomal and platelet assays. *Can J Physiol Pharmacol* 75: 1088–1095
57. Miralpeix M, Camacho M, López-Belmonte J, Canalias F, Beleta J, Palacios JM, Vila L (1997) Selective induction of cyclo-oxygenase activity in the permanent human endothelial cell line HUV-EC-C: biochemical and pharmacological characterization. *Br J Pharmacol* 121: 171–180
58. Warner TD, Giuliano F, Vojnovic I, Bukasa A, Mitchell JA, Vane JR (1999) Nonsteroid drug selectivities for cyclo-oxygenase-1 rather than cyclo-oxygenase-2 are associated with human gastrointestinal toxicity: a full *in vitro* analysis. *Proc Natl Acad Sci USA* 96: 7563–7568
59. Warner TD, Pairet M, Van Ryn J (2001) Test systems for inhibitors of cyclooxygenase-1 and cyclooxygenase-2. In: JR Vane, RM Botting (Eds): *Therapeutic roles of selective COX-2 inhibitors.* William Harvey Press, London. 76–94
60. Fenner H (1997) Differentiating among nonsteroidal antiinflammatory drugs by pharmacokinetic and pharmacodynamic profiles. *Semin Arthritis Rheum* 26 (Suppl 1) 28–33
61. Brooks P, Emery P, Evans JF, Fenner H, Hawkey CJ, Patrono C, Smolen J, Breedveld F, Day R, Dougados M et al. (1999) Interpreting the clinical significance of the differential inhibition of cyclooxygenase-1 and cyclooxygenase-2. *Rheumatology (Oxford)* 38: 779–788
62. Cullen L, Kelly L, Connor SO, Fitzgerald DJ (1998) Selective cyclooxygenase-2 inhibition by nimesulide in man. *J Pharmacol Exper Ther* 287: 578–582
63. Fahmi H, He Y, Zhang M, Martel-Pelletier J, Pelletier JP, Di Battista JA (2001) Nimesulide reduces interleukin-1beta-induced cyclooxygenase-2 gene expression in human synovial fibroblasts. *Osteoarthritis Cartilage* 9: 332–340
64. Di Battista JA, Fahmi H, Zhang M, Martel-Pelletier J, Pelletier J-P (2001) Differential regulation of interleukin-1β-induced cyclooxygenase-2 gene expression by nimeslide in human synovial fibroblasts. *Clin Exp Rheumatol*, 19 (Suppl 22): S3–S5
65. Taniguchi Y, Yokoyama K, Ikesue A, Noda K, Debuchi H, Nakamura T, Toda A, Shimeno H (1998) Inhibition by nimesulide of prostaglandin production in rat macrophages. *Drugs Exp Clin Res* 24: 17–27
66. Tool ATJ, Verhoeven AJ (1995) Inhibition of the production of platelet activating factor and of leukotriene B_4 in activated neutrophils by nimesulide due to an elevation of intracellular cyclic adenosine monophosphate. *Arzneim Forsch* 45: 1110–1114
67. Gray PA, Warner TD, Vojnovic I, Del Soldato P, Parikh A, Scadding GK, Mitchell JA (2002) Effects of non-steroidal anti-inflammatory drugs on the cyclooxygenase and lipoxygenase activity in whole blood from aspirin-sensitive asthmatics *versus* healthy donors. *Br J Pharmacol* 137: 1031–1038
68. Clish CB, Sun Y-P, Serhan CN (2001) Identification of dual cyclooxygenase-eicosanoid oxidoreductase inhibitors: NSAIDs that inhibit reductase/LTB_4 dehydrogenase 1. *Biochem Biophys Res Commun* 288: 868–874

69. Ueda N, Deutsch DG (2004) Biosynthesis and degradation of anandamide, an endogenous ligand of cannabinoid receptors. In: Curtis-Prior P (Ed): *The Eicosanoids*. Wiley, Chichester. 179–187
70. Fowler CJ, Holt S, Tiger G (2003) Acidic nonsteroidal anti-inflammatory drugs inhibit rat brain fatty acid amide hydrolase in a pH-dependent manner. *J Enzyme Inhib Med Chem* 18: 55–58
71. Kim J, Alger BE (2004) Inhibition of cyclooxygenase-2 potentiates retrograde endocannabinoid effects in hippocampus. *Nature Neurosci* 7: 697–698
72. Spencer AG, Woods JW, Arakawa T, Singer II, Smith WL (1998) Subcellular localization of prostaglandin endoperoxide H synthases-1 and -2 by immunoelectron microscopy. *J Biol Chem* 273: 9886–9893
73. Meade EA, Smith WL, DeWitt DL (1993) Differential inhibition of prostaglandin endoperoxide synthase (cyclooxygenase) isozymes by aspirin and other non-steroidal anti-inflammatory drugs. *J Biol Chem* 268: 6610–6614
74. Laneuville O, Breuer DK, Dewitt DL, Hla T, Funk CD, Smith WL (1994) Differential inhibition of human prostaglandin endoperoxide H synthases-1 and -2 by nonsteroidal anti-inflammatory drugs. *J Pharmacol Exp Ther* 271: 927–934
75. Copeland RA, Williams JM, Giannaras J, Nurnberg S, Covington M, Pinto D, Pick S, Trzaskos JM (1994) Mechanism of selective inhibition of the inducible isoform of prostaglandin G/H synthase. *Proc Natl Acad Sci USA* 91: 11202–11206
76. Gierse JK, Koboldt CM, Walker MC, Seibert K, Isakson PC (1999) Kinetic basis for selective inhibition of cyclo-oxygenases. *Biochem J* 339: 607–614
77. Walker MC, Kurumbail RG, Kiefer JR, Moreland KT, Koboldt CM, Isakson PC, Seibert K, Gierse JK (2001) A three-step kinetic mechanism for selective inhibition of cyclo-oxygenase-2 by diarylheterocyclic inhibitors. *Biochem J* 357: 709–718
78. FitzGerald GA (2003) COX-2 and beyond: Approaches to prostaglandin inhibition in human disease. *Nat Rev Drug Discov* 2: 879–890
79. Flower RJ (2003) The development of COX2 inhibitors. *Nat Rev Drug Discov* 2: 179–191
80. Hood WF, Gierse JK, Isakson PC, Kiefer JR, Kurumbail RG, Seibert K, Monahan JB (2003) Characterization of celecoxib and valdecoxib binding to cyclooxygenase. *Mol Pharmacol* 63: 870–877
81. Picot D, Loll PJ, Garavito RM (1994) The X-ray crystal structure of the membrane protein prostaglandin H2 synthase-1. *Nature* 367: 243–249
82. Luong C, Miller A, Barnett J, Chow J, Ramesha C, Browner MF (1996) Flexibility of the NSAID binding site in the structure of human cyclooxygenase-2. *Nat Struct Biol* 3: 927–933
83. Kurumbail RG, Stevens AM, Gierse JK, McDonald JJ, Stegeman RA, Pak JY, Gildehaus D, Miyashiro JM, Penning TD, Seibert K et al. (1996) Structural basis for selective inhibition of cyclooxygenase-2 by anti-inflammatory agents. *Nature* 384: 644–648
84. Berman HM, Westbrook J, Feng Z, Gilliland G, Bhat TN, Weissig H, Shindyalov IN, Bourne PE (2000) The Protein Data Bank. *Nucleic Acids Res* 28: 235–242

85. http://www.rcsb.org/pdb/
86. Marnett LJ, Kalgutkar AS (1998) Design of selective inhibitors of cyclooxygenase-2 as nonulcerogenic anti-inflammatory agents. *Curr Opin Chem Biol* 4: 482–490
87. Thuresson ED, Lakkides KM, Rieke CJ, Sun Y, Wingerd BA, Micielli R, Mulichak AM, Malkowski MG, Garavito RM, Smith WL (2001) Prostaglandin Endoperoxide H synthase-1. The functions of cyclooxygenase active site residues in the binding, positioning, and oxygenation of arachidonic acid. *J Biol Chem* 276: 10347–10359
88. Malkowski MG, Ginell SL, Smith WL, Garavito RM (2000) The productive conformation of arachidonic acid bound to prostaglandin synthase. *Science* 289: 1933–1937
89. Loll PJ, Picot D, Garavito RM (1995) The structural basis of aspirin activity inferred from the crystal structure of inactivated prostaglandin H2 synthase. *Nat Struct Biol* 2: 637–643
90. Kiefer JR, Pawlitz JL, Moreland KT, Stegeman RA, Hood WF, Gierse JK, Stevens AM, Goodwin DC, Rowlinson SW, Marnett LJ et al. (2000) Structural insights into the stereochemistry of the cyclooxygenase reaction. *Nature* 405: 97–101
91. Rieke CJ, Mulichak AM, Garavito RM, Smith WL (1999) The role of arginine120 of human prostaglandin endoperoxide H synthase-2 in the interaction with fatty acid substrates and inhibitors. *J Biol Chem* 274: 17109–17114
92. Mancini JA, Riendeau D, Falgueyret JP, Vickers PJ, O'Neill GP (1995) Arginine 120 of prostaglandin G/H synthase-1 is required for the inhibition by nonsteroidal anti-inflammatory drugs containing a carboxylic acid moiety. *J Biol Chem* 270: 29372–29377
93. Rowlinson SW, Kiefer JR, Prusakiewicz JJ, Pawlitz JL, Kozak KR, Kalgutkar AS, Stallings WC, Kurumbail RG, Marnett LJ (2003) A novel mechanism of cyclooxygenase-2 inhibition involving interactions with Ser-530 and Tyr-385. *J Biol Chem* 278: 45763–45769
94. Hochgesang GP, Rowlinson SW, Marnett LJ (2000) Tyrosine-385 is critical for acetylation of cyclooxygenase-2 by aspirin. *J Am Chem Soc* 122: 6514–6515
95. Kurumbail RG, Kiefer JR, Marnett LJ (2001) Cyclooxygenase enzymes: catalysis and inhibition. *Curr Opin Struct Biol* 11: 752–760
96. García-Nieto R, Pérez C, Gago F (2000) Automated docking and molecular dynamics simulations of nimesulide in the cyclooxygenase active site of human prostaglandin-endoperoxide synthase-2 (COX-2) *J Comp-Aided Mol Des* 14: 147–160
97. Pedretti A, Villa AM, Villa L, Vistoli G (1997) Interactions of some PGHS-2 selective inhibitors with the PGHS-1: an automated docking study by BioDock. *Farmaco* 52: 487–491
98. Fabiola GF, Pattabhi V, Nagarajan K (1998) Structural basis for selective inhibition of COX-2 by nimesulide. *Bioorg Med Chem* 6: 2337–2344
99. García-Nieto R, Pérez C, Checa A, Gago F (1999) Molecular model of the interaction between nimesulide and human cyclooxygenase-2. *Rheumatology* 38 (Suppl 1): 14–18
100. Dupont L, Pirotte B, Masereel B, Delarge J, Geczy J (1995) Nimesulide. *Acta Crystallogr* C51: 507–509

101. Allen FH, Bellard S, Brice MD, Cartwright BA, Doubleday A, Higgs H, Hummelink T, Hummelink-Peters BG, Kennard O, Motherwell WDS, Rodgers JR, Watson DG (1979) The Cambridge Crystallographic Data Centre: computer-based search, retrieval, analysis and display of information. *Acta Cryst* B35: 2331–2339
102. Rogge CE, Liu W, Wu G, Wang L-H, Kulmacz RJ, Tsai A-L (2004) Identification of Tyr504 as an alternative tyrosyl radical site in human prostaglandin H synthase-2. *Biochemistry* 43: 1560–1568
103. Futaki N, Takahashi S, Yokoyama M, Arai I, Higuchi S, Otomo S (1994) NS-398, a new anti-inflammatory agent, selectively inhibits prostaglandin G/H synthase/cyclooxygenase (COX-2) activity *in vitro*. *Prostaglandins* 47: 55–59
104. Harada Y, Kawamura M, Hatanaka K, Saito M, Ogino M, Ohno T, Ogino K, Yang Q (1998) Differing profiles of prostaglandin formation inhibition between selective prostaglandin H synthase-2 inhibitors and conventional NSAIDs in inflammatory and non-inflammatory sites of the rat. *Prostaglandins Other Lipid Mediat* 55: 345–358
105. Li CS, Black WC, Chan CC, Ford-Hutchinson AW, Gauthier JY, Gordon R, Guay D, Kargman S, Lau CK, Mancini J et al. (1995) Cyclooxygenase-2 inhibitors. Synthesis and pharmacological activities of 5-methanesulfonamido-1-indanone derivatives. *J Med Chem* 38: 4897–4905
106. Ouellet M, Percival MD (1995) Effect of inhibitor time-dependency on selectivity towards cyclooxygenase isoforms. *Biochem J* 306: 247–251
107. Guo Q, Wang LH, Ruan KH, Kulmacz RJ (1996) Role of Val509 in time-dependent inhibition of human prostaglandin H synthase-2 cyclooxygenase activity by isoform-selective agents. *J Biol Chem* 271: 19134–19139
108. Wong E, Bayly C, Waterman HL, Riendeau D, Mancini JA (1997) Conversion of prostaglandin G/H synthase-1 into an enzyme sensitive to PGHS-2-selective inhibitors by a double His513→Arg and Ile523→Val mutation. *J Biol Chem* 272: 9280–9286
109. Selinsky BS, Gupta K, Sharkey CT, Loll PJ (2001) Structural analysis of NSAID binding by prostaglandin H(2) synthase: time-dependent and time-independent inhibitors elicit identical enzyme conformations. *Biochemistry* 40: 5172–5180
110. Llorens O, Pérez JJ, Palomer A, Mauleón D (1999) Structural basis of the dynamic mechanism of ligand binding to cyclooxygenase. *Bioorg Med Chem Lett* 9: 2779–2784
111. Gierse JK, McDonald JJ, Hauser SD, Rangwala SH, Koboldt CM, Seibert K (1996) A single amino acid difference between cyclooxygenase-1 (COX-1) and -2 (COX-2) reverses the selectivity of COX-2 specific inhibitors. *J Biol Chem* 271: 15810–15814
112. Mancini JA, Vickers PJ, O'Neill GP, Boily C, Falgueyret JP, Riendeau D (1997) Altered sensitivity of aspirin-acetylated prostaglandin G/H synthase-2 to inhibition by nonsteroidal anti-inflammatory drugs. *Mol Pharmacol* 51: 52–60
113. Cignarella G, Vianello P, Berti F, Rossoni G (1996) Synthesis and pharmacological evaluation of derivatives structurally related to nimesulide. *Eur J Med Chem* 31: 359–364
114. Panara MR, Padovano R, Sciuii MG, Santini G, Renda G, Rotondo MT, Pace A, Patrono C, Patrignani P (1998) Effects of nimesulide on constitutive and inducible prostanoid biosynthesis in human beings. *Clin Pharmacol Ther* 63: 672–681

115. Kozak KR, Prusakiewicz JJ, Rowlinson SW, Schneider C, Marnett LJ (2001) Amino acid determinants in cyclooxygenase-2 oxygenation of the endocannabinoid 2-arachidonylglycerol. *J Biol Chem* 276: 30072–30077
116. Kalgutkar AS, Crews BC, Rowlinson SW, Marnett AB, Kozak KR, Remmel RP, Marnett LJ (2000) Biochemically based design of cyclooxygenase-2 (COX-2) inhibitors: facile conversion of nonsteroidal antiinflammatory drugs to potent and highly selective COX-2 inhibitors. *Proc Natl Acad Sci USA* 97: 925–930
117. Greig GM, Francis DA, Falgueyret JP, Ouellet M, Percival MD, Roy P, Bayly C, Mancini JA, O'Neill GP (1997) The interaction of arginine 106 of human prostaglandin G/H synthase-2 with inhibitors is not a universal component of inhibition mediated by nonsteroidal anti-inflammatory drugs. *Mol Pharmacol* 52: 829–838
118. Bhattacharyya DK, Lecomte M, Rieke CJ, Garavito M, Smith WL (1996) Involvement of arginine 120, glutamate 524, and tyrosine 355 in the binding of arachidonate and 2-phenylpropionic acid inhibitors to the cyclooxygenase active site of ovine prostaglandin endoperoxide H synthase-1. *J Biol Chem* 271: 2179–2184.
119. So OY, Scarafia LE, Mak AY, Callan OH, Swinney DC (1998) The dynamics of prostaglandin H synthases. Studies with prostaglandin H synthase 2 Y355F unmask mechanisms of time-dependent inhibition and allosteric activation. *J Biol Chem* 273: 5801–5807
120. Loll PJ, Picot D, Ekabo O, Garavito RM (1996) Synthesis and use of iodinated nonsteroidal antiinflammatory drug analogs as crystallographic probes of the prostaglandin H_2 synthase cyclooxygenase active site. *Biochemistry* 35: 7330–7340
121. Singh S, Shardra N, Mahajan L (1999) Spectrophotometric determination of pKa of nimesulide. *Int J Pharm* 176: 261–264
122. Julémont F, de Leval X, Michaux C, Damas J, Charlier C, Durant F, Pirotte B, Dogné JM (2002) Spectral and crystallographic study of pyridinic analogues of nimesulide: determination of the active form of methanesulfonamides as COX-2 selective inhibitors. *J Med Chem* 45: 5182–5185
123. Michaux C, Charlier C, Julemont F, Norberg B, Dogne J-M, Pirotte B, Durant F (2001) FJ6, *N*-methyl-*N*-(4-nitro-2-phenoxyphenyl)methanesulfonamide. *Acta Crystallogr* E57: 1012–1013
124. Hendlich M, Bergner A, Gunther J, Klebe G (2003) Relibase: design and development of a database for comprehensive analysis of protein–ligand interactions. *J Mol Biol* 326: 607–620
125. Penning TD, Talley JJ, Bertenshaw SR, Carter JS, Collins PW, Docter S, Graneto MJ, Lee LF, Malecha JW, Miyashiro JM et al. (1997) Synthesis and biological evaluation of the 1,5-diarylpyrazole class of cyclooxygenase-2 inhibitors: identification of 4-[5-(4-methylphenyl)-3-(trifluoromethyl)-1H-pyrazol-1-yl]benzenesulfonamide (SC-58635, celecoxib) *J Med Chem* 40: 1347–1365
126. Notredame C, Higgins D, Heringa J (2000) T-Coffee: A novel method for multiple sequence alignments. *J Mol Biol* 302: 205–217
127. Brooks PM, Day RO (1991) Nonsteroidal anti-inflammatory drugs. Differences and similarities. *N Eng J Med* 24: 1716–1723

128. Smith MJH (1978) Aspirin and prostaglandins: some recent developments. *Agents Actions* 8: 427–429
129. Kitchen EA, Dawson W, Rainsford KD, Cawston T (1985) Inflammation and possible modes of action of anti-inflammatory drugs. In: *Anti-Inflammatory and Anti-Rheumatic Drugs. Volume I Infalmmation Mechanisms and Actions of Traditional Drugs.* CRC Press, Boca Raton. 21–87
130. Rainsford KD (1996) Mode of action, uses, and side effects of anti-inflammatory drugs. In: Rainsford KD (Ed): *Advances in Anti-Rheumatic Therapy.* CRC Press, Boca Raton. 59–111
131. Dallegri F, Ottonello L (1997) Tissue injury in neutrophilic inflammation. *Inflamm Res* 46: 382–391
132. Diaz-Gonzales F, Sanchez-Madrid F (1998) Inhibition of leukocyte adhesion: an alternative mechanism of action for anti-inflammatory drugs. *Immunology Today* 19: 169–172
133. Tegeder I, Pfeilschifter J, Geisslinger G (2001) Cyclooxygenase-independent actions of cyclooxygenase inhibitors. *FASEB J* 224: 2057–2072
134. Tak PT, Bresnihan B (2000) The pathogenesis and prevention of joint damage in rheumatic arthritis. *Arthritis Rheum* 43: 2619–2633
135. Faurschou M, Borregaard N (2003) Neutrophil granules and secretory vesicles in inflammation. *Microbes Infect* 14: 1317–1327
136. Lindbom L, Werr J (2002) Integrin-dependent neutrophil migration in extravascular tissue. *Semin Immunol* 14: 115–121
137. Weiss SJ (1989) Tissue destruction by neutrophils. *N Engl J Med* 320: 365–376
138. Scapini P, Lapinet-Vera JA, Gasperini S, Calzetti F, Bazzoni F, Cassatella MA (2000) The neutrophil as a cellular source of chemokines. *Immunol Rev* 177: 195–203
139. Bevilacqua M, Vago T, Baldi G, Renesto E, Dallegri F, Norbiato G (1994) Nimesulide decreases superoxide production by inhibiting phosphodiesterase type IV. *Eur J Pharmacol* 268: 415–423
140. Dapino P, Ottonello L, Dallegri F (1994) The anti-inflammatory drug nimesulide inhibits neutrophil adherence to and migration across monolayers of cytokine-activated endothelial cells. *Respiration* 61: 336–341
141. Gomez-Gaviro MV, Gonzalez-Alvaro I, Dominguez-Jimenez C, Peschon J, Black RA, Sanchez-Madrid F, Diaz-Gonzalez F (2002) Structure-function relationship and role of tumor necrosis factor-converting enzyme in the down-regulation of L-selectin by non-steroidal anti-inflammatory drugs. *J Biol Chem* 277: 38212–38221
142. Capsoni F, Venegoni E, Minonzio F, Ongari AM, Maresca V, Zanussi C (1987) Inhibition of neutrophil oxidative metabolism by nimesulide. *Agents Actions* 21: 121–129
143. Capecchi PL, Ceccatelli L, Beermann U, Laghi Pasini F, Di Perri T (1993) Inhibition of neutrophil function *in vitro* by nimesulide. Preliminary evidence of an adenosine-mediated mechanism. *Arzneim Forsch* 43: 992–996
144. Mouithys-Mickalad AM, Zheng SX, Deby-Dupont GP, Deby CM, Lamy MM, Reginster JY, Henrotin YE (2000) *In vitro* study of anti-oxidant properties of non steroidal anti-in-

flammatory drugs by chemiluminescence and electron spin resonance (ESR) *Free Radic Res* 33: 607–621

145. Ottonello L, Dapino P, Pastorino G, Montagnani G, Gatti F, Guidi G, Dallegri F (1993) Nimesulide as a down-regulator of the activity of neutrophil myeloperoxidase pathway. *Drugs* 46: 29–33
146. Ottonello L, Dapino P, Scirocco MC, Bevilacqua M, Dallegri F (1995) Sulphonamides as anti-inflammatory agents: old drugs for new therapeutic strategies in inflammation? *Clin Sci* 88: 331–336
147. Dallegri F, Dapino P, Ottonello L, Guidi G (1991) Suppression of neutrophil chloramine production by nimesulide. *Drug Invest* 3:75–78
148. Ottonello L, Amelotti M, Barbera P, Dapino P, Mancini M, Tortolina G, Dallegri F (1999) Chemoattractant-induced release of elastase by tumour necrosis factor-primed human neutrophils: autoregulation by endogenous adenosine. *Inflamm Res* 48: 637–642
149. Zimmerli W, Sansano S, Wiesemberg-Böttcher I (1991) Influence of the anti-inflammatory compound flosulide on granulocyte function. *Biochem Pharmacol* 10: 1913–1919
150. Dallegri F, Ottonello L, Dapino P, Sacchetti C (1992) Effect of nonsteroidal anti-inflammatory drugs on the neutrophil promoted inactivation of alpha-1-proteinase inhibitor. *J Rheumatol* 19: 419–423
151. Dallegri F, Ottonello L, Dapino P, Bevilacqua M (1992) The anti-inflammatory drug nimesulide rescues alpha-1-proteinase inhibitor from oxidative inactivation by phagocytosis neutrophils. *Respiration* 59: 1–4
152. Ottonello L, Dapino P, Dallegri F (1993) Inactivation of alpha-1-proteinase inhibitor by neutrophil metalloproteinases. *Respiration* 60: 32–37
153. Kimura T, Iwase M, Kondo G, Watanabe H, Ohashi M, Ito D, Nagumo M (2003) Suppressive effect of selective cycloxygenase-2 inhibitor on cytokine release in human neutrophils. *Int Immunopharm* 3: 1519–1528
154. Sawada T, Hashimoto S, Tohma S (2000) Inhibition of L-leucine methyl ester mediated killing of THP-1, a human monocytic cell line, by a new anti-inflammatory drug, T614. *Immunopharmacology* 49: 285–294
155. Lardner A (2001) The effects of extracellular pH on immune function. *J Leukoc Biol* 69: 522–530
156. Kobayashi M, Tanaka T, Usui T (1982) Inactivation of lysosomal enzymes by the respiratory burst of polymorphonuclear leukocytes. Possible involvement of myeloperoxidase-H_2O_2-halide system. *J Lab Clin Med* 100: 896–907
157. Klempner MS, Styrt B (1983) Alkalinizing the intralysosomal pH inhibits degranulation of human neutrophils. *J Clin Invest* 72:1793–800
158. Klempner MS, Styrt B (1983) Alkalinization of the intralysosomal pH by clindamycin and its effects on neutrophil function. *J Antimicrob Chemother* 12 (Suppl C) 39–50
159. Yocum DE, Hempel S, Busse WW (1984) Regulation of the human polymorphonuclear leukocyte inflammatory response by inhibitors of arachidonic acid metabolism. *J Immunopharmacol* 6: 237–255

160. Styrt B, Klempner MS (1984) Inhibition of neutrophil oxidative metabolism by lysosomotropic weak bases. *Blood* 67: 334–342
161. Kurita N, Terao K, Brummer E, Ito E, Nishimura K, Miyaji M (1991) Resistance of Histoplasma capsulatum to killing by human neutrophils. Evasion of oxidative burst and lysosomal-fusion products. *Mycopathologia* 115: 207–213
162. Dri P, Presani G, Perticarari S, Alberi L, Prodan M, Decleva E (2002) Measurement of phagosomal pH of normal and CGD-like human neutrophils by dual fluorescence flow cytometry. *Cytometry* 48:159–166
163. Ivanov IT, Tzaneva M (2002) Direct cytotoxicity of non-steroidal anti-inflammatory drugs in acidic media: model study on human erythrocytes with DIDS-inhibited anion exchanger. *Pharmazie* 57: 848–851
164. Sinn H, Schrenk HH, Friedrich EA, Schilling U, Maier-Borst W (1990) Design of compounds having an enhanced tumour uptake, using serum albumin as a carrier. Part I. *In. J Radiat Appl Instrum B* 17: 819–827
165. Wosikowski K, Biedermann E, Rattel B, Breiter N, Jank P, Loser R, Jansen G, Peters GJ (2003) *In vitro* and *in vivo* antitumor activity of methotrexate conjugated to human serum albumin in human cancer cells. *Clin Cancer Res* 9: 1917–1927
166. Stossel TP (1992) The mechanical response of white blood cells. In: JI Gallin, IM Goldstein, R Sniderman (Eds): *Inflammation: Basic Principles and Clinical Correlates*. Raven Press, New York. 459–475
167. Ottonello L, Dapino P, Pastorino G, Dallegri F (1992) Inhibition of the neutrophil oxidative response induced by the oral administration of nimesulide in normal volunteers. *J Clin Lab Immunol* 37: 91–96
168. Bravo-Cuellar A, Garcia-Reyes G, Barba-Barajas M, Carranco-Lopez A, Dominguez-Rodriguez JR (2003) Modification by nimesulide administration of the phagocytic activity of polymorphonuclears in healthy subjects. *Biomed Pharmacol* 57: 434
169. Condliffe AM, Kitchen E, Chilvers ER (1998) Neutrophil priming: pathophysiological consequenbces and underlying mechanisms. *Clin Sci (Lond)* 94: 461–471
170. Hampton MB, Kettle AJ, Winterbourn CC (1998) Inside the neutrophil phagosome: oxidants, myeloperoxidase, and bacterial killing. *Blood* 92: 3007–3017
171. Reeves EP, Nagl M, Godovac-Zimmermann J, Segal AW (2003) Reassessment of the microbicidal activity of reactive oxygen species and hypochlorous acid with reference to the phagocytic vacuole of the neutrophil granulocyte. *J Med Microbiol* 52: 643–651
172. Belova LA (1997) Biochemistry of inflammatory processes and vascular injury. Role of neutrophils: a review. *Biochemistry (Mosc)* 62: 563–570
173. Wientjes FB, Segal AW (1995) NADPH oxidase and the respiratory burst. *Semin Cell Biol* 6: 357–365
174. Cassatella MA (1998) The neutrophil: one of the cellular targets of interleukin-10. *Int J Clin Lab Res* 28: 148–161
175. Binder R, Kress A, Kan G, Herrmann K, Kirschfink M (1999) Neutrophil priming by cytokines and vitamin D binding protein (Gc-globulin): impact on C5a-mediated chemotaxis, degranulation and respiratory burst. *Mol Immunol* 36: 885–892

176. Klebanoff SJ (1999) Myeloperoxidase. *Proc Assoc Am Physicians* 111: 383–389
177. Nauseef WM (1999) The NADPH–dependent oxidase of phagocytes. *Proc Assoc Am Physicians* 111: 373-382
178. Berridge MV, Tan AS (2000) High-capacity redox control at the plasma membrane of mammalian cells: trans-membrane, cell surface, and serum NADH-oxidases. *Antioxid Redox Signal* 2: 231–242
179. Lardner A (2001) The effects of extracellular pH on immune function. *J Leukoc Biol* 69: 522–530
180. Neumann NF, Stafford JL, Barreda D, Ainsworth AJ, Belosevic M (2001) Antimicrobial mechanisms of fish phagocytes and their role in host defense. *Dev Comp Immunol* 25: 807–825
181. Forman HJ, Torres M (2002) Reactive oxygen species and cell signaling: respiratory burst in macrophage signaling. *Am J Respir Crit Care Med* 166: S4–S8
182. Gougerot-Pocidalo MA, el Benna J, Elbim C, Chollet-Martin S, Dang MC (2002) Regulation of human neutrophil oxidative burst by pro- and anti-inflammatory cytokines. *J Soc Biol* 196: 37–46
183. Karlsson A, Dahlgren C (2002) Assembly and activation of the neutrophil NADPH oxidase in granule membranes. *Antioxid Redox Signal* 4: 49–60
184. Vignais PV (2002) The superoxide-generating NADPH oxidase: structural aspects and activation mechanism. *Cell Mol Life Sci* 59: 1428–1459
185. Quinn MT, Gauss KA (2004) Structure and regulation of the neutrophil respiratory burst oxidase: comparison with nonphagocyte oxidases. *J Leukoc Biol* 76: 760–781
186. Werner E (2004) GTPases and reactive oxygen species: switches for killing and signaling. *J Cell Sci* 117: 143–153
187. Chilvers ER, Rossi AG, Murray J, Haslett C (1998) Regulation of granulocyte apoptosis and implications for anti-inflammatory therapy. *Thorax* 53: 533–534
189. Haslett C (1997) Granulocyte apoptosis and inflammatory disease. *Br Med Bull* 53: 669–683
190. Haslett C, Savill JS, Whyte MK, Stern M, Dransfield I, Meagher LC (1994) Granulocyte apoptosis and the control of inflammation. *Philos Trans R Soc Lond B Biol Sci* 345: 327–333
191. Heasman SJ, Giles KM, Ward C, Rossi AG, Haslett C, Dransfield I (2003) Glucocorticoid-mediated regulation of granulocyte apoptosis and macrophage phagocytosis of apoptotic cells: implications for the resolution of inflammation. *J Endocrinol* 178: 29–36
192. Martin MC, Dransfield I, Haslett C, Rossi AG (2001) Cyclic AMP regulation of neutrophil apoptosis occurs via a novel protein kinase A-independent signaling pathway. *J Biol Chem* 276: 45041–45050
193. Rossi AG, Cousin JM, Dransfield I, Lawson MF, Chilvers ER, Haslett C (1995) Agents that elevate cAMP inhibit human neutrophil apoptosis. *Biochem Biophys Res Commun* 217: 892–899

194. Rossi AG, McCutcheon JC, Roy N, Chilvers ER, Haslett C, Dransfield I (1998) Regulation of macrophage phagocytosis of apoptotic cells by cAMP. *J Immunol* 160: 3562–3568
195. Walker A, Ward C, Dransfield I, Haslett C, Rossi AG (2003) Regulation of granulocyte apoptosis by hemopoietic growth factors, cytokines and drugs: potential relevance to allergic inflammation. *Curr Drug Targets Inflamm Allergy* 2: 339–347
196. Ward C, Walker A, Dransfield I, Haslett C, Rossi AG (2004) Regulation of granulocyte apoptosis by NF-kappaB. *Biochem Soc Trans* 32: 465–467
197. Krakstad C, Christensen AE, Doskeland SO (2004) cAMP protects neutrophils against TNF-{alpha}-induced apoptosis by activation of cAMP-dependent protein kinase, independently of exchange protein directly activated by cAMP (Epac). *J Leukoc Biol* 74: 641–647
198. Fadeel B, Kagan VE (2003) Apoptosis and macrophage clearance of neutrophils: regulation by reactive oxygen species. *Redox Rep* 8: 143–150
199. Jersmann HP, Ross KA, Vivers S, Brown SB, Haslett C, Dransfield I (2003) Phagocytosis of apoptotic cells by human macrophages: analysis by multiparameter flow cytometry. *Cytometry* 51A: 7–15
200. Kopperud R, Krakstad C, Selheim F, Doskeland SO (2003) cAMP effector mechanisms. Novel twists for an 'old' signaling system. *FEBS Lett* 546: 121–126
201. Borisenko GG, Iverson SL, Ahlberg S, Kagan VE, Fadeel B (2004) Milk fat globule epidermal growth factor 8 (MFG-E8) binds to oxidized phosphatidylserine: implications for macrophage clearance of apoptotic cells. *Cell Death Differ* 11: 943–945
202. Arroyo A, Modriansky M, Serinkan FB (2002) NADPH oxidase-dependent oxidation and externalization of phosphatidylserine during apoptosis in Me2SO-differentiated HL-60 cells. Role in phagocytic clearance. *J Biol Chem* 277: 49965–49975
203. Kagan VE, Gleiss B, Tyurina YY (2002) A role for oxidative stress in apoptosis: oxidation and externalization of phosphatidylserine is required for macrophage clearance of cells undergoing Fas-mediated apoptosis. *J Immunol* 169: 487–499
204. Tardieu D, Jaeg JP, Deloly A, Corpet DE, Cadet J, Petit CR (2000) The COX-2 inhibitor nimesulide suppresses superoxide and 8-hydroxy-deoxyguanosine formation, and stimulates apoptosis in mucosa during early colonic inflammation in rats. *Carcinogenesis* 21: 973–976
205. Wright CD, Kuipers PJ, Kobylarz-Singer D, Devall LJ, Klinkefus BA, Weishaar RE (1990) Differential inhibition of human neutrophil functions. Role of cyclic AMP-specific, cyclic GMP-insensitive phosphodiesterase. *Biochem Pharmacol* 40: 699–707
206. Vago T, Norbiato G, Baldi G, Chebat E, Bertora P, Bevilacqua M (1990) Respiratory-burst stimulants desensitize beta-2 adrenoceptors on human polymorphonuclear leukocytes. *Int J Tissue React* 12: 53–58
207. Pryzwansky KB, Kidao S, Merricks EP (1998) Compartmentalization of PDE-4 and cAMP-dependent protein kinase in neutrophils and macrophages during phagocytosis. *Cell Biochem Biophys* 28: 251–275

208. Tool AT, Mul FP, Knol EF, Verhoeven AJ, Roos D (1996) The effect of salmeterol and nimesulide on chemotaxis and synthesis of PAF and LTC_4 by human eosinophils. *Eur Respir J* (Suppl) 22: 141s–145s
209. Verhoeven AJ, Tool AT, Kuijpers TW, Roos D (1993) Nimesulide inhibits platelet-activating factor synthesis in activated human neutrophils. *Drugs* 46 (Suppl 1): 52–58
210. Kumar A, Jain NK, Kulkarni SK (2000) Analgesic and anti-inflammatory effects of phosphodiesterase inhibitors. *Indian J Exp Biol* 38: 26–30
211. Van Rensburg AJ, Theron AJ, Andreson R (1991) Comparison of the pro-oxidative interactions of flunoxaprofen and benoxaprofen with human polymophonuclear leucocytes *in vitro*. *Agents Actions* 33: 292–299
212. Ouellet M, Falgueyret JP, Percival MD (2004) Detergents profoundly affect inhibitor potencies against both cyclo-oxygenase isoforms. *Biochem J* 377: 675–684
213. Ouellet M, Percival MD (1995) Effect of inhibitor time-dependency on selectivity towards cyclooxygenase isoforms. *Biochem J* 306: 247–251
214. Van Antwerpen P, Dubois J, Gelbcke M, Neve J (2004) The reactions of oxicam and sulfoanilide non steroidal anti-inflammatory drugs with hypochlorous acid: determination of the rate constants with an assay based on the competition with para-aminobenzoic acid chlorination and identification of some oxidation products. *Free Radic Res* 38: 251–258
215. Van Antwerpen P, Neve J (2004) *In vitro* comparative assessment of the scavenging activity against three reactive oxygen species of non-steroidal anti-inflammatory drugs from the oxicam and sulfoanilide families. *Eur J Pharmacol* 496: 55–61
216. Uddin MJ, Rao PN, Knaus EE (2003) Design and synthesis of novel celecoxib analogues as selective cyclooxygenase-2 (COX-2) inhibitors: replacement of the sulfonamide pharmacophore by a sulfonylazide bioisostere. *Bioorg Med Chem* 11: 5273–5280
217. Kiyama R, Tamura Y, Watanabe F, Tsuzuki H, Ohtani M, Yodo M (1999) Homology modeling of gelatinase catalytic domains and docking simulations of novel sulfonamide inhibitors. *J Med Chem* 42: 1723–1738
218. Levin JI, Chen JM, Cheung K, Cole D, Crago C, Santos ED, Du X, Khafizova G, MacEwan G, Niu C, Salaski EJ, Zask A, Cummons T, Sung A, Xu J, Zhang Y, Xu W, Ayral-Kaloustian S, Jin G, Cowling R, Barone D, Mohler KM, Black RA, Skotnicki JS (2003) Acetylenic TACE inhibitors. Part 1. SAR of the acyclic sulfonamide hydroxamates. *Bioorg Med Chem Lett* 13: 2799–2803
219. Supuran CT, Casini A, Scozzafava A (2003) Protease inhibitors of the sulfonamide type: anticancer, antiinflammatory, and antiviral agents. *Med Res Rev* 23: 535–558
220. Levin JI, Du MT (2003) Sulfonate ester hydroxamic acids as potent and selective inhibitors of TACE enzyme. *Drug Des Discov* 18: 123–126
221. Skotnicki JS, DiGrandi MJ, Levin JI (2003) Design strategies for the identification of MMP-13 and Tace inhibitors. *Curr Opin Drug Discov Devel* 6: 742–759
222. Peyron P, Maridonneau-Parini I, Stegmann T (2001) Fusion of human neutrophil phagosomes with lysosomes *in vitro*: involvement of tyrosine kinases of the Src family and inhibition by mycobacteria. *J Biol Chem* 276: 35512–35517

223. Wilhelms OH, Linssen MJ, Lipponer L, Seilnacht W (1990) Nimesulide, indomethacin, BW 755 C, phenidon, mepacrin and nedocromil inhibit the activation of human and rat leucocytes. *Int J Tissue React* 12: 101–106
224. de Mello SB, Laurindo IM, Cossermelli W (1994) Action of the 4-nitro-2-phenoximethanesulphonanilide (nimesulide) on neutrophil chemotaxis and superoxide production. *Rev Paul Med* 112: 489–494
225. Dallegri F, Ottonello L, Bevilacqua M (1995) Possible modes of action of nimesulide in controlling neutrophilic inflammation. *Arzneimittelforschung* 45: 1114–1117
226. Ottonello L, Barbera P, Dapino P, Sacchetti C, Dallegri F (1997) Chemoattractant-induced release of elastase by lipopolysaccharide (LPS)-primed neutrophils; inhibitory effect of the anti-inflammatory drug nimesulide. *Clin Exp Immunol* 110: 139–143
227. Cronstein BN (1996) Molecular therapeutics. Methotrexate and its mechanism of action. *Arthritis Rheum* 39: 1951–1960
228. Cronstein BN (1997) The mechanism of action of methotrexate. *Rheum Dis Clin North Am* 23: 739–755
229. Cronstein BN, Montesinos MC, Weissmann G (1999) Salicylates and sulfasalazine, but not glucocorticoids, inhibit leukocyte accumulation by an adenosine-dependent mechanism that is independent of inhibition of prostaglandin synthesis and p105 of NFkappaB. *Proc Natl Acad Sci USA* 96: 6377–6381
230. Gadangi P, Longaker M, Naime D (1996) The anti-inflammatory mechanism of sulfasalazine is related to adenosine release at inflamed sites. *J Immunol* 156: 1937–1941
231. Baggott JE, Morgan SL, Ha T, Vaughn WH, Hine RJ (1992) Inhibition of folate-dependent enzymes by non-steroidal anti-inflammatory drugs. *Biochem J* 282: 197–202
232. Male D (2001) Cell migration and inflammation. In: Roitt I, Brostoff J, Male D (Eds): *Immunology*. Sixth Edition. Mosby, Edinburgh. 47–63
233. Auteri A, Saletti M, Blardi P, Di Perri T (1988) Action of a new non-steroid antiinflammatory drug, nimesulide, on activation of the complement system: an *in vitro* study. *Int J Tissue React* 10: 217–221
234. Eibl G, Bruemmer D, Okada Y, Duffy JP, Law RE, Reber HA, Hines OJ (2002) PGE_2 is generated by a specific COX-2 activity and increases VEGF production in COX-2-expressing human pancreatic cells. *Biochem Biophys Res Commun* 306: 887–897
235. Amano H, Hayashi I, Toshida S, Yoshimura H, Majima M (2002) Cyclooxygenase-2 and adenylate cyclase/protein kinase A signaling pathway enhances angiogenesis through induction of vascular endothelial growth factor in sponge implants. *Hum Cell* 15: 13–24
236. Wang YQ, Luk JM, Ikeda K, Man K, Chu AC, Kaneda K, Fan ST (2004) Regulatory role of vHL/HIF-1alpha in hypoxia-induced VEGF production in heaptic stellate cells. *Biochem Biophys Res Commun* 317: 358–362
237. Tamarat R, Silvestre JS, Durie M, Levy BI (2002) Angiotensin II angiogenic effect *in vivo* involves vascular endothelial growth factor- and inflammation-related pathways. *Lab Invest* 82: 747–756

238. Chen PY, Long QC (2004) Effects of cyclooxygenase 2 inhibitors on biological traits in nasopharyngeal carcinoma cells. *Acta Pharmacol Sin* 25: 943–949
239. Woolf CJ, Salter MW (2000) Neuronal Plasticity: increasing the gain in pain. *Science* 288: 1765–1769
240. Vanegas H, Schaible H-G (2001) Prostaglandins and cyclooxygenases in the spinal cord. *Prog Neurobiol* 64: 327–363
241. Dolan S, Nolan AM (1990) N-methyl D-aspartate induced mechanical allodynia is blocked by nitric oxide synthase and cyclooxygenase-2 inhibitors. *Neuroreport* 10: 449–452
242. Meller ST, Gebhart GF (1993) Nitric oxide (NO) and nociceptive processing in the spinal cord. *Pain* 52: 127–136
243. Hao JX, Xu XJ (1996) Treatment of a chronic allodynia-like response in spinally injured rats: effects of systemically administered nitric oxide synthase inhibitors. *Pain* 66: 313–319
244. Pardutz A, Krizbai I, Multon S, Vecsei L, Schoenen J (2000) Systemic nitroglycerin increases nNOS levels in rat trigeminal nucleus caudalis. *Neuroreport* 11: 3071–3075
245. Wu J, Fang L, Lin Q, Willis WD (2000) Fos expression is induced by increased nitric oxide release in rat spinal cord dorsal horn. *Neuroscience* 96: 351–357
246. Jurna I, Brune K (1990) Central effect of the non-steroid anti-inflammatory agents, indomethacin, ibuprofen, and diclofenac, determined in C fibre-evoked activity in single neurones of the rat thalamus. *Pain* 41: 71–80
247. Bianchi M, Panerai AE (2001) Anti-hyperalgesic effects of lornoxicam, piroxicam, meloxicam, ketorolac and aspirin in rats. *Br J Pharmacol* 133 (suppl 1): 51P
248. Tassorelli C, Greco R, Sandrini G, Nappi G (2003) Central components of the analgesic/antihyperalgesic effect of nimesulide: studies in animal models of pain and hyperalgesia. *Drugs* 63 (suppl 1): 9–22
249. Bianchi M, Broggini M (2002) Anti-hyperalgesic effects of nimesulide: studies in rats and humans. *Int J Clin Pract* 128 (suppl)11–19
250. Bianchi M, Panerai AE (1997) Formalin injection in the tail facilitates hindpaw withdrawal reflexes induced by thermal stimulation in the rat: effect of paracetamol. *Neurosci Lett* 237: 89–92
251. Biella G, Bianchi M, Sotgiu ML (1999) Facilitation of spinal sciatic neuron responses to hindpaw thermal stimulation after formalin injection in rat tail. *Exp Brain Res* 126: 501–508
252. Baba H, Kohno T, Moore KA, Woolf CJ (2001) Direct activation of rat spinal dorsal horn neurons by prostaglandin E_2. *J Neurosci* 21: 1750–1756
253. Samad TA, Moore KA, Sapirstein A, Billet S, Allchorne A, Poole S, Bonventre JV, Woolf CJ (2001) Interleukin-1beta-mediated induction of Cox-2 in the CNS contributes to inflammatory pain hypersensitivity. *Nature* 410: 471–475
254. Sorkin LS, Moore JH (1996) Evoked release of amino acids and prostanoids in spinal cords of anesthetized rats: changes during peripheral inflammation and hyperalgesia. *Am J Ther* 3: 268–275

255. Dirig DM, Konin GP, Isakson PC, Yaksh TL (1997) Effect of spinal cyclooxygenase inhibitors in rat using the formalin test and *in vitro* prostaglandin E_2 release. *Eur J Pharmacol* 331: 155–160
256. Yamamoto T, Sakashita Y (1998) COX-2 inhibitor prevents the development of hyperalgesia induced by intrathecal NMDA or AMPA. *Neuroreport* 9: 3869–3873
257. Bianchi M, Limiroli E, Ferrario P, Sacerdote P (2001) Anti-hyperalgesic effects of lornoxicam in the rat: behavioural and biochemical evidence. *Inflamm Res* 5 (suppl 3): S207
258. McCormack KJ (1994) Non-steroidal anti-inflammatory drugs and spinal nociceptive processes. *Pain* 59: 9–43
259. Brune K, Beck WS, Geisslinger G, Menzel-Soglowski S, Peskar BM, Peskar, BA (1991) Aspirin-like drugs may block pain independently of prostaglandin synthesis inhibition. *Experientia* 47: 257–261
260. McCormack KJ (1994) The spinal action of NSAIDs and the dissociation between anti-inflammatory and analgesic effects. *Drugs* 47: 28–45
261. Hay CH, Trevethick MA, Wheeldon A, Bowers JS, de Belleroche JS (1997) The potential role of spinal cord cyclooxygenase-2 in the development of Freund's complete adjuvant-induced changes in hyperalgesia and allodynia. *Neuroscience* 78: 843–850
262. Stein C, Millan MJ, Herz A (1988) unilateral inflammation of the hindpaw in rats as a model of prolonged noxious stimulation: alterations in behaviour and nociceptive thresholds. *Pharmacol Biochem Behav* 31: 315-324
263. Ferreira SH, Nakamura M (1979) III – Prostaglandin hyperalgesia: relevance of the peripheral effect for the analgesic action of opioid-antagonists. *Prostaglandins* 18: 201–208
264. Ferreira SH, Lorenzetti BB, De Campos DI (1990) Induction, blockade and restoration of a persistent hypersensitive state. *Pain* 42: 365–371
265. Kumazawa T, Mizumura K, Koda H (1993) Involvement of EP3 subtype of prostaglandin E receptors in PGE_2-induced enhancement of the bradykinin response of nociceptors. *Brain Res* 632: 321–324
266. Anderson GD, Hauser SD, McGarity KL, Bremer ME, Isakson PC, Gregory SA (1996) Selective inhibition of cyclooxygenase (COX)-2 reverses inflammation and expression of COX-2 and interleukin 6 in rat adjuvant arthritis. *J Clin Invest* 97: 2672–2679
267. Tassorelli C, Joseph SA (1995) Systemic nitroglycerin induces Fos immunoreactivity in brainstem and forebrain structures of the rat. *Brain Res* 682: 167–178
268. Tassorelli C, Joseph SA, Buzzi G, Nappi G (1999) The effect on the central nervous system of nitroglycerin – Putative mechanisms and mediators. *Prog Neurobiol* 57: 606–624
269. Tassorelli C, Joseph SA, Nappi G (1997) Neurochemical mechanisms of nitroglycerin-induced neuronal activation. *Neuropharmacology* 10: 1417–1424
270. Tassorelli C, Greco R, Morocutti A, Costa A, Nappi G (2001) Nitric oxide-induced neuronal activation in the central nervous system as an animal model for migraine: mechanisms and mediators. *Funct Neurol* 16 (suppl 4): 69–76

271. Tassorelli C, Greco R, Wang DC, Sandrini M, Sandrini G, Nappi G (2003) Nitroglycerin induces hyperalgesia in rats – a time-course study. *Eur J Pharmacol* 464: 159–162
272. Jones MG, Lever I, Bingham S, Read S, McMahon SB, Parsons A (2001) Nitric oxide potentiates response of trigeminal neurones to dural or facial stimulation in the rat. *Cephalalgia* 21: 643–655
273, Reuter U, Chiarugi A, Bolay H, Moskowitz MA (2002) Nuclear factor-kappaB as a molecular target for migraine therapy. *Ann Neurol* 51: 507–516
274. Torfgard K, Ahnler J, Axelsson KL, Norlander B, Bertler A (1989) Tissue distribution of glyceryl trinitrate and the effect on cGMP levels in rat. *Pharmacology & Toxicology* 64: 369–372
275. Mashimo T, Pak M, Choe H, Inagaki Y, Yamamoto M, Yoshiya I (1997) Effects of vasodilators guanethidine, nicardipine, nitroglycerin, and prostaglandin E_1 on primary afferent nociceptors in humans. *J Clin Pharmacol* 37: 330–335
276. Tjolsen A, Berge OG, Hunskaar S, Rosland JH, Hole K (1992) The formalin test: an evaluation of the method. *Pain* 51: 5–17
277. Taniguchi Y, Yokoyama K, Inui K, Deguchi Y, Furukawa K, Noda K (1997) Inhibition of brain cyclooxygenase-2 activity and the antipyretic action of nimesulide. *Eur J Pharmacol* 330: 221–229
278. Coderre TJ, Katz J, Vaccarino AL, Melzack R (1993) Contribution of central neuroplasticity to pathological pain: review of clinical and experimental evidence. *Pain* 52: 259–285
279. Sasaki M, Tohda C, Kuraishi Y (1998) Region-specific increase in glutamate release from dorsal horn of rats with adjuvant inflammation. *Neuroreport* 9: 3219–3122
280. DeGroot J, Zhow S, Carlton SM (2000) Peripheral glutamate release in the hind paw following low and high intensity sciatic stimulation. *Neuroreport* 14: 497–502
281. Lawand NB, McNearney T, Westlund KN (2000) Amino acid release into the knee joint: key role in nociception and inflammation. *Pain* 86: 69–74
282. Tegeder I, Niederberger E, Vetter G, Brauitigam L, Geisslinger G (2001) Effects of selective $COX\text{-}_1$ and -_2 inhibition on formalin-evoked nociceptive behaviour and prostaglandin E_2 release in the spinal cord. *J Neurochem* 79: 777–786
283. Beiche F, Geisslinger G, Goppelt-Struebe M (1988) Expression of cyclooxygenase isoforms in the rat spinal cord and their regulation during adjuvant-induced arthritis. *Inflamm Res* 47: 482–487
284. Maihofner C, Tegeder I, Euchenhofer C, deWitt D, Brune K, Bang R, Neuhunber W, Geisslinger G (2000) Localization and regulation of cyclo-oxygenase-1 and -2 and neuronal nitric oxide synthase in mouse spinal cord. *Neuroscience* 101: 1093–1108
285. Ferreira SH, Lorenzetti BB (1996) Intrathecal administration of prostaglandin E_2 causes sensitization of the primary afferent neuron via the spinal release of glutamate. *Inflamm Res* 45: 499–502
286. Kawamata T, Omote K (1999) Activation of spinal N-methyl-D-aspartate receptor stimulates a nitric oxide/cyclic guanosine 3,5-monophosphate/glutamate release cascade in nociceptive signaling. *Anesthesiology* 91: 1415–1424

287. Sakurada C, Sugiyama A, Nakayama M, Yonezawa A, Sakurada S, Tan-No K, Kisara K, Sakurada Y (2001) Antinociceptive effect of spinally injected L-NAME on the acute nociceptive response induced by low concentrations of formalin. *Neurochem Int* 38: 417–423
288. Kaube H, Hoskin HL, Goadsby PJ (1993) Intravenous acetylsalicylic acid inhibits central trigeminal neurons in the dorsal horn of the upper cervical spinal cord in the cat. *Headache* 33: 541–544
289. Sandrini M, Vitale G, Pini LA (2002) Central antinociceptive activity of acetylsalicylic acid is modulated by brain serotonin receptor subtypes. *Pharmacology* 65: 193–197
290. Willer JC (1990) Clinical exploration of nociception with the use of reflexologic techniques. *Neurophysiol Clin* 20: 335–356
291. Sandrini G, Arrigo A, Bono G, Nappi G (1993) The nociceptive flexion reflex as a tool for exploring pain control systems in headache and other pain syndromes. *Cephalalgia* 13: 21–27
292. Sandrini G, Alfonsi E, Bono G, Facchinetti F, Montalbetti L, Nappi G (1986) Circadian variations of human flexion reflex. *Pain* 25: 403–410
293. Sandrini G, Ruiz L, Capararo M, Garofoli F, Beretta A, Nappi G (1992) Central analgesic activity of ibuprofen. A neurophysiological study in humans. *Int J Clin Pharmacol Res* 12: 197–204
294. Willer JC (1985) Studies on pain: Effects of morphine on a spinal nociceptive flexion reflex and related pain sensation in man. *Brain Res* 331: 105–114
295. Sandrini G, Proietti Cecchini A, Alfonsi E, Nappi G (2001) The effectiveness of nimesulide in pain. A neurophysiological study in humans. *Drugs of Today* 37 (suppl B): 21–29
296. Sandrini G, Tassorelli C, Cecchini AP, Alfonsi E, Nappi G (2002) Effects of nimesulide on nitric oxide-induced hyperalgesia in humans – a neurophysiological study. *Eur J Pharmacol* 450: 259–262
297. Willer JC, De Broucker T, Bussel B, Roby-Brami A, Harrewyn JM (1989) Central analgesic effect of ketoprofen in humans: electrophysiological evidence for a supraspinal mechanism in a double-blind and cross-over study. *Pain* 38: 1–7
298. Cooke TDV (1985) Mechanisms of cartilage degradation: relation to choice of therapeutic agent. *Semin Arthritis Rheum* 15 (2 Suppl 1): 16–23
299. Burkhardt D, Ghosh P (1987) Laboratory evaluation of antiarthritic drugs as potential chondroprotective agents. *Semin Arthritis Rheum* 17 (2 Suppl 1): 3–34
300. Doherty M (1989) 'Chondroprotection' by non-steroidal anti-inflammatory drugs. *Ann Rheum Dis* 48: 619–621
301. Ghosh P, Wells C, Smith M, Hutadilok N (1990) Chondroprotection, myth or reality: An experimental approach. *Semin Arthritis Rheum* 19 (4 Suppl 1): 3–9
302. Rainsford KD, Rashad SY, Revell PA, Low FM, Hemingway AP, Walker FS, Johnson D, Stetsko P, Ying C, Smith F (1992) Effects of NSAIDs on cartilage proteoglycan and synovial prostaglandin metabolism in relation to progression of joint deterioration in osteoarthritis. In: Bálint G, Gömör B, Hadinka L (Eds): *Rheumatology, State of the Art.* Elsevier, Amsterdam. 177–183

303. Rashad S, Rainsford K, Revell P, Low F, Hemingway A, Walker F (1992) The effects of NSAIDS on the course of osteoarthritis. In: Bálint G, Gömör B, Hódinka L (Eds): *Rheumatology, State of the Art*. Elsevier, Amsterdam. 184–188
304. Jones AC, Doherty M (1992) The treatment of osteoarthritis. *Br J Clin Pharmacol* 33: 357–363
305. Brandt KD (1993) Should osteoarthritis be treated with nonsteroidal anti-inflammatory drugs? *Rheum Dis Clin North Am* 19: 697–712
306. Brandt KD (1993) NSAIDs in the treatment of osteoarthritis. Friends or foes? *Bull Rheum Dis* 42: 1–4
307. Jobanputra P, Nuki G (1994) Nonsteroidal anti-inflammatory drugs in the treatment of osteoarthritis. *Curr Opin Rheumatol* 6: 433–439
308. Brady SJ, Brooks P, Conaghan P, Kenyon LM (1997) Pharmacotherapy and osteoarthritis. *Baillieres Clin Rheumatol* 11: 749–768
309. Buckland-Wright C (1999) Evaluation of disease progression during nonsteroidal antiinflammatory drug treatment: imaging X-rays. *Osteoarthritis Cartilage* 7: 343–344
310. Dougados M (2001) The role of anti-inflammatory drugs in the treatment of osteoarthritis: a European viewpoint. *Clin Exp Rheumatol* 19 (6 Suppl 25): S9–14
311. Ding C (2002) Do NSAIDs affect the progression of osteoarthritis? *Inflammation* 26: 139–142
312. Dieppe P, Brandt KD (2003) What is important in treating osteoarthritis? Whom should we treat and how should we treat them? *Rheum Dis Clin North Am* 29: 687–716
313. Dieppe P, Bartlett C, Davey P, Doyal L, Ebrahim S (2004) Balancing benefits and harms: the example of non-steroidal anti-inflammatory drugs. *Br Med J* 329: 31–34
314. Tindall EA, Sharp JT, Burr A, Katz TK, Wallemark CB, Verburg K, Lefkowith JB (2002) A 12-month, multicenter, prospective, open-label trial of radiographic analysis of disease progression in osteoarthritis of the knee or hip in patients receiving celecoxib. *Clin Ther* 24: 2051–2063
315. Gandy SJ, Dieppe PA, Keen MC, Maciewicz RA, Watt I, Waterton JC (2002) No loss of cartilage volume over three years in patients with knee osteoarthritis as assessed by magnetic resonance imaging. *Osteoarthritis Cartilage* 10: 929–937
316. Gineyts E, Mo JA, Ko A, Henriksen DB, Curtis SP, Gertz BJ, Garnero P, Delmas PD (2004) Effects of ibuprofen on molecular markers of cartilage and synovium turnover in patients with knee osteoarthritis. *Ann Rheum Dis* 63: 857–861
317. Abadie E, Ethgen D, Avouac B, Bouvenot G, Branco J, Bruyere O, Calvo G, Devogelaer JP, Dreiser RL, Herrero-Beaumont G et al. Group for the Respect of Excellence and Ethics in Science (2004) Recommendations for the use of new methods to assess the efficacy of disease-modifying drugs in the treatment of osteoarthritis. *Osteoarthritis Cartilage* 12: 263–268
318. Murray RO (1976) Iatrogenic lesions of the skeleton. Caldwell lecture, 1975. *Am J Roentgenol* 126: 5–22
319. Allen EH, Murray RO (1971) Iatrogenic arthropathies. *Eur Assoc Radiol Practice. Excerpta Medica* 249: 204–210

320. Coke H (1967) Long-term indomethacin therapy of coxarthrosis. *Ann Rheum Dis* 26: 346–347
321. Goldie I (1978) Osteonekros och indomethacin *Läkartidningen* 75: 1275–1277
322. Hauge MF (1975) Hofteleddsartrose–indomethacin. *Tidsskr Nor Laegeforen* 95: 1594–1603
323. Serup J, Ovesen JO (1981) Salicylate-arthropathy. Accelerated coxarthrosis during long-term treatment with acetylsalicylic acid. *Schweiz Rundsch Med Prax* 70: 359–361
324. Rashad S, Revell P, Hemingway A, Low F, Rainsford K, Walker FS (1989) Effect of non-steroidal anti-Inflammatory drugs on the course of osteoarthritis. *Lancet* 2, 519–522
325. Mastbergen SC, Lafeber FP, Bijlsma JW (2002) Selective COX-2 inhibition prevents proinflammatory cytokine-induced cartilage damage. *Rheumatology* 41: 801–808
326. Hardy MM, Seibert K, Manning PT, Currie MG, Woerner BM, Edwards D, Koki A, Tripp CS (2002) Cyclooxygenase 2-dependent prostaglandin E_2 modulates cartilage proteoglycan degradation in human osteoarthritis explants. *Arthritis Rheum* 46: 1789– 1803
327. El Hajjaji H, Marcelis A, Devogelaer JP, Manicourt DH (2003) Celecoxib has a positive effect on the overall metabolism of hyaluronan and proteoglycans in human osteoarthritic cartilage. *J Rheumatol* 30: 2444–2451
328. Sylvia VL, Del Toro F Jr, Hardin RR, Dean DD, Boyan BD, Schwartz Z (2001) Characterization of PGE_2 receptors (EP) and their role as mediators of 1alpha,25-(OH)(2)D(3) effects on growth zone chondrocytes. *J Steroid Biochem Mol Biol* 78: 261–274
329. Miyamoto M, Ito, H, Kobayashi T, Yamamoto H, Kobayashi M, Maruyama T, Akiyama H, Nakamura T (2003) Simultaneous stimulation of EP2 and EP4 is essential to the effect of prostaglandin E_2 in chondrocyte differentiation. *Osteoarthritis Cartilage* 11: 644–652
330. de Brum-Fernandes AJ, Morisset S, Bkaily G, Patry C (1996) Characterization of the PGE_2 receptor subtype in bovine chondrocytes in culture. *Br J Pharmacol* 118: 1597–1604
331. Amin AR, Attur M, Patel RN, Thakker GD, Marshall PJ, Rediske J, Stukin SA, Patel IR, Abramson SB (1997) Superinduction of cyclooxygenase-2 activity in human osteoarthritic affected cartilage. *J Clin Invest* 99: 1231–1237
332. Amin AR, Attur M, Abramson SB (1998) Regulation of nitric oxide and inflammatory mediators in human osteoarthritic-affected cartilage: implication for pharmacological intervention. In: Rubanyl GW (Ed): *The Pathophysiological and Clinical Application of Nitric Oxide*. Harvard Academic Publishers, Boston. 397–412
333. Goldring MB, Berenbaum F, Buckwalter J (2004) The regulation of chondrocyte function by proinflammatory mediators: prostaglandins and nitric oxide. *Clin Orthop* 427 (Suppl): S37–S46
334. Clausen PA, Flechtenmacher J, Haeuselmann HJ, Kuettner KE, Aydelotte MB, Iyer AP (1996) Evidence of an eicosanoid contribution to IL-1 induction of IL-6 in human artcular chondrocytes. *Am J Ther* 3: 101–108

335. Woolley DE, Tetlow LC (2000) Mast cell activation and its relation to proinflammatory cytokine production in rheumatoid lesion. *Arthritis Res* 2: 65–74
336. Kerr JS, Stevens TM, Davis GL, McLaughlin JA, Harris RR (1989) Effects of recombinant interleukin-1 beta on phospholipase activity, phospholipase A_2 mRNA, and eicosanoid formation in rabbit chondrocytes. *Biochem Biophys Res Commun* 29: 1079–1084
337. Martel-Pelletier J, Mineau F, Fahmi H, Laufer S, Reboul P, Boileau C, Lavigne M, Pelletier JP (2004) Regulation of the expression of 5-lipoxygenase-activating protein/ 5-lipoxygenase and the synthesis of leukotriene B_4 in osteoarthritic chondrocytes: role of transforming growth factor beta and eicosanoids. *Arthritis Rheum* 50: 3925–3933
338. Amat M, Diaz C, Vila L (1998) Leukotriene A_4 hydrolase and leukotriene C_4 synthase activities in human chondrocytes: transcellular biosynthesis of leukotrienes during granulocyte-chondrocyte interaction. *Arthritis Rheum* 41: 1645–1651
339. Rouzer CA, Kargmann S (1988) Translocation of 5-lipoxygenase to the membrane in human leucocytes challenged with ionophore A23187. *J Biol Chem* 263: 10980–10988
340. Serhan CN (1997) Lipoxins and novel aspirin-triggered 15-epi-lipoxins (ATL): a jungle of cell–cell interactions or a therapeutic opportunity? *Prostaglandins* 53: 107–137
341. Brandwein SR (1986) Regulation of interleukin 1 production by mouse peritoneal macrophages. Effects of arachidonic acid metabolites, cyclic nucleotides, and interferons. *J Biol Chem* 261: 8624–8632
342. Brandwein SR (1990) Differential regulation of soluble interleukin 1 release and membrane expression by pharmacologic agents. *Agents Actions* 30: 381–392
343. Bonta IL, Elliott GR (1992) Non-steroidal anti-inflammatory drugs and the augmented lipoxygenase pathways: conceivable impact on joints. In: Rainsford KD, Velo GP (Eds): *Side-effects of Anti-inflammatory Drugs 3*. Kluwer Academic Publishers, Dordrecht. 269–274
344. Swierkosz TA, Mitchell JA, Warner TD, Botting RM, Vane JR (1995) Co-induction of nitric oxide synthase and cyclooxygenase: interactions between nitric oxide and prostanoids. *Br J Pharmacol* 114: 1335–1342
345. Sanchez de Miguel L, de Frutos T, Gonzalez-Fernandez F, del Pozo V, Lahoz C, Jimenez A, Rico L, Garcia R, Aceituno E, Millas I et al. (1999) Aspirin inhibits inducible nitric oxide synthase expression and tumour necrosis factor-alpha release by cultured smooth muscle cells. *Eur J Clin Invest* 29: 93–99
346. Wang ZY, Brecher P (1999) Salicylate inhibition of extracellular signal-related kinases and inducible nitric oxide synthase. *Hypertension* 34: 1259–1264
347. Rola-Pleszczynski M, Thivierge M, Gagnon N, Lacasse C, Stankova J (1993) Differential regulation of cytokine and cytokine receptor genes by PAF, LTB_4 and PGE_2. *J Lipid Mediat* 6:175–181
348. Harbrecht BG, Kim YM, Wirant EM, Shapiro RA, Billiar TR (1996) PGE_2 and LTB_4 inhibit cytokine-stimulated nitric oxide synthase type 2 expression in isolated rat hepatocytes. *Prostaglandins* 52: 103–116

349. Harizi H, Norbert G (2004) Inhibition of IL-6, TNF-alpha, and cyclooxygenase-2 protein expression by prostaglandin E_2-induced IL-10 in bone marrow-derived dendritic cells. *Cell Immunol* 228: 99–109
350. Fleming DC, Kelly RW (2004) Prostaglandins and the immune response. In: Curtis-Prior P (Ed): *The Eicosanoids*. Wiley, Chichester. 237–245
351. Rainsford KD (2004) Cytokines and eicosanoids in arthritis. In: Curtis-Prior P (Ed): *The Eicosanoids*. Wiley, Chichester. 347–358
352. Stamp LK, Cleland LG, James MJ (2004) Upregulation of synoviocyte COX-2 through interactions with T lymphocytes: role of interleukin 17 and tumor necrosis factor-alpha. *J Rheumatol* 31: 1246–1254
353. Arend WP (2001) Cytokine imbalance in the pathogenesis of rheumatoid arthritis: the role of interleukin-1 receptor antagonist. *Semin Arthritis Rheum* 30 (Suppl 2): 1–6
354. van Roon JA, Bijlsma JW, Lafeber FP (2002) Suppression of inflammation and joint destruction in rheumatoid arthritis may require a concerted action of Th2 cytokines. *Curr Opin Investig Drugs* 3: 1011–1016
355. Taylor PC (2003) Anti-cytokines and cytokines in the treatment of rheumatoid arthritis. *Curr Pharm Des* 9:1095–1106
356. Kato T, Xiang Y, Nakamura H, Nishioka K (2004) Neoantigens in osteoarthritic cartilage. *Curr Opin Rheumatol* 16: 604–608
357. Ghosh P, Cheras PA (2001) Vascular mechanisms in osteoarthritis. *Best Pract Res Clin Rheumatol* 15: 693–709
358. Burr DB (1998) The importance of subchondral bone in osteoarthrosis. *Curr Opin Rheumatol* 10: 256–262
359. Wilbrink B, Nietfeld JJ, den Otter W, van Roy JL, Bijlsma JW, Huber-Bruning O (1991) Role of TNF alpha, in relation to IL-1 and IL-6 in the proteoglycan turnover of human articular cartilage. *Br J Rheumatol* 30: 265–271
360. Neidel J, Zeidler U (1993) Independent effects of interleukin 1 on proteoglycan synthesis and proteoglycan breakdown of bovine articular cartilage *in vitro*. *Agents Actions* 39: 82–90
361. Sakkas LI, Platsoucas CD (2002) Role of T cells in the pathogenesis of osteoarthritis. *Arthritis Rheum* 46: 3112–3113
362. Haynes MK, Hume EL, Smith JB (2002) Phenotypic characterization of inflammatory cells from osteoarthritic synovium and synovial fluids. *Clin Immunol* 105: 315–325
363. Sakata M, Masuko-Hongo K, Nakamura H, Onuma H, Tsuruha JI, Aoki H, Nishioka K, Kato T (2003) Osteoarthritic articular chondrocytes stimulate autologous T cell responses *in vitro*. *Clin Exp Rheumatol* 21: 704–710
364. Haynes DR, Crotti TN (2003) Regulation of bone lysis in inflammatory diseases. *Inflammopharmacology* 11: 323–331
365. Martel-Pelletier J, Di Battista JA, Lajeunesse D, Pelletier JP (1998) IGF/IGFBP axis in cartilage and bone in osteoarthritis pathogenesis. *Inflamm Res* 1998 Mar; 47: 90–100

366. Rainsford KD, Omar H, Ashraf A, Hewson AT, Bunning RAD, Rishiraj R, Shepherd P, Seabrook RW (2002) Recent pharmacodynamic and pharmacokinetic findings on oxaprozin. *Inflammopharmacology* 10: 185–239
367. Dingle JT, Parker M (1997) NSAID stimulation of human cartilage matrix synthesis. *Clin Drug Invest* 14: 353–362
368. Glazer PA, Rosenwasser MP, Ratcliffe A (1993) The effect of naproxen and interleukin-1 on proteoglycan catabolism and on neutral metalloproteinase activity in normal cartilage *in vitro*. *J Clin Pharmacol* 33: 109–114
369. Rainsford KD (1992) Effects of anti-inflammatory drugs and agents that modify signal transduction signals or metabolic activities on interleukin 1-induced cartilage proteoglycan resorption *in vitro*. *Pharmacol Res* 25: 335–346
370. Rainsford KD, Ying C, Smith FC (1997) Effects of meloxicam compared with other NSAIDs on proteoglycan metabolism, synovial prostaglandin E_2, interleukins-1, 6 and 8 production, in human and porcine explants in organ culture. *J Pharm Pharmacol* 49: 991–998
371. Blot L, Marcelis A, Devogelaer JP, Manicourt DH (2000) Effects of diclofenac, aceclofenac and meloxicam on the metabolism of proteoglycans and hyaluronan in osteoarthritic human cartilage. *Br J Pharmacol* 131: 1413–1421
372. Choi JH, Choi JH, Kim DY, Yoon JH, Youn HY, Yi JB, Rhee HI, Ryu KH, Jung K, Han CK et al. (2002) Effects of SKI 306X, a new herbal agent, on proteoglycan degradation in cartilage explant culture and collagenase-induced rabbit osteoarthritis model. *Osteoarthritis Cartilage* 10: 471–478
373. El Hajjaji H, Marcelis A, Devogelaer JP, Manicourt DH (2003) Celecoxib has a positive effect on the overall metabolism of hyaluronan and proteoglycans in human osteoarthritic cartilage. *J Rheumatol* 30: 2444–2451
374. Fossati A (1999) Antiinflammatory effects of Seaprose-S on various inflammation models. *Drugs Exp Clin Res* 25: 263–270
375. Frean SP, Abraham LA, Lees P (1999) *In vitro* stimulation of equine articular cartilage proteoglycan synthesis by hyaluronan and carprofen. *Res Vet Sci* 67: 183–190
376. Ghosh P, Hutadilok N (1996) Interactions of pentosan polysulfate with cartilage matrix proteins and synovial fibroblasts derived from patients with osteoarthritis. *Osteoarthritis Cartilage* 4: 43–53
377. Hardy MM, Seibert K, Manning PT, Currie MG, Woerner BM, Edwards D, Koki A, Tripp CS (2002) Cyclooxygenase 2-dependent prostaglandin E2 modulates cartilage proteoglycan degradation in human osteoarthritis explants. *Arthritis Rheum* 46: 1789–1803
378. Henrotin YE, Labasse AH, Simonis PE, Zheng SX, Deby GP, Famaey JP, Crielaard JM, Reginster JY (1999) Effects of nimesulide and sodium diclofenac on interleukin-6, interleukin-8, proteoglycans and prostaglandin E_2 production by human articular chondrocytes *in vitro*. *Clin Exp Rheumatol* 17: 151–160
379. Lafeber FP, Beukelman CJ, van den Worm E, van Roy JL, Vianen ME, van Roon JA, van Dijk H, Bijlsma JW (1999) Apocynin, a plant-derived, cartilage-saving drug,

might be useful in the treatment of rheumatoid arthritis. *Rheumatology (Oxford)* 38: 1088–1093

380. Mastbergen SC, Lafeber FP, Bijlsma JW (2002) Selective COX-2 inhibition prevents proinflammatory cytokine-induced cartilage damage. *Rheumatology (Oxford)* 41: 801–808
381. Rainsford KD, Skerry TM, Chindemi P, Delaney K (1999) Effects of the NSAIDs meloxicam and indomethacin on cartilage proteoglycan synthesis and joint responses to calcium pyrophosphate crystals in dogs. *Vet Res Commun* 23: 101–113
382. Bassleer C, Magotteaux J, Geenen V, Malaise M (1997) Effects of meloxicam compared to acetylsalicylic acid in human articular chondrocytes. *Pharmacology.* 54: 49–56
383. Beluche LA, Bertone AL, Anderson DE, Rohde C (2001) Effects of oral administration of phenylbutazone to horses on *in vitro* articular cartilage metabolism. *Am J Vet Res* 62: 1916–1921
384. Rainsford KD, Davies A, Mundy L, Ginsburg I (1998) Comparative effects of azapropazone on cellular events at inflamed sites. Influence on joint pathology, leucocyte superoxide and eicosanoid production and actions on interleukin-1 – induced cartilage resorption correlated with drug uptake into cartilage *in vitro*. *J Pharma Pharmacol* 41: 322–330
385. Rainsford KD, Ying C, Smith FC (1996) Effects of 5-lipoxygenase inhibitors on interleukin-1 production by human synovial tissues in organ culture: comparison with interleukin-1 synthesis inhibitors. *J Pharm Pharmacol* 48: 45–50
386. Walker FS, Rainsford KD (1997) Do NSAIDS adversely affect joint pathology in osteoarthritis? In: KD Rainsford (ed.) Side effects of anti-inflammatory drugs – IV. Kluwer Academic Publishers, Dordrecht, 43–53
387. Steinmeyer J, Daufeldt S (1997) Pharmacological influence of antirheumatic drugs on proteoglycans from interleukin-1 treated articular cartilage. *Biochem Pharmacol* 53: 1627–1635
388. Torzilli PA, Tehrany AM, Grigiene R, Young E (1996) Effects of misoprostol and prostaglandin E2 on proteoglycan biosynthesis and loss in unloaded and loaded articular cartilage explants. *Prostaglandins* 52: 157–173
389. Wang B, Yao YY, Chen MZ (1998) Effects of indomethacin on joint damage in rat and rabbit. *Zhongguo Yao Li Xue Bao* 19:70–73
390. Solignac M (2004) Mechanisms of action of diacerein, the first inhibitor of interleukin-1 in osteoarthritis. *Presse Med* 33 (Pt 2): S10–S12
391. Botrel MA, Haak T, Legrand C, Concordet D, Chevalier R, Toutain PL (1994) Quantitative evaluation of an experimental inflammation induced with Freund's complete adjuvant in dogs. *J Pharmacol Toxicol Methods* 32: 63–71
392. Gilroy DW, Tomlinson A, Greenslade K, Seed MP, Willoughby DA (1998) The effects of cyclooxygenase 2 inhibitors on cartilage erosion and bone loss in a model of Mycobacterium tuberculosis-induced monoarticular arthritis in the rat. *Inflammation* 22: 509–519

393. Pelletier J-P, Martel-Pelletier J (1993) Effects of nimesulide and naproxen on the degradation and metalloproteinase synthesis of human osteoarthritic cartilage. *Drugs* 46 (Suppl 1): 34–39
394. Sanchez C, Mateus MM, Defresne M-P, Crielaard J-MR, Reginster J-YL, Henrotin YE (2002) Metabolism of human articular chondrocytes cultured in alginate beads. Long-term effects of interleukin 1β and nonsteroidal anti-inflammatory drugs. *J Rheumatol* 29: 772–782
395. Frean SP, Cambridge H, Lees P (2002) Effects of anti-arthritic drugs on proteoglycan synthesis by equine cartilage. *J Vet Pharmacol Ther* 25: 289–298
396. Amin AR, Abramson SB (1998) The role of nitric oxide in articular cartilage breakdown in osteoarthritis. *Curr Opin Rheumatol* 10: 263–268
397. Tung JT, Venta PJ, Caron JP (2002) Inducible nitric oxide expression in equine articular chondrocytes: effects of anti-inflammatory compounds. *Osteoarthritis Cartilage* 10: 5–12
398. Mathy-Hartert M, Deby-Dupont GP, Reginster J-Y, Ayache N, Pujol JP, Henrotin YE (2002) Regulation by reactive oxygen species of interleukin-1 beta, nitric oxide and prostaglandin E_2 production by human chondrocytes. *Osteoarthritis Cartilage* 10: 547–555
399. Barracchini A, Franceschini N, Amicosante G, Oratore A, Minisolat G, Pantaleoni G, Di Giulio A (1998) Can non-steroidal anti-inflammatory drugs act as metalloproteinase modulators? An *in vitro* study of inhibition of collagenase activity. *J Pharm Pharmacol* 50: 1417–1423
400. Bevilacqua M, Devogelaer J-P, Righini V, Famaey J-P, Manicourt D-H (2004) Effect of nimesulide on the serum levels of hyaluronan and stromelysin-1 in patients with osteoarthritis: a pilot study. *Int J Clin Pract* (Suppl 144): 13–19
401. Kullich WC, Niksic F, Klein G (2002) Effect of nimesulide on metallo-proteinases and matrix degradation in osteoarthritis: A pilot clinical study. *Int J Clin Pract* (Suppl) 128: 24–29
402. Pelletier JP, Mineau F, Fernandes J, Kiansa K, Ranger P, Martel-Pelletier J (1997) Two NSAIDs, nimesulide and naproxen, can reduce the synthesis of urokinase and IL-6 while increasing PAI-1, in human OA synovial fibroblasts. *Clin Exp Rheumatol* 15: 393–398
403. Pelletier JP, Di Battista JA, Zhang M, Fernandes J, Alaaeddine N, Martel-Pelletier J (1999) Effect of nimesulide on glucocorticoid receptor activity in human synovial fibroblasts. *Rheumatology (Oxford)* 38 (Suppl 1): 11–13
404. Di Battista JA, Fahmi H, He Y, Zhang M, Martel-Pelletier J, Pelletier JP (2001) Differential regulation of interleukin-1 beta-induced cyclooxygenase-2 gene expression by nimesulide in human synovial fibroblasts. *Clin Exp Rheumatol* 19: S3–S5
405. Di Battista JA, Zhang M, Martel-Pelletier J, Fernandes J, Alaaeddine N, Pelletier JP (1999) Enhancement of phosphorylation and transcriptional activity of the glucocorticoid receptor in human synovial fibroblasts by nimesulide, a preferential cyclooxygenase 2 inhibitor. *Arthritis Rheum* 42: 157–166

406. de Paulis A, Ciccarelli A, Marino I, de Crescenzo G, Marino D, Marone G (1997) Human synovial mast cells. II. Heterogeneity of the pharmacologic effects of antiinflammatory and immunosuppressive drugs. *Arthritis Rheum* 40: 469–478
407. Casolaro V, Meliota S, Marino O, Patella V, de Paulis A, Guidi G, Marone G (1993) Nimesulide, a sulfonanilide nonsteriodal anti-inflammatory drug, inhibits mediator release from human basophils and mast cells. *J Pharmacol Exp Therap* 267: 1375–1385
408. Hashimoto S, Takahashi K, Amiel D, Coutts RD, Lotz M (1998) Chondrocyte apoptosis and nitric oxide production during experimentally induced osteoarthritis. *Arthritis Rheum* 41: 1266–1274
409. Hashimoto S, Ochs RL, Komiya S, Lotz M (1998) Linkage of chondrocyte apoptosis and cartilage degradation in human osteoarthritis. *Arthritis Rheum* 41: 1632–1638
410. Hashimoto S, Ochs RL, Rosen F (1998) Chondrocyte-derived apoptotic bodies and calcification of articular cartilage. *Proc Natl Acad Sci USA* 95: 3094–3099
411. Pelletier J-P, Jovanovic DV, Lasaucoman V (2000) Selective inhibition of inducible nitric oxide synthase reduces progression of experimental osteoarthritis *in vivo*: Possible link with the reduction in chondrocyte apoptosis and caspase-3 level. *Arthritis Rheum* 43: 1290–1299
412. Miwa M, Saura R, Hirata S, Hayashi Y, Mizuno K, Itoh H (2000) Induction of apoptosis in bovine articular chondrocytes by prostaglandin E_2 through cAMP-dependent pathway. *Osteoarthritis Cartilage* 8: 17–24
413. Horton WE Jr, Feng L, Adams C (1998) Chondrocyte apoptosis in development, aging and disease. *Matrix Biol* 17: 107–115
414. Mukherjee P, Rachita C, Aisen PS, Pasinetti GM (2001) Non-steroidal anti-inflammatory drugs protect against chondrocyte apoptotic death. *Clin Exp Rheumatol* 19: S7–S11
415. Mukherjee P, Pasinetti GM (2002) Altered gene expression during nimesulide-mediated inhibition of apoptotic death in human chondrocytes. *Int J Clin Pract* (Suppl): 20–23
416. Maffei Facino R, Carini M, Aldini G, Saibene L, Morelli R (1995) Differential inhibition of superoxide, hydroxyl and peroxyl radicals by nimesulide and its main metabolite 4-hydroxynimesulide. *Arzneim Forsch* 45(II): 10–17
417. Facino RM, Carini M, Aldini G (1993) Antioxidant activity of nimesulide and its main metabolites. *Drugs* 46 (Suppl 1): 15–21
418. Maffei Facino R, Carini M, Aldini G, Saibene L, Macciocchi A (1993) Antioxidant profile of nimesulide, indomethacin and diclofenac in phosphatidylcholine liposomes (PCL) as membrane model. *Int J Tissue React* 15: 225–234
419. Zheng SX, Mouithys-Mickalad A, Deby-Dupont GP, Deby CM, Maroulis AP, Labasse AH, Lamy ML, Crielaard JM, Reginster JY, Henrotin YE (2000) *In vitro* study of the antioxidant properties of nimesulide and 4-OH nimesulide: effects on HRP- and luminol-dependent chemiluminescence produced by human chondrocytes. *Osteoarthritis Cartilage* 8: 419–425
420. Stanford SJ, Pepper JR, Mitchell JA (2000) Cyclooxygenase-2 regulates granulocyte-macrophage colony-stimulating factor, but not interleukin-8, production by human vascular cells: role of cAMP. *Arterioscler Thromb Vasc Biol* 20: 677–682

421. Ottino P, Bazan HE (2001) Corneal stimulation of MMP-1, -9 and uPA by platelet-activating factor is mediated by cyclooxygenase-2 metabolites. *Curr Eye Res* 23: 77–85
422. Turner MA, Vause S, Greenwood SL (2004) The regulation of interleukin-6 secretion by prostanoids and members of the tumor necrosis factor superfamily in fresh villous fragments of term human placenta. *J Soc Gynecol Investig* 11: 141–148
423. Ricote M, Li AC, Wilson TM, Kelly CJ, Glass CK (1998) The peroxisome proliferator-activated receptor-gamma is a negative regulator of macrophage activation. *Nature* 391: 79–82
424. Jiang C, Ting AT, Seed B (1998) PPAR-γ agonists inhibit production of monocyte inflammatory cytokines. *Nature* 391: 82–86
425. Fahmi H, Di Battista JA, Pelletier J-P, Mineau F, Ranger P, Martel-Pelletier J (2001) Peroxisome proliferators-activated receptor activators inhibit interleukin-1β-induced nitric oxide and matrix metalloproteinase 13 production in human chondrocytes. *Arthritis Rheum* 44: 595–607
426. Kalajdzic T, Faour WH, He QW, Fahmi H, Martel-Pelletier J, Pelletier JP, Di Battista JA (2002) Nimesulide, a preferential cyclooxygenase 2 inhibitor, suppresses peroxisome proliferator-activated receptor induction of cyclooxygenase 2 gene expression in human synovial fibroblasts: evidence for receptor antagonism. *Arthritis Rheum* 46: 494–506
427. Famaey J-P, Fontaine J, Reuse J (1975) Inhibiting effects of morphine, chloroquine, non-steroidal and steroidal anti-inflammatory drugs on electrically-induced contractions of guinea pig ileum and reversing effects of prostaglandins. *Agents Actions* 5: 354–358
428. Altura BM, Altura BY (1976) Vascular smooth muscle and prostaglandins. *Fed Proc* 35: 2360–2366
429. Vanhoutte PM, Rimele TJ, Rooke TW (1984) Calcium entry and the contraction of vascular smooth muscle. *Adv Cyclic Nucleotide Prot Phosph Res* 17: 569–573
430. Kurahashi K, Nishihashi T, Trandafir CC, Wang AM, Murakami S, Ji X (2003) Diversity of endothelium-derived vasocontracting factors – arachidonic acid metabolites. *Acta Pharmacol Sin* 24: 1065–1069
431. Malofiejew M, Blaszkiewicz Z (1979) The effect of nonsteroidal antiphlogistic drugs upon the spontaneous contractility of rats myometrium. *Gin Pol* 30: 299–304 (Polish)
432. Sawdy R, Knock GA, Bennett PR, Poston L, Aaronson PI (1998) Effect of nimesulide and indomethacin on contractility and the Ca^{2+} channel current in myometrial smooth muscle from pregnant women. *Br J Pharmacol* 125: 1212–1217
433. Baguma-Nibasheka M, Nathanielsz PW (1998) *In vivo* administration of nimesulide, a selective PGHS-2 inhibitor, increases *in vitro* myometrial sensitivity to prostaglandins while lowering sensitivity to oxytocin. *J Soc Gynecol Investig* 5: 296–299
434. Connolly C, McCormick PA, Docherty JR (1998) Effects of the selective cyclooxygenase-2 inhibitor nimesulide on vascular contractions in endothelium-denuded rat aorta. *Eur J Pharmacol* 352: 53–58
435. Slattery MM, Friel AM, Healy DG, Morrison JJ (2001) Uterine relaxant effects of cyclooxygenase-2 inhibitors *in vitro*. *Obstet Gynecol* 98: 563–569

436. Landen CN Jr, Zhang P, Young RC (2001) Differing mechanisms of inhibition of calcium increases in human uterine myocytes by indomethacin and nimesulide. *Am J Obstet Gynecol* 184:1100–1103
437. Karadas B, Kaya T, Guvenal T, Cetin M, Divrik I, Cetin A (2004) Comparison of the effects of nimesulide and 5,5-dimethyl-3-(3-fluorophenyl)-4-(4-methylsulphonyl) phenyl-2(5H)-furanone (DFU) on contractions of isolated pregnant human myometrium. *Eur J Obstet Gynecol Reprod Biol* 113: 172–177
438. Ross RG, Sathishkumar K, Naik AK, Bawankule DU, Sarkar SN, Mishra SK, Prakash VR (2004) Mechanisms of lipopolysaccharide-induced changes in effects of contractile agonists on pregnant rat myometrium. *Am J Obstet Gynecol* 190: 532–540
439. Karadas B, Kaya T, Bagcivan I, Kaloglu C, Guvenal T, Cetin A, Soydan AS (2004) Comparison of effects of cyclooxygenase inhibitors on myometrial contraction and constriction of ductus arteriosus in rats. *Eur J Pharmacol* 485: 289–282
440. White LR, Juul R, Cappelen J, Aasly J (2002) Cyclooxygenase inhibitors attenuate endothelin ET(B) mediated contraction in humnan temporal artery. *Eur J Pharmacol* 448: 51–57

Clinical applications of nimesulide in pain, arthritic conditions and fever

M. Bianchi[1], *G.E. Ehrlich*[2], *F. Facchinetti*[3], *E.C. Huskisson*[4], *P. Jenoure*[5], *A. La Marca*[3], *K.D. Rainsford*[6]

[1]Department of Pharmacology, Faculty of Medicine, University of Milan, Italy; [2]University of Pennsylvania, Philadelphia, PA, USA; [3]Clinica Ostetrica & Ginecologia, Via del Pozzo, 71, 41100 Modena, Italy; [4]14A Milford House, 7 Queen Anne Street, London, UK; [5]crossklinik am Merian Iselin Spital, Föhrenstrasse 2, 4009 Basel, Switzerland; [6]Biomedical Research Centre, Sheffield Hallam University, Howard Street, Sheffield, S1 1WB, UK

NSAIDs: The survivors from the laboratory

Signalling from pain

Pain remains the chief reason for medical consultation. It also remains the chief reason for self-medication. So what is the role of pain? While always unwelcome, pain protects against noxious stimuli (nociceptive pain) by encouraging withdrawal from and subsequent avoidance of the provoking agent (e.g., nettles, insects, fire, etc.) [1, 2]. This pain tends to be transient and is treated, if at all, locally. It obviously has a protective function, sometimes called adaptive, but the same response to modern medical and surgical interventions would not be as beneficial. The pain accompanying inflammation results from tissue damage. It can be acute or chronic and is almost always treated. Yet even this pain serves a teleologic purpose. It leads to discovery of the underlying cause and at the same time inhibition of movement of the affected part. As rest is anti-inflammatory, even if imperfectly so, this pain aids in the natural alleviation of its cause. Analgesic and anti-inflammatory medicines are particularly effective here. Neuropathic pain, an example of non-adaptive pain [3], results from a lesion of the nervous system, and generally requires stronger analgesics, not necessarily anti-inflammatory, for its control. Functional pain, resulting from a fault in central processing or from causes yet to be determined, challenges medical skills and is generally not helped by medication [4].

While the emphasis on NSAID use has been on rheumatologic conditions, they seem much more multipotential than that. Inflammation has lately been incriminated in the development of colonic polyps, and indeed, NSAIDs have become one preventive. Further, inflammation may play an important role in the development of atherosclerosis and probably other 'degenerative' conditions not previously thought of as successors to inflammation. Because of their versatility and generally good tolerability, NSAIDs have also become established treatments for dysmenorrhoea; one may hazard the guess that additional indications will emerge as well, and that NSAIDs will still be here, and needed, when most other current medicines have been superseded.

Control of pain

The attempts to control pain date back to antiquity and led to many herbal treatments [5]. Some of these were precursors to modern synthesised drugs, as for example willow bark and autumn madder to salicylates. The modern era of pharmacology, producing bioequivalent and bioavailable drugs stems from these antecedents. The most successful of the salicylates was surely acetyl salicylic acid, trademarked aspirin (though the trademark soon became a generic term) [6]. For much of the century succeeding its introduction in the late 1890s, aspirin remained the mainstay of treatment, especially of rheumatic pain, and led to the famous dictum of the early rheumatologists in Boston: "if aspirin doesn't work, give more of it." By the mid-1950s, as much as 10.8 g a day was advocated for recalcitrant rheumatoid arthritis (RA) [7]. Amidopyrine was a favourite analgesic of the 1930s, but agranulocytosis and aplastic anaemia were potential complications and it and some successor compounds lost favour (except in some Latin countries) [8]. In the 1950s, Geigy's phenylbutazone, both analgesic and anti-inflammatory, ushered in the modern era of non-steroidal anti-inflammatory drugs (NSAIDs). In rapid succession during the 1960s and 1970s came other NSAIDs, including the largest number derived from propionic acid (the first successful and safe one of which was ibuprofen) [9]. The pharmacologic rationale, inhibition of cyclooxygenase (COX), was a major breakthrough in our understanding of the road from membrane phospholipids through arachidonic acid to the cyclooxygenases, which, in time, were found to consist of at least two separate enzymes, COX-1 and COX-2, and probably even more [10].

Gastrointestinal and other untoward events

The limitations on aspirin use were always gastrointestinal intolerance, with gastric bleeding a major problem presented to emergency rooms [6]. Later, Reye's syndrome and other serious complications resulting from aspirin use led to the gradual supplanting of aspirin in the rheumatologic armamentarium and replacement by other NSAIDs. The first generation of these also potentially led to gastric erosions (at least observed by endoscopy; their significance remains controversial to this day, as few continued to clinically evident or diagnosed ulcers), and to a lesser extent, ulcers and haemorrhages [11]. While these gastric effects were the most common complications, renal, skin, hepatic, and haematological adverse events were also reported. This led to the development of COX-2 selective inhibitors, which in appropriate dosage had little effect upon the constitutive enzyme, COX-1, and targeted chiefly COX-2, which played a major role in continued inflammation [12]. Elsewhere in this book, the pharmacology of these compounds is discussed in depth (Chapter 4). However, the clinical reception brought to light some of the differences among the various drugs: though addressing similar targets, NSAIDs did not necessarily treat the same people equally or produce similar unwanted effects. As has previously been stated by Ehrlich [13], the NSAIDs treated overlapping, not concentric, circles of patients. They consistently rank among the best selling pharmaceutical products [14].

Efficacy or safety?

The current controversies regarding these NSAIDs question whether clinical efficacy or safety should govern the choice, and, to a lesser extent, whether cost should be considered as well [15]. Moreover, considerable controversy questions whether long-term administration is needed, or whether a short course can help sufficiently so that protracted treatment – at least for pain – is not necessary. This controversy rages particularly for osteoarthritis (OA), where some would treat initially with paracetamol (acetaminophen) and others at once with NSAIDs [16]. Though patients seem to prefer NSAIDs for OA [17, 18], the likelihood is that they have already tried paracetamol (available without prescription) before seeking the advice of a physician, and that long-term use may limit NSAIDs and not necessarily paracetamol, as some studies seem to suggest [19].

Purpose of this chapter

In this chapter, the clinical profile and uses of nimesulide are described and compared with other NSAIDs.

Osteoarthritis: A leading target for NSAIDs

As understanding of acute and chronic inflammation increases, and more drugs targeted at specific mechanisms are introduced, the likelihood remains that as one after another of these is replaced by newer compounds, NSAIDs will remain as mainstays, especially for OA. OA is ubiquitous, probably afflicts almost everyone by roentgenographic evidence but is only symptomatic in about 20% [20–22]. OA is the remote consequence of many insults to the joint and as such manifests at a time remote from the causation. The pain experienced is usually superimposed inflammation, which is probably why anti-inflammatory compounds are preferred. The old distinction between (osteo) arthrosis and osteoarthritis probably deserves to be resurrected to differentiate between those needing treatment and those for whom the joint changes are only casually discovered.

Development of osteoarthritis

Osteoarthritis is a disease of cartilage, characterised by progressive deterioration as the structure loses the natural ability to repair itself [21]. This process is accompanied by the release of enzymes and crystals leading to synovial inflammation [21]. It can be difficult to tell the difference between the histological appearance of osteoarthritis and rheumatoid arthritis (RA). There is sclerosis of bones and overgrowth also secondary to the change in cartilage.

There are two reasons for osteoarthritis. There is a very strong genetic influence and osteoarthritis can also be the result of an injury [22]. There are, therefore, many cases in which the disease appears to be entirely genetic. Examples would include the appearance of changes in the small joints of the hands of a 50-year-old woman or the hips of a 60-year-old man whose mother or father would have suffered the same problem at a similar age. There are a few cases in which the disease is entirely mechanical, developing for example in the knee after a meniscectomy or in any joint after a fracture. There are also cases in which both genetic and mechanical influences are evident; patients with generalised disease for example but with particularly severe changes in a joint, which has been exposed to undue stress. The evidence suggests multiple factors underlie development of osteoarthritis and the associators may vary according to whether it is localised in the knee or hip [22].

Osteoarthritis requiring some clinical intervention affects about 15% of the population, and so is a very common disease. It is an expensive disease, sufferers requiring drugs, operations and care.

Should NSAIDs be used for osteoarthritis? – efficacy

There is abundant evidence that anti-inflammatory drugs are more effective in OA than simple analgesics [23]. This is not surprising since there is also considerable evidence for the importance of inflammation in OA. This comes from clinical findings and histology as well as the effects of drugs. Osteoarthritic joints show cardinal signs of inflammation like warmth and swelling [21]. Morning stiffness, a characteristic symptom of inflammatory arthritis, is a regular complaint although it is of longer duration in diseases like RA. It is reduced by NSAIDs. Many studies have compared NSAIDs [23] and simple analgesics in OA, most showing a clear advantage for the anti-inflammatory. NSAIDs do much more than just relieve pain; they reduce stiffness and improve function [23].

Compliance is also a problem in this age group. Thus a simple dosage schedule helps.

Should NSAIDs be used for osteoarthritis? – tolerability

Traditional NSAIDs may cause serious gastric problems including ulceration and its complications, perforation and bleeding [23], with elderly, female patients particularly vulnerable to these problems. They are also more likely to be fatal in the frail elderly and in patients with other medical problems. A fit young man will probably recover from a brisk gastric bleed but a sick old woman will not.

While it would be a shame to deny elderly patients the potential benefits of NSAIDs for their OA, it is obviously sensible to look for the safest possible drug. Selective COX-2 enzyme inhibitors like nimesulide with their superior gastric safety present a huge potential advantage. Experimental evidence predicting a favourable gastric tolerance is discussed in Chapter 6 of this book.

Choice

Unlike some other products, the older NSAIDs remain useful and were employed even after newer ones (e.g., coxibs) arrived. The dictum of Alexander Pope (Essay on Criticism, 1711) "Be not the first by whom the new is tried nor yet the last to lay the old aside" may not fully describe the life cycle of NSAIDs. Some very successful NSAIDs, reaching the end of their patent life, have been reformulated in lower doses for over-the-counter (OTC) sale. While the lower dosage loses much of the anti-inflammatory action of these drugs, they remain popular favourites.

RA seems to have become less severe in the past 50 years [24]. Formerly, patients with marked deformities, even ankylosis, required prolonged hospitalisation, surgery, and a sequence of so-called remittive drugs; today, the hospitalisations and the parade of drugs are ancient history. What accounts for the change? Of course, we have available many newer targeted compounds, and they are often administered earlier in the course. However, almost all patients also have access to NSAIDs, by prescription or OTC, and it may well be that these have made a difference in the outcomes.

Even more to the point is the treatment of OA. Most OA probably needs little or no treatment, as was stated above, as it is generally asymptomatic or only mildly and intermittently symptomatic. Which OA goes on to become very painful, requiring treatment cannot as yet be predicted, but it is this OA that needs to be addressed. Earlier guidelines promulgated by a committee of the American College of Rheumatology [25], following the publication by Brandt's group [26] recommended paracetamol as the first line of defence. But the study by Bradley et al. [26] treated OA for a relatively short period, and was succeeded by a number of studies that continued the comparisons for longer periods of time or questioned patients as to their preferences. In most of these, NSAIDs were preferred to paracetamol [16, 17, 27–32]. Recent large scale trials and meta-analysis studies [27–32] have now clearly shown that NSAIDs give superior analgesia and relief of joint symptoms over paracetamol in OA, thus providing a definitive answer to the controversy about the relative efficacy of this drug in relation to NSAIDs.

There has also been a backlash against paracetamol, a derivative of phenacetin, now regarded as a toxic compound but for many years used as an analgesic. A recent editorial in the New York Times [33] even gave voice in the lay press to these concerns and reported a death rate of some 500 annually in the United States alone. Moreover, the comparisons of paracetamol and NSAIDs were based predominantly on controlled studies, and the populations admitted to such studies hardly resemble those presenting to clinical practices, as those taking other drugs or suffering from concurrent diseases are generally excluded from enrolment [34].

In the studies, some patients reported more ultimate discontinuations with NSAIDs, but whether these resulted from untoward effects or because the drugs were more successful in alleviating the symptoms is not clear. At any rate, a subsequent revision of the recommendations gave pride of place to NSAIDs [35], although these needed to be looked at critically: most patients (probably) took paracetamol first or concomitantly, on their own, so that the groups were not pure. Moreover, the favourable responses were reported by those with the more severe symptoms during the wash-out periods, so it was not clear if they were commenting on the feeling of relief alone or reporting an actual difference of effectiveness [36]. Current commentaries accept both conclusions, and it is left to the physician and the individual patient to decide the best approach for each case on an individual basis. It is unlikely that protracted treatment for osteoarthritic complaints is necessary, so that relatively short courses will suffice to contain the intermittent worsening of symptoms. Under these circumstances, some of the limitations of untoward events are minimised, although gastric events often appear early in the course of treatment, but may not become severe and clinically relevant unless such treatment continues. Those most at risk are patients who have had prior untoward reactions to such medicines, those who have pre-existing conditions that could be worsened, those taking other medications with which interactions can develop (including nutraceuticals and herbals) and the elderly (which include many of those who have OA). A recent study concluded that some patients with prior reactions to other NSAIDs might yet be able to take some preferential

COX-2 inhibitors, such as nimesulide or meloxicam [37]. Nevertheless, when judiciously administered and monitored, the various NSAIDs have made a remarkable difference in the quality of life, especially in life space and life content.

Nimesulide in the treatment of osteoarthritis

In considering applications for therapy with nimesulide, we must consider the properties of the drug, efficacy and tolerability but also the characteristics of the patient. Patients with OA are often elderly; they tend to have other diseases and to be taking other drugs. They are vulnerable. Is nimesulide the right drug for these patients? The evidence in support of this is evaluated here.

Nimesulide – efficacy

There is now a very large worldwide experience of nimesulide in OA, confirming its efficacy in treating this condition. Two dose-finding studies were initially performed in patients with OA. Dreiser and Riebenfeld [38] studied 24 patients with osteoarthritis of the hip. They compared 100 mg nimesulide b.i.d. with 200 mg b.i.d. in a crossover trial with a week of placebo between the two doses. Each treatment was given for 1 week. There was a significant reduction in pain scores and improved articular function in the nimesulide-treated groups compared with placebo. Global efficacy was assessed as good in 21% of patients on placebo, 35% of patients taking 100 mg twice daily of nimesulide and 50% of those taking 200 mg twice daily. The authors concluded that 100 mg twice daily was the minimum effective dose.

In a multicentre trial undertaken in France [39] 392 patients were divided into four groups which received placebo or nimesulide in doses of 50, 100 or 200 mg twice daily. Treatment continued for 1 month. The physician's assessment of efficacy was rated as excellent in 42% of patients on placebo, 59% of the 50 mg nimesulide group and 72% of the 100 and 200 mg groups. This study showed dose-effect relationships with nimesulide. It supports the view that the optimal dose of nimesulide for efficacy is 100 mg twice daily.

Table 1 shows a summary of trials in OA in which nimesulide has been investigated for effects on painful symptoms in patients with OA of the knee and/or hip in comparison with either placebo or comparator NSAIDs. These data show that nimesulide was superior to placebo and, in all but two trials equivalent to comparator NSAIDs [40–46]. Two trials, one comparing nimesulide with rofecoxib or celecoxib [42] and another with piroxicam [38] showed that nimesulide was superior in comparison with these drugs for pain relief in osteoarthritis.

A study by Huskisson et al. in 1999 [43] compared nimesulide with diclofenac, the latter of which has long been a market leader. This was an active control equiv-

Table 1 – Summary of studies showing comparative efficacy of oral nimesulide in relief of painful symptoms in osteoarthritis

Patient characteristics (% of patients and arthritic site)	Trial design (duration of treatment)	Dosage (no. of patients evaluated)	VAPS (mean baseline/ end scores)	Efficacy* (%)	Relative Efficacy	Reference
OA 100% knee	R, mc, db, pc, pl (4 weeks)	NIM 50 mg bid (97) NIM 100 mg bid (98) NIM 200 mg bid (97) PLA (100)	6.2/2.5 6.1/2.3 6.1/3.0 6.3/3.2	59 72 72 42	NIM>PLA	Bourgeois et al. (1994) [39]
OA 100% knee	r, mc, db, pc, pl (2 weeks)	NIM 100 mg bid (30) PLA (97)	6.0/3.2 6.5/5.2	62 17	NIM>PLA	Dreiser & Riebenfeld (1993) [38]
OA 90% knee, 10% hip	r, mc, db, pl (3 weeks)	NIM 100 mg bid (29) PIR 10 mg od (30)	7.2/3.6 6.9/2.7	63 59	NIM>PIR	" "
OA 60% knee, 40% hip	r, mc, db, pl (8 weeks)	NIM 100 mg bid (28) KET 100 mg bid (27)	6.8/2.9 6.9/3.3	58 39	NIM ≡ KET	" "
OA 100% hip and/ or knee	r, db, pl (4 weeks)	NIM gr 100 mg bid (20) NAP gr 250 mg bid (27)	not stated	75 58	NIM ≡ NAP	Fossaluzza & Montagnani (1989) [40]
OA 100% knee	r, mc, db, pl (12 weeks)	NIM 100 mg bid (100) ETO 300 mg bid (99)	7.5/3.4 7.5/3.5	75 58	NIM ≡ ETO	Lûcker (1994) [47]

Table 1 – (continued)

Patient characteristics (% of patients and arthritic site)	Trial design (duration of treatment)	Dosage (no. of patients evaluated)	VAPS (mean baseline/ end scores)	Efficacy* (%)	Relative Efficacy	Reference
OA 100% knee	r, db, xo (1 week)	NIM 100 mg od (31) CEL 200 mg od (31) ROF 25 mg od (31)	5.7/3.1 5.6/3.6 5.8/3.6	75 73	NIM>CEL, ROF	Bianchi & Broggini (2003) [42]
OA 100% hip and/ or knee	r, mc, db, pl (24 weeks)	NIM 100 mg bid (135) DIC 50 mg tid (144)	5.4/4.0 6.0/4.0	79 86	NIM ≡ DIC	Huskisson (1999) [43]
OA knee or hip	r, db (6/12 months)	NIM 100 mg bid (183) NAP 250 mg am & NAP 500 mg pm (187)		69 62	NIM ≡ NAP	Kriegel (2001) [49]
OA 100% hip and/ or knee	r, mc, db, pl (4 weeks)	NIM 100 mg bid (44) DIC 50 mg tid (45)	7.2/3.8 6.9/3.7	72 74	NIM ≡ DIC	Porto et al. (1998) [44]
OA 100% hip	r, db, pl (4 weeks)	NIM 100 mg bid (52) NAP 500 mg bid (51)	7.2/3.1 7.1/2.7	87 82	NIM ≡ NAP	Quattrini & Paladin (1995) [46]

* *Global clinical efficacy rated by investigator as % of patients with very good/good clinical response. Overall efficacy for patients completing 12 weeks of therapy for nimesulide (69%) versus etodolac (62%) was similar.*
Abbreviations and symbols: bid = twice daily; db = double blind; CEL = celecoxib; DIC = diclofenac; ETO = etodolac; KET = ketoprofen; NAP = naproxen; NIM = nimesulide; PIR = piroxicam; ROF = rofecoxib; +ret = retard form of nimesulide; gr = granules; mc = multicentre; OA = osteoarthritis; PLA = placebo; pc = placebo-controlled, od = once daily; pl = parallel; r = randomised; VAPS = visual analogue pain score; xo = crossover ≡ indicates similar efficacy; > = indicates statistically significant greater efficacy than comparator ($p < 0.05$).
Modified and updated from [63].

alence study designed to show that nimesulide was as effective as diclofenac, which it was. 279 patients with OA of the hip or knee received either nimesulide 100 mg twice daily or diclofenac 50 mg three times daily. Global efficacy and the Lequesne Functional Index were the primary efficacy measures. Global pain scores and Lequesne Functional Index values were reduced by about 15–20% by both drugs at 2 weeks and remained constant thereafter to the end of the study at 24 weeks. Patients did not continue to take drugs in the long-term if they were not effective, and so it was interesting to see that 65% of patients given nimesulide and 68% of those given diclofenac completed 6 months of treatment.

Porto et al. [44] also found nimesulide 100 mg twice daily and diclofenac 50 mg three times daily equally effective in a parallel group study in 83 patients with OA of the hip or knee, measuring pain and functional impairment. A trial in China [45] compared nimesulide 100 mg twice daily and diclofenac 50 mg three times daily in 60 patients with OA of the knee. Nimesulide was significantly more effective than diclofenac after both 7 and 21 days of treatment. The efficacy of nimesulide was assessed as good or excellent by 85% of patients taking nimesulide and 47% of those taking diclofenac.

Quattrini and Paladin [46] compared nimesulide 100 mg twice daily and naproxen 500 mg twice daily for 4 weeks and found them equally effective.

A multicentre study in Germany [47] compared nimesulide 100 mg twice daily with etodolac 300 mg twice daily in 199 patients with OA of the knee. Both groups showed significant improvements in variables like pain and the Lequesne Functional Index. Both patients and physicians assessed the results as good or excellent in 80% of patients taking nimesulide and 64% of those taking etodolac.

Another multicentre study which was performed in Italy [48] compared nimesulide and flurbiprofen, both given by suppository. Nimesulide was given in a dose of 200 mg twice daily and flurbiprofen in a dose of 100 mg twice daily. All efficacy variables improved significantly with both treatments. Both patients and physicians rated the efficacy as good or excellent in more than 80% of cases.

In some of these trials the effects of nimesulide were studied for relatively short periods of time. A long-term study was undertaken to compare nimesulide (100 mg twice daily) with naproxen (250 mg in the morning and 500 mg at night) in a multicentre, double-blind, parallel group, double-dummy, active equivalence study lasting 1 year in patients with OA of the hip (27.3% or 28.9% for each treatment group respectively) or knee (72.7% or 71.1% respectively) [49]. The intensity of pain, joint stiffness and physical function were determined by Visual Analogue Scale (VAS) at 2, 4, 8, 12, 18, 26, 42 and 52 weeks and entered into the relevant sections of the WOMAC osteoarthritis index (Version VAS 3.0). The Lequesne Functional Index of the knee or hip was determined at each visit. Global efficacy and tolerability was assessed on a four-point scale (ranging from 1 = excellent to 4 = poor) by both investigator and patient at 6 and 12 months.

The median values of the WOMAC pain sum-scores in intention-to-treat population were almost identical with the two treatments at both 6 and 12 months with the two drug treatments. The mean percentage changes from baseline at 6 months with nimesulide were 22.5% and with naproxen 22.4%. At 12 months these changes were 22.5% and 19.9% respectively. Global efficacy was similar for both the investigators' and patients' assessments. Their respective assessments at 6 months for ratings of good or excellent were 59.3% and 57.0% of patients on nimesulide and 56.3% and 52.7% on naproxen. At 12 months these respective rating were 68.8% and 65.6% for those that received nimesulide and 69.7% and 65.7% for those on naproxen.

In a meta-analysis carried out by Wober [50] of six trials (see [38, 43, 44, 46, 47]) nimesulide was compared with other NSAIDs for efficacy and safety in patients with OA. In these studies nimesulide was taken 100 mg twice daily for 2 weeks compared with piroxicam, ketoprofen, naproxen, etodolac and diclofenac. Based on Mann-Whitney statistical analysis of the efficacy parameters he concluded that nimesulide was as efficacious as the comparator drugs. Similar outcomes were observed by this author in two other studies in patients with extra-articular rheumatism. Based on results of the meta-analysis of the adverse reactions (principally symptomatic reactions in the gastrointestinal tract) analysed by the Cochran-Mantel-Haenzsel-Pooling procedure there were no differences among the treatment groups. However, there were fewer dropouts from treatment with nimesulide than either the comparator drugs or placebo [50].

Post-marketing surveillance [51] in 22,938 patients with OA showed good or excellent efficacy in 76% of cases taking 100 or 200 mg of nimesulide twice daily for up to 3 weeks. An open study in France showed good or excellent efficacy in 77% of 132 patients taking 100 mg twice daily for 3 months.

A placebo-controlled study [52] confirmed the efficacy of nimesulide 100 mg twice daily in 40 elderly patients with OA; there were significant improvements in pain, stiffness and functional impairment.

All these studies say essentially the same thing. Nimesulide at a daily dose of 100 mg b. i. d. is at least as effective as the traditional NSAIDs taken at their recommended daily doses, which are widely used for patients with OA.

Much interest has been shown recently in the effects in rheumatic conditions of the new category of COX-2 selective NSAIDs (known as 'coxibs'), especially in view of their claims for better gastrointestinal tolerability compared with established or so-called unselective COX-inhibitory NSAIDs (e.g., diclofenac, naproxen). It is, therefore, of interest to compare the effects of these drugs with that of nimesulide in the treatment of OA. Recently, Bianchi and Broggini [42] performed a very interesting study comparing nimesulide with celecoxib and rofecoxib (the latter drug was withdrawn from the market worldwide on 29 September 2004 because of unacceptably high risks of myocardial infarction and stroke [53]). The study was designed to assess particularly the analgesic efficacy of the three compounds. 31 patients with OA of the knee received all three treatments for 7 days in random order in a Latin square design. Pain was measured for 3 h on the first and last days. It was interesting to see that the state of the patients before treatment and the effect of the treatment were very similar on day 1 and day 7. However nimesulide was more effective than celecoxib or rofecoxib both on days 1 and 7; it also exerted a more rapid analgesic effect, which was evident 15 min after administration. Good or very good analgesic efficacy was reported at the end of the week of treatment by 53.4% of patients on nimesulide, 50% on rofecoxib and 46.7% on celecoxib. Nimesulide treatment was the first choice in 40%, rofecoxib also in 40% and celecoxib in 20% of these patients. An assessment of the analgesic responses of nimesulide compared with the coxibs is discussed in the section "Comparison of Analgesic Properties of Nimesulide with Coxibs".

Some studies have been performed in South America and India examining the effects of nimesulide using non-Helsinn preparations and for which in some cases little data is available on the bioequivalence or safety parameters of these preparations. Thus, Roy and co-workers [54] compared the effects of nimesulide 100 mg daily with piroxicam 20 mg daily in a randomised, double-blind trial in 90 patients with OA of the knee focussing on evidence for chondroprotection as determined by magnetic resonance imaging (MRI). Both treatments resulted in significant improvement in severity indices and physicians' and patients' assessment of global arthritic condition at 4 weeks and a reduction in joint tenderness at 8 weeks. Functional activity was improved in 64% of patients on nimesulide and 74.5% on piroxicam. No differences were found in efficacy or tolerability between the two treatments. After 6 months of therapy MRI scans of the knees of 10 patients showed no differences in articular cartilage and associated joint structures compared with baseline from both the treatments. The latter is perhaps hardly surprising since the extent of joint damage would be expected to be considerable with the patients recruited to the study and any reversal of joint damage at this stage would be unlikely. While the numbers of patients that were examined by MRI is small this study probably shows that there is possibly no deterioration in joint structures with the drug treatments, although more extensive studies are required to establish if there is protection, reversal of deterioration in joint structure where there is evidence of improvement in joint mobility. A similar study was performed from the same study centre [55], but with a placebo control in 49 patients with OA of the knee. Functional parameters were improved to a greater extent with nimesulide (72.2%) than with piroxicam (44.4%) at 8 weeks. No differences were observed in the articular cartilage at 24 weeks of treatment with either drug.

A beta-cyclodextrin formulation of nimesulide 400 mg b.i.d. (= 100 mg b.i.d. nimesulide) was compared with naproxen 500 mg b.i.d. in an 'on-demand' drug treatment randomised, double-blind, multicentre design that extended for 2 weeks and 5.5 months in patients with OA of the knee or hip [56]. Similar pain relief on movement, morning stiffness and values of the Lequense Index were observed with both treatments, but fewer gastrointestinal symptomatic events were observed with the beta-cyclodextrin nimesulide preparation than with naproxen. It could be argued that the 'on-demand' use of the drugs could create conditions where the intake of the drugs is not known and therefore is not a study in drug equivalence, but this situation is closer to the real world usage of drugs by patients with OA.

In a study in Uruguay, Estevez and co-workers [57] compared the effects of once daily treatment with nimesulide 200 mg (Nodo®) and diclofenac 100 mg (Voltaren®, sustained release) for 91 days (following a 1 week washout) and measured the plasma concentrations of the drugs at 7, 49 and 91 days. After 2 weeks there was improvement in indices of pain with both drug treatments and this progressively over 91 days of treatment. This coincided with a progressive increase in plasma concentrations of the drugs suggestive of drug accumulation.

Nimesulide – tolerance and safety in OA patients

The studies in Table 1 and discussed above have examined the adverse effect profile of nimesulide. In most cases the adverse events have been symptomatic gastrointestinal reactions and with nimesulide have been similar to or slightly better than comparator drugs. Compared with diclofenac in the active control equivalence study [43], the overall incidence of adverse events was similar in the two groups, 65% of patients taking nimesulide and 68% of patients taking diclofenac reporting one or more adverse event. However, more patients in the diclofenac group had adverse gastrointestinal events, 47% of those taking diclofenac compared to 36% of those taking nimesulide, a statistically significant difference. Global evaluation showed excellent tolerance in 37% of patients taking nimesulide and 24% taking diclofenac. No serious haematological or biochemical abnormalities occurred in either group. Porto et al. [44] comparing the same drugs found excellent or good tolerance assessed by the physician in 84% of patients taking nimesulide and 79% of those taking diclofenac. Endoscopies were carried out in this study. Ulcers developed in one patient on nimesulide (2.4%) and three on diclofenac (7.3%). In the study by Gui-Xin and co-workers in China [45], adverse events occurred in 13% of patients taking nimesulide and 29% of those taking diclofenac. Gastrointestinal events occurred in 6.7% of patients taking nimesulide and 30% of those on diclofenac, a statistically highly significant difference. Few patients in either group had abnormal laboratory findings suggestive of liver abnormalities.

In the one-year 'active' control study comparing nimesulide with naproxen in 370 patients with OA [49], gastrointestinal side effects were less common with nimesulide than with naproxen. Gastrointestinal adverse events were reported in 47.5% of patients taking nimesulide and 54.5% of those taking naproxen, concluding that nimesulide was as effective but with fewer gastrointestinal adverse events. Quattrini and Paladin [46] recorded four adverse events in each of the two groups, receiving naproxen or nimesulide. They were mainly gastrointestinal and either mild or moderate.

In the comparison with etodolac [47], 39 patients on nimesulide had side effects compared with 34 on etodolac; 59% of those occurring with nimesulide were gastrointestinal compared to 64% with etodolac – so these are essentially comparable. In the meta-analysis [50], nimesulide had a superior benefit–risk ratio to the other drugs with a comparable safety and tolerability to placebo, especially regarding gastrointestinal adverse events. In a direct comparison with placebo [52], four patients in the nimesulide group and two in the placebo group had adverse events, all mild.

In the French open study [39], adverse events occurred in 33% of patients and were mostly mild or moderate in severity. There were no laboratory abnormalities. Adverse events occurred in only 9.4% of patients in the post-marketing surveillance in 22,938 cases of OA [51]. They were usually mild and rarely required a in dosage or cessation of treatment. The drop-out rate in this study was only 3.5%.

In the comparison with celecoxib and rofecoxib [42], good or excellent tolerance was reported by 76.7% of patients taking both nimesulide and rofecoxib and by 70% of patients on celecoxib following 1 week's treatment.

Two studies have looked at the economic consequences of better gastric tolerance. Using data from meta-analysis, Liaropoulos [58] calculated that in Greece, nimesulide was 56% cheaper than diclofenac. Using similar data for France, Italy and Spain, Tarricone [59] found that nimesulide saved between 1.5 and 3.6 Euros per patient in a 15-day treatment period.

The adverse events in trials from nimesulide in OA [60] and in spontaneous reports [61] highlighted the three types of adverse event which occurred with nimesulide that are also observed with other NSAID comprising allergic skin reactions, liver injury and gastric complaints (see also Chapter 6). The latter showed a lower incidence with nimesulide than with other NSAIDs. Skin and hepatic reactions were comparable with that of other NSAIDs. There are a number of confounding variables which made it difficult to be sure about the cause in many of these cases, including other pre-existing diseases and other predrugs being taken by the patient. Overall, the benefits from relief of pain and inflammation compared with risks due to adverse reactions with nimesulide are in favour of the drug.

Conclusions

There is a very large experience of the use of nimesulide in OA from around the world. The studies clearly show that nimesulide is at least as effective as other

NSAIDs with which it has been compared but with less gastric adverse events. It has a convenient dosing schedule of 100 mg twice daily and is an ideal drug for use in OA. All drugs have side effects and even with a better-tolerated drug like nimesulide, caution and vigilance are required to ensure the safety of patients who are often vulnerable.

Miscellaneous rheumatic conditions

Rheumatoid arthritis

Several pilot or preliminary investigations were performed in small patient numbers during the 1980s in uncontrolled studies in which nimesulide 400–800 mg/day was shown to relieve painful symptoms in patients with RA (reviewed in [62, 63]). These and the early clinical investigations at Riker as part of the development of nimesulide (Chapter 1) included some Phase I/II studies in patients with RA. These studies showed that nimesulide provided effective relief of pain and joint symptoms in patients with RA. However, the doses of the drug were relatively high being in some of the Riker studies up to 800 mg/day, so it was not surprising that increase in plasma levels of liver enzymes occurred in some of the patients.

Recently, Balabanova and co-workers [64] undertook a multicentre open clinical trial of nimesulide 200–400 mg/day in 52 patients with RA. Articular signs and pain symptoms were recorded at 4 and 8 weeks after initiation of treatment. The drug resulted in improvement or marked improvement in 84.6% of patients. Side effects occurred in 15.3% of patients which were reversible upon cessation of the drug.

In the reports of adverse drug reactions attributed to nimesulide (Chapter 6) it is apparent that the drug has been prescribed to a considerable number of patients with RA even though the drug is not recommended for use in this condition. In some cases the doses have exceeded the recommended daily doses for the treatment of OA and musculoskeletal pain. The question arises whether higher doses of nimesulide are required for effective relief of pain and joint symptoms in RA as indicated in these studies if under these conditions there would be an increase in side effects, e.g., in the liver as a consequence?

Psoriatic arthritis

Psoriatic arthritis comprises a heterogeneous group of arthritic conditions that present with associated psoriasis [65]. Psoriatic arthritis is present in some 2–3% of the population and so has considerable clinical significance [65]. Assessment of the outcome of patients with this condition has only recently received much attention [66]. Currently management of this condition is directed towards controlling the progressive radiological evidence of erosions and is usually treated with immunosuppressive drugs or more recently with biologics along with NSAIDs of which most have been shown to relieve joint symptoms but probably have little effect on the psoriatic symptoms [65].

Sarzi-Puttini et al. [67] undertook a randomised, double-dummy, placebo-controlled, dose-ranging study in 80 patients with psoriatic arthritis who received 100, 200 or 400 mg/day nimesulide for 4 weeks. Pain (assessed on a visual analogue scale), tender and swollen joints were reduced in all three nimesulide treated groups compared with baseline to the end of therapy, while in the placebo group there was no change. Overall pain and morning stiffness were reduced by 200 and 400 mg/day nimesulide but not by 100 mg/day compared with placebo. Paracetamol escape medication was used by more patients that received placebo than those that had nimesulide. Side effects (in 15%) of patients were mild in all treatment groups but gastric pain in one patient that received 200 mg/day nimesulide was such that the patient withdrew from therapy.

Gout

Although gout is not a recognised indication for application of nimesulide, its effects have recently been studied by Barskova and co-workers [68]. These authors treated 20 male patients with established gout (mean duration of disease 8.1 years) with nimesulide 100 mg b.i.d. for 14 or 21 days. Joint swelling index, supra-articular skin hyperaemia, articular index and pain on rest and movement were determined on the day of initiating treatment and at 5, 14 and 21 days after initiating treatment. Nimesulide caused rapid improvement in joint parameters of pain and inflammation and this was evident at 5 days of treatment. The ESR and seromucoid levels were also significantly reduced but there was no alteration in plasma levels of uric acid, glucose or liver enzymes. One patient developed urticaria. These preliminary results deserve further investigation.

The analgesic properties of nimesulide in inflammatory pain

Onset of analgesia

Recent studies in patients with inflammatory arthritis in whom COX-2 mRNA and protein were measured along with COX-2-derived PGE_2 in both synovial tissues and fluid and in the whole blood assay showed that nimesulide in contrast to diclofenac has a rapid onset of action in reducing production of PGE_2 which is regarded as a surrogate mediator of analgesia [69]. Thus, in a pharmacological

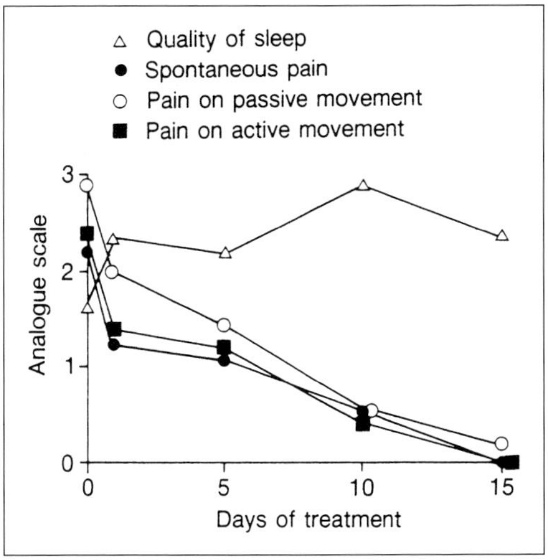

Figure 1
Effects of nimesulide 100 to 300 mg/day on some clinical parameters of efficacy in 11 patients with osteoarthritis of the cervical spine. Rapid onset is evident within a day of treatment with nimesulide of the relief of spontaneous pain, pain on passive and active movement together with improved quality of sleep. This progressively improves over 15 days of treatment with the drug. From [70]. Permission: The Journal of International Medical Research.

sense nimesulide can be considered to have rapid actions within 0.5–1.0 h in chronically inflamed joints.

The duration of functional pain relief in OA, i.e., attributable to spontaneous pain, as well as the pain on passive and active movement was investigated in patients with OA of the cervical spine. A study by Reiner [70] is instructive in as much as it shows that with dose-adjustment the onset of analgesia in this spinal inflammatory/degenerative condition is quite rapid with the indices of pain being reduced by half within the first day of treatment with 100 mg nimesulide (Fig. 1). With increasing dosage of nimesulide up to 300 mg/day adjusted according to patients needs pain relief progresses to the extent that by 15 days the indices of pain relief are almost zero (Fig. 1). Thus, these studies show that initially there is rapid onset of analgesia with nimesulide followed by a sustained period where the drug progressively acts presumably on deep inflammatory pain.

Comparison of analgesic properties of nimesulide with coxibs

The analgesic effects of nimesulide have been compared with the coxibs in both experimental and clinical settings. From a pharmacological point of view, a body of data exists showing that nimesulide belongs to the group of preferential COX-2 inhibitors [71, 72]. In this section we focus attention on comparisons of nimesulide with other NSAIDs with similar pharmacodynamic characteristics (at least with regard to the inhibition of COX-2 rather than COX-1).

Experimental studies

The analgesic responses to nimesulide in various animal and human models are discussed in Chapter 4. Here we consider comparisons of nimesulide with other COX-2 inhibitors in models of hyperalgesia as a prelude to consideration of their therapeutic responses in clinical pain states.

The effects in models of hyperalgesia of nimesulide, celecoxib, and rofecoxib have been assessed by using two animal models and in a human model of inflammatory hyperalgesia [73–75]. In animal studies [75], each drug was administered intraperitoneally (i.p.) at its previously defined ED_{50} for the anti-inflammatory effect in the rat (i.e., the inhibition of carrageenan-induced hind paw oedema measured by plethysmometry). In the first animal study, nimesulide (2.9 mg/kg) totally prevented the development of thermal hind paw hyperalgesia induced by the injection of formalin in the tail. In this model of centrally-mediated hyperalgesia, celecoxib (12.7 mg/kg) reduced the hyperalgesia significantly but not completely, whereas rofecoxib (3.0 mg/kg) was ineffective. In the second animal study [75], nimesulide was significantly more effective than celecoxib and rofecoxib in reducing the mechanical hind paw hyperalgesia induced by the intraplantar injection of Freund's Complete Adjuvant (FCA). It is important to point out that the latter represents a reliable and widely used experimental model of monoarthritis [73].

In the human model, after oral administration in patients with RA all drugs reduced the inflammatory hyperalgesia to mechanical stimuli applied to a middle phalange joint [75]. However, only the effect of nimesulide was already evident 15 min after treatment. Moreover, nimesulide (100 mg) proved to be significantly more effective than rofecoxib (25 mg).

Clinical data

Meaningful response in OA patients treated with nimesulide has been demonstrated in a considerable number of studies (see previous section). Here we focus on the pain parameters that are influenced by nimesulide compared with other COX-2 inhibitors including the coxibs.

In comparison with other COX-2 inhibitors, the efficacy and tolerability of nimesulide (200 mg/day) were compared with those of etodolac (600 mg/day) in the chronic treatment of patients with OA of the knee. In this study, both the beneficial and unwanted effects of the two drugs were generally comparable, although overall judgments of the efficacy by both the physicians and the patients were in favour of nimesulide [47].

More recently, a study was performed to examine the analgesic efficacy of nimesulide, celecoxib and rofecoxib in patients with knee OA [42]. This was a prospective, randomised, double-blind, intra-patient Latin square design trial comparing three COX-2 selective inhibitors at indicated doses for the treatment of knee OA, over a period of 3 weeks. Using this design, each drug was tested against all the others and was administered equally either as first, second, or third in the sequence to the same number of patients. Enrolled patients were randomly assigned to treatment with nimesulide 100 mg p.o., celecoxib 200 mg p.o., or rofecoxib 25 mg p.o. Each drug was given in a single oral administration for 7 days. Only the following concomitant treatment was allowed: one 500 mg paracetamol tablet, once a day, 12 h after the administration of one of the tested drugs. No other rescue medication was allowed during the study.

As patients with OA have pain that typically increases with activity and is particularly evident after a period of inactivity, special attention was devoted to the onset of the action against pain connected with movement after the drug administration in the morning. The intensity of pain was recorded at baseline and 15, 30, 60, 120, and 180 min after drug consumption.

The overall analgesic efficacy in the first hours after drug administration was determined by total pain relief over 3 h (TOPAR3).

At the end of each week of treatment patients answered questions about analgesic efficacy on a five-point categorical scale: none, mild, moderate, good, very good. At the end of the study, each patient was asked about which of the three forms of treatment he or she would opt for as a continuation of the therapy. For tolerability assessment, at the end of each period of treatment (7 days) patients replied to questions about the overall tolerability of the treatment on a five-point categorical scale: very poor, poor, fair, good, very good. Before treatment, all the patients recorded a score >40, the basal values ranging from 42–95. These VAS scores indicate that the patient would have recorded at least moderate pain on a standard four-point categorical scale.

Although all the drugs induced a reduction in pain intensity, the analgesic efficacy of nimesulide was clearly superior to that of the other two NSAIDs (Tab. 2).

In fact, a single dose of nimesulide 100 mg provided greater therapeutic benefit than celecoxib 200 mg and rofecoxib 25 mg over a 3 h period. This difference in TOPAR3 values was evident both on the first and on the last day of a week-long treatment (Fig. 2). In addition, it is particularly worth underlining that the analgesic action of nimesulide was more rapid than that exerted by the other drugs tested. Indeed, only in the group of patients treated with nimesulide was the

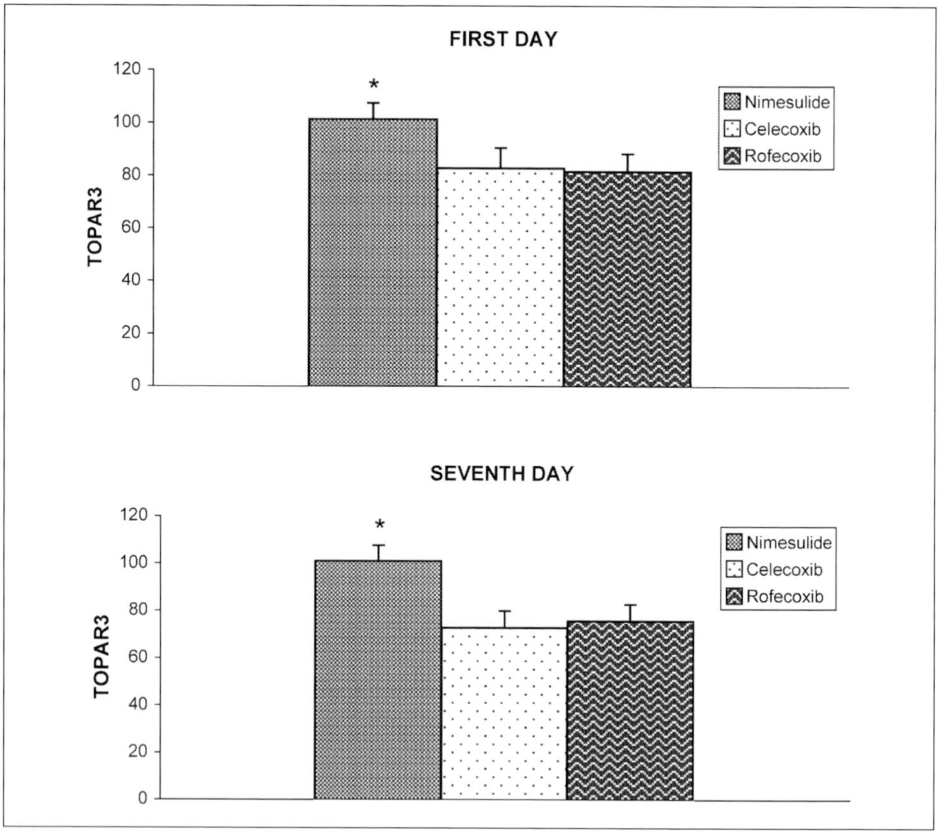

Figure 2
*Overall analgesic effects (expressed as TOPAR3) of nimesulide (100 mg), celecoxib (200 mg), and rofecoxib (25 mg) on the first day (upper panel) and on the last day (lower panel) of treatment in patients with knee OA. TOPAR3 represents the sum of pain relief scores over 3 hours, and was derived by adding up time-weighted pain relief scores (expressed as the difference between the value recorded at baseline and that recorded at each time point after drug administration) over a period of 3 hours [42]. * = $P < 0.05$ vs celecoxib and rofecoxib (One-way ANOVA followed by Bonferroni's t test).*

mean VAS values measured 15 and 30 min after consumption significantly different from those measured in basal conditions (Fig. 3).

This observation seems to be of particular importance if we consider that a rapid decrease of pain intensity will make a considerable difference in the ability of patients with OA to carry out their normal everyday activities. The percentage of patients who reported good or very good analgesic efficacy was 53.4% in the nimesulide group, 46.7% in the celecoxib group, and 50% in the rofecoxib group.

Table 2 – Percentage of patients with OA of the knee who achieved at least 50% reduction in pain score, compared with basal value, after treatment with celecoxib (200 mg), nimesulide (100 mg) or rofecoxib (25 mg) [42]

Day 1						
Time	**15′**	**30′**	**60′**	**120′**	**180′**	**12 h**
Celecoxib	0	3.3	23.3	20	16.6	16.6
Nimesulide	0	6.6	50	60	66.6	46.6
Rofecoxib	0	3.3	33.3	36.6	33.3	33.3
Day 7						
Time	**15′**	**30′**	**60′**	**120′**	**180′**	**12 h**
Celecoxib	0	0	26.6	30	20	13.3
Nimesulide	3.3	3.3	36.6	56.6	56.6	40.0
Rofecoxib	0	0	36.6	30.0	30.0	20.0

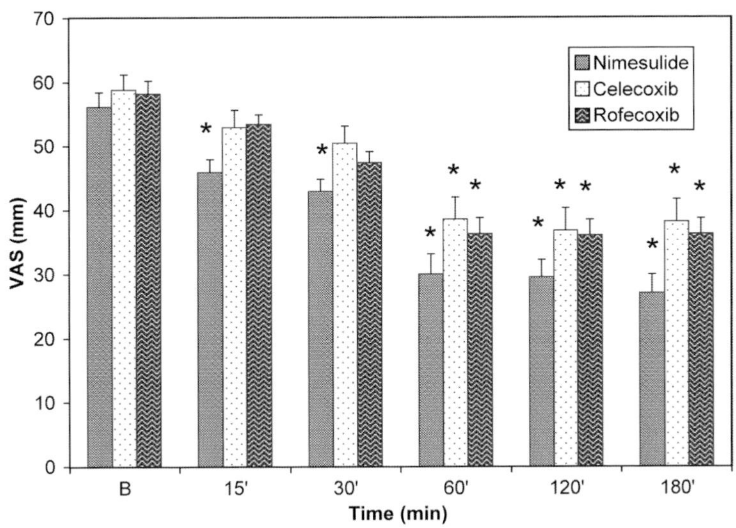

Figure 3
*Pain intensity as recorded by the patient on a 100-mm Visual Analogue Scale (VAS) from 15 to 180 minutes after drug administration at the first day of treatment with nimesulide, celecoxib and rofecoxib. Each bar represents means ± SEM of 30 patients with knee OA. * = P < 0.05 vs baseline (One-way ANOVA followed by Dunnett's t test) [42].*

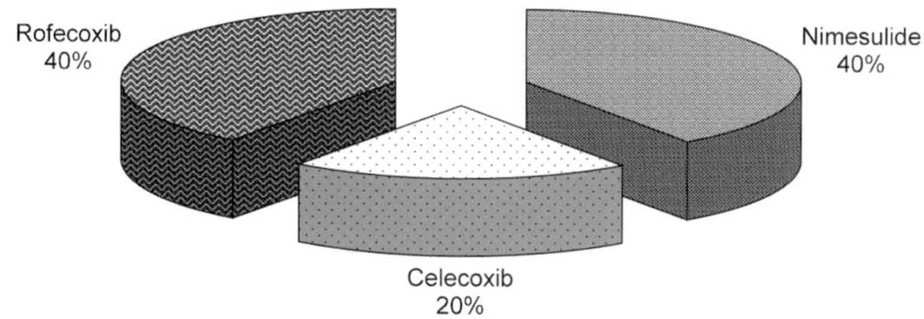

Figure 4
Percentage of patients with knee OA who chose nimesulide, celecoxib or rofecoxib for a continuation of the analgesic therapy at the end of the study. Total number = 30 (100%) [42].

The percentage of patients who reported good or excellent tolerability were 76.7% in the nimesulide-treated group, 70% in the celecoxib-treated group, and 76.7% in the group of patients treated with rofecoxib. No patient withdrew from the study for serious adverse events.

At the end of the study, the percentage of patients who expressed their preference for nimesulide treatment was 40%. The same percentage of patients expressed their preference for rofecoxib. The percentage of patients who expressed their preference for celecoxib was 20% (Fig. 4).

Thus, in this study on patients with knee OA nimesulide proved to be significantly more effective in providing symptomatic relief than celecoxib and rofecoxib. Furthermore, nimesulide provided more rapid relief of pain connected with walking than the other two drugs tested in this study.

From this comprehensive analysis of available data emerges that nimesulide represents an effective agent for the treatment of joint pain, with particular reference to the rapid onset of its analgesic effect.

Nimesulide in the treatment of primary dysmenorrhoea and other gynaecological conditions

Pelvic pain and pain in dysmenorrhoea

Pelvic pain is a common and significant disorder of women. Pelvic pain is estimated to have a prevalence of 3.8% in women aged 15–73, which is higher than the prevalence of migraine (2.1%) and is similar to that of asthma (3.7%) or back pain (4.1%) [76]. In primary care practice, 39% of women complain of pelvic pain [77,

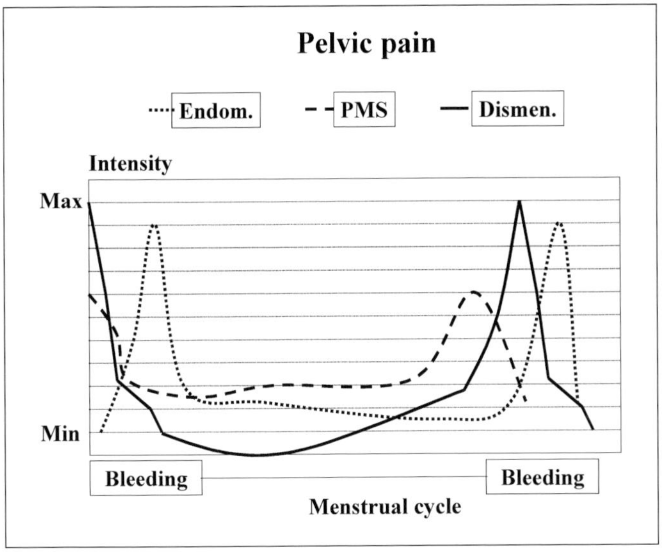

Figure 5
Variation in the occurrence of pelvic pain in different gynecological conditions, i.e. endometriosis (Endom) and premenstrual syndrome (PMS) from primary dysmenorrhoea (Dysmen).

78] and it is estimated to account for 10% of all referrals to gynaecologists. Pelvic pain represents the indication for 12% of all hysterectomies and over 40% of gynaecologic diagnostic laparoscopies [79]. Direct costs of healthcare for chronic pelvic pain in the United States are estimated at $880 million per year, and both direct and indirect costs may total over $2 billion per year [78]. At an individual level, pelvic pain leads to years of disability and suffering, with loss of employment, marital discord and divorce, and numerous untoward and unsuccessful medical misadventures. Clearly, pelvic pain is an important issue in the healthcare of women.

Although definitions vary, chronic abdominal pain may be considered any pain that has been present, continuously or intermittently, for at least 6 months. Recurrent or intermittent pain may either be cyclic or non-cyclic in nature (see Fig. 5). Pain with a specific, identifiable physiological cause is often referred to as 'organic' pain; pain without a clear identifiable cause and/or pain that appears to be exacerbated by psychosocial factors is frequently referred to as 'functional' pain. Among the cyclic pelvic pain primary dysmenorrhoea is the commonest problem in young women [77].

Primary dysmenorrhoea

Definition, prevalence and diagnosis

Primary dysmenorrhoea is usually defined as cramping pain in the lower abdomen occurring near the onset of menstruation in the absence of any identifiable pelvic disease [76–87]. It is distinguished from secondary dysmenorrhoea, which refers to painful menses resulting from pelvic pathology such as endometriosis. Prevalence rates are as high as 90%. Initial presentation of primary dysmenorrhoea typically occurs in adolescence and is a common cause of absenteeism and reduced quality of life in women.

Primary dysmenorrhoea is highly prevalent among adolescent girls. The prevalence of dysmenorrhoea has been extensively examined in teenagers [80]. A majority of adolescents report experiencing dysmenorrhoea and about 15% of adolescents describe their dysmenorrhoea as severe to require treatment. This supports the widely held idea that dysmenorrhoea is related to the establishment of ovulatory menstrual cycles.

Dysmenorrhoea is the major cause of activity restriction and school and work absence in adolescent girls. In a questionnaire study of 182 US high school girls, 59% reported that cramps caused them to be less active, 45% reported missing school or work due to cramps, and 40% reported missing class in the past year due to cramps [81]. In a sample of Swedish schoolgirls ages 14–19 years, 15% reported being unable to participate in normal activities, 10% reported school absence, and 5% reported staying in bed due to dysmenorrhoea [82]. Among 54 Norwegian factory workers aged up to 19 years, 24% reported being absent from work in the previous 6 months [83].

In a prospective cohort study, menstrual diary data have been collected during the first year of university from 165 college entrants aged 17–19 years [83]. During the study, 1,396 bleeding episodes were observed. Menstrual pain led to "ever-missing any activity" in 42% and "ever-missing school" in 25% of subjects. Of the reported pain episodes, 10% were associated with missing any activity, 4% were associated with missing school, and 10% were associated with staying in bed.

In a larger, representative sample of US adolescents aged 12–17 years, 14% frequently missed school because of cramps [80]. Those with severe cramps (50%) were more likely to miss school than those with mild cramps (17%), and African–American girls (24%) were more likely than Caucasian girls (12%) to miss school due to cramps after adjustment for socioeconomic status. Some authors have estimated that dysmenorrhoea is the single greatest cause of lost working hours and school absence in adolescent girls, although no systematic studies have prospectively examined the impact of dysmenorrhoea on quality of life or cost [84].

A diagnostic evaluation is unnecessary in patients with typical symptoms and no risk factors for secondary causes. Primary dysmenorrhoea usually presents during adolescence, within 3 years of menarche [76]. It is unusual for symptoms to start within the first 6 months after menarche. Affected women experience sharp, intermittent spasms of pain, usually centred in the suprapubic area. Pain may radiate to the back of the legs or the lower back. Systemic symptoms of nausea, vomiting, diarrhoea, fatigue, fever, headache or light headedness are fairly common. Pain usually develops within hours around the start of menstruation and peaks as the flow becomes heaviest during the first or the second day of the cycle.

A focussed history collection and physical examination are usually sufficient to establish the diagnosis of primary dysmenorrhoea [85–87]. The history reveals the typical cramping pain with menstruation, and the physical examination is completely normal. Secondary causes of dysmenorrhoea must therefore be excluded [87].

The most important causes of secondary dysmenorrhoea include endometriosis, adenomyosis, malformation of Mullerian ducts, ovarian cysts, pelvic varicocele, pelvic inflammatory disease, uterine fibroids, contraceptive intrauterine devices, and stenosis of the cervical channel [84, 85].

With a typical history and a lack of abnormal findings on routine pelvic examination, further diagnostic evaluation is not required. In many instances, it is preferable to confirm the diagnosis "ex adjuvantibus" through a therapeutic trial of NSAIDs [88, 89]. At least partial relief of pain with NSAID therapy is so predictable in women with primary dysmenorrhoea that failure to respond should raise doubts about the diagnosis.

Etiology

The etiology of primary dysmenorrhoea is not precisely understood, but most symptoms can be explained by the action of uterine prostaglandins, namely PGF2α. During endometrial sloughing, the disintegrating endometrial cells release PGF2α as menstruation begins. PGF2α stimulates myometrial contractions, ischaemia and sensitisation of nerve endings. Pain is produced through three mechanisms, all of which are mediated by the effect of prostaglandins on pelvic tissue. Indeed, the increased production of prostaglandins gives rise to increase and/or abnormal uterine contractility. Moreover, such uterine activity reduces uterine blood flow and favours ischaemia or hypoxia, leading to pain. Furthermore, cyclic endoperoxides, the intermediates in the biosynthesis of prostaglandins, have direct pain-producing properties through sensitisation of the pain fibres.

The clinical evidence for this theory is quite strong. Women who have more severe dysmenorrhoea have higher levels of PGF2α in their menstrual fluid. These levels are highest during the first two days of menses, when symptoms peak [88]. In addition, several studies have documented the impressive efficacy of NSAIDs, which act through prostaglandin synthetase inhibition [89]. Some studies have also implicated increased levels of leukotrienes and vasopressin, but these connections are not yet well established.

Nimesulide compared with other NSAIDs in the clinical management of primary dysmenorrhoea

Most patients with primary dysmenorrhoea show subjective improvement upon treatment with NSAIDs [88–90] and successful pain relief ranged 64–100% of subjects, according to various reports. These familiar drugs have a record of efficacy demonstrated by numerous studies over the past 15 years. Table 3 gives a summary of the responses from nimesulide compared with placebo or other NSAIDs in primary dysmenorrhoea and pelvic inflammatory disease [91–100]. These data show that nimesulide is superior to placebo and some other NSAIDs (i.e. diclofenac, naproxen, mefenamic acid) with the exception of piroxicam and methoxybutropate for which it was equivalent.

Oral contraceptives provide another effective and well-studied choice of treatment, especially in women desiring birth control (Tab. 4). Oral contraceptives are effective in about 90% of patients with primary dysmenorrhoea. For the approximately 10% who do not respond to the above options, a host of alternatives exists, ranging from laparoscopic surgery to acupuncture, although with much less evidences supporting their use. Again, it is important to underline that lack of pain relief should increase suspicion of a secondary cause of dysmenorrhoea.

The most appropriate first-line choice of therapy in most women with primary dysmenorrhoea is an NSAID. Such class of medications work through the inhibition of the production and release of prostaglandins, also at uterine level. As previously mentioned prostaglandins are responsible for the painful uterine contractions and associated systemic symptoms of primary dysmenorrhoea, such as nausea and diarrhoea. The choices of specific agents are numerous. Response to NSAIDs usually occurs within 30–60 min. Since individual response may vary, it may be prudent to try a second agent of a different class if the pain is not relieved with the first agent after one or two menstrual cycles.

Nimesulide has gained attention recently for its selective properties as an inhibitor of prostaglandin production in reproductive target tissues [98–99]. COX-1 derived prostaglandins (PGs) induce progesterone withdrawal (luteolysis), while COX-2 derived PGs, inhibited by nimesulide, induce uterine activity. When the uterine smooth muscle contracts, expression of COX-2 transcript is elevated. It is well known that increased PGs production as shown in endometrial tissue and indicated by high menstrual blood PGs concentration is the main factor in the pathology of primary dysmenorrhoea [88, 89].

The reduction in PG production obtained with NSAIDs allows the conversion of uterine smooth muscle function from painful anoxic contractures to painless

Table 3 – Effects of nimesulide compared with placebo or other NSAIDs in relief of pain in dysmenorrhoea or pelvic inflammatory disease

Treatment and Dosage (mg/d)	No. of Patients	Relative Efficacy	Reference
NIM 200 PIR 200	18 18	NIM = PIR	Bacarat et al. (1991) [93]
NIM 200 PLA	18 15	NIM > PLA	Chiantera et al. (1993) [94]
NIM 200 PLA	20 19	NIM > PLA	Di Leo et al. (1988) [91]
NIM 200 MET 1200	30 30	NIM = MET	Melis et al. (1997) [97]
NIM 100 NAP 500	6 6	NIM > NAP	Pirhonen & Pulkkinen (1995) [96]
NIM 200 DIC 150	30 30	NIM > DIC	Rinaldi & Cymbalista (1994) [95]
NIM 200 FEN 200 MEF 1500	20 20 20	NIM > FEN NIM > MEF	Lopez Rosales & Cisneros Lugo (1989) [92]
NIM 200 PLA	14 14	NIM > PLA	Pulkkinen (1987) [98, 99]
NIM 100-300 DIC 150	152 156	NIM > DIC	Facchinetti (2001) [100]

Abbreviations: DIC = diclofenac; FEN = fentiazac; MEF = mefenamic acid; MET = methoxybutropate; NAP = naproxen; NIM = nimesulide; PLA = placebo; PIR = piroxicam; = indicates no statistically significant difference in efficacy; > denotes statistically greater efficacy compared with comparator drug ($p < 0.05$).
Modified and updated from [63].

contractions. The recent characterisation of COX-1 as a constitutively expressed isoenzyme, and of COX-2, as an inducible isoenzyme, gives the rationale for the comparison between the COX-2 inhibitor nimesulide and the nonselective COX inhibitors in the treatment of primary dysmenorrhoea [98–100].

The effects of nimesulide on both prostaglandin content in the menstrual blood and the overall intrauterine perfusion was investigated. The concentrations

Table 4 – Comparison of different treatments for primary dysmenorrhoea

Type of treatment	Effectiveness	Advantages or Limitations	Balance of the treatment
Nimesulide	Very good	Less GI adverse effects than other NSAIDs	First line treatment
NSAIDs	Very good	Gastric adverse effects	First line treatment
Oral Contraceptives	Very good	Every day pill assumption Pill-related adverse effects	Second line treatment
Other treatments (Vitamins, NO donors, magnesium)	Quite good	Additional medication is usually required	Studies are required to confirm the efficacy, dosage and duration of the treatment
Complementary medicine	(too few controlled studies to reach any conclusion)		

of $PGF_{2\alpha}$ (which causes uterine contraction) and PGE_1 or PGE_2 (which have relaxing/contractile effects) were reduced by 80% and 60% respectively [93, 94]. Following nimesulide treatment a slight decrease in active pressure and a gradual normalisation of resting pressure and frequency of pressure cycles were observed. Hence, it seems that nimesulide transforms smooth muscle from a pathological state of dysmenorrheic contracture to a state of eumenorrheic physiological contractions. Moreover, nimesulide treatment seems to be associated with a reduction in vascular resistance of uterine arteries [98, 99].

In a recent multicentre double-blind study 308 women were randomised in two groups to receive up to 3 tablets/day of nimesulide or diclofenac, for the first 3 days of the menstrual cycles [100]. Abdominal pain was the primary endpoint and it was evaluated before and every 30 min after the first drug administration through a visual analog scale. Both drugs progressively and significantly decreased pain which was reduced by 82% (nimesulide) and 79% (diclofenac) at the second hour. However, nimesulide showed faster activity than diclofenac starting from 30 min with a reduction of 35% *versus* 27% at both the first and second cycle of treatment. Headache and back pain were significantly and equally improved by both treatments. Tolerability was good with both drugs. However, 16 of 155 and 7 of 149 patients reported gastric side effects with diclofenac and nimesulide, respectively [100].

Nimesulide has a favourable tolerability profile, since the incidence of adverse reactions is equal or slightly higher than that of placebo. Moreover, in comparison with other NSAIDs the prevalence of adverse reactions for nimesulide, mostly gastrointestinal was lower than that observed in those patients receiving other NSAIDs [93, 97].

Given its fast analgesic action and efficacy in relieving pelvic-cramp related symptoms on the one hand, and the absence of significant adverse reactions on the other hand, nimesulide should be the preferred treatment for primary dysmenorrhoea.

Other treatments (Tab. 4) include (a) oral contraceptives (OC) as second line of treatment for most patients, unless birth control is also desired [101], (b) transdermal nitroglycerine patches (which are probably less effective than NSAIDs) [102, 103], (c) some possible benefits from a low dietary intake of omega-3 fatty or magnesium supplements [104], and/or (d) complementary medicine including acupuncture [105]. None of the treatments (b) or (c) alone surpasses that of NSAIDs and with its favourable tolerability, nimesulide has a place as a first line therapy (Tab. 4).

Conclusions

Dysmenorrhoea is the most common gynaecologic complaint among young women. Despite progress in understanding the physiology of dysmenorrhoea and the availability of effective treatments, many women do not seek medical advice or are under-treated. Dysmenorrhoea in young women is usually primary (functional), and it is associated with normal ovulatory cycles and no pelvic pathology. In the pathogenesis of dysmenorrhoea, prostaglandins and arachidonic acid metabolites play an important role, being elevated in women with dysmenorrhoea. Penetration of excess prostaglandins into general circulation fully accounts for the systemic symptoms of dysmenorrhoea (nausea, vomiting, diarrhoea, headache, etc.). Rational treatment of dysmenorrhoea with nimesulide is directed at elimination of the excess prostaglandin action.

NSAIDs in sports medicine

Introduction

Some 10 years ago a nationwide epidemiological survey in France of more than 7,000 consultations for injury due to involvement in sports was conducted among more than 150 sports injury physicians. The objective of this survey was to provide some insight into the diagnostic and prescribing habits of practitioners who have to deal with the problems of injuries and overuse lesions during a sporting activity [106]. This survey was instructive in providing information on the topography of lesions and their nature, as well as insight into the aetiopathogenesis and the importance of stopping work and sport as part of therapy. This study also provided information on the therapeutic habits of the doctors questioned. Although, as expected, the respondents showed a multiple pragmatic approach to therapy it was hardly surprising that in 7,282 responses, systemic NSAIDs (50%) and local NSAIDs (51%) were by far the most frequent therapies prescribed. This finding is further supported by the admittedly modest use of infiltrations (5%) often also used for anti-inflammatory purposes. Since the survey was conducted in France, it is not surprising to find a fairly high percentage (19%) of responses mentioning mesotherapy, a surface injection technique where the therapeutic agent, a cocktail of injectable solutions, contains a large proportion of injectable non-steroidal anti-inflammatories.

It is, therefore, concluded that NSAIDs are the drug of choice available to doctors to treat patients suffering from sport injuries [107–108].

Inflammation

The various aspects of the responses of the musculoskeletal system to sports injuries have been comprehensively reviewed elsewhere [109–112]. In sports injuries it is usual to distinguish between the acute event or injury ('macrotrauma') and the more chronic lesion or overuse lesion ('microtrauma'). These two concepts overlap, as shown in the Figure 6A. These two categories comprise acute and overstraining lesions (Tab. 5). The pathophysiology of these two types of lesion has not yet been fully determined.

In the case of acute injuries, it is clear that the injurious force causes tissue lesions of the articular capsule, tendons, muscle fibres, cartilage or other structures depending on the type of lesion [110–112]. These lesions are generally accompanied by blood capillary tears with local bleeding. A multitude of functional and structural reactions will then occur and these form the basis of the inflammatory reaction.

Table 5 – Categories of sport injuries

Acute lesions	Overstraining lesions
Contusions	Tendinopathies
Partial or total tears of muscles, tendons or ligaments	Stress fractures
	Compartment syndromes
Dislocations	Osteoarthritis
Fractures	Bursitis

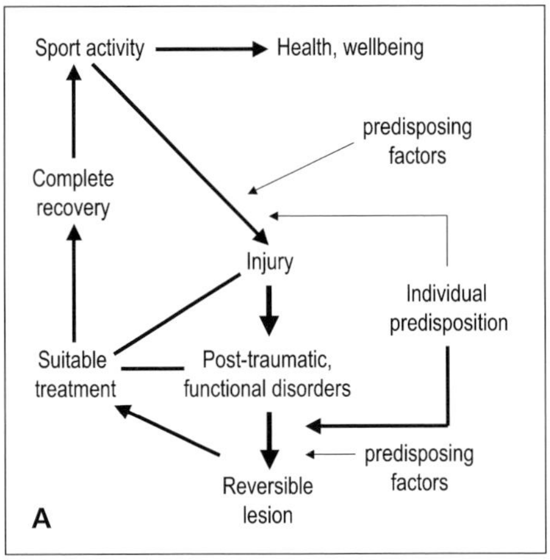

Figure 6
Causes of and responses to sports injuries. Predisposing factors principally involving overload reactions (Figure 6A) and pathophysiological changes and responses to repair and therapeutic modalities (Figure 6B).

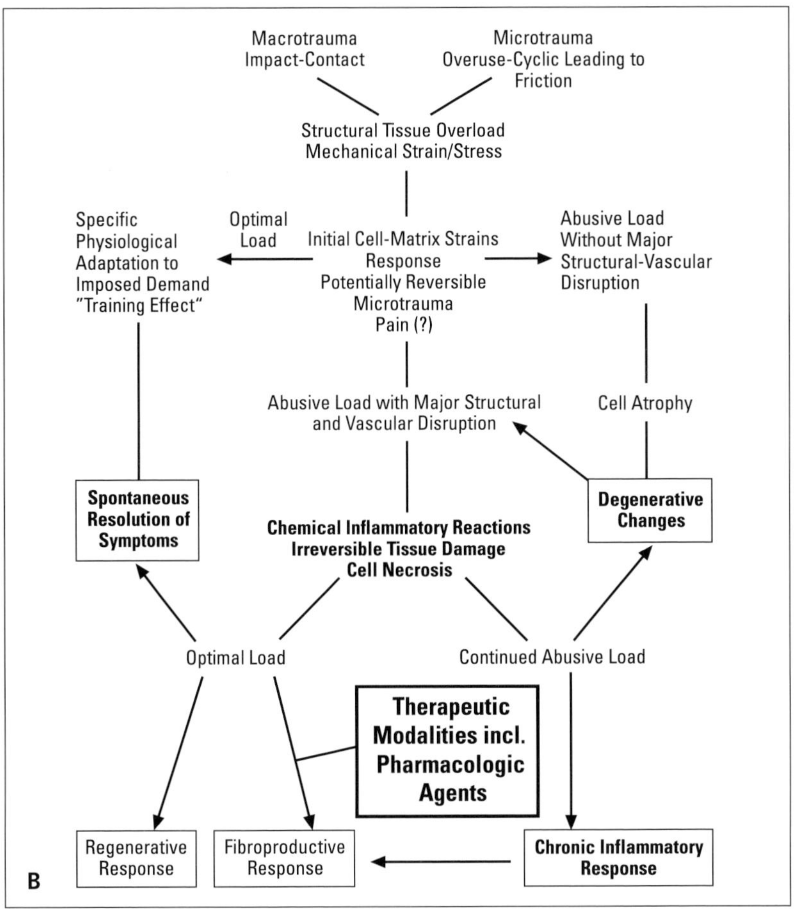

It is important to appreciate that these inflammatory reactions are a physiological response, brought about by the tissue lesion, which forms part of the process of healing. Healing can be obtained by regeneration where the damaged tissue is replaced by functionally and morphologically similar tissue, or by repair, where the injured tissue will be replaced by granulation tissue which will organise itself into a scar. Thus, the response of tissues to a lesion is to cause inflammation, regardless of the cause of the lesion. The process is complex and not yet understood in detail. It involves many types of inflammatory cells, joint and tissue destructive enzymes and other physiologically active substances, and it may take varied forms.

In the lesions incurred during sports, acute injuries or overuse lesions, the trigger of the inflammatory response is probably the degradation products of the damaged tissue. This will set off a cascade of sequences with associated healing (Fig. 6B). Although inflammation is essential to healing, it may be self-perpetuating, thus becoming chronic. This may cause new destructive damage to surrounding tissue. It may thus be important to control this reaction before it magnifies and this is where the use of NSAIDs is worthwhile.

It has not been demonstrated that in every sports injury, particularly microtraumas, there is an inflammatory reaction. Many authors have been able to demonstrate the absence of cells and other inflammation mediators, for example in many forms of tendon inflammation, e.g., the Achilles tendon or the patellar tendon. A plausible explanation for these findings is that the classic inflammatory process is triggered only if sufficient tissue and microvascular injury is present.

The use of nimesulide in sports medicine

There is a wide choice of NSAIDs for use in sport injuries. Currently there are about 50 different preparations of NSAIDs available many distinguished from one another in clinical response or their adverse effect profile, although there may be differences at an interindividual level [108–110].

Several reports have been published showing the efficacy of nimesulide in various types of soft tissue conditions including those from sports injury [113–124] (see review of earlier literature in [62]).

In a randomised double-blind study comparing the effects of 100 mg of nimesulide twice daily with placebo in 60 sprained ankles, Dreiser and Riebenfeld [113] clearly demonstrated the superiority of the active product over the placebo. On day 4, three treated patients (10%) stopped the treatment due to the disappearance of the symptoms, while 11 patients (37%) in the control group stopped due to aggravation. Not only was the absolute efficacy superior in the treated group, but also the time taken to obtain this result was also shorter. Overall and in particular gastrointestinal tolerability was of the same order of magnitude in both groups. The difference was statistically significant in favour of the treated group.

In another double-blind study, which was multicentre, Lecomte et al. [114] were able to demonstrate that the efficacy of oral nimesulide in the treatment of tendonitis and bursitis related to involvement in sport, was similar to that of oral naproxen – used for comparison and which has been in widespread use for many years. This study compared the reduction in pain recorded on a visual analogue scale (VAS) as well as pain during movement against specific resistance in the affected joint, as well as side effects. The group studied consisted of 201 patients, 101 that received nimesulide and 100 treated with naproxen. The distribution of disorders was very similar in the two groups, as were the characteristics of the groups as regards lifestyle, age and morphology. The authors found similar real efficacy in both groups, but without statistically significant differences; the same applied to side effects, with more frequent gastric disturbances in the naproxen group.

In a multicentre double-blind study, oral nimesulide was compared with oral naproxen in the treatment of minor injuries resulting from involvement in sport [115]. A total of 660 patients suffering from minor lesions such as contusions, tendonitis, pulled muscles and strains were divided into two comparable groups, one receiving 300 mg of nimesulide daily, and the other 750 mg of naproxen daily. The evaluation criteria were judged as much by the patient as by the attending doctor and concerned mainly efficacy in reducing pain and tolerability. After 7 days of examinations, the authors concluded that the two products had similar properties in relation to oedema and pain intensity, since both parameters improved significantly in both cases. As regards tolerability more patients that took naproxen had gastric side effects compared with those on nimesulide, but the difference was not statistically significant.

Nimesulide 200 mg has been found to be as effective as diclofenac 150 mg in the relief of pain and swelling from soft tissue injuries [116]. Jenoure et al. [117] reported the effects of using nimesulide compared with diclofenac in daily sports injury practice in a specialist sports injury clinic, and also in a randomised, double-blind study conducted on a multicentre basis with colleagues practising in sports medicine. The aim of the study was to compare the efficacy of nimesulide 100 mg twice daily with that of diclofenac 75 mg (a well-known reference standard) twice daily taken orally in the treatment of acute injuries arising from involvement in sport. A total of 343 patients were investigated within 48 h of the accident affecting mainly joints or muscles. They were monitored over a week of drug treatment. Although it was not possible to demonstrate any difference in efficacy between the two products in improvement of symptoms during the study period, nimesulide appeared to have better tolerability. This is not unimportant bearing in mind the global use of NSAIDs, although in sports medicine, the duration of treatment with these products is generally short, and the patients are usually in good health.

Saillant and co-workers [118] conducted a randomised, double-blind, multicentre study in France in 293 patients of either sex with ankle sprain due to sport activities who received nimesulide 100 mg tablets or ketoprofen 100 mg capsules b.i.d., with the respective identical capsule or tablet placebo, for 7 days. Paracetamol was permitted as a rescue medication. Pain on active or passive movement, pain intensity recorded on 100 mm VAS scales, pain on palpation, joint swelling and ability to stand on the affected foot were recorded at entry and on the 2nd and 7th days of treatment. Both drug treatments produced virtually identical effects and the intake of rescue medication (required in 2–3% of patients was similar). The global judgement of efficacy rated as "very good" or "good" by patients was 83.9% in those that received nimesulide and in 77.8% of those on ketoprofen. Physician's ratings on the same scale were similar, being 82.5% for nimesulide and 75.8% for ketoprofen. There were similar numbers of responders and non-responders in pain relief for both treatments (> 92%). Thus in all respects the two drugs had identical benefit and this was substantial over the treatment period.

In Table 6 studies summarise the effects of nimesulide in comparison with other NSAIDs or placebo for the treatment of pain in acute musculoskeletal injuries and tendonitis/bursitis; some of these may have been attributed to sport [113–124].

These studies therefore demonstrate that nimesulide is a molecule with anti-inflammatory effects entirely comparable with those of "classic" NSAIDs with, however, tolerability that tends to be better, in particular in terms of gastrointestinal symptoms.

Topical nimesulide in acute musculoskeletal injuries

The popularity of topical preparations (ointments, gels) for both self-treatment as well as by prescription for treating acute musculoskeletal injuries, including those sustained during sport, is well established. The application of different formulations of nimesulide for these states has been reported by a number of authors.

As reviewed in Chapter 2, a gel formulation of 3% nimesulide (90 mg) when applied to the outer part of the shaven right thigh three times daily for 8 days in healthy volunteers was absorbed to about 1% of that from an oral dose of the drug [125]. Using this same formulation at the same dose two double-blind, multicentre placebo-controlled trials were undertaken following 7 days treatment t.i.d. in 105 patients with benign ankle sprains [126] and in 103 patients with acute tendonitis of the upper limb [127]. Despite a relatively high rate of placebo response (54.3% in the ankle sprains and 34% in tendonitis groups, respectively), nimesulide treatment showed significant and pronounced improvement in 100 mm VAS scores (82% in the ankle sprain group and 60% in the tendonitis group, respectively). Nimesulide treatment was judged by investigators to be "very good" or "good" in 96% compared with placebo 47% in the ankle sprain study and

Table 6 – Effects of nimesulide compared with placebo or other NSAIDs on relief of painful symptoms in patients with various acute injuries or conditions

Treatment and Dosage (mg/d)	No. of Patients	Relative Efficacy	Reference
Acute musculo-skeletal Injury			
NIM 300 NAP 750	330 330	NIM ≡ NAP	Calligaris et al. (1993) [115]
NIM 200 DIC 150	14 20	NIM ≡ DIC	Costa et al. (1995) [119]
NIM 200 SER 15	18 20	NIM > SER	Di Marco et al. (1989) [120]
NIM 200 PLA	30 30	NIM > P	Dreiser & Riebenfeld [113]
NIM 200 SER 15	17 17	NIM > SER	Gusso & Innocenti (1989) [121]
NIM 200 DIC 150	14 20	NIM ≡ DIC	Ribamar et al. (1995) [116]
NIM 200 KET 200	154 153	NIM ≡ KET	Saillant et al. (1997) [118]
NIM 200 DIC 100	29 32	NIM ≡ DIC	Zarraga Corrales et al. (1992) [123]
Bursitis/tendonitis			
NIM 200 NAP 1000	101 100	NIM ≡ NAP	Lecomte et al. (1994) [114]
NIM 200 DIC 150	62 60	NIM ≡ DIC	Wober et al. (1999) [124]
Thrombophlebitis			
NIM 200 DIC 100	30 30	NIM ≡ DIC	Agus et al. (1993) [193]
NIM 200 DIC 100	23 24	NIM ≡ DIC	Ferrari et al. (1993) [194]
NIM 200 SER 15	30 30	NIM > SER	Zanetta et al. (1988) [195]

Table 6 – (continued)

Treatment and Dosage (mg/d)	No. of Patients	Relative Efficacy	Reference
Ear, nose and throat disorders			
NIM 200 SEA 60	195 195	NIM > SEA	Bianchini et al. (1993) [171]
NIM 200 FLU 300	30 28	NIM ≡ FLU	Cadeddu et al. (1988) [177]
NIM 200 DIC 150	30 30	NIM ≡ DIC	Gananca et al. (1990) [187]
NIM 200 NAP 500	29 31	NIM ≡ NAP	Miniti & Dieb Miziara (1991) [181]
NIM 200 DIC 150	30 30	NIM ≡ DIC	Munhoz et al. (1990) [182]
NIM 200 NAP 500	27 26	NIM > NAP	Nouri & Monti (1993) [173]
NIM 200 FEP 400	20 20	NIM ≡ FEP	Passali et al. (1988) [183]
NIM 400 PR FLU 200 PR	48 47	NIM ≡ FLU	Rossi et al. (1991) [178]
Urogenital disorders			
NIM 200 PLA	40 40	NIM > PLA	Lotti et al. (1993) [196]
NIM 200 BRO 240	20 20	NIM < BRO	Lotti et al. (1993) [196]

Abbreviations: BRO = bromeline; DIC = diclofenac; FEN = fentiazac; FEP = feprazone; FLU = flurbiprofen; KET = ketoprofen; MEF = mefenamic acid; NAP = naproxen; NIM = nimesulide; PLA = placebo; PIR = piroxicam; SEA = seaprose S; SER = serrapeptase (serratiopeptidase) d = day; ≡ indicates no statistically significant difference in efficacy; > denotes statistically greater efficacy compared with comparator drug ($p < 0.05$).
Modified and updated from [63].

75.5% compared with placebo 22% in the tendonitis study. Patients' assessments were similar to those of the investigators. In the ankle sprain study nimesulide had above-mentioned rating of 96% c.f. placebo 51%. In the tendonitis study the scores for "very good" or "good" were 77.6% and 2%, respectively. Two patients reported minor skin reactions – one in each group in the ankle sprain group while 10 reports of adverse reactions in nine patients were recorded in the tendonitis study. In five patients that had nimesulide there were skin reactions c.f. two in placebo, which required discontinuation of treatment in one patient in each group. Three patients on nimesulide had nausea or heartburn.

In another multicentre, double-blind study that extended over 14 days treatment t.i.d. with a 3% gel containing 90 mg nimesulide was compared in 111 patients with that of the same mass of gel containing 30 mg diclofenac in 109 patients that had tendonitis of the upper limb [128]. The 100 mm scale VAS responses were identical and showed a progressive statistically-significant decline in pain at days 7 and 15. Improvement in pain, functional disability, active joint movement and reduced sleep disturbance was reflected in the response time to the drug treatments. At the end of treatment 75% of patients on nimesulide and 76% on diclofenac had shown significant improvement. The time of onset of improvement was 6.4 days (range 1–14) in the nimesulide group and 6.9 days (range 2–15) in the diclofenac group, with the difference being not statistically significant. Moreover, the consumption of the rescue analgesic, paracetamol, was the same in both groups. Investigators judged the nimesulide treatment to be "very good" or "good" in 55% of patients compared with 60% in the diclofenac group, the difference being not statistically significant. The patients' ratings of the treatments using the same criteria were 51% in the nimesulide group and 50% in the diclofenac group.

Adverse reactions were reported in 17.1% patients that received nimesulide compared with 13.8% in the diclofenac group; the difference being not statistically significant. The most frequent adverse events were dry skin, erythematous rash and pruritus that were present in 45% of patients that received nimesulide and 40% of the diclofenac group that reported adverse reactions.

Another similarly designed multicentre, double-blind randomised trial was performed to compare the effects of another popular topical NSAID, ketoprofen 3.0% gel 90 mg formulation with that of 3% nimesulide gel 90 mg for a total of 7 days in 120 patients with mild ankle sprains [129]. VAS scores (100 mm scale) were similar with the two drugs over the 7 day time period and were not statistically significant. Decreased in joint oedema also occurred over the same period and the difference was not statistically significant.

Efficacy was judged by investigators to be "very good" or "good" in 87.1% of patients that had nimesulide and 89.7% that had ketoprofen at day 7. The same rating judged by patients was 79% of those that received nimesulide and 77.6% on ketoprofen. The intake of rescue analgesic, paracetamol, was the same in both groups. Only one case of dry skin was observed in the nimesulide group.

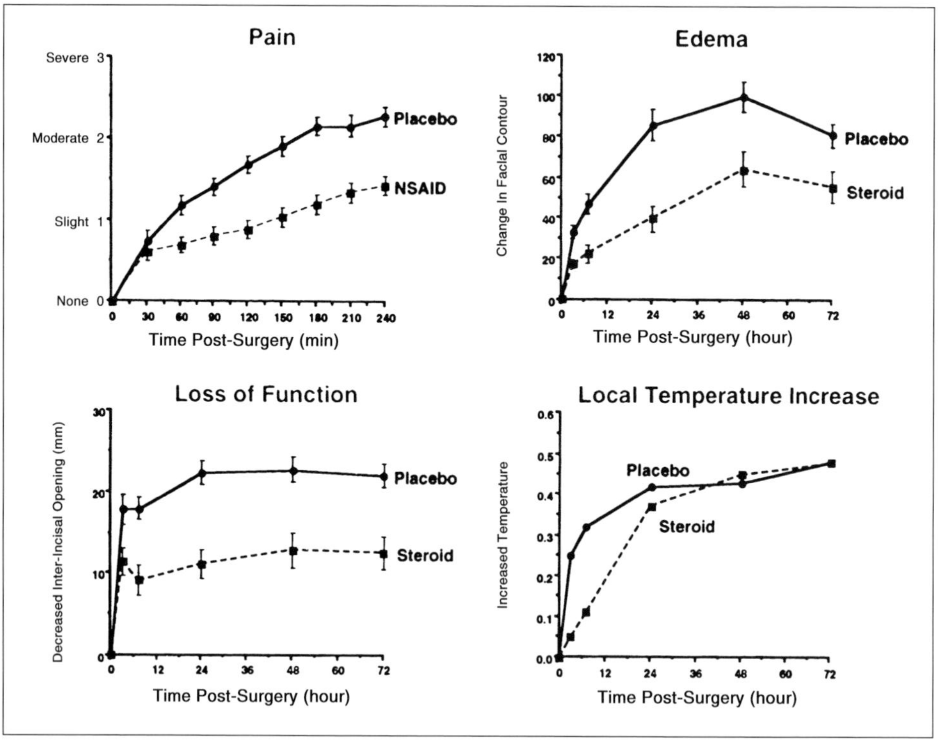

Figure 7

Cardinal signs of inflammation during the development of acute pain in the oral surgical model. Demonstration of the ability to measure pain, oedema, loss of function and local temperature changes and the responses to prototypic drugs (corticosteroid, the NSAID, ketoprofen). From [138]. Reproduced with permission of Kluwer Academic Publishers, Dordrecht, The Netherlands.

Post-marketing experience up to December 2004 of the 3% nimesulide gel (Helsinn) following sales of approximately 2.4 million units in 14 countries has revealed a total of three adverse reaction reports [130].

The results show that nimesulide gel is an efficacious treatment for pain relief and has comparable efficacy in treatment of the pain associated with acute musculoskeletal conditions with that of two commonly used NSAIDs, diclofenac and ketoprofen, formulated in the same gel system as used in the nimesulide 3% gel. Mild skin reactions which are relatively frequent with topical NSAIDs were also found to occur in a few patients that receive these NSAID gel formulations.

Two studies have shown the effectiveness of another topical formulation of nimesulide not of Helsinn origin (whose pharmaceutical characteristics were not described) in treating acute musculoskeletal conditions [131, 132]. In another study by Sengupta et al. [133] a gel formulation with an unspecified composition but containing 100 mg nimesulide was compared following topical administration with that of gel formulations of diclofenac and piroxicam as the same dose in the Hollander acute pain model induced in the forearm of volunteers. The pain response was determined by VAS, placebo related ratings on a ten-point scale and TOTPAR. Overall pain relief was faster from nimesulide than with the other two drugs with peak analgesia being observed at 120 min and this was correlated with plasma concentrations of the drug.

These studies show that gel formulations of nimesulide are effective when topically applied for *acute* pain relief. The question of their long-term utility in *chronic* musculoskeletal conditions, e.g., OA of the knee, is still to be resolved.

Acute pain models and conditions

Oral surgical model

Oral and other acute surgical pain models are considered useful for quantitative determination of the analgesic activities of NSAIDs as well as opioid and non-opioid analgesics in humans [134–139]. In extraction of third molars there is appreciable trauma to the dental alveolar cavities and surrounding inflamed tissues [138, 139]. The cardinal signs of inflammation can be assessed with time following surgery (Fig. 8) [138, 139]. There is accompanying production of PGE_2, endorphin, bradykinin and other proinflammatory molecules in the oral surgical extraction site (Fig. 9) [139–142]. The production of PGE_2 in the oral surgical extraction site is inhibited by NSAIDs in parallel with the reduction in pain symptoms (Fig. 10) [138, 139, 141, 142].

Recent studies with COX-2 specific NSAIDs (coxibs) suggest that suppression of COX-2 products is coincident with pain suppression and that there is effective analgesia with these drugs [143–150]. The inference that the inhibition of COX-2 alone underlies analgesia may be true for coxibs. However, there are also indications from studies with various COX-1 and COX-2 inhibitors that inhibition of COX-1 derived prostanoids may also contribute to the initial stages of analgesia in the periphery from non-selective NSAIDs or COX-1 inhibitors [148, 151–153].

Effects of post-operative nimesulide in oral surgery

A considerable number of studies have been reported showing the efficacy of nimesulide in controlling postoperative pain following dental surgery. Among

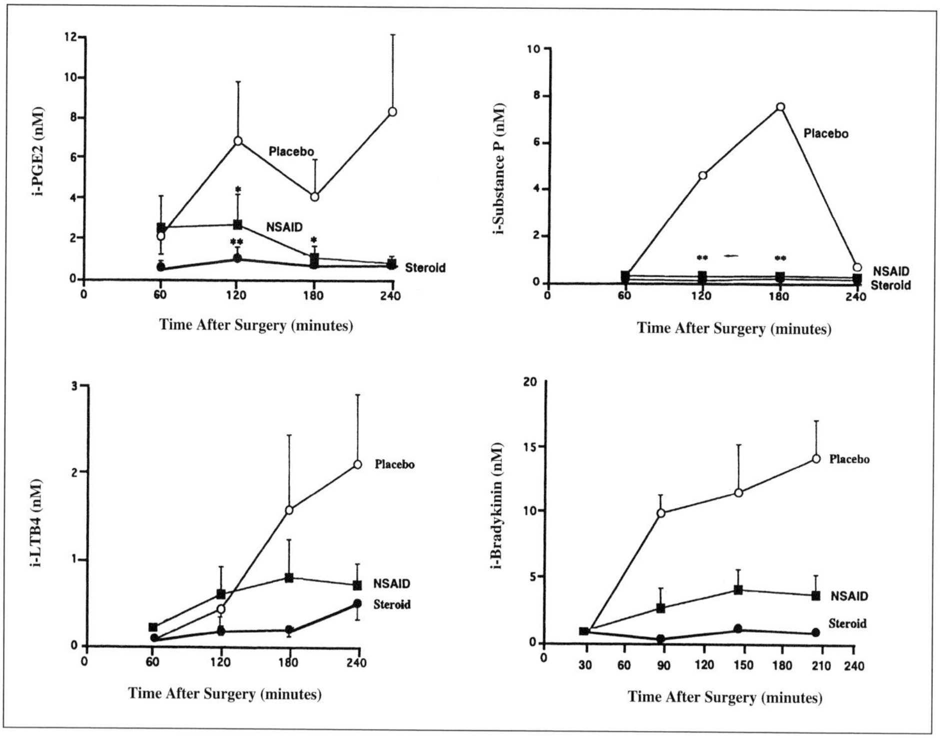

Figure 8
Production of the inflammatory mediators, PGE_2, substance P, LTB_4 and bradykinin, in samples collected by microdialysis from surgical extraction sites during 3rd molar surgery. The inhibitory effects are shown of an NSAID (ketoprofen) and a steroid on production of these inflammatory mediators. The inflammatory mediators were measure by immunoassay. Data of [141]. From [138]. Reproduced with permission of Kluwer Academic Publishers, Dordrecht, The Netherlands.

the early investigations was a study by Cornaro in 1983 [154] who studied the effects of nimesulide 200 mg/d compared with placebo in 49 patients who had undergone oral surgery for various conditions. Overall, pain relief judged by being "excellent" or "good" was found in 64% of patients treated with nimesulide compared with 25% in those given placebo. One patient withdrew from therapy with nimesulide and seven on placebo because of lack of efficacy.

Salvato and co-workers [155] compared the effects of 6 days treatment with nimesulide 200 mg/d, *Serratio* peptidases 15 mg/d or no pain therapy in 100 patients who had undergone tooth extraction or surgery for osteolysis. All patients received amoxicillin 1,500 mg/d. Reduction in pain and inflammation was rated

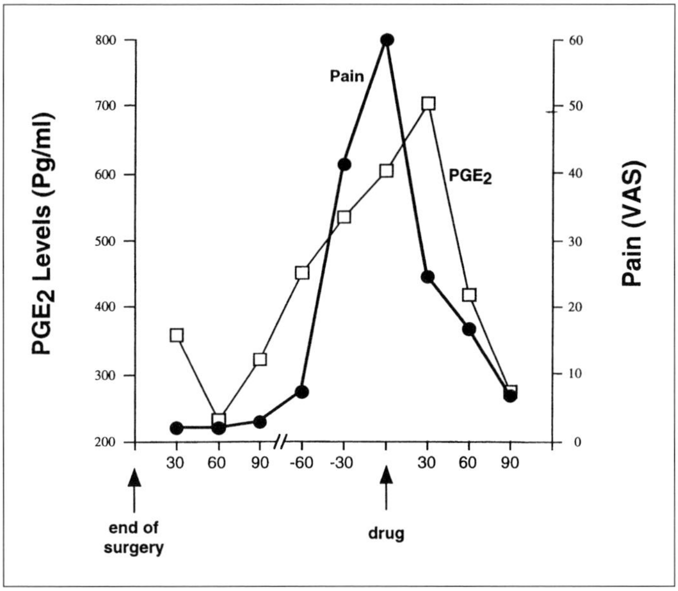

Figure 9
Relation between PGE_2 measured (using immunoassay) in samples collected by microdialysis from oral surgery extraction sites and parallel subjective assessments of pain intensity (measured by a visual analogue scale, VAS). Intake of the NSAID, ketoprofen, (shown as 'drug') caused a parallel reduction in pain and PGE_2 levels. From [138]. Reproduced with permission of Kluwer Academic Publishers, Dordrecht, The Netherlands.

to be "excellent" or "good" in 95% patients that received nimesulide, 65% of those given the peptidase preparation and in 25% of the non-treatment group. Nimesulide had faster onset of analgesia than the other treatments. In a similar study of 100 patients who had undergone dental surgery for tooth extractions or apical granulomas, Bucci et al. [156] showed the effectiveness of nimesulide in patients that received bacampicillin. A limited study by Moniaci [157] showed that nimesulide 100 mg twice daily had faster onset of analgesia than that from ketoprofen 200 mg/d for 14 days in patients who had undergone surgery for temporomandibular pain or extraction of third molars.

Using the third molar surgery trial design and pain assessment and quantitation developed by Cooper and Beaver in 1976 [134], Ragot and co-workers [158] showed in a randomised double-blind placebo-controlled trial in 134 patients that pain intensity difference (PID) and pain relief (PAR) scores from intake of a single dose of 100 or 200 mg nimesulide or 250 mg niflumic acid were significantly greater than placebo over the 6 h period of the study (Fig. 10a, b). PID

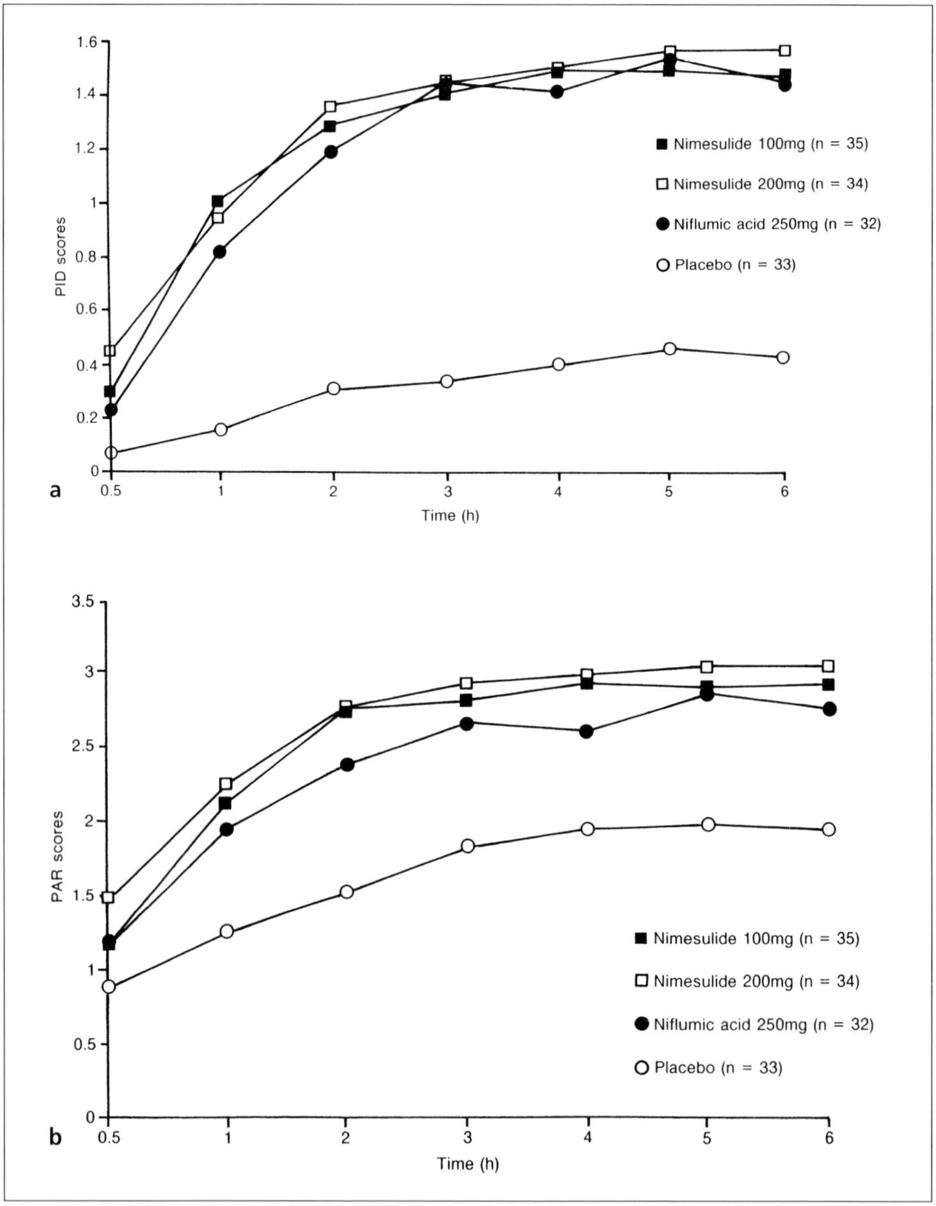

Figure 10
(a) Mean values for pain relief (PAR) scores in patients undergoing extraction of impacted third molars. (b) Mean values for pain intensity difference (PID) scores in patients undergoing extraction of impacted third molars. Pain intensity scores were adjusted for missing values and rescue analgesic administration. From [158]. Reproduced with permission of the publishers of Drugs.

scores in the drug-treated patients were about four-fold greater than in those that received placebo (Fig. 10a). PAR scores in the NSAID treated patients were about twice those on placebo (Fig. 10b). There were no significant differences between these drug treatments in PID at hourly intervals over 6 h, the sum of PID (SPID), the PAR scores over 6 h or the total PAR (TOTPAR). This is of interest in the case of nimesulide since it shows that there is no advantage in taking the higher dose of 200 mg compared with the 100 mg dose of the drug.

In a study in 51 adult patients that underwent maxillofacial surgery Ferrari Parabita et al. [159] compared the analgesic effects of nimesulide 100 mg twice daily with 250 mg naproxen twice daily, both taken as granulated formulations in water. There was no placebo treatment group presumably because of ethical or recruitment difficulties in such a study. Antibiotic treatments were allowed. Pain was graded on a visual analogue scale (VAS) at different periods during the day. Ranges of symptoms were rated on a four-point scale including difficulty in chewing and swallowing, swelling, hyperaemia, muscle contraction and impairment of sleep. There were no significant differences between nimesulide and naproxen treatments in VAS pain scores over 6 days of treatment, although the responses obtained were slightly greater with nimesulide than naproxen. Sleep quality was good in both groups with slightly more than half the patients in both groups reporting no pain. Swelling was completely resolved by day 6 in 85% patients that received nimesulide and in 56% of those that had naproxen. Hyperaemia was reduced in 92% of patients that had nimesulide and in 64% that received naproxen; these differences being statistically significant. Muscle contraction was not present in 96% of patients that had nimesulide and in 60% of those that received naproxen, and again these differences were statistically significant. Chewing and swallowing also improved in both groups. There were no adverse reactions recorded in the two treatment groups.

Pierleoni and co-workers [160] compared the effects of 5 days rectal suppository treatment with nimesulide 200 mg twice daily with that of ketoprofen 100 mg twice daily in a double-blind study (without placebo control) in 46 patients who underwent surgical removal of impacted molars. Efficacy was determined by assessment of "spontaneous" pain (quantified by the patient on the Scott-Huskisson VAS from 0 mm (for no pain), to 100 mm (for maximal pain), swelling, hyperaemia, pain upon mastication, night pain, ability to swallow and quality of sleep. The intensity of the symptoms was read on a four-point scale of increasing intensity. Both treatments caused reduction in the VAS spontaneous pain over the 5 day period with nimesulide showing slightly greater (but not statistically significant) pain relief than ketoprofen. The error in VAS data also progressively declined over the 5 day treatment period and both treatments had low VAS values near zero by this time. There was a trend towards increased pain in the morning compared with that in the evening. Other symptoms showed improvements although the responses varied among the patients on the two drug treatments. The investigator judged nimesulide to be excellent or good in 21/23 patients compared with 15/23 that received ketoprofen.

A β-cyclodextrin inclusion formulation of nimesulide 100 mg single dose was compared by Scolari et al. [161] for its analgesic effects with that of nimesulide 100 mg single dose in a randomised double-blind multicentre study in 148 outpatients who had undergone dental surgery. Pain intensity was evaluated on a VAS scale 30–360 min after ingestion of the drug and pain relief on a categorical scale over the same time period.

While the reduction in pain intensity from β-cyclodextrin-nimesulide was significantly greater than nimesulide itself over the first 60 min of treatment and pain relief significantly faster the overall assessments rated excellent or good were 95% with the former and 92% with the latter. The translation of such small differences into clinical practice may not be so pronounced.

A large randomised, multicentre placebo-controlled double-blind study by Ragot and co-workers in 469 patients (of whom 431 were evaluable) who had undergone molar tooth extraction compared the effects of single doses of 100 or 200 mg nimesulide with 500 mg mefenamic acid [162]. Rescue medication (paracetamol) was allowed and the quantities consumed by the different groups were recorded. There appeared to be two placebos – one matched for sachets of nimesulide taken in water and the other capsules to match the mefenamic acid formulation. Yet the puzzling feature about this report was that the data from the two placebo treatments were grouped together for the statistical analyses, without comparison been made within the two placebo groups. It can only be assumed that the two placebo treatments produced the same placebo responses. Pain intensity and pain relief were rated on a 100 mm VAS and on a four-point verbal scale of increasing severity or relief respectively.

The PID and SPID values were calculated from the former and PID at 1 h ≥ 1 values was used to determine the numbers of responders and non-responders respectively for each of the treatments.

The percentage of responders in the 100 and 200 mg nimesulide groups was 77.7% and 74.5%, respectively, whereas the mefenamic acid group had 43.4% and placebo group(s) combined had 16.5%. These differences in pain responses to nimesulide are quite striking and likewise the lack of differences between the two doses of the drug. The PID values over the period of 0.5–6 h of the study and cumulative or SPID values showed similar responses with the greatest pain relief being shown with the 100 and 200 mg doses of nimesulide, there being no difference between the two doses, and both these being about twice those achieved with mefenamic acid. Excellent or good pain relief was achieved in 87/110 (79.1%) of patients that received 100 mg nimesulide, in 91/112 (81.3%) that had 200 mg nimesulide, in 51/98 (52%) who had mefenamic acid compared with 33/103 (32%) who received placebo. The percentage of patients who required additional paracetamol use was 27% in the nimesulide groups, 57% in the mefenamic acid group and 70% in the placebo group. The low placebo response is striking in view of the large number of patients in the placebo group who took paracetamol. The use of paracetamol by the other NSAID groups paralleled the overall analgesic response to the NSAIDs.

This study is instructive in showing the extent of the *acute* pain relief from nimesulide compared with mefenamic acid and placebo which was quite low. The 'delta' or difference in P&D and SPID and overall assessment of pain relief was good from the nimesulide treatments. Also, of interest is the lack of any differences in pain relief from the two doses of nimesulide. This has been observed in some other respects and generally the 100 mg dose of the drug provides sufficient analgesia. The rate of onset of analgesia from nimesulide is also quite rapid.

Other acute surgical pain

Patients suffering from pain and inflammation following general, orthopaedic, urological or gynaecological interventions have also been employed in studies to investigate the pain-relieving properties of nimesulide. Thus, Stefanoni and co-workers [163] performed a randomised, double-blind study in 20 patients who had undergone mastectomy or quadrectomy and another 20 who had surgery for inguinal hernia comparing the effects of suppositories of either nimesulide 200 mg three times daily or diclofenac sodium 100 mg for 3 days. Assessment of the efficacy of the treatments was determined by recording the pain at rest as well as the pain on active and passive movement (using the Scott-Huskisson VAS), and swelling, hyperaemia and pyrexia. There were no significant differences in the pain responses and all these declined over the 3 days of the study. Likewise the other parameters declined and the over responses were such that there were virtually no evident symptoms in most of the patients at 3 days. Two patients that received diclofenac had rashes, one of which required treatment with antihistamines and the other had an erythematous rash. No other adverse events were recorded.

Schmökel and co-workers performed a double-blind study in 53 patients who had undergone various surgical procedures, mostly orthopaedic but there were some who had hernias and facial plastic surgery [164]. They received suppositories twice daily of 200 mg nimesulide or 500 mg paracetamol for variable periods according to the patients' needs. Pain relief and other signs of inflammation were recorded on an arbitrary four-point categorical scale. Analgesic efficacy over a 6 h period and in the first and second day following treatment was similar with the two drugs. There does not appear to have been any records of relief of other inflammatory symptoms with the two treatments. Adverse events were recorded in four patients in each of the two groups, mostly diarrhoea or mild CNS reactions.

In a second open label study in 17 patients, nimesulide was also found to have pain relief and reduction in other symptoms of inflammation [164].

Ramella and co-workers [165] undertook a randomised double-blind study in 40 patients who underwent saphenectomy or inguinal hernioplasty who received nimesulide 200 mg three times daily or diclofenac 100 mg three times daily administered rectally for a total of 3 days. No other medications were allowed and the study was not placebo controlled. The efficacy of the treatments was assessed by evaluation of fever, pain, spontaneous or on active or passive movement, oedema or hyperaemia of soft tissues. Fever was assessed by recording body temperature four times daily. Pain intensity was measured four times daily using the Scott-Huskisson VAS. Oedema and hyperaemia were assessed daily by the physician as being absent, mild, moderate or severe. The pain scores from spontaneous active movement or passive movement declined over the 3 day period and by the third day had virtually achieved no values indicating that there was almost complete pain relief. There appeared to be no differences between the two drug treatments. There was a significant reduction in oedema with both treatments along with the mild fever which was observed in 11 nimesulide treated patients and 13 in diclofenac treated patients and had resolved after 2 days of therapy.

Binning [166] recently reported a study in 94 patients who underwent knee arthroscopy who following the operation were randomised in a double-blind trial to receive nimesulide 100 mg b.i.d., naproxen sodium 500 mg b.i.d. or placebo for up to 3 days. The summed pain intensity (SPID) and total pain relief (TOTPAR) scores up to 6 h showed that nimesulide was superior to placebo and naproxen. These results also were paralleled by the intake of rescue medication that was taken by half those patients that received placebo and in those that received naproxen between the number of placebo and nimesulide patients. This study is interesting for showing that nimesulide had a faster onset of action than naproxen. As this drug was taken as the sodium salt it would have on pharmacokinetic grounds been expected to act rapidly.

A recent study of postoperative inflammatory events in 100 patients that had undergone coronary bypass surgery who received 100 mg nimesulide b.i.d, and another 100 who received 250 mg naproxen b.i.d routinely, the pain relief and plasma levels of interleukin-6 (IL-6), soluble tumour necrosis factor-α-receptor-I (sTNF-RI) and C-reactive protein (CRP), as well as the ESR and white cell count were determined [167]. There were no differences between the levels of IL-6, TNF-RI, CRP or ESR between the two treatment groups and comparable pain relief was achieved with both drugs. The levels of IL-6 and TNF-RI increased in the period after the operation then fell to basal levels thereafter. Unfortunately, these authors failed to include a control group which might have received either paracetamol alone or as a rescue medication. There were no gastrointestinal reactions observed in the nimesulide group in contrast to that in 7% of patients that received naproxen.

A recent study by McCrory and Fitzgerald [168] showed that nimesulide gave added pain relief in combination with the narcotic, morphine, following surgery for thoracotomy. This study was undertaken in 30 patients with adenocarcinoma (mean age 63 years) who had undergone thoracotomy followed by intrathecal morphine 0.5–1.0 mg p.r.n., then randomised to receive no NSAID, ibuprofen or nimesulide in an open label manner. They monitored cerebrospinal fluid (CSF) levels of 6-keto-$PGF_{1\alpha}$ and *ex vivo* whole blood production of TxB_2 or LPS-stimulated PGE_2 as surrogate measurement of COX-1 and COX-2 activity respectively, and also pain was measured on a ten-point visual analogue scale. Nimesulide treatment reduced COX-2 but not COX-1 activity while ibuprofen reduced COX-1 but not COX-2 activity in whole blood *ex vivo*. Nimesulide reduced CSF levels of 6-keto-$PGF_{1\alpha}$ while ibuprofen had no effect. Pain relief at rest and after coughing was greater with nimesulide than from ibuprofen in the period up to 48 h following the operation. The results show that pain relief from nimesulide was related to reduction in CSF levels of COX-2 derived PGI_2 (in the CSF) and PGE_2 in the blood and that this accounts for the improved analgesia seen with this drug compared with ibuprofen and lower requirements for opiate analgesia.

Overall, these studies have shown that in acute surgical pain, nimesulide has comparable activity with that of other NSAIDs. While in some cases the numbers of patients in these studies is relatively small the results are nonetheless clear-cut and show conclusively that nimesulide has rapid pain relieving activities.

Otorhinolaryngological and upper respiratory tract inflammation

The throat pain associated with tonsillitis and other painful throat conditions has been considered to be a useful model for determining analgesic activity and the speed of onset of analgesia from NSAIDs and paracetamol [169]. A considerable number of studies have been undertaken comparing the effects of nimesulide in ear, nose and throat (ENT) infections as well as upper respiratory tract infections, bronchitis or laryngotracheitis (Tab. 6) [170–187]. Some of the earlier studies that were published up to 1988 have been comprehensively reviewed and evaluated by Ward and Brogden [62]. In essence they show that there is a time-dependent improvement in many of the clinical symptoms when standard doses of 100 or 200 mg nimesulide are given twice daily (Fig. 11), often in combination with antibiotics. The usual time of treatment has been up to 7 days. Many of these studies were performed in small patient groups and in view of the wide variability and symptomology it is not surprising that there has been some variability in response but overall the efficacy of nimesulide is quite striking in these respiratory tract and ENT infections.

A larger multicentre study was undertaken by Ottaviani and co-workers [170] in 940 male and female patients aged 15–77 years in a non-comparative study in patients with otorhinolaryngal infections. The lack of an adequate control group

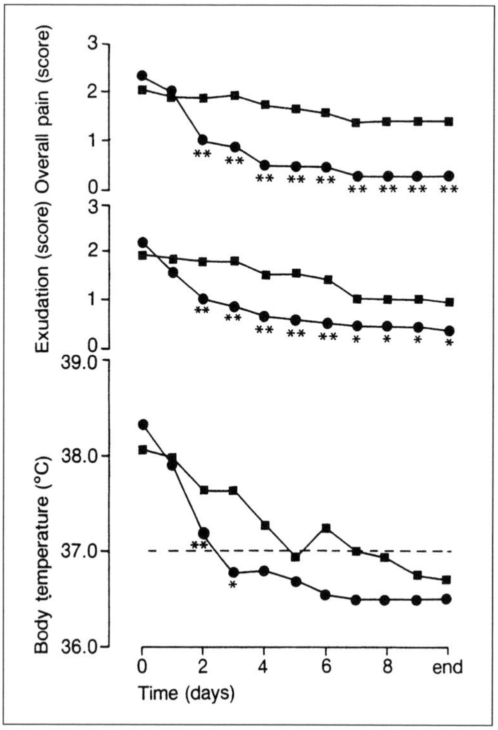

Figure 11
*Effects of nimesulide 200 mg/day (●) or benzydamine 150 mg/day (■) on overall pain, exudation and body temperature in 50 patients with otorhinolaryngological inflammatory disease (after [172]). * = p<0.05, ** = p<0.01. From [172]. Reproduced with permission of the publishers of The Journal of International Medical Research.*

either of a comparative drug or placebo has obviously limited the interpretation of this particular study. However, nimesulide 100 mg twice daily taken in a granular formulation for a mean of 10 days showed reduction in signs and symptoms graded on a four-point categorical scale of increasing severity.

In a study of 200 professional or amateur divers of either sex aged 18–54 years, Bianchini and co-workers [171] undertook a double-blind comparison of the effects of 100 mg nimesulide taken twice daily compared with that of Seaprose S™ (a proteolytic enzyme complex frequently used in ENT treatments in Italy) taken as tablet formulations for 1 week for the relief of symptoms attributed to non-bacterial inflammatory disorders of the ear, nose and throat. Patients were evaluated at baseline and at 3 and 7 days of treatment in which the signs and symptoms recorded were congestion, oedema, exudate formation, cough, hoarseness, pain, nasal obstruction, rhinorrhoea, sneezing, headache, a sensation of obstructive ear, deafness, autophony, difficulty in compensation and vertigo. The intensity of these symptoms and signs was graded on a four-point categorical scale of increasing severity. At the end of the treatment both patients and physicians evaluated the effectiveness of the therapies based on the scale of very good, good, moderate and no effect. Of the 200 patients entered into the study only 195 were of value for statistical analysis since four patients were excluded because of concomitant drug intake and one in the nimesulide group because of nausea. One patient that received Seaprose S™ experienced mild ortocherium but continued treatment. Both the treatments reduce the symptoms but there was a statistically significant difference in favour of nimesulide in respect of relief of pharyngeal congestion, nasal obstruction and congestion, rhinorrhoea, headache, earache, deafness, autophony, sensation of obstructive ear, ear congestion, ear oedema and ear exudate formation. The overall evaluation by physicians favoured nimesulide treatment showing very good or good effectiveness in 92.7% of patients assessed by physicians compared with that with Seaprose which was rated in the same way by 78.4% of patients. The patients' assessments were for nimesulide 93% finding the treatment very good or good compared with that of Seaprose 74%.

In a Phase III double-blind trial in patients with "non-bacterial" acute inflammation of the ear, nose and throat, Nouri and Monti [173] compared the effects of nimesulide 100 mg or naproxen 500 mg given twice daily for 5–10 days depending on the patient requirements. No use of antitussive preparations, expectorants or other anti-inflammatory, analgesic, antipyretic drugs was permitted. Those patients that required antibiotic treatment were excluded from the study. Efficacy of the treatments was evaluated by recording the intensity of local and referred pain, the quantity of secretion and a degree of swelling. These parameters were graded on a five-point categorical scale of increasing severity. Signs and symptoms were assessed by the physician at the initial clinical examination and daily thereafter where possible. A total of 53 patients were evaluable, most of them suffering from rhinopharyngitis, otitis or sinusitis. Both treatments resulted in a reduction of pain and local symptoms over 7 days of treatment. The relief of pain intensity was significantly greater over a 5 day period of treatment with nimesulide compared with that of patients that received naproxen and likewise the relief of inflammatory symptoms favoured nimesulide. In all cases there was quite rapid relief of symptoms of fever. In the global assessment by the physicians' nimesulide was considered effective in 92.5% of patients compared with that of 61.6% that received naproxen. In a review of three studies previously published, Bellussi and Passali [174] evaluated the effectiveness of nimesulide compared with feprazone, nimesulide plus ambroxol *versus* ambroxol alone (to control infections) and nimesulide in otitis media. These studies were undertaken in relatively small groups of patients ranging from 40–62 per trial and overall showed that (a) there was an advantage in combining nimesulide treatment with an antibiotic, e.g., ambroxol, when taken over a 7–10 day period.

In another review of the effectiveness of nimesulide in the treatment of chronic bronchitis, Sofia and co-workers [175] noted that the effects of nimesulide on the functions of neutrophils and other components of inflammation that are of significance in bronchitis along with hypersecretion of mucous were thought to be the basis of the improved effectiveness of nimesulide in sputum viscosity compared with that of peptidase treatment or tiopronin over 1–3 weeks of treatment. Furthermore, at 3 weeks of treatment nimesulide resulted in a reduction in the bronchio-alveolar lavage fluid fractions. These results suggest that nimesulide may reduce the symptoms associated with inflammation of the airways and mucus hypersecretion in bronchitis.

In a multicentre double-blind randomised control trial in 316 patients with acute otitis externa or acute otitis media or exacerbations of chronic otitis media, nimesulide 100 mg in an inclusion complex with a total mass of 400 mg with β-cyclodextrin was compared with 700 mg morniflumate taken twice daily for up to 10 days [176]. Patients were aged between 15 and 65 years and upon enrolment the patients had an intensity of otalgia represented by a score of ≥50 mm on a 100 mm visual analogue scale without the immediate need for antibiotic treatment. If needed, antibiotics were allowed for approximately 3 h after the first dose of study medication. The drugs were given as a sachet taken orally with water every 12 h. A total of 10 patients in the nimesulide β-cyclodextrin group and 12 in the morniflumate group were excluded because of violations. Both treatments led to a reduction in VAS scores over the first 3 h following drug administration that were not significantly different from one another. In those patients that were defined as responders on a basis of having a reduction in pain of >50% of the value at baseline within the first 3 h of administration, a total of 56% on nimesulide β-cyclodextrin were responders compared with that of 47.4% in the morniflumate group. The difference wasn't statistically significant. A range of secondary clinical symptoms related to inflammation, pain and temperature, were found to be decreased significantly by both nimesulide β-cyclodextrin as well as morniflumate and the differences between these two treatments were not statistically significant. As mentioned previously, the inclusion of nimesulide in the β-cyclodextrin formulation complex has been considered to have faster onset of action than that of nimesulide alone although the differences between the two may not be striking in the clinical context.

Nimesulide has been shown to be effective in relieving symptoms of upper respiratory tract infections and associated fever and treating upper respiratory tract infections in children [178–186] (reviewed in [188]) and has been found to have a satisfactory safety profile [185, 186]. Since the risks of allergic reactions from nimesulide in the respiratory tract and intolerance in aspirin-intolerant asthma patients appears low with this drug [189, 190], nimesulide may be given to patients with upper respiratory tract infections and ENT conditions with relative safety.

As discussed in Chapter 6 asthmatic reactions are uncommon with nimesulide. Although these cannot be completely ruled out the relative risk of a reaction occurring with nimesulide is obviously much lower than that of many other NSAIDs.

Nimesulide 200 mg/d has also been shown in a number of studies to be effective in treating symptoms of acute rhinitis especially in combination with antihistamines, e.g., terfenadine 120 mg/d [191], or nimesulide 100 mg/d alone or in combination with cetirizine 10 mg/d [192].

Clearly, the multiple anti-inflammatory mechanisms of nimesulide contribute to its effectiveness in treating a wide range of respiratory and ENT infections.

Miscellaneous conditions

Nimesulide has been found to have good analgesic activity in several different painful conditions with pronounced local inflammatory reactions including thrombophlebitis [193–195], urinogenital disorders [196], prostato-vesiculitis [197], mastalgia and carpal tunnel syndrome [198, 199] (Tab. 6).

Antipyretic effects

In many of the above-mentioned studies the relief of symptoms of fever has been observed following treatment with nimesulide and this has been noted in a studies examining the antipyretic effects of either orally administered nimesulide 100 or 200 mg or that taken by suppositories [178, 176] (reviewed by Ward and Brogden [62] and Davis and Brogden [188]).

A number of clinical trials have examined the mode of action and relative antipyretic efficacy of nimesulide compared with paracetamol or NSAIDs [200–205]. In a double-blind crossover trial in 18 patients of both sexes aged between 42 and 87 years (medium 72 years) presenting with fever above or equal to 38 °C (axillary) who were hospitalised for treatment, received single oral doses of either nimesulide 100 mg, aspirin 500 mg or dipyrone 500 mg taken orally in a variable sequence of treatments [200]. Axillary temperatures and pulse rates were measured immediately before administration and subsequently at 30–360 min thereafter. Nimesulide and dipyrone showed a marked reduction in body temperature to achieve near normal values at 240 and 360 min. Aspirin only achieved a reduction in fever at this period to about half the extent of the two former drugs. It could be argued that the dose of aspirin may have been suboptimal, for generally two or three 325 mg tablets of aspirin are normally administered to achieve antipyretic effects. There were no statistically significant differences between nimesulide and dipyrone treatments.

In contrast to this study (which was in a relatively small patient group that received drugs taken orally), the antipyretic effect of nimesulide 200 mg suppositories was compared with that of diclofenac 100 mg in a placebo-controlled trial by Reiner et al. [201]. This study was undertaken in 81 inpatients of both sexes ranging from 18–90 years with a mean of 65 years. In comparison with placebo the body temperatures following the two drug treatments were reduced at 60 min and declined rapidly to near normal values at 360 min; statistical significance being achieved in data from 90 min onwards. It was observed that both drug treatments led to a decrease in heart rate and systolic arterial pressure in comparison with placebo. This quite sizeable study shows the effectiveness of nimesulide in comparison with the standard diclofenac formulation to be pronounced in the treatment of fever. The use of suppositories is of particular interest especially as incapacitated or elderly patients may not be able to take oral formulations of the drug.

In a study of 39 elderly inpatients of both sexes (aged 65 years or more) in a geriatric ward admitted for rehabilitation after stroke or orthopaedic surgery, who presented with either viral or bacterial infections of the upper or lower respiratory tract were randomly assigned to receive nimesulide 200 mg or paracetamol 500 mg suppositories three times daily for two consecutive days [206] (see also [207]). On the third day therapy was withdrawn in order to determine if there was control of hyperpyrexia. Of 18 patients that received nimesulide, one was excluded because of being unable to complete the study because of an adverse reaction and most of these had influenza symptoms including pharyngitis or pneumonitis. Within the first three to four treatments with nimesulide, fever had started to be reduced and was at near normal levels by the end of the first day and continued to decline to the third day of treatment. No hyperpyrexia was observed on the third day. Similar results were observed with paracetamol in 21 patients that received the drug. At the 6 a.m. period the mean temperature in the paracetamol group on the second day was still greater than 37 °C in 10 patients, suggesting that in about half the patients there were still febrile symptoms. At that period only 23% of the nimesulide group were febrile. There was no rebound on the third day with either of the treatments. Heart rate and diastolic blood pressure did not vary significantly although there was a marginal reduction in systolic pressure observed during the second and third day of treatments in both groups.

Use of nimesulide as an antipyretic in children has been reported in a number of studies [62, 188, 203–205]. An important consideration for children is to know the safe and effective dosage and to know at what plasma concentration antipyretic effects are apparent. Thus, Ugazio et al. [204] showed that a dose of 50 mg nimesulide taken as granules to hypoglycaemic children produced plasma concentrations of 3.5 mg/l within 2 h of oral administration which declined progressively over the following 12 h. The 4′-hydroxy-metabolite appeared in the plasma at 0.5 h then progressively increased to peak at 9 h. In a randomised trial (which was not blinded) in 100 hospitalised children greater antipyretic effects were observed with an oral suspension of nimesulide 5 m/kg/d (which is comparable to the dose employed in the pharmacokinetic study) compared with that of paracetamol 26 mg/kg/d over 3–9 days of treatment.

Headache

The symptoms of non-migraineous and migraine-type headaches and the response to aspirin and other NSAIDs has been reviewed elsewhere [208]. In most of the clinical trials that have been undertaken in acute conditions involving inflammation of the airways or ear, nose and throat conditions clinical symptoms involving headache have been improved with the drug [62, 188].

In a double-blind, parallel, placebo-controlled study in 30 patients with menstrual migraine, nimesulide 100 mg three times daily was taken for 10 days starting from the beginning of the symptoms of migraine, and then through a further two menstrual cycles, during which it was found that pain intensity and duration were significantly better than placebo [209]. The daily dose of 300 mg nimesulide is quite high but perhaps this is needed for relief of migraine in contrast to other less severe headaches.

In a pharmaco-epidemiological study in a specialist headache clinic in northern Italy, the most used drug was nimesulide, with 'tryptans' and anti-depressants being also used prophylactically [210]. It is assumed that since the patients in this study were being treated in a specialist centre and taking a cocktail of drugs that they were quite severe cases of this condition.

Cancer pain

Pain during the onset and progress of cancer has represented a major challenge for the physician. Among the problems that are presented for this severely debilitating manifestations of cancer is the problem of patient variability and inevitable decline of general wellbeing associated with the onset of chronic pain [208]. The now well-established World Health Organization guidelines provide for three-step analgesia in which NSAIDs are employed in the first step [208, 211, 212]. NSAIDs are often used alone or in combination with opioids for the treatment of cancer pain [211, 212]. Most often patients receive oral NSAIDs but the use of rectally administered drug holds particular advantages especially as there is often frequent intolerance to intake of oral formulations of NSAIDs.

In 64 patients with pain associated with advanced cancer Corli et al. [213] compared the effectiveness of oral nimesulide 300 mg/d or oral diclofenac 150 mg/d compared with rectal nimesulide 400 mg/d and rectal diclofenac 200 mg/d in non-blinded but in patients who were randomly assigned to these treatments (Fig. 12). These were given for 1 week in patients who did not have any impaired renal function, coagulopathy, positive history or gastropathy or NSAID intolerance. The efficacy of each treatment was evaluated by daily recordings of the Integrated Pain Score of Ventafridda and co-workers [214] and sleep duration. Adverse events were also recorded daily. After the first day of treatment and up to 7 days of treatment all the treatments gave reductions in integrated pain scores by about half the initial values. The responses obtained with the tablet formulations appeared to be slightly greater although not significantly different compared with that of the suppository formulations (Fig. 12). Both drug formulations showed marked reduction in integrated pain scores on the first day of the treatments then maintained this reduced level of pain for the 7 days of the trial (Fig. 12). Some of the patients that received the oral formulations of the drugs developed gastric symptoms but overall nimesulide suppositories were the best tolerated among the treatments.

In a study in 68 patients with advanced cancer who were undergoing therapy in the first step of the standard protocol provided by the WHO for pain control [215], nimesulide 200 mg was compared with that of naproxen 500 mg both given twice daily and the pain was evaluated using the integrated pain score of Ventafridda [214]. Patients were treated up to 14 days and adverse events were recorded. Of the 22/34 patients that received nimesulide and 21/34 that received naproxen, the integrated pain score was reduced from baseline in 65% and 70% respectively. There was a statistically significant reduction by 1 week of therapy compared with baseline of both the treatments and although there was a slight

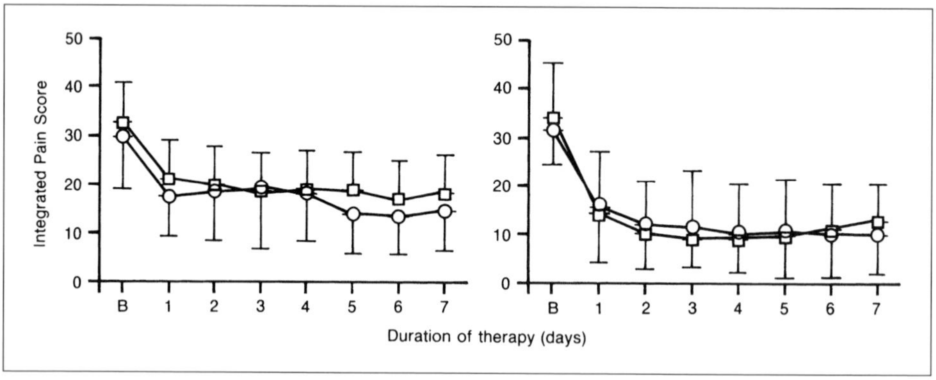

Figure 12
Effects of nimesulide (○) and diclofenac (□) in either suppository (left panel) or tablet (right panel) formulations on the integrated Pain Score (± SD) in patients with cancer-related pain. B = baseline. From [213]. Reproduced with permission of the publishers of Drugs.

difference in favour of naproxen in the first week both had integrated pain scores down to a value of 10 by 2 weeks with the difference not being significant. Both drugs showed gastrointestinal symptoms (gastric pain, nausea and hyperchlorhydria and vomiting). Clearly in comparison with the study undertaken by Corli et al. [213] it would seem to be preferable to institute pain control with suppositories of nimesulide or other NSAIDs for adequate pain control without gastric symptoms. In a study by Gallucci et al. [216] cancer patients who were also treated on the first step of the WHO analgesic ladder with nimesulide 200 mg b.i.d. appeared identical to that of naproxen 500 mg b.i.d. Similar results were found from another study comparing these two drugs [217].

Adverse events encountered in clinical trials

Case reports of adverse events have been noted in a number of the studies that have been reviewed in this chapter. Because of the relatively small numbers involved in some of the studies it is not being considered worthwhile to report these individually. However, a comprehensive analysis of all the adverse events reported in all the clinical trials is presented in Chapter 6 to which the reader is referred.

Conclusions

In comparison with conventional NSAIDs (with COX-1 as well as COX-2 inhibitory effects) and the coxibs, nimesulide has been shown in a large number of studies to be equivalent to, or in some cases more effective in relieving pain and inflammatory signs and symptoms. Recent evidence suggesting that nimesulide may have fast onset of action in acute pain may be an advantage for the drug in certain clinical situations. Nimesulide has proven to be an effective drug in comparison with other NSAIDs including the coxibs.

References

1. Hunt SP, Mantyh PW (2001) The molecular dynamics of pain control. *Nat Rev Neurosci* 2: 83–91
2. Scholz J, Woolf CJ (2002) Can we conquer pain? *Nat Rev Neurosci* 5: suppl: 1062–1067
3. Koltzenburg M, Scadding J (2001) Neuropathic pain. *Curr Opin Neurol* 14: 641–647
4. Woolf CJ (2004) Pain: Moving from symptom control toward mechanism-specific pharmacologic management. *Ann Intern Med* 140: 441–451

5. Weiner DK, Ernst E (2004) Complementary and alternative approaches to the treatment of persistent musculoskeletal pain. *Clin J Pain* 20: 244–245
6. Rainsford KD (ed) (2004) *Aspirin and Related Drugs*. CRC Press/Taylor & Francis, London and New York
7. Waine H (1956) Management of rheumatoid arthritis. *Arch Intern Med* 98: 332–339
8. International Agranulocytosis and Aplastic Anemia Study (1986) Risks of agranulocytosis and aplastic anemia. A first report of their relation to drug use with special reference to analgesics. *JAMA* 256: 1749–1757
9. Rainsford KD (ed) (1999) *Ibuprofen. A Critical Bibliographic Review*. Taylor & Francis, London
10. Wolfe F, Anderson J, Burke TA, Arguelles LM, Pettitt D (2002) Gastroprotective therapy and risk of gastrointestinal ulcers: risk reduction by COX-2 therapy. *J Rheumatol* 29: 467–473
11. Mason DH, Bernstein J, Bortnichak EA, Ehrlich GE (1990) Spontaneous reporting of adverse drug reactions: Diclofenac sodium and four other leading NSAIDs. *IM* 11: 1–8
12. Needleman P, Isakson PC (1997) The discovery and function of COX-2. *J Rheumatol* 24: 6–8
13. Ehrlich GE (1977) Guidelines for anti-inflammatory drug research. *J Clin Pharm* 17: 697–703
14. Woolf AD, Pfleger B (2003) Burden of major musculoskeletal conditions. *Bull WHO* 81: 646–656
15. Scott DL, Kingsley G (2004) Translating research into practice: Acetaminophen in osteoarthritis revisited. *J Rheumatol* 31: 199–202
16. Wegman A, der van Windt D, van Tulder M, Stalman W, de Vries T (2004) Nonsteroidal antiinflammatory drugs or acetaminophen for osteoarthritis of the hip or knee? A systematic review of evidence and guidelines. *J Rheumatol* 31: 344–354
17. Wolfe F, Zhan S, Lane N (2000) Preference for nonsteroidal anti-inflammatory drugs over acetaminophen by rheumatic disease patients: a survey of 1799 patients with osteoarthritis, rheumatoid arthritis, and fibromyalgia. *Arthritis Rheum* 43: 378–385
18. Pincus T, Swearingen C, Cummins P, Callahan LF (2000) Preference for nonsteroidal anti-inflammatory drugs *versus* actaminophen and concomitant use of both types of drugs in patients with osteoarthritis. *J Rheumatol* 28: 1020–1027
19. Boutron I, Tubacj F, Girandeau B, Ravaud P (2003) Methodological differences in clinical trials evaluating nonpharmacological and pharmacological treatments of hip and knee osteoarthritis. *JAMA* 290: 1062–1070
20. Ehrlich GE (2003) The rise of osteoarthritis. *Bull WHO* 81: 630
21. Buchanan WW, Kean WF, Kean R (2003) History and current status of osteoarthritis in the population. *Inflammopharmacology* 11: 301–316
22. Buchanan WW, Kean WF (2002) Osteoarthritis I: epidemiological risk factors and historical considerations. *Inflammopharmacology* 10: 5–21

23. Buchanan WW, Kean WF (2002) Osteoarthritis IV: clinical therapeutic trials and treatment. *Inflammopharmacology* 10: 79–155
24. Hakala P, Nieminen O, Kolvisto O (1994) More evidence from a community based series of better outcomes in rheumatoid arthritis. Data on the effect of multidisciplinary care on the retention of functional ability. *J Rheumatol* 21: 1432–1437
25. Hochberg MC, Altman RD, Brandt KD, Clark BM, Dieppe PA, Griffin MR, Moskowitz RW, Schnitzer TJ (1995) Guidelines for the medical management of osteoarthritis. Part II. Osteoarthritis of the knee. *Arthritis Rheum* 38: 1541–1546
26. Bradley JD, Brandt KD, Katz BP, Kalasinski LA, Ryan SI (1991) Comparison of an antiinflammatory dose of ibuprofen, an analgesic dose of ibuprofen, and acetaminophen in the treatment of patients with osteoarthritis of the knee. *N Engl J Med* 325: 87–91
27. Case JP, Baliunas AJ, Block JA (2003) Lack of efficacy of acetaminophen in treating symptomatic knee ostearthritis: a randomised, double-blind, placebo-controlled comparison with diclofenac sodium. *Arch Intern Med* 163: 169–178
28. Wegman AC, van der Windt DA, de Haan M, Deville WL, Fo CT, de Vries TP (2003) Switching from NSAIDs to paracetamol: a series of n of 1 trials for individual patients with osteoarthritis. *Ann Rheum Dis* 62: 1156–1161
29. Pope JE, Prashker M, Anderson J (2004) The efficacy and cost effectiveness of N of 1 studies with diclofenac compared to standard treatment with nonsteroidal antiinflammatory drugs in osteoarthritis. *J Rheumatol* 31: 140–149
30. Pincus T, Kock G, Lei H, Mangal B, Sokka T, Moskowitz R, Wolfe F, Gibofsky A, Simon L, Zlotnick S et al. (2004) Patient Preference for Placebo, Acetaminophen (paracetamol) or Celecoxib Efficacy Studies (PACES): two randomised, double-blind, placebo controlled, crossover clinical trials in patients with knee or hip osteoarthritis. *Ann Rheum Dis* 63: 931–939
31. Boureau F, Schneid H, Zeghari N, Wall R, Bourgeois P (2004) The IPSo study: ibuprofen, paracetamol study in osteoarthritis. A randomised comparative clinical study comparing the efficacy and safety of ibuprofen and paracetamol analgesic treatment of osteoarthritis of the knee or hip. *Ann Rheum Dis* 63: 1028–1034
32. Wolfe F, Michaud K, Burke TA, Zhau SZ (2004) Longer use of COX-2 specific inhibitors compared to nonspecific nonsteroidal anti-inflammatory drugs: A longitudinal study of 3639 patients in community practice. *J Rheumatol* 31: 355–358
33. Lee WM (2004) Aches, pains and warning labels. *NYTimes* 2004: 153; 17 March, p. A25
34. Pencharz JN, Grigoriadis E, Jansz GF, Bombardier C (2002) A critical appraisal of clinical practice guidelines for the treatment of lower limb osteoarthrtitis. *Arthritis Res* 4: 36–44
35. American College of Rheumatology Subcommittee on Osteoarthritis Guidelines (2000) Recommendations for the medical management of osteoarthritis of the hip and knee. 3000 update. *Arthritis Rheum* 43: 1905–1915
36. Dieppe P (2001) From protocols to principles, from guidelines to toolboxes: aids to good management of osteoarthritis. *Rheumatology* 40: 841–842

37. Senna GE, Passalacqua G, Dama A, Crivellaro M, Schiabboli M, Bonadonna P, Caconica GW (2003) Nimesulide and meloxicam are a safe alternative drugs for patients intolerant to nonsteroidal anti-inflammatory drugs. *Allerg Immunol* (Paris) 35: 393–396
38. Dreiser RL, Riebenfeld D (1993) Nimesulide in the treatment of osteoarthritis. Double-blind studies in comparison with piroxicam, ketoprofen and placebo. *Drugs* 46: 191–195
39. Bourgeois P, Dreiser RL, Lequesne MG, Macciocchi A, Monti T (1994) Multi-centre double-blind study to define the most favourable dose of nimesulide in terms of efficacy/safety ratio in the treatment of osteoarthritis. *Eur J Rheum Inflamm* 14: 39–50
40. Fossaluzza V, Montagnani G (1989) Efficacy and tolerability of nimesulide in elderly patients with osteoarthritis: double-blind trial *versus* naproxen. *J Int Med Res* 17: 295–303
41. Rabasseda X (1996) Nimesulide: a selective cyclooxygenase 2 inhibitor antiinflammatory drug. *Drugs of Today* 32: 365–384
42. Bianchi M, Broggini M (2003) A randomized, double blind, clinical trial comparing the efficacy of nimesulide, celecoxib and rofecoxib in osteoarthritis of the knee. *Drugs* 63 (Suppl 1) 37–46
43. Huskisson EC, Macciocchi A, Rahlfs VW, Bernstein RM, Bremner AD, Doyle DV, Molloy MG, Burton AE (1999) Nimesulide *versus* diclofenac in the treatment of osteoarthritis of the hip or knee: an active control equivalence study. *Curr Ther Res* 60: 253–265
44. Porto A, Reis C, Perdigoto R, Leone R, Goncalves M, Freitas P, Macciocchi A (1998) Gastroduodenal tolerability of nimesulide and diclofenac in patients with osteoarthritis. *Curr Ther Res* 59: 654–665
45. Gui-Xin Q, Macciocchi A (1997) Trial on the efficacy and tolerability of nimesulide *versus* diclofenac in the treatment of osteoarthritis of the knee. Helsinn Healthcare, Report No. TSD 7530
46. Quattrini M, Paladin S (1995) A double blind study comparing nimesulide and naproxen in the treatment of osteoarthritis of the hip. *Clin Drug Invest* 10: 139–146
47. Lucker PW, Pawlowski C, Friederich I (1994) Double-blind randomized multicentre clinical study evaluating the efficacy and tolerability of nimesulide in comparison with etodolac in patients suffering from osteoarthritis of the knee. *Eur J Rheum Inflamm* 14: 29–38
48. Renzi G, Della Marchina M (1989) Nimesulide: comparative, multi-centre, double blind evaluation *versus* Flurbiprofen concerning its therapeutic potential in rectal form in osteoarticular diseases. Helsinn Healthcare, Report No. TSD 5438
49. Kriegel W, Korff KJ, Ehrlich JC, Lehnhardt K, Macciocchi A (2001) Double-blind study comparing long-term efficacy of the COX-2 inhibitor nimesulide and naproxen in patients with osteoarthritis. *Int J Clin Pract* 55: 510–514
50. Wober W (1999) Comparative efficacy and safety of nimesulide and diclofenac in patients with acute shoulder and a meta-analysis of controlled studies with nimesulide. *Rheumatology* 38: 33–38

51. Pochobradsky MG, Mele G, Beretta A, Montagnani G (1991) Post-marketing surveillance of nimesulide in the short term treatment of osteoarthritis. *Drugs Exptl Clin Res* 17: 197–204
52. Blardi P, Gatti F, Auteri A, Di Perri T (1992) Effectiveness and tolerability of nimesulide in the treatment of osteoarthritic elderly patients. *Int J Tiss Reac* 14: 263–268
53. Topol EJ (2004) Failing the public health – Rofecoxib, Merck, and the FDA. *N Engl J Med* 351: 1707–1709
54. Roy V, Gupta U, Sharma S, Dhaon BK, Singh NP, Gulati P (1999) Comparative efficacy and tolerability of nimesulide and piroxicam in osteoarthritis with special reference to chondroprotection. *J Indian Med Assoc* 97: 442–445
55. Sharma S, Rastogi S, Gupta V, Rohtagi D, Gulati P (1999) Comparative efficacy and safety of nimesulide *versus* piroxicam in osteoarthritis with special reference to chondroprotection. *Am J Ther* 6: 191–197
56. Fioravanti A, Storri L, DiMartino S, Bisogno S, Oldani V, Scotti A, Marcalongo R (2002) A randomised, double-blind, multicenter trial of nimesulide-beta-cyclodextrin *versus* naproxen in patients with osteoarthritis. *Clin Ther* 24: 504–519
57. Estevez F, Amoro G, Giusti M, Lasalvia L, Havranek H (1993) Diclofenac vs. nimesulida en artrosis. Niveles plasmaticos y eficacia clinica. *Medicina* (Buenos Aires) 53: 307–314
58. Liaropoulos L (1999) Economic comparisons of nimesulide and diclofenac and the incidence of adverse events in the treatment of rheumatic disease in Greece. *Rheumatology* 38 (Suppl 1): 39–46
59. Tarricone R, Martelli E, Parazzini F, Darba, J, Le Pen C (2001) Economic evaluation of nimesulide *versus* diclofenac in the treatment of osteoarthritis in France, Italy and Spain. *Clin Drug Invest* 21: 453–464
60. Rainsford KD (1998) An analysis from clinico-epidemiological data of the principal adverse events from the Cox-2 selective NSAID, nimesulide, with particular reference to hepatic injury. *Inflammopharmacology* 6: 203–221
61. Conforti A, Leone R, Moretti U, Mozzo F, Velo G (2001) Adverse drug reactions related to the use of NSAIDs with a focus on nimesulide. *Drug Safety* 24: 1081–1090
62. Ward A, Brogden RN (1988) Nimesulide: a preliminary review of its pharmacological properties and therapeutic efficacy in inflammation on pain States. *Drugs* 36: 732–753
63. Anonymous (1998) *In Focus Nimesulide* (Aulin) Helsinn Healthcare SA. Adis International Ltd, Milan
64. Balabanova RM, Belov BS, Chichasova NV, Oliunin IuA, Badokin VV, Pchelintseva AO, Fedina TP, Gukasian DA, Lopatina NE, Korshunov NI et al. (2002) Nimesil effectiveness in rheumatoid arthritis. *Klin Med (Mosk)* 80: 49–52
65. Mease PJ (2004) Recent advances in the management of psoriatic arthritis. *Curr Opin Rheumatol* 16: 366–370
66. Taylor WJ (2004) Assessment of outcome in psoriatic arthritis. *Curr Opin Rheumatol* 16: 350–356

67. Sarzi-Puttini P, Santandrea S, Boccassini L, Panni B, Caruso I (2001) The role of NSAIDs in psoriatic arthritis: evidence from a controlled study with nimesulide. *Clin Exp Rheumatol* 19 (Suppl 22): S17–S20
68. Barskova VG, Iakunina IA, Nasonova VA. (2003) Nimesil treatment of gouty arthritis. *Ter Arkh* 75: 60–64
69. Duffy T, Belton O, Bresnihan B, FitzGerald O, FitzGerald D (2003) Inhibition of PGE_2 by nimesulide compared with diclofenac in the acutely inflamed joint of patients with arthritis. *Drugs* 63 (Suppl 1): 31–36
70. Reiner M (1982) Nimesulide in the short-term treatment of osteoarthrosis: a pilot study for assessing minimal effective dose. *J Int Med Res* 10: 92–98
71. Flower RJ (2003) The development of COX-2 inhibitors. *Nature Rev* 2: 179–191
72. Bianchi M (2004) Are all NSAIDs other than "coxibs" really equal? *Trends Pharmacol Sci* 25: 6–7
73. Stein C, Millan MJ, Herz A (1988) Unilateral inflammation of the hindpaw in rats as a model of prolonged noxious stimulation: alterations in behavior and nociceptive thresholds. *Pharmacol Biochem Behav* 31: 445–451
74. Bianchi M, Panerai AE (1997) Formalin injection in the tail facilitates hindpaw withdrawal reflexes induced by thermal stimulation in the rat: effect of paracetamol. *Neurosci Lett* 237: 89–92
75. Bianchi M, Broggini M (2002) Anti-hyperalgesic effects of nimesulide: studies in rats and humans. *Int J Clin Practice* Suppl 128: 11–19
76. Zondervan KT, Yudkin V, Vessey MP, Dawes MG, Barlow DH, Kennedy SH (1999) Prevalence and incidence in primary care of chronic pelvic pain in women: Evidence from a national general practice database. *Br J Obstet Gynaecol* 106: 1149–1155
77. Jamieson DJ, Steege JF (1996) The prevalence of dysmenorrhoea, dyspareunia, pelvic pain, and irritable bowel syndrome in primary care practices. *Obstet Gynecol* 87: 55–58
78. Mathias SD, Kuppermann M, Liberman RF, Lipschutz RC, Steege JF (1996) Chronic pelvic pain: Prevalence, health-related quality of life, and economic correlates. *Obstet Gynecol* 87: 321–327
79. Howard FM (1993) The role of laparoscopy in chronic pelvic pain: Promise and pitfalls. *Obstet Gynecol* Surv 48: 357–387
80. Klein JR, Litt IF (1981) Epidemiology of adolescent dysmenorrhoea. *Pediatrics* 68: 661–665
81. Harlow SD, Park M (1996) A longitudinal study of risk factors for the occurrence, duration and severity of menstrual cramps in a cohort of college women. *Br J Obst Gyn* 103: 1134–1138
82. Svanberg L, Ulmsten U (1981) The incidence of primary dysmenorrhoea in teenagers. *Arch Gynecol* 230: 173–177
83. Andersch B, Milsom I (1982) An epidemiologic study of young women with dysmenorrhoea. *Am J Obstet Gynecol* 144: 655–658
84. Ylikorkala O, Dawood MY (1978) New concepts in dysmenorrhoea. *Am J Obstet Gynecol* 130: 833–835

85. Smith RP (1997) *Gynecology in Primary Care*. Williams & Wilkins, Baltimore, 389–404
86. Campbell MA, McGrath PJ (1999) Non-pharmacologic strategies used by adolescents for the management of menstrual discomfort. *Clin J Pain* 15: 313–318
87. Malinak, LR, Buttram V, Elias S, Simpson JL (1980) Heritage aspects of endometriosis. II. Clinical characteristics of familial endometriosis. *Am J Obstet Gynecol* 137: 332–337
88. Dawood MY (1993) Nonsteroidal antiinflammatory drugs and reproduction. *Am J Obstet Gynecol* 169: 1255–1265
89. Dawood MY (1988) Nonsteroidal anti-inflammatory drugs and changing attitudes toward dysmenorrhoea. *Am J Med* 8: 23–29
90. Johnson J (1988) Level of knowledge among adolescent girls regarding effective treatment for dysmenorrhoea. *J Adoles Health* 9: 398–402
91. Di Leo S, Meli MT, Dal Pra ML (1988) Esperienza clinica con nimesulide nella patologia flogistica ginecologica. *Minerva Ginecol* 40: 119–124
92. Lopez Rosales C, Cisneros Lugo JH (1989) Nimesulide en el tratamiento de la dismenorrea primaria. Evaluación clínica comparativa con acido mefenamico y fentiazac. *Ginecol Obstet Mexico* 57: 196–201
93. Bacarat EC, Alves da Motta EL, de Lima GR (1991) Avaliação clínica da eficácia e tolerabilidade do nimesulide versus piroxicam na terapêutica da dismenorréia primária. Rev Bras Ginecol 101: 467–470
94. Chiantera A, Tesauro R, Di Leo S, Meli S, Scaricabarozzi I (1993) Nimesulide in the treatment of pelvic inflammatory diseases. A multicentre clinical trial conducted in Campania and Sicily. *Drugs* 46 (Suppl 1): 134–136
95. Rinaldi JF, Cymbalista N (1994) Avaliação comparativo do nimesulide *versus* diclofenaco sódico em pacientes com doençã inflamatória pélvica aguda. *Arq Bras Med* 68: 229–232
96. Pirhonen J, Pulkkinen M (1995) The effect of nimesulide and naproxen on the uterus and ovarian arterial blood flow velocity. A Doppler study. *Acta Obstet Gynecol Scand* 74: 549–553
97. Melis GB, Paoletti AM, Mais V, Ajossa S, Guerriero S (1997) Studio clinico controllato sull'efficacia e tollerabilità del metoxibutropato verso nimesulide in campo ginecologico. *Minerva Ginecol* 49: 409–415
98. Pulkkinen MO (1987) Alterations in intrauterine pressure, menstrual fluid prostaglandin F levels, and pain in dysmenorrheic women treated with nimesulide. *J Clin Pharmacol* 27: 65–69
99. Pulkkinen MO (2001) Is there a rationale for the use of nimesulide in the treatment of dysmenorrhoea? *Drugs of Today* 37: 31–38
100. Facchinetti F, Piccinini F, Sgarbi L, Renzetti D, Volpe A (2001) Nimesulide in the treatment of primary dysmenorrhoea: a double-blind study *versus* diclofenac. *Drugs of Today* 37: 39–45
101. Ekstrom P, Juchnicka E, Laudanski T (1989) Effect of an oral contraceptive in primary dysmenorrhoea-changes in uterine activity and reactivity to agonists. *Contraception* 40: 39–43

102. The Transdermal Nitroglycerine/Dysmenorrhoea Study Group (1997) Transdermal nitroglycerine in the management of pain associated with primary dysmenorrhoea: a multinational pilot study. *J Int Med Res* 25: 41–44
103. Facchinetti F, Sgarbi L, Piccinini F, Volpe A (2002) A comparison of glyceryl trinitrate with diclofenac for the treatment of primary dysmenorrhoea: an open, randomized, cross-over trial. *Gynecol Endocrinol* 16: 39–43
104. Harel Z, Bhiro FM, Kottenhahn RK (1996) Supplementation with omega-3 polyunsaturated fatty acids in the management of dysmenorrhoea in adolescents. *Am J Obstet Gynecol* 174: 1335–1338
105. Helms JM (1987) Acupuncture for the management of primary dysmenorrhoea. *Obstet Gynecol* 69: 51–56
106. Anonymous (date unknown) *Enquête épidémiologique nationale sur plus de 7000 consultations de traumatologie sportive*. Menarini France, p. 78
107. Gardinet PF (1983) Anti-inflammatory medications. *The Physician & Sport Medicine* 9: 71–73
108. Almekinders LC (1999) Anti-inflammatory treatment of muscular injuries in sport. An update of recent studies. *Sports Med* 28: 383–388
109. Leadbeater WB, Buchwalter JA, Gordon SL (eds) (1990) *Sports-Induced Inflammation.* American Academy of Orthopaedic Surgeons, Park Ridge, Illinois
110. Garrick J (ed) (2004) *Sports Medicine 3*. American Academy of Orthopaedic Surgeons, Park Ridge, Illinois
111. Scott A, Khan KM, Roberts CR, Cook JL, Duronio V (2004) What do we mean by the term "inflammation"? A contemporary basic science update for sports medicine. *Br J Sports Med* 3: 372–380
112. Herring SA, Nilson KL (1987) Introduction to overuse injuries. *Clinics in Sports Med* 2: 225–239
113. Dreiser RL, Riebenfeld D (1993) A double-blind study of the efficacy of nimesulide in the treatment of ankle sprain in comparison with placebo. *Drugs* 46 (Suppl 1): 183–186
114. Lecomte J, Buyse H, Taymans J, Monti T (1994) Treatment of tendonitis and bursitis: a comparison of nimesulide and naproxen sodium in double-blind parallel trial. *Eur J Rheumatol Inflamm* 14: 29–32
115. Calligaris A, Scaricabarozzi I, Vecchiet L (1993) A multicentre double-blind investigation comparing nimesulide and naproxen in the treatment of minor sport injuries. *Drugs* 46: 187–190
116. Ribamar Bacelar Costa J, Carlos Lima Rehfeld L, Antônio Gonçalves da Silva S (1995) Traumatic sprains and strains. Prospective blind study to compare the efficacy and safety of nimesulide and diclofenac. *Rev Bras Med* 366–372
117. Jenoure P, Gorschewsky O, Ryf C, Steigbugel Werzel C, Frey W, Voisin D (1998) Randomised, double-blind, multicentre study of nimesulide vs. diclofenac in adults with other sport injuries. *J Clin Res* 1: 343–356

118. Saillant G, Magni E, Macciocchi A et al. (1997) Double-blind, controlled, multicentre study comparing the efficacy and tolerability of nimesulide vs. ketoprofen in sports trauma (Phase III). Helsinn Healthcare, Report No. TSD 5873
119. Costa RJB, Rehfeldt LCL, Gonclaves da Silva SA (1995) O traumatismo osteoarticular. Estudo terapeutico, cego simples, randomico, comparativo entre nimesulide e diclofenaco potassico. *Rev Bras Med* 52: 366–373
120. Di Marco C, Fesi P, Letizia GA (1989) Studio controllato sull'attività terapeutica e sulla tollerabilità della nimesulide nel trattamento della patologia algoflogistica post-traumatica. *Minerva Ortopedica* 40: 111–116
121. Gusso MI, Innocenti M (1989) Nimesulide (Fans dell'ultima generazione) nel trattamento della patologia algoflogistica post-traumatica. *Ortopedia e Traumatologia Oggi* 9: 162–168
122. Lederman R, Rubenstein J (1982) Estudo controlado con Nimesulide a Ibuprofen no tratamiento de processos inflammatorios extraarticulares. *A Folio Medica* 84 (Suppl 1): 301–306
123. Zarraga Corrales J, López Torres D, Cruz Cruz M (1992) Estudio comparativo de nimesulide vs diclofenac en el tratamiento del traumatismo agudo de partes blandas. *Invest Med Int* 19: 133–141
124. Wober W, Ralfs VW, Buchl N, Grassle A, Macciocchi A (1998) Comparative efficacy and safety of the non-steroidal anti-inflammatory drugs nimesulide and diclofenac in patients with acute subdeltoid bursitis and bicupital tendonitis. *Int J Clin Pract* 52: 169–175
125. Eicher MG et al. (1999) Pharmacokinetic profile and transdermal penetration of nimesulide in male healthy volunteers after single and multiple epicutaneous administration of a new 3% gel formulation. Helsinn Healthcare, Report No. TSD 7781.2E
126. Saillant G (1998) Multicentre, double-blind, parallel group trial to compare the efficacy and tolerability of nimesulide 3% gel *versus* placebo in the treatment of benign ankle sprains. Helsinn Healthcare, Report No. TSD 7218 E
127. Bourgeois P (1998) Multicentre, double-blind, parallel group trial to compare the efficacy and tolerability of nimesulide 3% gel *versus* placebo in the treatment of acute tendinitis of the upper limb. Helsinn Healthcare, Report No. TSD 7219 E
128. Jenoure PJ (1999) Multicentre, double-blind, parallel group study to compare the efficacy and tolerability of 3% nimesulide gel *versus* diclofenac in the treatment of acute tendinitis of the upper limb. Helsinn Healthcare, Report No. TSD 7552 E
129. Rodriguez Alvarez J (2000 & 2001) Multicentre, double-blind, parallel group study comparing the efficacy and safety of 3% nimesulide gel *versus* ketoprofen gel in the treatment of mild ankle sprain. Helsinn Healthcare, Reports No. TSD 7318 & TSD 7318.2
130. Jenoure PJ (2001) Nimesulide gel. Expert report on the clinical documentation. Helsinn Report October 2001, *and* Personal Communication, Helsinn Healthcare SA, January, 2005
131. Dhaon BK, Farooque MF, Sharma DR, Bhutani S (1998) Open labelled clinical evaluation of local application of nimesulide transdermal gel in painful musculoskeletal conditions. *Indian J Orthop* 32: 75–78

132. Dhaon BK, Singh OP, Gupta SP, Maini L, Sharma DR, Bhutani S (2000) Efficacy and safety of nimesulide transdermal gel *versus* diclofenac and piroxicam gel in patients with acute musculoskeletal condition. *Indian J Orthop* 34: 288–292
133. Sengupta S, Velpandian T, Kabir SR, Gupta SK (1998) Analgesic efficacy and pharmacokinetics of topical nimesulide gel in healthy human volunteers: double-blind comparison with piroxicam, diclofenac and placebo. *Eur J Clin Pharmacol* 54: 541–547
134. Cooper SA, Beaver WT (1976) A model to evaluate mild analgesics in oral surgery outpatients. *Clin Pharmacol Ther* 20: 241–250
135. Cooper SA (1984) Five studies on ibuprofen for post-surgical dental pain. *Am J Med* 77A: 70–77
136. Valanne J, Korttila K, Ylikorkala O (1987) Intravenous diclofenac sodium decreases prostaglandin synthesis and postoperative symptoms after general anaesthesia in outpatients undergoing dental surgery. *Acta Anaesthesiol Scand* 31: 722–727
137. Wuolijoki E, Oikarinen VJ, Ylipaavalniemi P, Hampf G, Tolvanen M (1987) Effective postoperative pain control by preoperative injection of diclofenac. *Eur J Clin Pharmacol* 32: 249–252
138. Dionne RA (1998) Evaluation of analgesic mechanisms and NSAIDs for acute pain using the oral surgery model. In: Rainsford KD, Powanda MC (eds): *Safety and Efficacy of Non-Prescription (OTC) Analgesics and NSAIDs*. Kluwer Academic Publishers, Dordrecht, 105–117
139. Dionne RA, Cooper SA (1999) Use of ibuprofen in dentistry. In: Rainsford, KD (ed): *Ibuprofen. A Critical Bibliographic Review.* Taylor and Francis, London, 407–430
140. Swift JQ, Garry MG, Roszkowski MT, Hargreaves KM (1993) Effect of flurbiprofen on tissue levels of immunoreactive bradykinin and acute postoperative pain. *J Oral Axillofac Surg* 51: 112–116
141. Roszkowski MT, Swift JQ, Hargreaves KM (1997) Effect of NSAID administration on tissue levels of immunoreactive prostaglandin E_2, leukotriene B_4, and (S)-flurbiprofen following extraction of impacted third molars. *Pain* 73: 339–345
142. Gordon SM, Brahim JS, Rowan J, Kent A, Dionne RA (2002) Peripheral prostanoid levels and nonsteroidal anti-inflammatory drug analgesia: replicate clinical trials in a tissue injury model. *Clin Pharmacol Ther* 72: 175–183
143. Khan AA, Brahim JS, Rowan JS, Dionne RA (2002) *In vivo* selectivity of a selective cyclooxygenase 2 inhibitor in the oral surgery model. *Clin Pharmacol Ther* 72: 44–49
144. Dionne KA, Khan AA, Gordon SM (2001) Analgesia and COX-2 inhibition. *Clin Exp Rheumatol* 19 (Suppl 25): S63–S70
145. Chang DJ, Fricke JR, Bird SR, Bohidar NR, Dobbins TW, Geba GP (2001) Rofecoxib *versus* codeine/acetaminophen in postoperative dental pain: a double-blind, randomized, placebo- and active comparator-controlled clinical trial. *Clin Ther* 23: 1446–1455
146. Desjardins PJ, Grossman EH, Kuss ME, Talwalker S, Dhadda S, Baum D, Hubbard RC (2001) The injectable cyclooxygenase-2-specific inhibitor parecoxib sodium has analgesic efficacy when administered preoperatively. *Anesth Analg* 93: 721–727

147. Edwards JE, Moore RA, McQuay HJ (2004) Individual patient meta-analysis of single-dose rofecoxib in postoperative pain. *BMC Anesthesiol* 4: 3
148. Chang DJ, Desjardins PJ, Chen E, Polis AS, McAvoy M, Mockoviak SH, Geba GP (2002) Comparison of the analgesic efficacy of rofecoxib and enteric-coated diclofenac placebo-controlled clinical trial. *Clin Ther* 24: 490–503
149. Jackson ID, Heidemann BH, Wilson J, Power I, Brown RD (2004) Double-blind, randomized, placebo-controlled trial comparing rofecoxib with dexketoprofen trometamol in surgical dentistry. *Br J Anaesth* 92: 675–680
150. Chen LC, Elliott RA, Ashcroft DM (2004) Systematic review of the analgesic efficacy and tolerability of COX-2 inhibitors in post-operative pain control. *J Clin Pharm Ther* 29: 215–229
151. Ballou LR, Botting RM, Goorha S, Zhang J, Vane JR (2000) Nociception in cyclooxygenase isozyme-deficient mice. *Proc Natl Acad Sci USA* 97: 10272–10276
152. Mazario J, Gaitan G, Herrero JF (2001) Cyclooxygenase-1 vs. cyclooxygenase-2 inhibitors in the induction of antinociception in rodent withdrawal reflexes. *Neuropharmacology* 40: 937–946
153. Torres-Lopez JE, Ortiz MI, Castaneda-Hernandez G, Alonso-Lopez R, Asomoza-Espinosa R, Granados-Soto V (2002) Comparison of the antinociceptive effect of celecoxib, diclofenac and resveratrol in the formalin test. *Life Sci* 70: 1669–1676
154. Cornaro G (1983) A new non-steroidal anti-inflammatory drug in the treatment of inflammations due to parodontal surgery. *Curr Ther Res* 33: 982–989
155. Salvato A, Zambruno E, Ventrini E, Savio G (1984) Sperimentazione clinica di un nuovo antiedemigeno orale: nimesulide. *Giorn Stomat Ortognat* 3: 184–191
156. Bucci E, Mignogna MD, Bucci P (1987) Aulin: una nuova e moderna terapia nel trattamento delle infiammazioni in odontostomatologia. *Min Stomatol* 36: 101–103
157. Moniaci D, Mozzati M, Anglesio Fariua G, Giacometti E (1988) Valutazone della-efficacia e della tolerabilita della nimesulide in alcune patologie odostomalogiche. *Stomatologica* 37: 291
158. Ragot, JP, Monti T, Macciochi A (1993) Controlled clinical investigation of acute analgesic activity of nimesulide in pain after oral surgery. *Drugs* 46 (Suppl 1): 162–167
159. Ferrari Parabita G, Zanetti U, Scalvini F, Rossi D, Scaricabarozzi I (1993) A controlled clinical study of the efficacy and tolerability of nimesulide vs. naproxen in maxillo-facial surgery. *Drugs* 46 (Suppl 1): 171–173
160. Pierleoni P, Tonelli P, Scaricabarozzi I (1993) A double-blind comparison of nimesulide and ketoprofen in dental surgery. *Drugs* 46 (Suppl 1): 168–170
161. Scolari G, Lazzarin F, Fornaseri C, Carbone V, Rengo S, Amato M, Cicciu D, Braione D, Argentino S, Morgantini A et al. (1999) A comparison of nimesulide beta cyclodextrin and nimesulide in postoperative dental pain. *Int J Clin Pract* 53(5): 345–348
162. Ragot J-P, Giorgi M, Marinoni M, Macchi M, Mazza P, Rizzo S, Garramone R, Monti T (1994) Acute activity of nimesulide in the treatment of pain after oral surgery- double-blind, placebo and mefenamic acid controlled study. *Europ J Clin Res* 5: 39–50

163. Stefanoni G, Saccomanno F, Scaricabarozzi I, Volontieri G, Persiani L, Boselli L, Beretta P, Giroda M (1990) Efficacia clinica della nimesulide in confronto a diclofenac sodio nella prevenzione e nel trattamento della sintomatologia algico-flogistica postchirurica. *Minerva Chirurgica* 45: 1469–1475
164. Schmoekel W, Bisaz E, Choendle S (1985) Nimesulide suppositories in traumatology. Study on the analgesic and anti-inflammatory activity. *Praxis* 74: 1460–1463
165. Ramella G, Costagli V, Vetere M, Capra C, Casella G, Sogni A, Scaricabarozzi, I (1993) Comparison of nimesulide and diclofenac in the prevention and treatment of painful inflammatory postoperative complications of general surgery. *Drugs* 46: 159–161
166. Binning AR (2004) Nimesulide in the treatment of acute pain: double-blind comparative study in a post-operative setting. *Abstracts of the Satellite Symposium on Nimesulide. "The Control of Pain for a Better Compliance of the Patients"*. 3rd World Congress of Pain. 21–25 September 2004, Barcelona
167. Alotti N, Bodo E, Gombocz K, Gabor V, Rashed A (2003) Management of postoperative inflammatory response and pain with nimesulide after cardiac surgery. *Orv Hetil* 144: 2353–2357
168. McCrory CR, FitzGerald DJ (2004) Spinal prostaglandin formation and pain perception following thoracotomy: a role for cyclooxygenase-2. *Chest*, 125: 1321–1327
169. Boureau F (1998) Multicentre study of the efficacy of ibuprofen compared with paracetamol in throat pain associated with tonsillitis. In: Rainsford KD, Powanda MC (eds): *Safety and Efficacy of Non-Prescription (OTC) Analgesics and NSAIDs*. Kluwer Academic Publishers, Dordrecht 119–121
170. Ottaviani A, Mantovani M, Scaricabarozzi I (1993) A multicentre clinical study of nimesulide in inflammatory diseases of the ear, nose and throat. *Drugs* 46 (Suppl 1): 96–99
171. Bianchini G, Scaricabarozzi I, Montecorboli U, Ceccarelli A, Chiesa F, Ditri L, Mazzer G, Maroni R, Viola M, Roggia F et al. (1993) Double-blind study of nimesulide in divers with inflammatory disorders of the ear, nose and throat. *Drugs* 46 (Suppl 1): 100–102
172. Milvio C (1984) Nimesulide for the treatment of painful inflammatory process in the ear, nose and throat areas: a double-blind controlled study with benzydamine. *J Int Med Res* 12: 327–332
173. Nouri E, Monti F (1993) Nimesulide granules for the treatment of acute inflammation of the ear, nose or throat. *Drugs* 46 (Suppl 1): 103–106
174. Bellussi L, Passali D (1993) Treatment of upper airways inflammation with nimesulide. *Drugs* 46 (Suppl 1): 107–110
175. Sofia M, Molino A, Mormile M, Stanziola A, Scaricabarozzi I, Carratu (1993) Nimesulide in the treatment of chronic bronchitis. *Drugs* 46 (Suppl 1): 111–114
176. Passali D, Balli R, Scotti A, Oldani V (2001) Controlled, double-blind, randomized comparison of nimesulide β-cyclodextrin and morniflumate in acute otitis. *Curr Ther Res* 62: 153–166

177. Cadeddu L, Piragine F, Puxeddu P, Scornavacche V, Sellari Franceschini S (1988) Comparison of nimesulide and flurbiprofen in the treatment of non-infectious acute inflammation of the upper respiratory tract. *J Int Med Res* 16: 466–473
178. Rossi M, Monea P, Lomeo G et al. (1991) Studio clinico sull'efficacia e la tollerabilità della nimesulide in formulazione supposte in patologie algico-infiammatorie otorinolaringoiatre. *Minerva Med* 82: 845–853
179. D'Apuzzo V, Monti T (1992) Pilot study of the antipyretic and analgesic activity of nimesulide paediatric suppositories. *Drugs Exp Clin Res* 18: 63–68
180. Cappella L, Guerra A, Laudizi L, Cavazzuti GB (1993) Efficacy and tolerability of nimesulide and lysine-acetylsalicylate in the treatment of paediatric acute upper respiratory tract inflammation. *Drugs* 46 (Suppl 1): 222–225
181. Miniti A, Dieb Miziara I (1991) Estudo comparativo de nimesulide versus naproxeno em pacientes con faringo-amigdalites. *Arqu Bras Med* 65: 511–514
182. Munhoz MSL, Ganaca MM, Munhoz MLGS (1990) Estudo comparativo de Mesulide (nimesulide) vs diclofenac potásico en afecciones otorrinolaringológias. *Rev Bras Med* 7: 591–594
183. Passali D, Bellussi L, Ciferri G et al. (1988) Prospectiva terapeutica nelle otiti medie secretive: nimesulide. *Otorinolaringol* 38: 169–175
184. Polidori G, Titti G, Pieragostini P, Comito A, Scaricabarozzi I (1993) A comparison of nimesulide and paracetamol in the treatment of fever due to inflammatory diseases of the upper respiratory tract in children. *Drugs* 46 (Suppl 1): 231–233
185. Salmon Rodriguez LE, Arista Viveros HA, Lujan ME, Maciel RM, Trujillo CL, Lopez E (1993) Assessment of the efficacy and safety of nimesulide vs. naproxen in paediatric patients with respiratory tract infections. A comparative single-blind study. *Drugs* 46 (Suppl 1): 226–230
186. Ulukol B, Koksal Y, Cin S (1999) Assessment of the efficacy and safety of paracetamol, ibuprofen and nimesulide in children with upper respiratory tract infections. *Eur J Clin Pharmacol* 55: 615–618
187. Gananca MM, Munhoz MSL, Caovilla HH (1990) Comparative study of nimesulide *versus* potassium diclofenac in acute otitis media. *Rev Bras Med* 47: 373–376 (Article in Portuguese)
188. Davis R, Brogden RN (1994) Nimesulide. An update of its pharmacodynamic and pharmacokinetic properties, and therapeutic efficacy. *Drugs* 48: 431–454
189. Bianco S, Robuschi M, Petrigni G, Scuri M, Pieroni M, Refini RM, Vaghi A, Sestini (1993) Efficacy and tolerability of nimesulide in asthmatic patients intolerant to aspirin. *Drugs* 46: 115–120
190. Bavbek S, Çelik G, Ediger D, Mungan D, Demirel YS, Mysyrhgil Z (1999) The use of nimesulide in patients with acetylsalicylic acid and nonsteroidal anti-inflammatory drug intolerance. *J Asthma* 36: 657–663
191. Andri L, Senna GE, Betteli C, Givanni S, Andri G, Scaricabarozzi I (1992) Combined treatment of allergic rhinitis with terfenadine and nimesulide, a non-steroidal antiinflammatory drug. *Allerg Immunol (Paris)* 24: 313–314

192. Kotwani A, Puri R, Gupta U (2001) Efficacy of nimesulide alone and in combination with cetirizine in acute allergic rhinitis. *J Assoc Physicians India* 49: 518–522
193. Agus GB, de Angelis R, Mondani P, Moia R (1993) Double-blind comparison of nimesulide and diclofenac in the treatment of superficial thrombophlebitis with telethermographic assessment. *Drugs* 46 (Suppl 1): 200–203
194. Ferrari E, Pratesi C, Scaricabarozzi I, Trezzani R (1992) Clinical study of the therapeutic efficacy and tolerance of nimesulide in comparison with sodium diclofenac in the treatment of acute superficial thrombophlebitis. *Minerva Cardioangiol* 40: 455–460 (Article in Italian)
195. Zanetta M, Martelli E, Corsi G (1988) The use of nimesulide in the treatment of thrombophlebitis of the lower limbs. *Minerva Angiol* 13: 49–52 (Article in Italian)
196. Lotti T, Mirone V, Imbimbo C, Corrado F, Corrado G, Garofalo F, Scaricabarozzi I (1993) Controlled clinical studies of nimesulide in the treatment of urinogenital inflammation. *Drugs* 46 (Suppl 1): 144–146
197. Canale D, Turchi P, Giorgi PM, Scaricabarozzi I, Menchini-Fabris GF (1993) Treatment of abacterial prostato-vesiculitis with nimesulide. *Drugs* 46 (Suppl 1): 147–150
198. Gabbrielli G, Binazzi P, Scaricabarozzi I, Massi GB (1993) Nimesulide in the treatment of mastalgia. *Drugs* 46 (Suppl 1): 137–139
199. Panagariya A, Sharma AK (1999) A preliminary trial of serratiopeptidase in patients with carpal tunnel syndrome. *J Assoc Physicians India* 47: 1170–1172
200. Reiner M, Massera E, Magni E (1984) Nimesulide in the treatment of fever: a double-blind crossover clinical trial. *J Int Med Res* 12: 102–107
201. Reiner M, Cereghetti S, Haeusermann M, Monti T (1985) Antipyretic activity of nimesulide suppositories: double-blind *versus* diclofenac and placebo. *International J Clin Pharmacol* 23: 673–677
202. Cunietti E, Monti M, Vigano A, Aprile ED, Saligari A, Scafuro E, Scaricabarozzi I (1993) Nimesulide in the treatment of hyperpyrexia in the aged. *Drug Res* 2: 160–162
203. Lecomte J, Monti T, Pochobradsky MG (1991) Antipyretic effects of nimesulide in paediatric practice: a double-blind study. *Curr Med Res Opin* 12: 296–303
204. Ugazio AG, Guarnaccia S, Berardi M, Renzetti I (1993) Clinical and pharmacokinetic study of nimesulide in children. *Drugs* 46 (Suppl 1): 215–218
205. Kapoor SK, Sharma J, Batra B, Paul E, Anand K, Sharma D (2002) Comparison of antipyretic effect of nimesulide and paracetamol in children. *Indian Pediartr* 39: 437–477
206. Cunietti E, Monti M, Viganò A, Aprile ED, Saligari A, Scafuro E, Scaricabarozzi I (1993) Nimesulide in the treatment of hyperpyrexia in the aged. *Arzneimmitel-Forsch* 43: 160–162
207. Cunietti E, Monti M, Viganò A, D'Aprile E, Saligari A, Scafuro E, Scaricabarozzi I (1993) A comparison of nimesulide vs. paracetamol in the treatment of pyrexia in the elderly. *Drugs* 46 (Suppl 1): 124–126
208. Rainsford KD (2004) Salicylates in the treatment of acute pain. In: Rainsford KD (ed): *Aspirin and Related Drugs*. CRC Press, Boca Raton (Florida), 587–618

209. Giacovazzo M, Gallo MF, Guidi V, Rico R, Scaricabarozzi I (1993) Nimesulide in the treatment of menstrual migraine. *Drugs* 46 (Suppl 1) 140–141
210. Ferrari A, Pasciullo G, Savino G, Cicero AF, Ottani A, Bertolini A, Sternieri E (2004) Headache treatment before and after the consultation of a specialist centre: a pharmacoepidemiological study. *Cephalgia* 24: 356–362
211. McNicol E, Strassels S, Goudas L, Lau J, Carr D (2004) Nonsteroidal anti-inflammatory drugs, alone or combined with opioids, for cancer pain: a systematic review. *J Clin Oncol* 22: 1975–1992
212. Lucas LK, Lipman AG (2002) Recent advances in pharmacotherapy for cancer pain management. *Cancer Pract* 10 (Suppl 1): S14–S20
213. Corli O, Cozzolino A, Scaricabarozzi I (1993) Nimesulide and diclofenac in the control of cancer-related pain. Comparison between oral and rectal administration. *Drugs* 46 (Suppl 1): 152–155
214. Ventafridda V, Toscani F, Tamburini M, Corli O, Gallucci M (1990) Sodium naproxen vs. sodium diclofenac in cancer pain control. *Arzneimittel- Forsch* 40: 1132–1138
215. Toscani F, Gallucci M, Scaricabarozzi I (1993) Nimesulide in the treatment of advanced cancer pain. Double-blind comparison with naproxen. *Drugs* 46 (Suppl 1): 156–158
216. Gallucci M, Toscani F, Mapelli A, Cantarelli A, Veca G, Scaricabarozzi I (1992) Nimesulide in the treatment of advanced cancer pain. Double-blind comparison with naproxen. *Arzneimittelforschung* 42: 1028–1030
217. Cantarelli A, Giannunzio D, Ligorio L, Mapelli A, Veca G, Gallucci M, Toscani F (1991) Comparison of nimesulide and naproxen sodium in the control of cancer pain. *Minerva Anestesiol* 57: 1103–1104 (Article in Italian)

Adverse reactions and their mechanisms from nimesulide

I. Bjarnason[1], F. Bissoli[2], A. Conforti[3], L. Maiden[1], N. Moore[4], U. Moretti[3], K.D. Rainsford[5], K. Takeuchi[1], G.P. Velo[6]

[1]Department of Medicine, Guy's, King's and St Thomas' Medical School, University of London, London, UK; [2]Divisione di Medicina, Clinica S Gaudenzio, Novara, Italy; [3]Università di Verona, Istituto di Farmacologia, Policlinico Borgo Roma, 37134 Verona, Italy; [4]Department of Pharmacology, Université Victor Segalen, Bordeaux, France; [5]Biomedical Research Centre, Sheffield Hallam University, Howard Street, Sheffield, S1 1WB, UK; [6]Ospedale Policlinico, Via delle Menegone 10, 37134 Verona, Italy

Introduction

The pattern of adverse drug reactions (ADRs) in different organ systems from the NSAIDs is essentially similar [1–7]. The main distinctions are in the quantitative differences that exist in the occurrence or frequency of ADRs among the different groups, especially those more frequently occurring in the gastrointestinal (GI) tract, liver and to some extent the kidney [1–7] (Fig. 1). Some drugs do have a propensity to cause rare ADRs, e.g., agranulocytosis and aplastic anaemia with phenylbutazone [1, 2]; Stevens Johnson and Lyell's Syndromes and other severe skin reactions with isoxicam and piroxicam [8, 9]. The difficulty is to quantify many of the individual reactions especially when it comes to population studies [2]. Here the main issue is to establish the exposure of a known population to individual drugs and to know if individual members of the population are taking other drugs or have conditions that might contribute to, or be major confounding factors in the development of ADRs [1, 2, 8–10].

In the case of nimesulide, the consensus reviewed here is that the drug has a relatively low propensity to produce severe GI reactions in comparison with other NSAIDs. Severe renal, cardiovascular and skin reactions are relatively rare. Liver reactions (hepatitis, cholestatic jaundice and liver failure) while having attracted attention in the period from 2001–2003 following a number of reports in Finland, were recently evaluated by the European Medicines Evaluation Agency (now the European Medicines Agency) and recent published reports, and found to be no more frequent than with other NSAIDs.

In this chapter the evidence of the safety of nimesulide compared with other NSAIDs has been reviewed from information derived from:

a) **Spontaneous adverse drug reaction (ADR) reports** recorded in the Helsinn Drug Safety Unit and supplemented by information from literature reports.

Nimesulide – Actions and Uses, edited by K. D. Rainsford

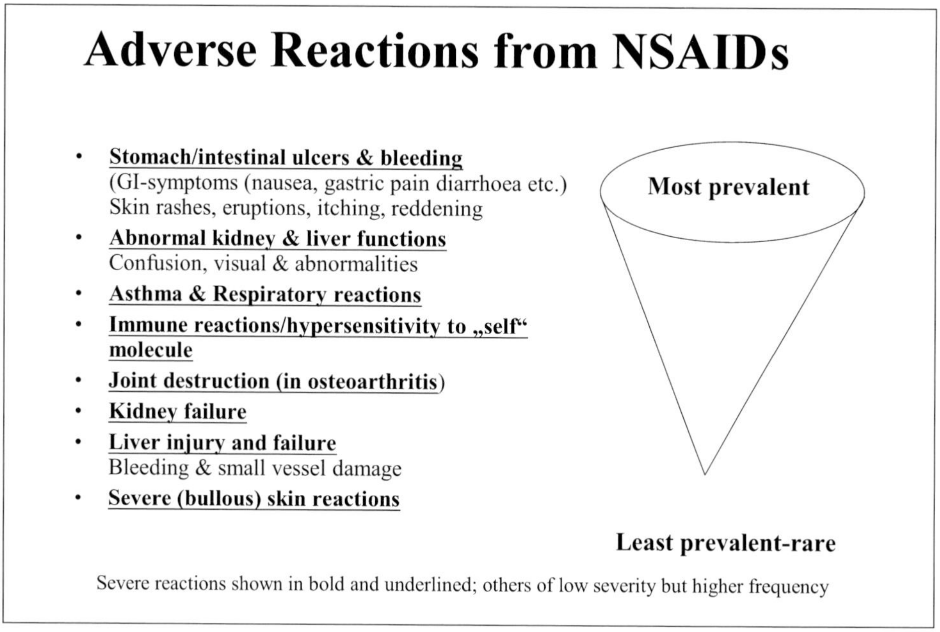

Figure 1
Conceptual view of the occurrence of adverse reactions in various organ systems from NSAIDs in relation the frequency (right side) and severity either in terms of morbidity or mortality. The more severe reactions are shown in bold and underlined.

Each report has been assessed for the quality of the information provided therein, any confounding factors (other drugs or diseases that might have precipitated the ADR) and causality.

b) **Epidemiological and population studies** principally those where there was analysis of the serious upper GI reactions and hepatic events. Information principally comprising clinical reports about renal, cutaneous and allergic reactions was also assessed. Most of the evidence from upper GI and hepatic events was derived from regional pharmaco-epidemiological studies some of which are retrospective in study design.
c) **Clinical trial data** involving prospective investigations in randomised, double-blind trials in normal human volunteers and patients most of whom had arthritic conditions.
d) **Clinical investigations** in normal healthy or patient volunteers designed to investigate the mode of actions of the drug in humans.
e) **Mechanistic studies** in animal models *in vivo* or *ex vivo* and in insolated cellular models *in vitro* as well as biochemical investigations. These studies serve to

show how nimesulide has in some cases unique cellular and molecular properties in comparison with various other NSAIDs that probably explains its safety features (e.g., low GI reactions).

Thus, the safety profile of nimesulide has been evaluated in various organ systems according to the accepted criteria for determining the causative and mechanistic basis of adverse reactions observed with the NSAIDs, with the clinical significance and risk/benefit ratios also being assessed. The reader is referred to the detailed summary at the end of this chapter in which the major points that have emerged from the different types of studies and investigations are included.

Nimesulide safety profile from spontaneous reporting

Spontaneous reporting is relevant for signal generation but cannot give a true incidence rate due to the lack of a definite denominator (number of patient exposed) and to the under-reporting. Furthermore, the ADRs causality assessment is often difficult due to the presence of confounding factors (e.g., patient's pre-existing clinical conditions, concurrent diseases, concomitant drugs). In addition, the delay between the occurrence of the ADR and the date of its reporting is a measure of notoriety bias: when alerts appear, or when publications of other events bring the drug to the public's attention, older cases tend to be more frequently reported.

Nimesulide has been marketed by Helsinn Healthcare's partners since 1985, initially in Italy, then extending progressively to over 50 countries by mid 2004. In this period, 3,249 adverse events have been reported in 2,005 patients from more than 415 million[1] treatment courses used.

An analysis of the adverse reactions from nimesulide has been undertaken from data held on file at Helsinn Healthcare SA (Lugano, Switzerland). The data was examined for number and characteristics of adverse reactions, and patients.

Also, a detailed analysis has been undertaken to assess the quality of the ADR report and from this determine more precisely the likelihood of the reaction being attributed to the drug and the factors (other drug(s), disease(s) or environmental) that may have contributed to the development of the nimesulide-associated adverse event. It is important to note that these data do not always record the ADRs in those countries where generic formulations are sold.

Spontaneous reports reported to the company directly or reports sent to the company from the regulatory authorities according to regulations have been con-

[1] *Assuming a nimesulide 200 mg/day as "daily dose" (equivalent to nimesulide β-cyclodextrin 800 mg/day) and a mean treatment period of 15 days.*

sidered. Data from the World Health Organization (WHO) Monitoring Service in Uppsala (Sweden) of adverse events in different body systems attributed to nimesulide were examined to check that these cases had been recorded in the database. However, because of the stochastic nature of such reporting and the fact that it does not represent a comprehensive database for all reports the data should be treated with caution as specified by the WHO. It should also be noted that these data may include ADR reports from those countries where generic or other non-Helsinn brands of the drug are marketed (e.g., Greece, Portugal, India, Italy, South America).

Overall pattern of adverse event reports

Figure 2 shows the total number of ADR reports received since nimesulide was introduced in 1985. Figure 3 shows the trends in ADR reports over the past half decade. In general, there is a consistent pattern of ADRs paralleling the number of treatment courses, with the exception of a peak in events that occurred during the first half of 2002 coinciding with occurrence of a "spike" of reporting of hepatic reactions in Finland, as detailed in Figure 4.

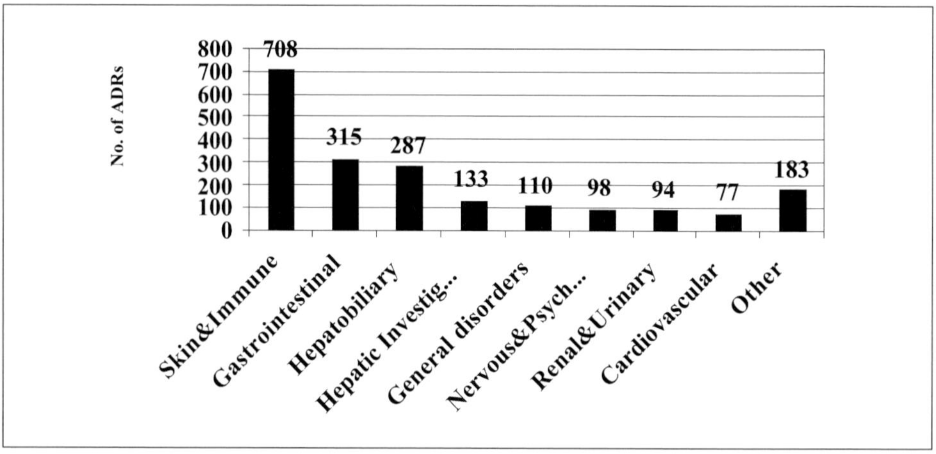

Figure 2
Total number of all adverse drug reactions (ADRs) (both serious and non-serious) attributed to intake of nimesulide reported worldwide since the drug was introduced in 1985 up to June 2004. Graph from the ADR database of Helsinn Healthcare SA.

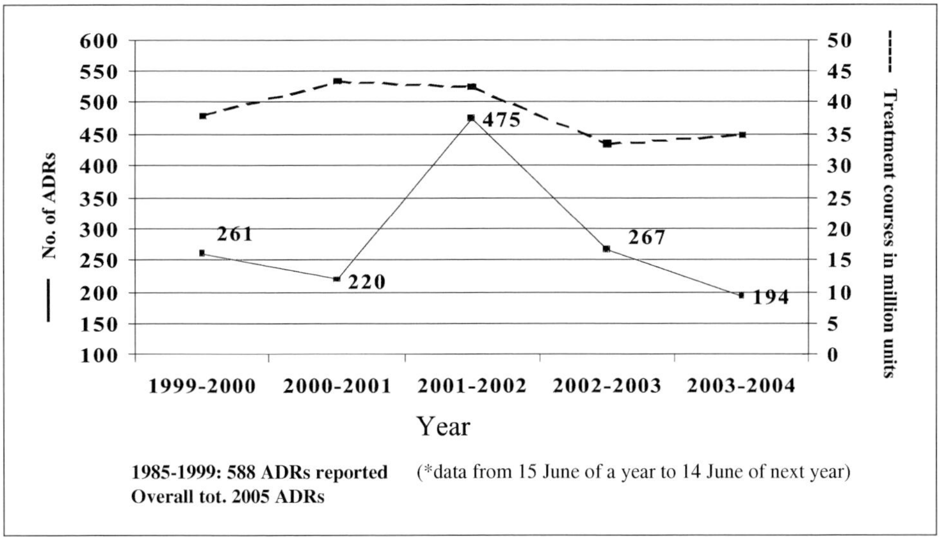

Figure 3

Total number of all ADRs from nimesulide over the last 5 years in comparison with treatment courses in the 5-year period until mid-2004.

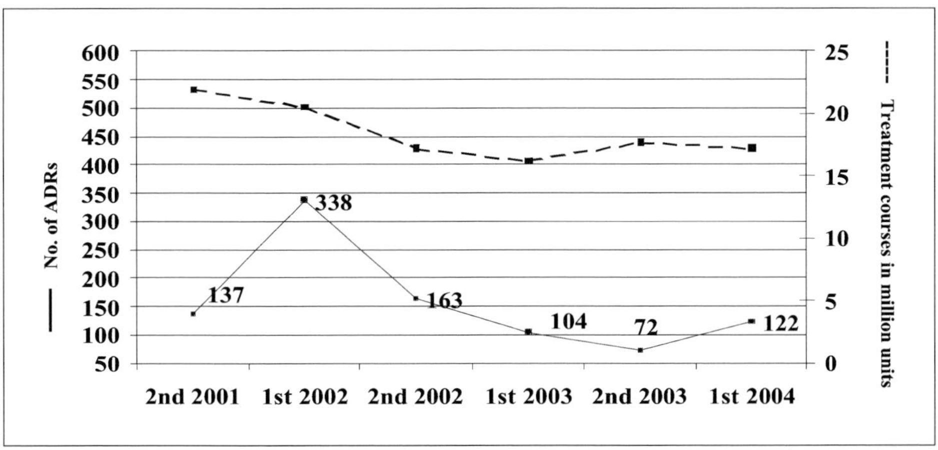

Figure 4

Adverse reactions attributed to nimesulide worldwide in the last six semesters until mid-2004. A peak of ADRs occurred during the first half of 2002.

Characteristics of the adverse reactions

Most of the reports were of skin disorders (35.3%), followed by GI events (15.7%), hepatobiliary (14.3%) and hepatic investigations (abnormal laboratory tests, 6.6%) (Tab. 1).

Overall 63 out of 2005 were fatal (3%), being 5% of hepatic cases (23/420) and 4.4% of GI (14/315).

Figure 5 shows the numbers of ADRs classified according to system organ class (SOC), reported in the major countries in the world where nimesulide is marketed.

The mean age of patients in whom reactions were reported varied widely between reactions, with patients reported having skin, allergic, central nervous system (CNS) or respiratory reactions being significantly younger than patients complaining of hepatic or cardiac reactions, or abnormal investigations. Signif-

Table 1 – Case reports of adverse reactions from nimesulide (system organs classified according to MedDRA dictionary)

Body/Organ Systems	Number of Cases	Percent of Total
Skin and immune	708	35.3
Gastrointestinal disorders	315	15.7
Hepatobiliary disorders	287	14.3
Hepatic investigations	133	6.6
General disorders	110	5.5
Nervous system & psychiatric disorders	98	4.9
Renal and urinary disorders	94	4.7
Blood and lymphatic system disorders	43	2.1
Vascular disorders	56	2.8
Injury and poisoning	36	1.8
Respiratory, thoracic and mediastinal disorders	34	1.7
Cardiac disorders	21	1.1
Pregnancy, puerperium and perinatal conditions/reproductive/congenital	20	1.0
Endocrine disorders[§]	18	0.9
Ear or eye disorders	18	0.9
Investigations	14	0.7
Number of Cases	2005	–

[§] *Includes musculoskeletal, metabolism and nutrition, infections and neoplasms.*

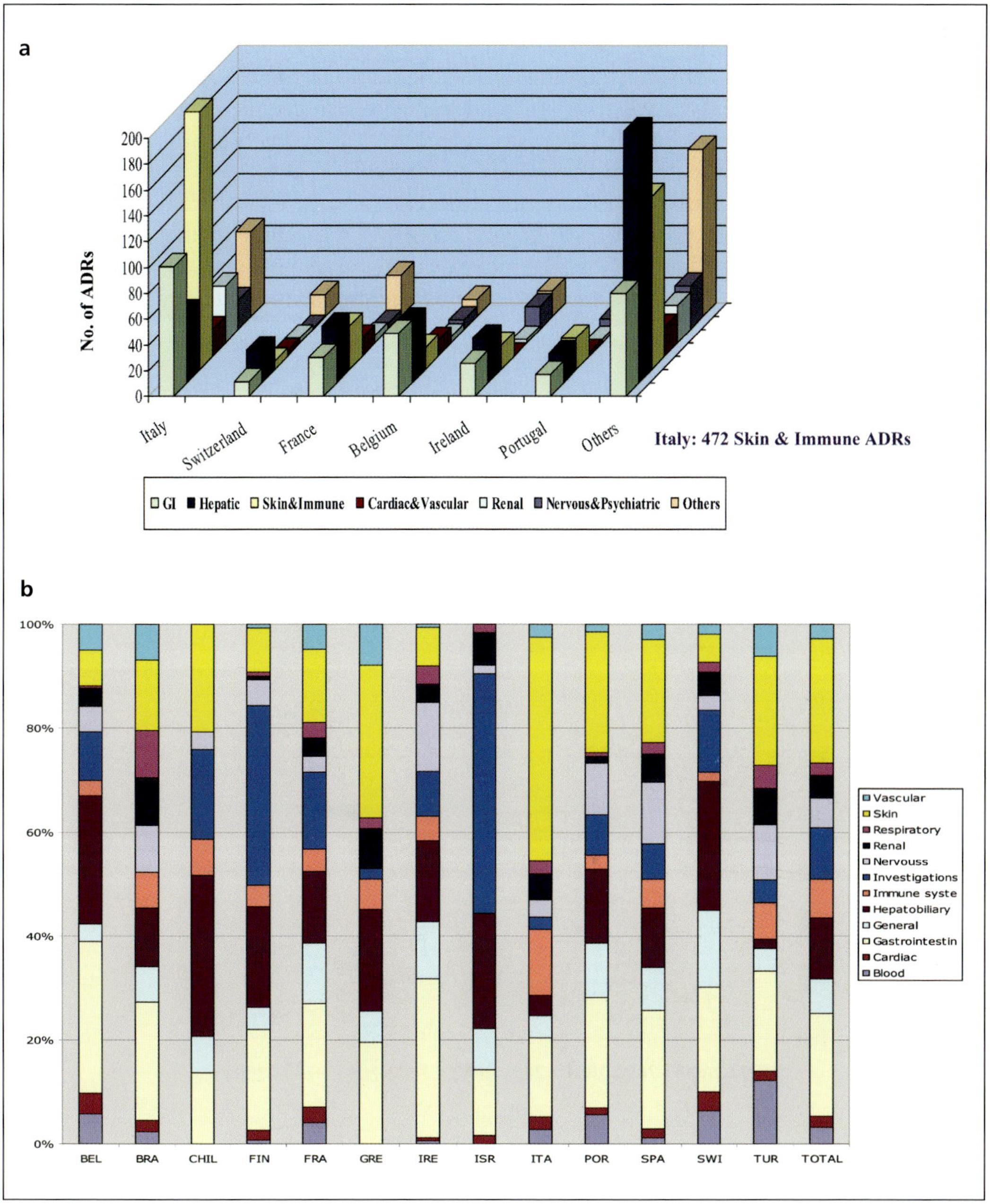

Figure 5

Number of ADRs classified according to System Organ Class (SOC), reported in the major EU countries where nimesulide is marketed.

icantly, more males suffered from GI reactions, while more females had hepatobiliary reactions.

Onset delays (time from beginning of treatment to onset of reaction) were also significantly different between organ systems, with for instance skin disorders having a mean onset delay of 7 days, compared to 90 days for hepatobiliary disorders, a feature noted previously with diclofenac [11].

This adverse reaction profile is not different in nature from that of other NSAIDs, although the proportion of hepatic reports is at the upper end of the range of reports seen with all other NSAIDs, for reasons that will be discussed further.

In Figures 6 and 7 the distribution of serious and non-serious ADR case reports is shown, respectively, for the past 5 years. The total numbers of reports in the GI, hepatic and skin and immune systems are shown in Figures 8, 9 and 10, respectively. To some extent these show a relatively constant level of reporting overall. This trend is evident in relation to sales of the drug (Figs 8–10).

The variations in the numbers of reports in different countries depend on date of marketing, and place of nimesulide on market, i.e., number of users, and indications. In decreasing order the number of events reported were Italy (782 reports), followed by Spain (171), France (165), Belgium (152), Finland (146), Ireland (121), Portugal (90), Turkey (88) and Switzerland (71) in the period up to June 2004.

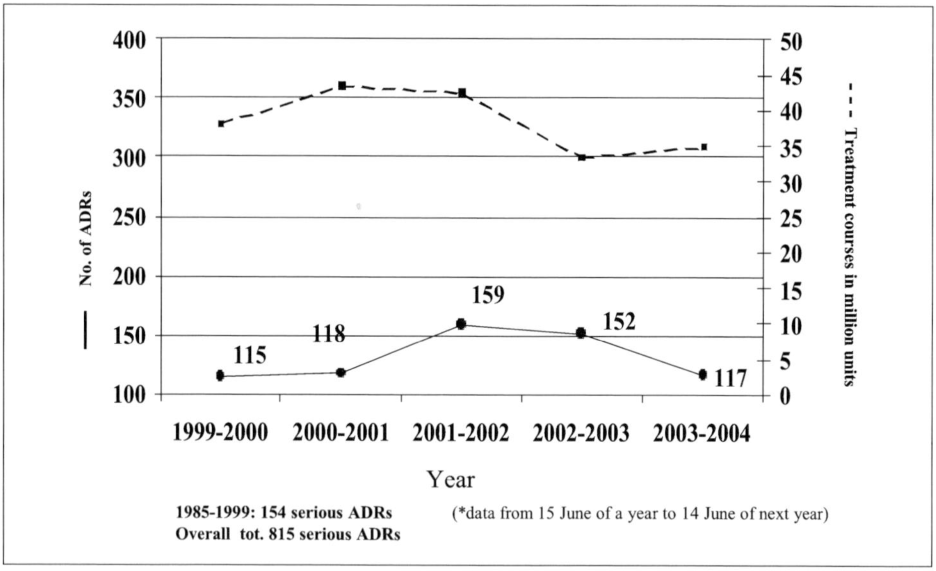

Figure 6

Serious adverse reactions attributed to nimesulide worldwide in comparison with treatment courses in the 5-year period until mid-2004.

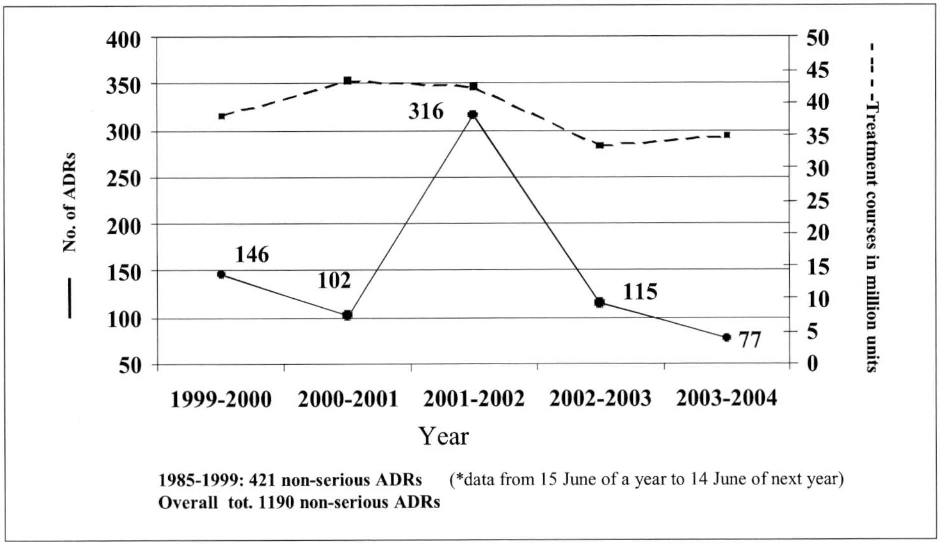

Figure 7
Non-serious ADRs from nimesulide over the last 5 years in comparison with treatment courses in the 5-year period until mid-2004.

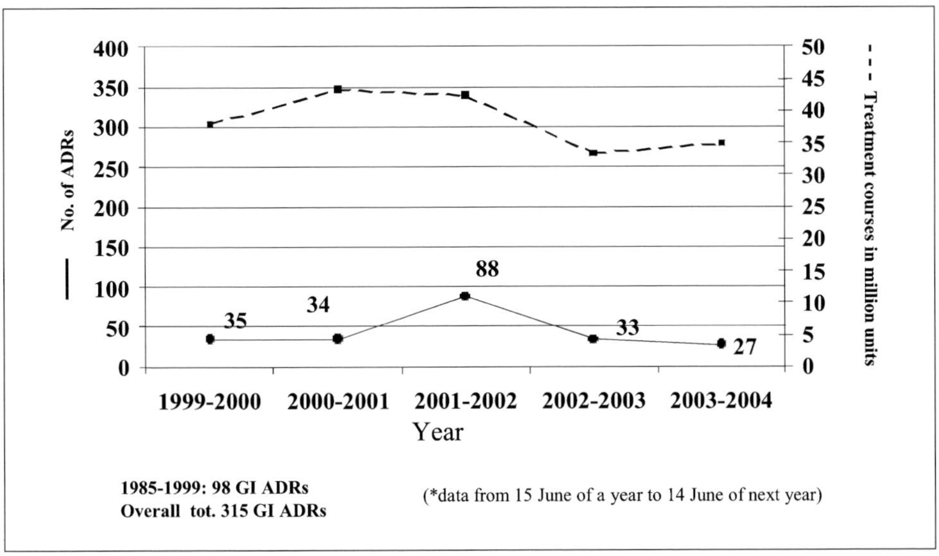

Figure 8
Total number of GI ADRs from nimesulide over the last 5 years in comparison with treatment courses in the 5-year period until mid-2004.

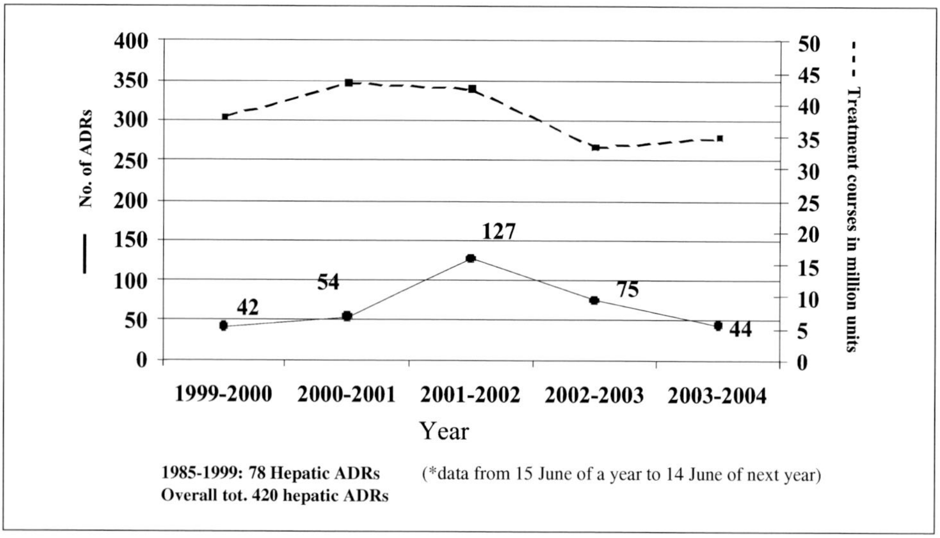

Figure 9

Total number of hepatic ADRs from nimesulide over the last 5 years in comparison with treatment courses in the 5-year period until mid-2004.

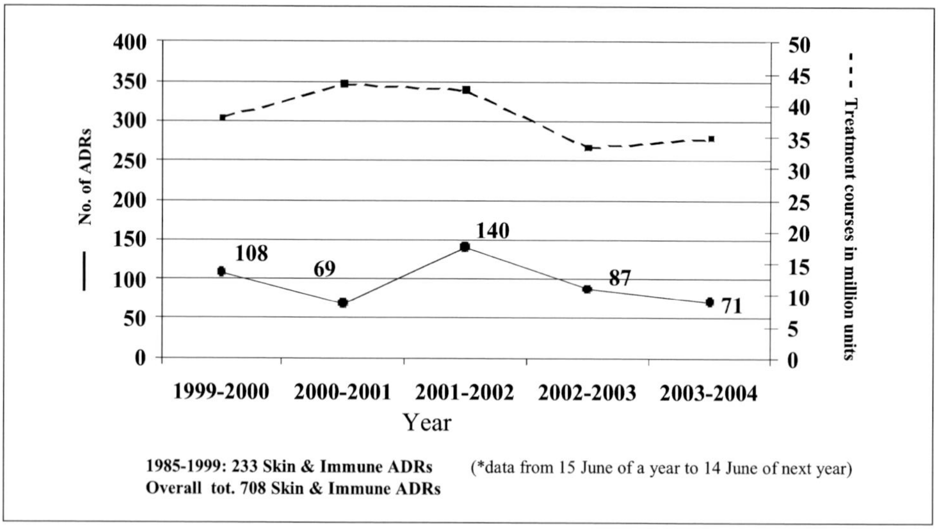

Figure 10

Total number of skin and immune ADRs from nimesulide over the last 5 years in comparison with treatment courses in the 5-year period until mid-2004.

There are significant differences in ADRs patient age by country. These variations may be due to the ages of patients and may, in part, reflect differences in the populations using the drug. For instance, greater use in paediatric patients that is evident in countries like Brazil will obviously influence the average age of patients being reported. Indication and duration of treatment should differ accordingly, which contributes to the different adverse reaction profiles between countries.

Reporting patterns show large variations between countries (Fig. 5). Whereas GI reports both serious and non-serious represent about 20% (range 13–32%) of all events across countries, there are major differences between countries for instance for hepatobiliary reactions and investigations (mostly attributed to elevated plasma levels of liver enzymes), which represent 69% of all reports in Finland, 68% in Israel but only 6% in Turkey or Italy. In contrast, skin reactions represent 60% in Italy, 29% of reactions in Greece, compared with only 11% in Belgium or Finland and none in Israel. These differences could be related to different patient susceptibilities related to genetic or cultural factors, or to different indications and usage patterns.

Analysing these patterns can help identify some of the origins of the occurrence of peaks in reports (notoriety). For instance for nimesulide, three main events occurred related to suspicions of hepatotoxicity: the publication of a case series in Belgium in 1998 [12], the suspension of the drug from the Israeli market in 1999 for a few months [13, 14], other publications from Ireland and Spain since 1999 [13, 15–22], and the temporary suspension of the drug in 2002 in Finland and Spain [23]. It is interesting to see that these instances were preceded or accompanied by the reporting of older, sometimes even undated reactions, some as much as 10 years old.

A pattern of “spiking” of reports that appears in total numbers of ADRs (Figs 3 and 4) and notably in non-serious ADRs (Fig. 7) is also evident with GI, hepatic and skin reactions though to a slightly lesser extent (Figs 7–9, respectively), with a peak during the period of the second semester 2001–first semester 2002. This coincided with publicity and alerts by drug regulatory authorities especially concerning hepatoxicity and publication of a considerable number of accumulated reports, some extending over the past decade. Hepatotoxicity is a feature common to many NSAIDs [24–29]. The hepatic risk associated with nimesulide was thoroughly investigated by the European Medicines Evaluation Agency in 2002–2003, and nimesulide was found not to carry a greater risk than other NSAIDs [30].

The patterns of ADRs and factors underlying their development in those reports up until 1999 have been analysed in published reports [10, 31]. With particular reference to hepatic reactions attributed to nimesulide it was found that in many of the case reports there was evidence of concomitant intake of many drugs that are known to be hepatotoxic including antibiotics, paracetamol, certain NSAIDs (diclofenac, sulindac), statins and oestrogenic steroids [10].

Causality assessment and quality of information

Recently, an analysis of ADR reports has been performed according to a system where the reports were graded according to the quality of information provided in the report and information on case details including information on those factors may influence the development of the hepatic events [32]. Data on ADRs in the hepatic, digestive, renal and skin body systems were analysed with respect to (a) likelihood of association graded on the basis of **A** (most likely), **B** (possible) or **O** (zero or unlikely), (b) age and (c) gender based on the reported classification of serious and non-serious cases.

The reports of all events were subject to quality assessment (termed "discriminant analysis") in which case reports were graded **α** where there is adequate information to be confident about the report having a reasonable degree of reliability, **β** where there is information provided that enables some association with nimesulide, but where there is some information or data missing, and **O** (zero) where the report is so poor or without substantial information to enable confidence to be ascribed to the accuracy of the report or the information provided therein. The cases of hepatic events were analysed in depth to establish what confounding factors were evident that may have influenced the development of the reaction(s) in this body system.

Slightly less than half of the total numbers of serious reports were given a **α**-rating and slightly less were **β** rated, while about 5–10% of serious cases can be considered to be of zero quality. The ratings of hepatic events are more variable since these are predominantly in the **β** category.

It seems possible to separate the serious case reports and ascribe credibility to about half of these. The evaluation of the hepatic risk associated with nimesulide showed that this drug does not carry a greater risk than other NSAIDs [30].

Nimesulide safety from epidemiological and population studies

Gastrointestinal adverse reactions

GI adverse reactions are certainly the most frequent reactions related to NSAIDs. In most cases they are mild and reversible upon cessation of the drug, but sometimes they can be serious and lead to patient deaths. Several studies have been published on severe upper GI complications, including upper GI bleeding associated with NSAIDs and they have shown wide differences in the risk associated to single drugs [33–40].

Epidemiological studies have been reviewed recently and the data have been pooled to give a more definitive estimate of risks [36]. In this research, case-control or cohort studies on non-aspirin NSAIDs have been selected. They included data on bleeding, perforation or other serious upper GI tract events resulting in hospital admission or referral to a specialist, and had the possibility to calculate relative risk. The authors identified 852 papers using Medline but only 18 original articles were included in the meta-analysis following the inclusion criteria. Ibuprofen was associated with the lowest risk (RR 1.9; CI[2] = 1.6–2.2), especially at dose <2,400 mg, followed by diclofenac, sulindac, naproxen sodium, indomethacin, ketoprofen and piroxicam. Nimesulide was not considered in this study, since it was not included in almost all the selected articles. Many of the epidemiological studies that include nimesulide come from Italy.

Among the first studies on nimesulide was that published by Traversa and co-workers [37]. This was a case-control study including 600 outpatients, selected in an Italian hospital, with a confirmed endoscopic diagnosis of ulcer and erosion and 6,000 community controls matched by age and sex. The prescription history was retrieved through a computerised prescription monitoring system. The drug with the highest risk was ketorolac (adjusted odd ratio, OR = 4.2; CI = 1.9–9.4). Nimesulide (OR = 1.2; CI = 0.6–2.5) was not different from the other NSAIDs. The work, as explained by the authors, presented some methodological bias: among these the impossibility to know the days of therapy within the month in which a prescription was filled. This probably led to an underestimation of odds ratios.

A large epidemiological study by Menniti-Ippolito and co-workers [38] was based on the drug prescription database of the Umbria region in Italy. A cohort and a nested case-control study were carried out. All residents who had been given at least one NSAID prescription in 1993 and 1994 were identified and rate ratios of hospitalisation for gastroduodenal ulcer with or without complications in the current, recent or past period according to exposure to different NSAIDs were estimated. The highest rate ratio (RR), adjusted for age and sex, for lesions of any severity was related to piroxicam (RR = 4.6; CI = 3.2–6.7). Nimesulide had the lowest rate ratio (RR = 2.0; CI = 1.1–3.4); this value was not far from the value for naproxen or diclofenac even when only cases with haemorrhage or perforation were considered (RR = 2.5; CI = 1.2–5.3). When the exposure period was restricted to 15 days from the date of prescription the rate ratios rose for all NSAIDs but not for nimesulide (RR = 2.1; CI = 0.8–5.1). The number of cases however was low (only five cases for nimesulide).

A larger study by Garcia Rodriguez and co-workers [39] used the regional computerised records of hospitalisation and drug prescriptions; the authors selected 1,505 cases of upper GI bleeding and evaluated the exposures to NSAIDs, calcium antagonists and other antihypertensive drugs. 20,000 controls were randomly selected in the reference population.

---

[2] *CI denotes 95% Confidence Interval.*

The annual background incidence rate of hospitalisation for upper GI bleeding was estimated as 1 per 1,000 persons. The relative risk associated for any NSAIDs was 4.4 (CI = 3.7–5.3). Ibuprofen was the NSAID with the lowest risk (RR = 2.1; CI = 0.6–7.1), followed by diclofenac (RR = 2.7; CI = 1.5–4.8) and ketoprofen (RR = 3.2; CI = 0.9–11.9). Piroxicam (RR = 9.5; CI = 6.5–13.8) and ketorolac (RR = 24.7; CI = 9.6–63.5) were associated to the highest risks; nimesulide had a relative risk of 4.4 (CI = 2.5–7.7), which overlapped that of those drugs with the lowest risk. Age, sex and previous history of bleeding were identified as risk factors, whereas duration of use seems not to be relevant, as observed in other studies [36]. There were relatively few cases included in the study and there were a wide range of confidence intervals. Furthermore it has been noted that the random nature of controls who had no prior exposure to the drug lead to a great variability of odds ratio. Another possible bias related to the methodology was the evaluation of drug exposures through the regional database of prescriptions. A few of the NSAIDs are available as non-prescription or "over-the-counter" (OTC) drugs. However, it should be considered that in some cases a prescription-only NSAID is given without a prescription (so called "under the counter drugs"). This may lead to an underestimation of the risk.

Recently, Laporte and co-workers [40] performed a large multicentre population-based case-control study. All patients older than 18 years admitted to the 18 participating hospitals in Italy and Spain with haematemesis or melaena and with a primary diagnosis of acute upper GI bleeding (UGIB) were considered for inclusion. In total 12,804 potential cases were identified: 4,309 patients fulfilled the primary inclusion criteria. Up to three hospital controls were randomly selected for each case, matched according to centre, time from admission, sex and age. The controls were patients admitted with acute clinical disorders thought to be unrelated to the intake of analgesics or NSAIDs. Secondary exclusion criteria were the same for cases and controls and led to 2,813 cases and 7,193 controls for analysis. Specially trained monitors interviewed with a structured questionnaire the potential cases and controls as soon as possible after admission. The interview covered general demographic and habit information, clinical history, and drug history, on a daily basis for the 21 days before admission (and on general basis from 22 days to 3 months).

The annual incidence of UGIB was 401 per million inhabitants older than 15 years. Age, sex and previous history of GI bleeding were confirmed as risk factors. Table 2 shows the odds ratio estimates for the most commonly used NSAIDs in the week before. The odds ratio estimates associated with NSAIDs ranged from 1.4 (CI = 0.6–3.3) for aceclofenac, to 24.7 (CI = 8.0–77.0) for ketorolac. 48 cases and 46 controls were exposed to nimesulide in the week before the index day: nimesulide had an odds ratio of 3.2 (CI = 1.9–5.6) similar to ibuprofen and diclofenac and much lower than other classical NSAIDs like aspirin, naproxen or piroxicam. Rofecoxib was associated to a high-risk estimate even if the number

Table 2 – Upper gastrointestinal bleeding odds ratios for single NSAIDs taken in the week before hospitalization. Data from Laporte et al. (2004) [40]

Drug	No. of Cases/ No. of Controls	Odds Ratios (95% CI)
Aceclofenac	15/30	1.4 (0.6–3.3)
Aspirin	591/403	8.0 (6.7–9.6)
Dexketoprofen	16/8	4.9 (1.7–13.9)
Diclofenac	100/98	3.7 (2.6–5.5)
Ibuprofen	60/58	3.1 (2.0–4.9)
Indomethacin	29/16	10.0 (4.4–22.6)
Ketoprofen	16/9	10.0 (3.9–25.8)
Ketorolac	33/6	24.7 (8.0–77.0)
Meloxicam	14/11	5.7 (2.2–15.0)
Naproxen	52/27	10.0 (5.7–17.6)
Nimesulide	48/46	3.2 (1.9–5.6)
Piroxicam	119/40	15.5 (10.0–24.2)
Rofecoxib	10/10	7.2 (2.4–23.0)
Other NSAIDs	34/33	3.6 (2.0–6.8)

95% CI = 95% Confidence Interval.

of exposed were low both in cases and controls, leading to a wide confidence interval.

For all drugs, the risk of GI bleeding increased with dose. Nimesulide had an odds ratio of 3.0 (CI = 1.6–5.5) at a dose lower than 200 mg/day compared to an odds ratio of 7.0 (CI = 2.2–22.7) at a dose equal or greater than 200 mg/day.

The following are the strengths of this study:

a) It was population-based, and this enabled the estimate of the public health impact of NSAID-induced UGIB in terms of incidence and attributable risk.
b) The sample size calculations were based on the estimated prevalence of use of the drugs of interest, and the study had therefore enough statistical power to estimate individual risks associated with a number of drugs.
c) The accuracy in selection of both cases and controls. Information on the clinical course leading to hospital admission was carefully examined, and the index day was established blindly with respect to drug exposures, thus avoiding exposure misclassification.
d) Detailed information on the patients' medical history was carefully collected by specially trained monitors, and this enabled controlling for confounding factors.

e) The numbers of drug exposures were evaluated considering actual use as referred by the patients, while other studies have considered a surrogate indicator of exposure, i.e., data obtained by prescription databases.
f) A conditional model was used for the estimates of risk, considering also for potential confounders.

Hepatic reactions

Hepatic reactions are well known with many NSAIDs. They are generally idiosyncratic reactions related to an individual susceptibility [41–46]. Hepatic reactions are rare (typically 1–5 among 100,000 exposed [41, 42]) and can widely differ from transient and asymptomatic increase of hepatic enzymes to serious cases of liver dysfunctions and hepatic injury. The molecular mechanisms underlying this toxicity are as yet unclear (see later section on "Mechanisms of toxicity").

Nimesulide has also been associated with serious hepatic reactions including acute hepatitis and fulminant liver failure [12–32, 44–46]. An increased risk of hepatotoxicity with nimesulide was suggested by spontaneous reports in Finland [32] and Spain [47]. Data from spontaneous reporting in larger patient populations in Italy and France did not confirm this signal [48, 49]. However, spontaneous reporting data cannot give true incidence rates due to the lack of a definite denominator (number of exposed) and because of underreporting. Furthermore, the causality assessment for hepatic reactions is often difficult to evaluate due to the presence of confounding factors like concurrent illnesses, alcohol intake or other concomitant hepatotoxic substances.

Recently, a large cohort study was reported by Traversa and co-workers [50]. This retrospective cohort study considered all the patients receiving a NSAID prescription in the years 1997–2001 in the Umbria region in Italy. Potential cases were all admissions to hospital for acute non-viral hepatitis. A nested case-control study was also carried out to control for potential confounders. 176 cases of hepatopathy occurring during current use of a NSAID were included in the final analysis (incidence 29.8 per 100,000 person years). The risk of hepatopathy among patients exposed to NSAIDs was small. Compared with the incidence for past use an adjusted rate ratio of 1.4 (CI 1.0–2.1) has been estimated. No fulminant hepatitis or liver injury related deaths were observed.

The risk of hepatopathy among current users of nimesulide was slightly higher than that of other NSAIDs (rate ratio, RR = 1.3; CI = 0.7–2.3) even if not significant. When cases with ALT increase above 5 × UNL were included in the analysis the rate ratio among users of nimesulide was higher (RR = 1.9; CI = 1.1–3.8). One possible bias of the study is that drug exposure was estimated through a prescription database, with no information on indications or dosage. However, the study was very large, covering the prescriptions made in a population of around 835,000 inhabitants for a 5 year period. The increase of risk for hepatotoxicity associated with nimesulide was low. Other studies report that the risk for nimesulide is comparable to that of most other currently used NSAIDs [42, 51]. There are however exceptions, such as sulindac, where the incidence is approximately 10-fold higher [42]. On the other hand only a few cases have been reported for coxibs, which have been introduced only recently [52, 53].

As for the other non-steroidals, the molecular mechanism underlying the hepatotoxicity related to nimesulide remains unclear. However, the absolute risk of developing hepatic adverse effects including fulminant hepatitis is very low.

Cutaneous and allergic reactions

Cutaneous reactions are often reported during a treatment with NSAIDs [54]. Most of them are mild and include pruritus, urticaria and morbilliform rashes. More severe reactions including Stevens-Johnson syndrome and toxic epidermal necrolysis (Lyell's syndrome) may occur, even if the absolute risk is low [55]. An analysis made in three international independent databases showed that oxicams have the highest risk for developing these serious reactions compared to the other more often used NSAIDs [56]. Recently, some cases of toxic epidermal necrolysis related to celecoxib have been reported [57–59].

Data from spontaneous reporting systems indicate that cutaneous reactions also occur with for nimesulide, however serious cutaneous reactions are few [60]. Some cases of Stevens-Johnson syndrome, toxic epidermal necrolysis [29, 60] and fixed eruptions [60] have been related to nimesulide, even if the causality assessment for the involved reports is sometime confounded by the presence of concomitant drugs or other predisposing conditions.

Many cutaneous adverse effects attributed to NSAIDs are pseudo-allergic reactions like urticaria, angioedema, asthma and anaphylaxis, often related to pyrazolone derivatives, aspirin or diclofenac [61, 62]. Prevalence of urticaria and/or angioedema by NSAIDs has been estimated in 0.1–0.3% of exposed patients [63]. Cross-sensitivity with other non-steroidals is often present [64]. Pseudo-allergic cutaneous reactions are more frequent in COX-1 selective drugs. This fact suggests that COX-1 inhibition has a relevant pathogenetic role in these reactions. However, anaphylactoid reactions have been observed with coxibs [67–69] indicating that COX-2 inhibition may be associated with this condition. The protective effect by leukotriene receptors inhibitors in cutaneous reactions related to NSAIDs maybe related to the overproduction of peptide-leukotrienes caused by COX-2 inhibition and diversion of arachidonic acid through the 5-lipoxygenase pathway [65, 66].

The relatively favourable tolerability of nimesulide in patients with NSAID intolerance has been demonstrated in a large number of clinical studies [70–73]. Furthermore, it has been shown to have antihistamine activity and this might confer a potential protective advantage in allergic conditions [74].

Renal adverse events

Renal toxicity is another known adverse effect to NSAIDs [75]. Renal prostaglandin inhibition by these drugs may alter renal function, especially in elderly patients, where low effective circulating volume is often present [75], or in patients with pre-existing renal diseases. However, the large spectrum of NSAIDs induced nephropathies like tubular, interstitial or tubulointerstitial nephritis, nephrotic syndromes may be a response to a state of hypersensitivity against these drugs [76]. The risk of renal adverse effect has been reported to increase with the number of NSAIDs. A recent study on the French pharmacovigilance database showed that in comparison with reports that did not mention any use of NSAIDs, the odds ratios for acute renal failure associated with the use of a single NSAID and two or more NSAIDs were respectively 3.2 (95% CI: 2.5–4.1) and 4.8 (95% CI: 2.6–8.8) [77].

Data from spontaneous reporting system [31, 32, 78] and from single published case reports [79, 80] suggested that the use of nimesulide could be associated with an increased risk of nephrotoxicity. However epidemiological studies trying to demonstrate and quantify this risk are lacking.

Cardiovascular events

Epidemiological evidence suggests that use of NSAIDs may be associated with increased risk of cardiovascular events including congestive heart failure, increased hypertension and myocardial infarction [81–83]. There is also extensive clinico-pharmacological data suggesting that NSAIDs may antagonise the effects of diuretics, beta-blockers, angiotensin-converting enzyme (ACE) inhibitors and other drugs used to treat cardio- and cerebro-vascular conditions [2, 83]. If they affect platelet aggregation or other components of blood coagulation it is possible they could exacerbate the actions of anticoagulants. Some NSAIDs also affect the actions of warfarin and other cardiovascular drugs by affecting their pharmacokinetics [2].

The sudden withdrawal of rofecoxib (Vioxx®; Merck) from the market worldwide in September 2004 followed substantial evidence that this drug is associated with a markedly greater risk of myocardial infarction and other serious vascular conditions and increased fatalities [84]. There followed evidence has showed this increased risk of serious cardiovascular adverse events was evident with celecoxib (Celebrex®; Pfizer) especially at high dose (400 and 600 mg daily) and also that this may be a problem with valdecoxib [85–88]. The evidence was extensively reviewed at a special joint meeting of the US Food and Drug Administration's (FDA) Arthritis Advisory and Drug Safety and Risk Management Committees on 16–18 February 2005 [86]. They concluded there is an increased risk of cardiovascular complications from all these coxibs marketed in the USA. This may be more pronounced in the elderly [89]. Speculation has been aroused that this may extend to the other coxibs and possibly COX-2 selective drugs in general (e.g., meloxicam) [87–89]. The evidence including that reported to the FDA's meeting suggests that naproxen, diclofenac and possibly ibuprofen may not have the high risk of serious cardiovascular events and high mortality from these conditions as seen with the three coxibs [86–88].

In addition to increasing risks of cardiovascular events, rofecoxib has been found to increase systolic blood pressure and this while evident with celecoxib is probably of lower severity [86–88, 90, 91]. The consequences of the withdrawal of rofecoxib from the market have sparked an extensive ethical debate [84, 87, 88]. Whether some of the newer coxibs (e.g., lumaricoxib) have similar risks of cardiovascular adverse events has yet to be resolved but the indications are (especially from the celecoxib data) that this might be a class effect [86–88].

The cardiovascular risk of rofecoxib and possibly other coxibs has been suggested to be related to their inhibitory effects on COX-2 and possibly creating an imbalance between partial inhibition of prostacyclin synthesis without affecting the production of the pro-aggregatory, thromboxane A_2 [92]. In contrast to the expected sparing of COX-1 induced thromboxane production, rofecoxib has been found to be a marked inhibitor of platelet aggregation [93], so leading to a puzzling situation because this effect might be regarded as a desirable property for reducing cardiovascular risk. Other studies suggest that endothelial cell proliferation and apoptosis may be reduced by celecoxib and not by rofecoxib [94]. It has also been suggested that rofecoxib may form a reactive metabolite leading to oxidative damage of low density lipoprotein (LDL) and phospholipids so initiating a cycle of cellular injury especially in atheromatous areas of the vasculature [86]. Thus, the situation regarding understanding the basis of the increased risk of cardiovascular events from the coxibs is unresolved.

Cardiovascular events associated with nimesulide

In view of these events with rofecoxib (which has a long plasma half life) as well as celecoxib (which has complex liver metabolism) and with both these drugs clearly differing in chemistry and pharmacology from nimesulide, it is important to examine the data on the cardiovascular reactions from nimesulide and assess its risk. From a pharmacological viewpoint nimesulide only slightly diminished thromboxane and prostacyclin generation [95, 96]. These properties may confer a degree of selectivity of the drug on components of prostanoid metabolism that are relevant to control of platelet aggregation. Thus, sparing of prostacyclin production as well as arachidonic acid induced thromboxane production, while controlling agonist-induced platelet aggregation may be beneficial in enabling prostacyclin product to persist and at the same time reducing the apparent risk from platelet stickiness especially in at-risk cardiovascular and rheumatic patients that have greater platelet adhesive properties than normal subjects [89, 92–95].

The clinical trials and ADR monitoring of nimesulide at Helsinn has revealed relatively few reports of patients with serious cardiovascular events. Thus, of 4,815 subjects that received nimesulide in 40 clinical trials, for which individual data could be obtained, 28 non-serious adverse events in the cardiac disorders system organ class and one case of phlebitis (a vascular disorder) were reported. Hypertension, oedema, palpitation and tachycardia were the most frequent cardiac events occurring in the clinical trials.

In an observational cohort study performed in Eire, no serious or non-serious adverse events in the cardiac and vascular disorders system organ classes were reported in 3,807 patients that had received nimesulide (Helsinn; data on file).

Post-marketing data available indicate that there have been relatively few cardiovascular adverse reactions (classified as both cardiac disorders and vascular disorders) reported to Helsinn Healthcare from all the sources, from August 1985 until 30 September 2004.

These serious cases include principally atrial fibrillation and/or cardiac failure that occurred in most patients that had pre-existing cardiovascular diseases. All patients (but one, whose outcome is unknown) recovered. The non-serious cardiac cases were mostly tachycardia and palpitations and all recovered.

The case reports belonging to the vascular reactions include cases of purpura and haemorrhage, hypertension, hypotension and vasculitis. Two fatal cases have been reported. One was a case of haemorrhagic shock (which was considered to be unrelated to nimesulide intake) and another of haemorrhage and blood dyscrasia (which was considered to be unassessable due to the very poor information available about the case).

The patient's gender or age (i.e., if ≥65 years) did not appear to be a relevant risk factor, and neither did the duration of treatment.

In summary, of the serious cardiovascular ADRs reported only 16 of these can be considered clinically relevant. This is considered to indicate a low risk of cardiovascular events especially in relation to the drug having been available for some two decades with more than 415 million of treatment courses available.

Meta-analysis and systematic reviews of adverse reactions from clinical trials

A meta-analysis was undertaken recently to evaluate the overall occurrence of adverse reactions in relation to drug efficacy that were reported in clinical trials attributed to NSAIDs including nimesulide in patients with rheumatoid and osteoarthritis [97]. The studies had been reported in Chinese journals comprising safety and efficacy trials in 19 articles published from 1990–2001. The total number of patients in the ADR analysis was relatively small (totalling 2,925) and likewise that in the efficacy studies (1,732). The rates for efficacy were (expressed as percentage with CI) for nimesulide, 79.8% (75.7–84.0), nabumetone, 66.7% (61.9–71.4), meloxicam, 68.4% (59.2–77.6), naproxen, 64.5% (59.8–69.1), ibuprofen 77.2% (70.7–83.8), diclofenac, 77.1% (69.2–85.0), and oxaprozin, 65.8% (59.5–72.0), respectively. Thus, it can be concluded that all the NSAIDs were about equally effective.

The rates of ADRs (95% CI) were for nimesulide 20.2% (16.0–24.3), nabumetone 16.3% (12.5–20.0), meloxicam 10.2% (4.2–16.2), naproxen 29.2% (14.7–18.8), diclofenac 19.3% (11.9–26.7) and oxaprozin, 12.7% (8.9–1.7), respectively.

These studies show considerable heterogeneity of the ADRs in relation to efficacy. On the basis of percentage data nimesulide would appear to be effective compared with other NSAIDs, with intermediate ratings for ADRs.

A systematic review of the GI toxicity induced by non-steroidal anti-inflammatory drugs was published recently by Hooper and co-workers [98]. In this study a total of 51 randomised controlled trials (28,178 participants) were reviewed in which the "COX-2 selectives" (etodolac, meloxicam, nabumetone and nimesulide) with non-selective NSAIDs ("COX-1 or conventional"). The results indicated that the COX-2 selectives significantly reduce the risk of symptomatic ulcers and probably the risk of serious GI complications.

The favourable GI safety profile of nimesulide is in agreement with what have been published in many other clinical trials [99–102].

Gastrointestinal tolerance of nimesulide compared with other NSAIDs: Clinical studies

Introduction

NSAIDs are the most prescribed of the antirheumatic drugs and some are now widely available as OTC medicines. However, there is continuing concern about their GI adverse effects [33–40, 98] that principally affects the gastric [1–3, 103, 104] and small intestinal [105, 106] mucosa. The pathogenesis of the intestinal damage is complex [104–109]. Gone are the days when theory dictated that inhibition of cyclooxygenase (COX) with decreases in mucosal prostaglandins accounted for both gastric and intestinal damage [104–109]. Rather the damage is initiated by an interaction between inhibition of the two COX enzymes and topical effects (defined as a COX independent action that requires mucosal contact of the drug from the luminal side) [106]. Hence dual inhibition of COX-1 and COX-2 causes GI damage and this damage is made worse by the topical effect [106, 109]. More recently it has been shown that NSAID-enteropathy can be caused by inhibition of COX-2 and the topical effect (without a concomitant decrease in mucosal prostaglandins) and similar to the above this damage is made worse by inhibition of COX-1 [109, 110].

The biochemical consequences of these actions of conventional NSAIDs are incompletely understood and speculative. Nevertheless, and contrary to the COX dogma, the picture is emerging that COX-2 has a much more important role in maintaining intestinal integrity that was previously recognised [109–111]. The precise nature of this action remains uncertain, but in the small bowel it may involve maintenance of oral tolerance [111]. One of the main consequences of COX-1 inhibition is the impaired regulation of microvascular blood flow leading to a state of relative hypoxia at times of increased oxygen need [106]. The topical effect encompasses two processes. Firstly there is a NSAID interaction with surface membrane phospholipids, and secondly there is uncoupling of mitochondrial oxidative phosphorylation within the absorptive cells [106, 110, 112]. Either of these effects can cause cellular damage and the uncoupling, in particular, will increase oxygen requirements. If so, NSAID-induced GI damage can be viewed as a mucosal weakening caused by the combination of COX-2 inhibition and the relative hypoxia caused by COX-1 inhibition and/or the topical effect. The mucosal is then further damaged by luminal aggressive factors gaining access to the mucosa [106]. This framework is supported by much experimental evidence and also explains the fact that selective COX-1 inhibition or absence by itself does not lead to damage (there are at least two other mechanisms for regulating microvascular blood flow), why the topical effect can disrupt intestinal integrity with increased permeability and low grade inflammation and why selective COX-2 absence or inhibition in experimental animals is associated with some ileo-caecal inflammation and ulcers [107].

Types of gastrointestinal investigations

Endoscopic observation of GI mucosal injury is among the most direct evidence for NSAID-associated injury in the GI tract. These side effects are best described in the terms of the study type and location of the damage. Accordingly, the upper GI side effects are classified as:

1) **Short-term (1–2 weeks) endoscopy studies in volunteers:** In general, all conventional acidic NSAIDs cause gastric damage when taken short-term. The damage is usually expressed according to the Lanza score, but there are numerous variations of this scale. There are also good visual analogue scales and homemade damage scales that lump together unrelated composite outcome measures in order to exaggerate differences between drugs. The mechanism of the short-term gastric damage in man is controversial, but against a background of COX-1 and COX-2 inhibition there is a significant correlation be-

tween the damage and the acidity of the drug as shown in Table 3 [106, 110, 112]. Damage in these studies predicts intolerance when taken long-term while tolerance does not necessarily predict longer-term safety.

2) **Long-term (3–6 months) endoscopy studies in patients:** These studies all show an ulcer prevalence of 10–30% with the various conventional NSAIDs [113]. There is however no clear pecking order here in their ability to cause damage. The importance of these ulcers is controversial. Some write the damage off as "endoscopy ulcers" that are unlikely to cause any significant problems [114] while others suggest that this damage should be viewed with the same gravitas as *Helicobacter pylori* associated peptic ulcers [113] as the natural history of the two may not differ significantly.
3) **Serious outcome studies:** Most of these studies are population based and assess the prevalence of life threatening complication such as bleeding and perforation. While some NSAIDs are clearly associated with more frequent serious side effects than others there is some lack of conformity in the pecking order for these drugs as shown in Table 4. If so it is difficult to understand why such emphasis is placed on these studies. Sometimes they indeed give misleading information due to channelling of high-risk patients to safer alternatives or are interpreted in strange ways [115–117]. More recently the serious outcomes have been assessed by prospective studies [118–120], but even these have their critics as the data has been extrapolated from a group of patients that is at low risk for complications to the high risk population (age over 65, previous ulcer history, concomitant use of steroid or anticoagulants, etc.) [121]. A recent and welcome development is to specifically study the high-risk patients [122].

The lower gastrointestinal side effects are similarly assessed as:

1) **Short-term studies (ranging from single doses to 1–4 weeks ingestion) in volunteers:** All conventional NSAIDs increase small intestinal permeability in 80–90% of subjects (within 24 h) [123–129] and this leads to intestinal inflammation within 10 days of ingestion [102]. Equally important, while all conventional NSAIDs are associated with similar changes in intestinal permeability, the prevalence of these changes is maintained long-term [130]. Furthermore, unlike the short-term endoscopy studies, safety short-term for the small intestine translates into long-term tolerability [104, 112, 130, 131].
   Misoprostol reduces but does not abolish the permeability changes induced by NSAIDs in the short-term [120, 125] while H_2 receptor antagonists have no significant effect [132]. In separate studies it is clear that celecoxib does not cause macroscopic small bowel when given for 2 weeks, as assessed by wireless capsule enteroscopy, while naproxen + PPI does [133] (55% of subjects which is quite comparable with the prevalence of small bowel inflammation when studied with faecal markers of inflammation [102]).

Table 3 – Anti-inflammatory drugs, pKa and gastric damage [106, 112]

Drug	drug dose (mg)	Number	Duration of ingestion	pKa	Lanza score (stomach)
Aspirin				3.5	
	3900	5	7		3.4
buffered	3900	5			4
enteric coated	3900	5			0.6
	2600	15	7/14		3.5/3.8
	3900	10	1		2.1
	2600	5	7		2.8
	2600	10	7		3.6
	3900	30	6		3.5
	3900	31	7		3.6
	2600	14	15		3.4
	3900	30	7		3.1
	3600	17	7		3.47
	3600	19	14		2.68
	1200	8	SD		2.88
	2600	21	3		1.4
Naproxen				4.2	
	750	20	7		1.6
	500	15	7/14		1.1/2.3
	1000	12	7		1.92
	1000	30	7		2.25
	1000	30	7		1.8
	1000	19	14		1.5
	1000	36	14		2.4
	1000	15	5		1.53
	1000	12	7		2.25
enteric coated	1000	12	7		1.08
	1000	16	5		2.13
Flurbiprofen				4.2	
	100	10	7		0.8
	150	10	7		1.2
	200	10	7		2.2
	300	10	7		1.5
	400	20	7		2.5
	500	10	7		2.6

Table 3 – (continued)

Drug	drug dose (mg)	Number	Duration of ingestion	pKa	Lanza score (stomach)
Indomethacin				4.5	
capsules	150	20	7		1.8
	200	12	7		2.25
Sulindac				4.7	
	300	15	7		0.93/0.93
Etodolac				4.7	
	600	12	7		0.17
	1000	12	7		0.25
Ibuprofen				5.2	
	2400	10	1		0.2
	2400	15	7		1.3
	2400	12	7		0.92
	3200	30	7		1.8
	2400	51	7		1.88
Fenbufen				5.7	
	1000	20	7		0.7
capsules	1000	20	7		0.6
Ketoprofen				5.9	
	75	24	7		2.38
Nimesulide				6.4	
	200	35	14		0.7
Flosulide				7.0	
	40	19	14		0.4
Paracetamol				7.0	
	1500	12	14		0.1
	3900	15	7		0.1
	4000	24	7		0.25
Placebo		24	14	7.0	0.42
		51	7		0.24
Dipyrone				8.5	
	1500	12	14		0.4
	1500	12	14		0.25
	3000	12	14		0.92
Rofecoxib				>8.5	
	250	51	7		0.27

Table 4 – Serious outcome toxicity ranking of NSAIDs

Drug/Author	Kaufman [174]	Henry [35]	Langman [176]	Rodrigues [177]	Henry [178]	MacDonald [179]
Aspirin					10	
Azapropazone			1		1	2
Diclofenac	6	7	6	7	9	4
Diflunisal		1		8	7	
Fenbufen						11
Fenprofen	4			5	11	1
Ibuprofen	7	8	7	9	12	8
Indomethacin	5	3	4	4	5	9
Ketoprofen	1	2	2	2	2	5
Nabumetone						10
Naproxen	3	4	5	3	6	6
Mefenamic acid						7
Piroxicam	2	5	3	1	3	3
Sulindac		6		6	8	
Tolmetin					4	

2) **Long-term (≥3 month ingestion of NSAIDs) cross sectional studies:** NSAID-induced small bowel inflammation (NSAID-enteropathy) is evident in 50–65% of patients, irrespective of the particular NSAID, sex or age [124, 130, 134]. The same drugs that increase intestinal permeability short-term lead to the long-term permeability and inflammatory changes [130]. Half of those affected have discrete small bowel ulcers or erosions on enteroscopy and the other half have haemorrhagic spots [135]. The occult complications of NSAID-enteropathy (evident in most of those with inflammation) include sustained low-grade bleeding and protein loss. In some patients this may contribute to iron deficiency anaemia and hypoalbuminaemia, respectively [136–138]. The complications of NSAID-enteropathy, namely bleeding and protein loss can be reduced by co-administration of sulphasalazine [137], metronidazole [138, 139] or misoprostol [140].
3) **Serious outcomes:** Long-term NSAID ingestion is associated with small bowel perforation [141] (sometimes detected only at autopsy [142]), overt bleeding [143] and "diaphragm" like strictures [144, 145] that may require surgery [145–151].

The main contention is to whether the overall prevalence of the serious complications originating from the small bowel approximates that from the stomach (1–2%

annual incidence of serious outcomes). Detailed analyses of the serious outcome studies associated with NSAIDs show that in the MUCOSA study 95 (40%) and 147 (60%) suspicious complication events were upper and lower GI tract events, respectively [118]. Secondly a re-analysis of VIGOR [119] showed that the relative prevalence of the serious outcomes from gastric and small bowel lesions was 60% and 40%, respectively [152]. An identical conclusion was reached when CLASS was analysed in a similar manner [153].

The small bowel toxicity of NSAIDs has not been considered important as the stomach damage for marketing purposes. The reasons for this may be that many of the NSAID "opinion leaders" are armchair epidemiologists and the complexities of the techniques for assessing the small bowel damage is beyond many of them. However, the prevalence and severity of the effects of NSAIDs on the small bowel now demands in depth investigations to establish if these drugs also affect the intestinal mucosal integrity.

Gastrointestinal studies with nimesulide

Nimesulide has however many properties that are in theory predictive of good GI tolerability including a pKa (6.5) which is close to neutrality [112]. Its selectivity for COX-2 is evident from a standardised selectivity assay (the William Harvey Human Modified Whole Blood Assay) [154]; this method out performs other assay systems [155, 156] as it relates the relative inhibitory effects of the drugs to their levels in serum or plasma. However, the most compelling evidence for selectivity comes from an endoscopic study where it is shown that nimesulide given at therapeutic doses did not inhibit gastric COX-1 significantly as assessed by prostaglandin production rates in gastric biopsies [100] (platelet aggregation and serum thromboxane levels were also unaffected). Rofecoxib has also been found to be without effects on COX-1-derived gastric mucosal PGE_2 production coincident with little gastric mucosal irritancy being observed endoscopically [157]. In contrast, lumiracoxib has been found to decrease gastric COX-1 activity by about 30% [158].

The published studies reviewed here have shown the favourable GI tolerability of nimesulide in human volunteers and patients with arthritis conditions. They show that this drug exhibits low GI mucosal injury when examined in comparison with other NSAIDs using the above-mentioned standard systems for investigating GI injury in short- and long-term studies in both the upper and lower GI tract.

Endoscopy studies

The first study compared the gastric tolerability of nimesulide (100 mg twice a day) and indomethacin (50 mg three times a day) when taken for 12–15 days in patients (n = 16/group) requiring anti-inflammatory analgesics [159]. An unconventional gastric damage score system was used (Grade 0 = normal, Grade 1 = hyperaemia, Grade 2 = hyperaemia and oedema, Grade 4 = erosive gastritis and Grade 5 = ulcer). In the nimesulide group 9 were normal (56%), 4 (25%) had grade 1–3 and 3 (19%) had erosions. Corresponding figures were 2 (12%), 5 (31%) and 8 (50%) for indomethacin with one patient (6%) having ulcers. Nimesulide was significantly better tolerated than indomethacin.

A volunteer study showed that nimesulide (100 mg twice a day for 2 weeks) was associated with Lanza grade 3 (>10 erosions) and 4 (ulcer) in 1 (3%) subject while naproxen (500 mg twice a day for 2 weeks) the corresponding figure was 20 (57%) [159]. The same study showed that nimesulide did not affect prostaglandin E_2 generation in gastric biopsies significantly while naproxen did show that at the doses given nimesulide does not inhibit gastric COX-1.

Marini and Spotti assessed the effect of nimesulide 10 mg and 200 mg taken twice a day for 7 days as compared with placebo in dyspeptic patients (n = 10/group) [160]. A Lanza type of scoring system was used and while there was no significant difference between nimesulide and placebo one patient on high dose nimesulide developed an ulcer (the other 19 were normal or showed hyperaemia and/or oedema).

In a large study of almost 100 patients with osteoarthritis the gastric damage with nimesulide (100 mg twice a day) while not being significantly different from diclofenac (50 mg three times a day) showed fewer numbers of ulcers (nimesulide 2% compared with diclofenac 7%) [161].

It would therefore seem that nimesulide has a relatively good level of gastric tolerance in these short-term endoscopy studies. When compared with other NSAIDs and selective COX-2 inhibitors in Table 3 it is clear that nimesulide is associated with no more damage than other COX-2 selective agents such as etodolac, flusolide and rofecoxib (and celecoxib). However etoricoxib which is a selective acidic (pK_a 4.5), COX-2 inhibitor is associated with 4.4–5.3 as many ulcers compared with placebo [162, 163].

Small bowel studies

Small bowel tolerability studies are not required at present for registration purposes despite the fact that conventional NSAIDs frequently cause clinically significant small bowel damage. Shah et al. showed that nimesulide (100 mg twice a day for 2 weeks) did not increase small intestinal permeability significantly or cause small bowel inflammation while naproxen (500 mg twice a day) did [101]. This is similar to that found with rofecoxib [127] and celecoxib [129] neither of which increases small bowel permeability. In keeping with the suggestion that the physicochemical properties of NSAIDs underlie the permeability changes, meloxicam, another putative COX-2 selective agent, increased intestinal permeability [129].

The clinical implication of this short-term study with nimesulide is uncertain, but to date all conventional NSAIDs that increase small bowel permeability in the short-term are associated with NSAID-enteropathy when taken long-term.

NSAIDs and inflammatory bowel disease

Apart from being implicated in colitis [163, 164] early reports suggested that NSAID may be therapeutically useful in patients with ulcerative colitis [165, 166] but subsequent studies suggest a detrimental effect of NSAIDs [167]. Indeed most clinicians are of the opinion that NSAID may cause relapse of quiescent inflammatory bowel disease (IBD) [167–169]. Those patients who are prone to relapse do so within a few days of receiving NSAID. The British National Formulary indeed cautions against NSAID use in IBD patients [170]. However, many patients with IBD have disease associated arthritis, ankylosing spondylitis, osteoporosis related fractures, etc., that necessitates NSAID administration.

There has been no systemic study on the effect of NSAIDs on the inflammatory process in patients with IBD, the observations on relapse rates being clinical rather than investigative. However a recent study compared the effect of naproxen (500 mg twice a day), nimesulide (100 mg twice a day) and paracetamol (1 g three times a day) on clinical and laboratory indices of intestinal inflammation (faecal calprotectin concentrations) [171–173] when ingested for 4 weeks. Naproxen was associated with clinical relapse in 25% of patients taking the drug and this was associated with concomitant increased intestinal inflammation. The effects of nimesulide and paracetamol did not differ significantly; one (5%) patient in each group had a clinical relapse of disease.

In conclusion, nimesulide has a favourable GI side effect profile in comparison with conventional NSAIDs and although parity with rofecoxib and celecoxib seems likely in this respect there is insufficient data to fully establish this from long-term studies at present.

Clinical aspects of nimesulide-related hepatic reactions from published case reports

As previously reported, the widespread use of NSAIDs has led to the recognition that unwanted GI effects can be common and severe. The risk of liver injury is a generally less relevant problem: the incidence of serious gastroduodenal lesions (bleeding and perforation) among users of NSAIDs [174–179] is almost 10 times higher than liver damage [9].

The possibility of drug-induced liver damage has been described for over a thousand drugs [180–192]. The frequency of clinical hepatotoxicity is difficult to determine, but most of the drugs cause liver injury infrequently, typically 1–5 among 100,000 exposed [41, 42, 189]. A variety of clinical presentations of drug-induced liver damage may be seen, ranging from asymptomatic mild biochemical abnormalities to acute or chronic illnesses mimicking almost every kind of liver disease [190].

Hepatotoxicity is a rare but potentially serious adverse effect of NSAIDs [191]. Borderline elevations in one or more liver function tests (LFTs) have been reported in up to 15% of patients treated with NSAIDs during clinical trials; elevated LFTs usually return to pre-treatment levels during continued treatment with the NSAIDs, but a few patients develop clinically significant liver injury, which requires prompt discontinuation of the NSAIDs for the prevention of worsening of hepatic disease and avoidance of liver failure [191, 192].

The published case reports are considered effective in description of the events [193, 194]. Main data from published case reports regarding nimesulide-related liver damage are summarised here.

The first cases of liver damage related to nimesulide were published in 1997, from Argentina [195], where the drug (not of Helsinn origin) was marketed in 1986, Italy [196], where the drug was marketed in 1985, and Belgium [197], where the drug was marketed in 1996. After then other published case reports followed [12–22, 198, 199, 203, 208, 211, 212, 214, 216, 220]. The reports from Argentina and Uruguay were from nimesulide preparations made locally and which have subsequently been found to contain substantial impurities (KD Rainsford, unpublished studies).

Data are available in 41 sufficiently well-documented cases in 30 females (73.2%) and 11 males (26.8%) and are considered here in detail. Their age covers a range of 17–83 years (mean: 57.2 years), with 17 cases (41.5%) above 65 years without difference between males (range 18–83, mean 59.3 years, above 65 years: 4/11 – 36%) and females (range 17–81, mean 56.5 years, above 65 years: 12/30 – 40%). Daily doses have been for all cases within the recommended range: only one case [19] received more than usual recommended dose (200–400 mg/day, for more than 5 months, and recovered), indicating a not dose-related effect.

The treatment duration up to the event (latency) is known in 40 cases (F 30, M 10): range 3–190 days (males 3–180, females 4–190), usually shorter in men (mean: males 33.9, females 56.0 days, as reported in Table 5). Prior use of nimesulide appeared to shorten the latency, both in males and in females, as shown in the data in 11 out of 41 cases, summarised in Table 5.

There does not appear to be a relationship between blood eosinophilia (>5%) and/or eosinophils presence in liver tissue, as markers of hypersensitivity, and latency (Tab. 5).

Table 5 – Factors associated with hepatic events from nimesulide

FACTOR(s)	No. of Cases	Proportion	Ref. No.
1. Period of treatment duration until the event (latency):			
≤1 week	10/40	6F 4M	14, 15, 17,
1–2 weeks	5/40	3F 2M	19, 20–22,
2–4 weeks	5/40	4F 1M	44, 195–199,
>4 weeks	20/40	17F 3M	203, 206, 208,
		[TOTAL:	211, 212, 214,
		75% F 25% M]	216–220
2. Latency and prior treatment with nimesulide:			
• Previous treatment without ADR (Latency range 4–12 days)	4	3F (4, 11 & 12 days) 1M (7 days)	12, 198*, 15, 219
• Previous treatment with previous ADR		1F (<6 days)	20, 198*
(Latency <6, 11days)	2	1M (11days)	
• No prior exposure to nimesulide (Latency range 21–105 days)	5	4F (21–105 days) 1M (35 days)	12
3. Latency and eosinophilia:			
• Eosinophils in liver sections			
Present (latency 5–60 days)	8/12		14, 15, 17,
Not present (latency 7–105 days)	4/12		19, 20–22,
• Blood eosinophilia			44, 195–199,
Present (latency 4–60 days)	8/28		203, 206, 208,
Not present (latency 5–190 days)	20/28		211, 212, 214,
• Blood eosinophilia and tissue eosinophils			216–220
Both present (latency 12–60 days)	4/11		
Not present (latency 7–105 days)	4/11		
Tissue eosinophils but no blood eosinophilia	3/11		

Table 5 – (continued)

FACTOR(s)	No. of Cases	Proportion	Ref. No.
4. Liver Function Tests/Pathology:			
• Alanine transaminase (ALT)			
>5 × upper limit of normal (ULN)	33/37 (89%)		14, 15, 17, 19, 20–22, 44, 195–199, 203, 206, 208, 211, 212, 214, 216–220
>10 × ULN	25/37 (68%)		
• Aspartate transaminase (AST)			
>5 × ULN	30/37 (81%)		
>10 × ULN	21/37 (57%)		
• Acute liver injury categories:			
Hepatocellular	25/33 (76%)		
Cholestatic	6/33 (18%)		
Mixed (of above)	2/33 (6%)		

* *Case reports from non-Helsinn nimesulide preparations sold in Uruguay and Argentina believed to have contained impurities.*

History

The analysed cases have no history of blood transfusions, other risk factors for viral diseases, alcohol abuse, hepatitis (except old hepatitis A in one case) or other liver diseases. Previous drug allergy is known in two cases (diclofenac [198]; amoxicillin [16]). One patient [17] suffered from allergy to dust mites and pollens.

Osteoarthritis (16 cases), hypertension (6 cases) and obesity (3 cases) were the most frequently reported concomitant diseases. One case respectively of Paget's disease of the bone, rheumatoid arthritis, lupus erythematosus, undefined connective tissue syndrome, pancreatic cancer, diabetes, and post-surgical hypothyroidism, psoriasis are published. One case occurred in the first quarter of pregnancy [17]: this patient recovered and had a normal delivery.

Clinical presentation

The great majority of the published cases were clinically symptomatic: jaundice is reported in 31/41 (76%), right upper quadrant pain in 9/41 (22%), pruritus in 7/41 (17%). Other commonly reported symptoms are fever, general malaise, asthenia, anorexia, nausea, vomiting.

Liver function tests (LFTs)

In liver function tests (LFTs) high bilirubin level (range: 1.3–43 × UNL) is reported in 33/37 cases (89%). Transaminases were higher than twice upper limit of normal range (UNL) in all patients, with ALT and AST higher than five times UNL in 89% and 81%, and higher than 10 per UNL in 68% and 57%, respectively. ALT/AST was >1 in 25/37 (68%). The majority (76%) can be biochemically classified [221, 222] as hepatocellular liver injury. Cholestatic or mixed cases are less frequent (Tab. 5).

Histology

Data are reported in 20/41 (49%) cases. They can be summarised as follows:

a) Hepatocellular necrosis: 13 cases, from perivenular to massive. A concomitant inflammatory infiltration (mainly portal and/or perivenular) is reported in 10, from mild (4) to moderate (2) or severe (3), not detailed in one; eosinophils are present in six, absent in three and not reported in four. Steatosis is reported in two cases (mild, moderate). Regenerating nodules are described in one case [211]. Vasculitis of hepatic vein branches is described in one case [219].
b) Cholestatic hepatitis: five cases, with cholestasis (canalicular, hepatocytes) and inflammatory infiltration ranging from mild to marked. Eosinophils are reported in two cases. In no case steatosis, regenerating nodules or vasculitis are reported.
c) Pure cholestasis: marked cholestasis without necrosis is reported in two males; inflammatory infiltration is absent in one, and mild, with eosinophils, in the other [12].

Outcome: 31/41 cases (76%) recovered after about 2 weeks–7 months; two other patients recovered after transplantation. Three of the recovered cases had other events: acute renal failure [199], haematemesis – gastric and duodenal ulcers [211], melaena from duodenal ulcer [22]. One of the successfully transplanted cases had a concomitant anaemia [218]. One patient died due to a pancreatic cancer [12]. Seven fatal nimesulide-related cases occurred (one after transplantation). In four of them the treatment continued for 2 weeks [21, 204], about 1 month [13] and about 4 months [17]) despite the appearance of the ADR, and another one resumed the treatment despite a previous nimesulide-related liver injury [198]. The histology is known in four cases in which three cases showed hepatocellular necrosis and one case with cholestatic hepatitis.

Clinical and pathologic data regarding published case reports of nimesulide-related liver damage do not differ from what is published in the literature regarding drug-induced liver injury [190, 200–202, 204, 205, 207, 209, 210, 213, 215, 222–225].

The majority of clinically symptomatic drug-induced (and NSAIDs-induced) liver injury is acute, with signs, symptoms and LFTs indicating mainly hepatocellular damage, cholestasis, mixed pattern of cytotoxic and cholestatic injury, or steatosis. The cytolytic injury is clinically similar to viral hepatitis and has markedly elevated serum aminotransferases (8- to 2,000-fold elevation) and mildly elevated serum alkaline phosphatase level (<three-fold); the patient is anicteric or with variable degrees of jaundice. Cytotoxic hepatocellular injury can lead to fulminant hepatic failure, with high fatality rate without transplantation.

Acute cholestatic injury often resembles extrahepatic obstructive jaundice. Patients rarely feel ill, with the most common symptoms being pruritus and jaundice; serum aminotransferases are only mildly elevated (usually less than eight-fold), while alkaline phosphatase are increased up to 3- to 10-fold. The prognosis is better than for hepatocellular injury, although fatalities have been reported. The cholestatic hepatitis pattern is clinically similar to viral hepatitis, however jaundice is greater than would be expected from the degree of liver injury and laboratory data of cholestasis are evident.

The case reports regarding nimesulide are representative of the above-described patterns. Acute steatosis, not described with nimesulide, leads to clinical features similar to Reye's syndrome or acute fatty liver of pregnancy: jaundice is usually mild and serum aminotransferases are lower than that seen in cytotoxic injury (8- to 20-fold elevation) with mildly elevated serum alkaline phosphatase level (<three-fold). Although the biochemical features do not appear as severe as those seen in hepatocellular disease, the illness can be severe and the prognosis poor with high mortality.

From a morphological point of view [224, 225] the hepatitis-like injury may be indistinguishable from acute viral hepatitis, with ballooning and apoptosis of hepatocytes and a predominantly lymphocytic inflammatory response; occasionally there are prominent eosinophils, suggesting a drug rather than a virus: many drugs may cause this type of injury. The acute combined hepatocellular-cholestatic injury generally corresponds to the clinical syndrome of cholestatic hepatitis, and it is characterised by hepatocellular injury and parenchymal inflammation, along with a significant degree of intrahepatic cholestasis: it may be caused by nearly all drugs that can cause either hepatitis-like injury or cholestasis. A few drugs, such as anabolic and contraceptive steroids, produce pure canalicular bile stasis. Many drugs produce predominantly canalicular bile stasis, but usually with a mild degree of hepatocellular injury.

Minor degrees of microvesicular steatosis are very common in drug-induced liver injury; clinically relevant acute steatosis, characterised by severe hepatitis and even hepatic failure, can be seen with a few drugs, but it is not described for nimesulide. No case of hepatic granulomas has been associated with nimesulide. This reaction has been described with more than 60 drugs, but for many of these there are only isolated case reports, so that the actual etiologic relationships are obscure. Drug-induced vascular lesions are uncommon, but there are several types of vascular diseases that may have an acute onset (necrotising angiitis, veno-occlusive disease, hepatic vein thrombosis): one case of vasculitis of hepatic vein branches is described with nimesulide.

It is to be noted that the diagnosis of drug-related liver injury cannot be made on morphologic grounds alone or on the basis of any specific laboratory test, and a number of scales have been developed that attempts to codify causality assessment into objective criteria [223]. Data regarding involvement in men and women follows the known but not explicated fact that women are more frequently involved than men in drug-induced liver injury [188].

The great majority (31/41 patients: 76%) recovered; two other patients recovered after transplantation. Four out of seven fatal published cases perhaps could have been prevented by interrupting the treatment at the beginning of ADR, and another one by not re-using the drug after a previous hepatic ADR.

In summary, the clinical and pathological characteristics of nimesulide-related liver damage reported in published case reports follow the above-summarised features of drug-induced hepatotoxicity. Acute cases only (hepatocellular necrosis, mixed and more rarely cholestatic type) are reported. No cases showing granuloma or microvesicular steatosis or chronic hepatic injury are published. The majority of published cases is clinically symptomatic and has relevant derangement of liver function tests. This can be due in part to the known tendency to publish new events or unusual presentations or the most severe cases.

Hepatic adverse events reported in Finland

The "spike" of case reports of hepatic ADRs noted in 2001–2002 (Fig. 9) was related to a considerable number of reports from Finland. This led to an investigation of the factors that may have been responsible for the sudden appearance of these cases in Finland [225] (Figs 11–16). As seen in Figure 11 the greatest numbers of serious and non-serious ADRs reported from Finland were in the hepatic system. Indeed, as a percentage of the total, the number of serious hepatic events (35.8%) is three-fold higher than that reported worldwide (11.5%). Thus, there is a striking difference in the number of adverse events in serious and non-serious cases reported in the hepatic system in Finland contrasted with that worldwide. The reasons underlying this are not clear but it is evident from inspection of the database that a large number of cases have appeared within the period of 2001–2002. The distribution of digestive, skin and renal serious and non-serious cases is relatively speaking lower, accounting for the predominance of hepatic events.

A statistical technique developed by Weber [226] employed determining the ratio of an event in one body system with respect to those in the skin since it was

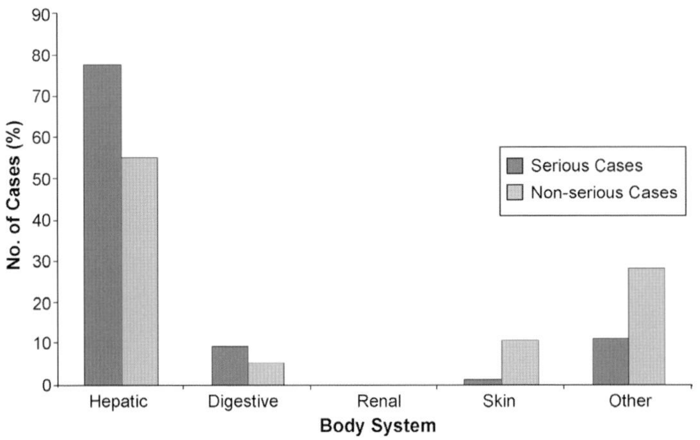

Figure 11
Total of all adverse reactions attributed to nimesulide that were reported in Finland since the drug was marketed there in 1998.

considered that skin reactions would be the most frequently reported. This procedure was an attempt to "standardise" data on ADRs to enable some basis of comparison. The accuracy of this procedure has not been ascertained but it is the intuitive view that using skin values as a denominator it may be possible to compare the occurrence adverse reactions in various body systems. Thus, taking the ratio of serious cases reported in the hepatic body system compared with those in the skin worldwide indicates that there is about a 2.4-fold greater number of serious cases reported in the hepatic compared with those in the skin. In contrast there is a 49-fold greater number of serious adverse events reported in the hepatic body system compared with those in the skin in Finland. In terms of the total of all adverse events whether serious or non-serious those in the hepatic system worldwide are approximately 1.5-fold greater in the hepatic compared with skin while there is a 10-fold difference in the numbers of hepatic reactions overall in serious and non-serious cases compared with those in skin reported from Finland. These data are indicative of there being some kind of signal of this pattern of serious adverse events in the hepatic body system from Finland.

There appears to be a slightly greater number of serious cases of hepatic reactions in elderly subjects (Fig. 13a) and in females (Fig. 13b). However, when age and gender are considered of the reported serious reactions from all adverse events in Finland indicated in Figure 14a there is a predominance in under 65 years old females. The distribution of reported ADRs in males with respect to age shows

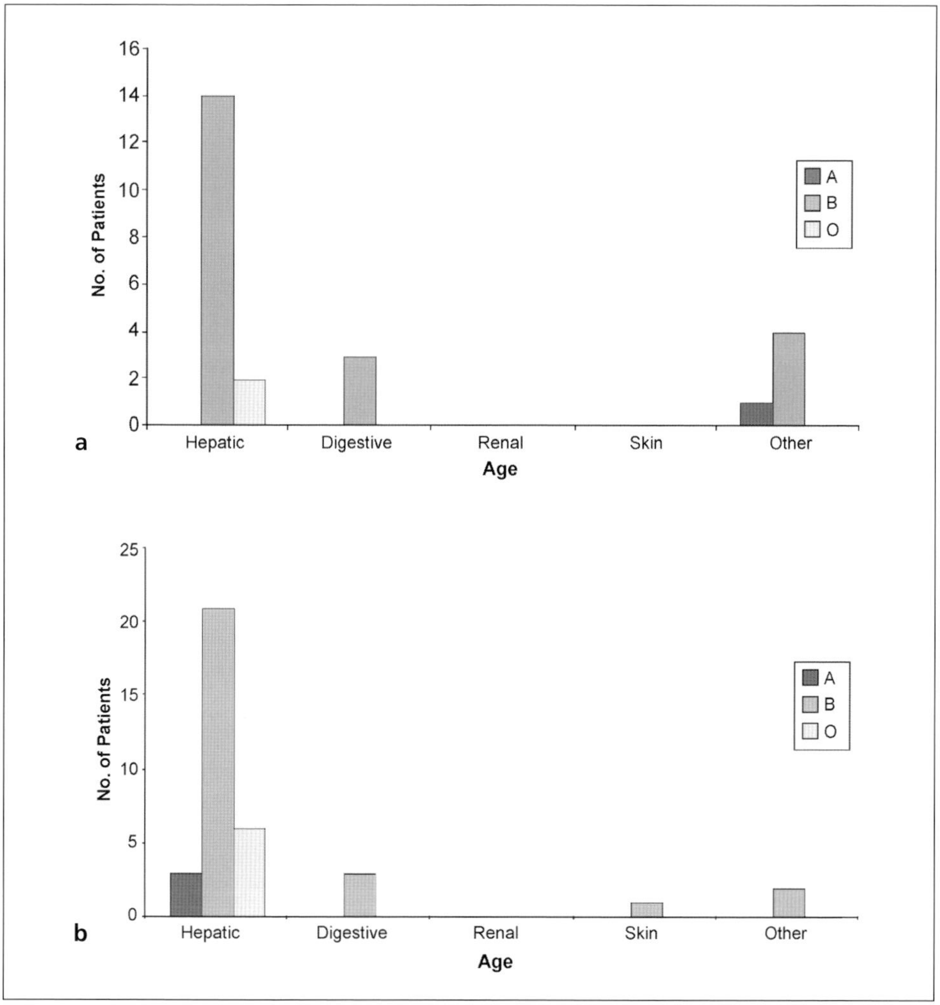

Figure 12

*(a) Serious adverse reactions, that were reported in Finland in the major organ systems of α-rated cases of adverse events assessed according to their probability of being associated with nimesulide [**A** = definite; **B** = probable or possible; **O** = unlikely or unknown). Most of the definite cases that were determined to be serious were in the hepatic and digestive system. However, there were only 17.9% of the total cases in the hepatic system that were definitely associated with the drug. None of these cases were found to be in the probable/possible category. (b) Non-serious adverse reactions that were reported in Finland in the major organ systems of **β**-rated cases of adverse events assessed according to their probability of being associated with nimesulide [**A** = definite; **B** = probable or possible; **O** = unlikely or unknown].*

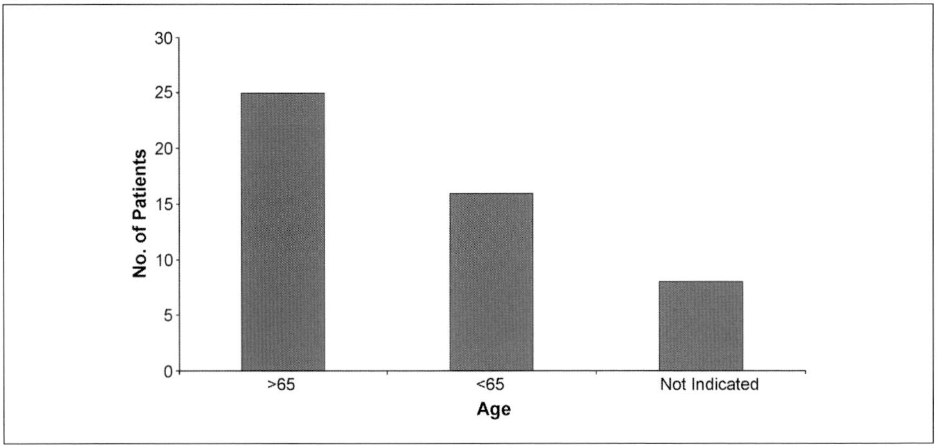

Figure 13a
Distribution of serious hepatic adverse events attributed to nimesulide that were reported in Finland by age.

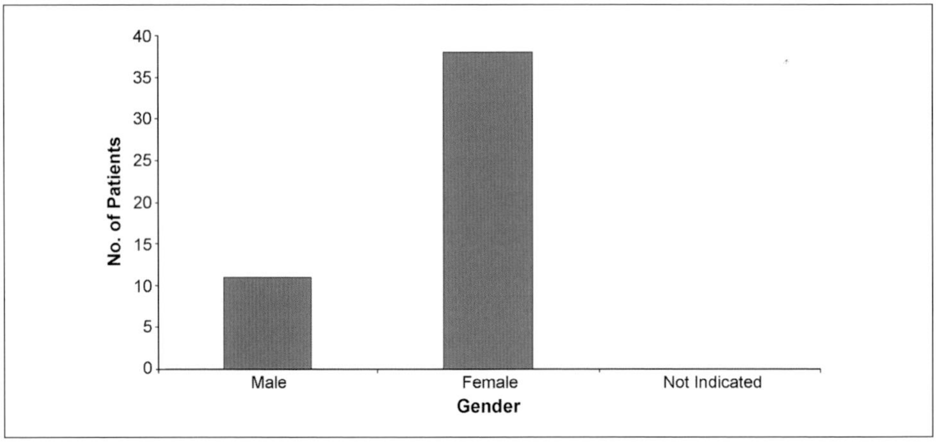

Figure 13b
Distribution of serious hepatic adverse events attributed to nimesulide that were reported in Finland by gender.

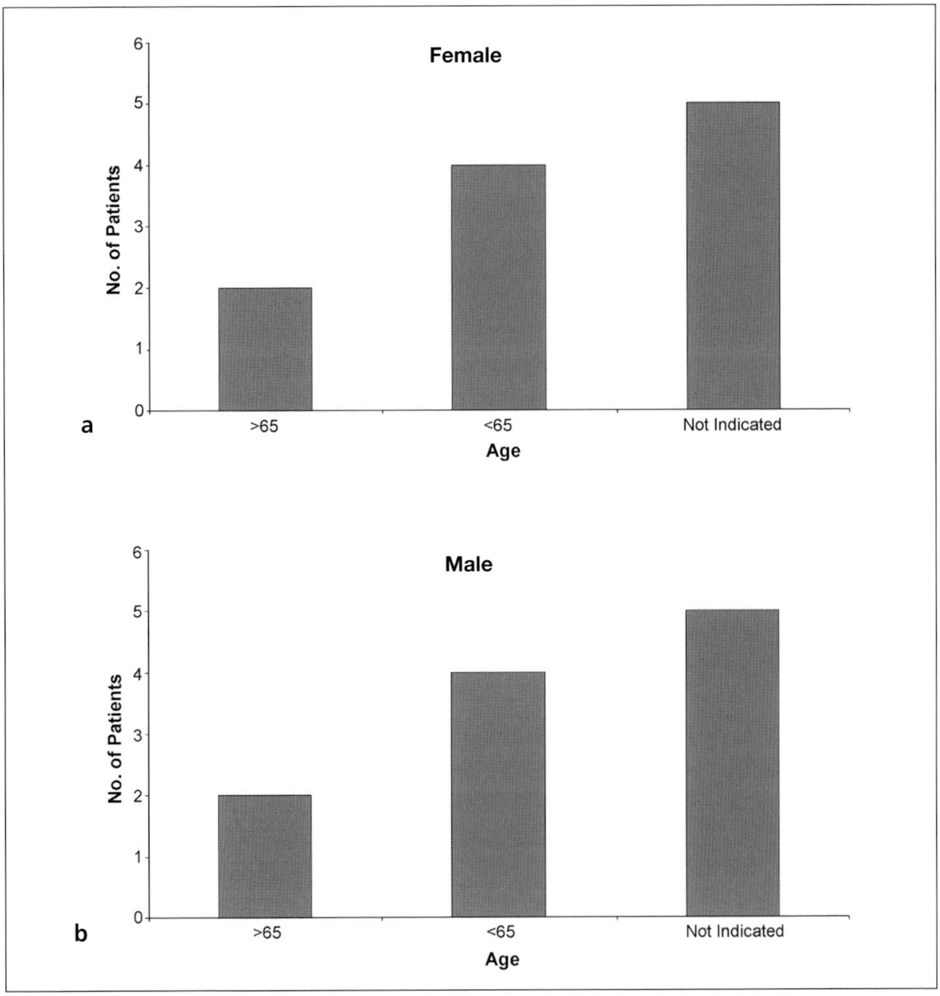

Figure 14

Distribution of serious hepatic adverse events attributed to nimesulide that were reported from females (Fig. 14a) or males (Fig. 14b) aged >65 or <65 years in Finland. The data in males is from relatively small numbers and so has limited significance. However, the predominance of females aged <65 years in which serious adverse events in the hepatic system were reported indicates that this is the group wherein special circumstances may prevail predisposing this group of patients to adverse reactions.

that these also predominate in under 65 year olds but it should be pointed out the numbers here are small and there is an appreciable number of cases where the age has not been clearly indicated in male subjects (Fig. 14b).

Using the case report rating system based on "discriminant analysis" (as employed in the earlier section "Causality Assessment and Quality of Information" [32] the predominant numbers of serious cases reported from the hepatic body system are of **B** association (possible) in the **α** (adequate information) and **β** (some information or data missing) quality rated cases (Figs 12a and 12b). There are only three cases reported from the hepatic body system where there is a clear indication of an association and these are from **β** rated cases. In attributing causality to nimesulide intake alone it must be stated that predominance of **B** association cases indicates that there is a substantial amount of information that is missing in the case reports from Finland. In these studies it was observed that many of the patients have indications of confounding factors that may have affected the onset of the development of hepatic reactions reported in Finland. A substantial number of patients have obviously taken hepatotoxic drugs or drugs concurrently which have the potential to substantially influence drug metabolism of nimesulide or of one another. Polypharmacy, especially in the use of agents to relieve pain, is clearly evident in a number of these patients in many cases there are conditions the patient has had, which have also influenced the condition (for example diabetes, chronic hepatitis, rheumatoid arthritis and other rheumatic conditions). A total of six patients had history of alcoholism or had severe alcoholic liver disease and since this comprises 12.2% of the total this is a remarkably high incidence of alcoholism in those patients in which serious hepatic events has been recorded.

Biopsy data

A review of the reports from liver biopsies reported was also undertaken in the hepatic cases that were reported worldwide and the ascription to cause based on the **A** (most likely), **B** (possible) or **O** (zero or unlikely) categories for causality, the temporality of drug intake and a brief note of the potential factors that could have been implicated in the development of the adverse events. A total of 42 cases have been reported for which biopsy information was available but in one case the biopsy was reported to have failed so there is a total of 41 cases that are evaluable. Of these, 12 cases have been reported in elderly females (29.3%) while two have been reported in elderly males (4.9%). In under 65 year olds, 30 cases were reported in females (73.2%) while 11 were reported in males (26.8%). In 32 cases there was a clear **A** or **B** attribution to nimesulide being a factor in the development of the hepatic reaction(s). In only 23 of these cases was there a clear indication of temporal intake of the drug established.

Compared with these data a total of 13 cases with biopsy data were reported from Finland [225]. The histological reports state that there are features of hepatitis that comprise centri-lobular necrosis in a few case with additionally some cholestatic changes, and eosinophilic or inflammatory infiltrates. In a number of cases there was from the clinical history evidence of an autoimmune condition, viral infection of the liver or other state that could have contributed to the development of the inflammatory reactions. There are a few cases of toxic hepatitis, fulminant liver failure with associated jaundice that contributed to the pathology.

The general impression is that nimesulide-associated hepatitis is the predominant histological feature with associated centrilobular necrosis and inflammatory changes encountered in case reports of hepatic reactions with this drug. However, given that there have been a substantial number of confounding factors particularly intake of known hepatotoxic drugs and hepatic changes from disease(s) or conditions affecting the liver it is difficult to establish what role nimesulide has played in the initiation of these events.

In an investigation of the factors underlying the development of liver reactions ascribed to nimesulide in Finland [225] it was found from interviewing rheumatologists and other clinical experts in that country that:

a) Female rheumatic patients frequently employ hormone replacement therapy (HRT) or other forms of oestrogenic steroids – these are well known to elicited hepatic reactions [227–232].
b) A considerable number of patients with rheumatoid arthritis (RA) have been prescribed nimesulide. Since elevation of transaminases was noted in RA patients in the early studies on nimesulide at very high doses conducted by Riker in the USA who received high doses of the drug [233] it is possible that this patient group may be susceptible to hepatic reactions from NSAIDs, a feature which is known with aspirin and some other NSAIDs [234–241]. It has also been noted that RA is not an approved indication for nimesulide and that the prescribing doses of NSAIDs is much higher in Finland than in other Nordic countries [242] so there may have been a tendency to over-prescribe nimesulide as well as that observed with other NSAIDs.

In the analysis of the factors in patients with serious liver reactions in Finland it was noted that alcoholic liver disease, and intake of statins, paracetamol, diclofenac and oestrogenic steroids were reported. Statins are known to produce hepatic reactions [243–245]. The issue of alcoholic liver disease is important for Finland as in all Nordic countries, especially in women [246–263].

To obtain some insight into possible underlying features of the population genetics or environmental factors leading to liver toxicity in general in the Finnish population it is worth noting the following population based factors:

1) The genetic associations in Finland (as well as in some other European countries) of intrahepatic cholestasis of pregnancy [264–270] which often appears not only during pregnancy but also in women taking oral contraceptives.

Among the genetic associations connected with oestrogens is that involving the multidrug resistance-3 (MDR-3) and bile salt transporter polymorphisms [266, 271–275]. There are also reported associations with apolipoprotein E alleles in women with this condition [276].

2) Polycystic liver disease with genetic associations in the Finnish population [277–280].
3) Hepatic lipase abnormalities with genetic associations [281–283] and those affecting metabolism of statins [284].
4) A polymorphism for the CD14 receptor for endotoxin that may be important in alcoholic liver disease [285] and other hepatic conditions involving reactions to bacterial infections.
5) Polymorphisms in cytochrome P_{450} 2E1 that could have importance for impaired metabolism of ethanol in patients with alcoholic liver disease [286].
6) Cytochrome P_{450} polymorphisms involving abnormalities of liver metabolism of cholesterol [286, 287], hormone steroids [288–290] and xenobiotics [291–294].
7) Various other liver diseases and conditions as well as environmental effects on the liver in the Finnish population [295–300].

It is clear that there are a considerable number of factors in addition to the previously commented notoriety bias determining a clustering of reports in Finland that may have accounted for the unusually high numbers of serious hepatic cases reported in the period of 2001–2002 from that country. Further studies are warranted to establish the mechanisms underlying these case reports.

Benefit/risk assessments

The effectiveness of nimesulide in each of its indications has been adequately demonstrated in clinical trials comparing its efficacy with placebo and with other NSAIDs. It has a rapid onset of action as an analgesic and is at least as effective as other NSAIDs as described in Chapter 5. This is supported by its very extensive adoption by clinicians in managing the various conditions.

The reporting rate of hepatic AEs for nimesulide is extremely low and similar to other NSAIDs. In clinical studies, there were no cases reported with hepatitis or hepatic failure among more than 64,000 nimesulide-treated patients (Helsinn, data on file).

The occurrence of hepatic reactions does not extensively modify the benefit/risk profile of nimesulide. This is not a new phenomenon, nor is it restricted to nimesulide among drugs used to treat similar conditions. What has changed recently is a relatively high incidence of reporting in certain countries, likely associated with notoriety bias. In many cases, there is insufficient information to confirm or refute a causal association; many are complicated by concomitant medicines known to cause liver injury or other risk factors. Some of the reported cases of hepatitis would qualify only as liver injury according to CIOMS definitions. Some would not even meet those criteria.

Considering the class effects of NSAIDs, such as serious renal, hypersensitivity and skin reactions, we see a very low spontaneous reporting rate for nimesulide with a frequency comparable with those for the other drugs used in the same indications. For the newer COX-2 inhibitors, there is also recent evidence for an increased risk of cardiovascular adverse reactions that appears not to be shared by nimesulide.

Considering risk overall, the use of anti-inflammatory drugs is associated with a distinct incidence of fatalities and the major hazard is upper GI perforation, ulceration and bleeding. The reporting frequency for PUB for nimesulide is extremely low, estimated at 0.4 per million patients/treatment courses. This appears lower than other NSAIDs. The positive benefit/risk profile has been demonstrated in respect to its efficacy and tolerability in clinical trials and long-established use.

Mechanisms of toxic reactions

In the earlier studies by Swingle and co-workers [301, 302] the acute gastric lesions from nimesulide in rats were compared with their anti-oedemic activity in the carrageenan injected paw model, to derive a therapeutic index (TI) (see Chapter 4). These studies showed that nimesulide had the highest TI (LD_{50} ulcers/ ED_{50} anti-oedemic activity = 260) compared with naproxen (TI = 190), ibuprofen (TI = 68), flufenamic acid (TI = 20), aspirin (TI = 11) and indomethacin (TI = 7). Thus, on a comparative basis nimesulide is amongst those NSAIDs with the lowest ulcerogenicity in the stomach. Studies in stressed rats given 100 mg/kg nimesulide [303] and unpublished studies in unstressed pigs given 100 mg/kg nimesulide [304] have shown this drug produces little if any GI damage.

Gastrointestinal injury and bleeding

The clinical and epidemiological data discussed in the previous sections shows that nimesulide in comparison with other NSAIDs has a low risk for developing GI ulcers and bleeding. In laboratory animal models these observations are largely confirmed and it is only at exceptionally high, sometimes supra-therapeutic doses, that gastric lesions develop in animals, and then under specific conditions. It is possible this constitutes a toxicological reaction to nimesulide distinct from being evident within the therapeutic ranges for anti-inflammatory, analgesic and antipyretic activities in the same species as used in GI ulcer studies.

Of the initial investigations performed in rats to investigate the acute gastric ulcerogenic effects of nimesulide (then R-805) it was found that a single oral dose of 100 mg/kg to fasted and cold-stressed rats did not lead to any lesions, ulcers of morphological signs of mucosal injury in the stomach or upper GI tract [303, 304]. This dose of 100 mg/kg nimesulide is at least 10–20 times that required for achieving therapeutic effects in rats in standard models of anti-inflammatory, analgesic or antipyretic activity (Chapter 4). Most other conventional NSAIDs exhibited gastric lesions, some haemorrhagic, in this model with the exception of well-established low ulcerogenic drugs (e.g., azapropazone, nabumetone) [303, 305–310]. The cold-stressed rat and the chronic pig models of NSAID-induced gastric injury have been found to be highly reproducible and sensitive predictors of gastric mucosal injury in humans [303, 305–310].

Tanaka and co-workers [311] compared the GI ulcerogenic effects of nimesulide with ibuprofen, indomethacin and piroxicam in rats that had been fasted for a total of 48 h. This is an unusually long period of fasting and would not normally be acceptable ethically today and it would be expected that exceptional nutritional and other physio-pathologic stress would have been inflicted on these animals. The extent of mucosal injury in the GI tract was determined using the Evan's blue dye labelling technique that measures the permeability and highlights areas of mucosal cell injury.

The results showed that nimesulide 30–300 mg/kg p.o. produced dose-related injury to the mucosa in the stomach comparable with that of the same doses of ibuprofen, indomethacin 1.0–10 mg/kg and piroxicam 3–30 mg/kg. The lowest dose of nimesulide 30 mg/kg did not produce statistically significant mucosal injury (assessed by the Evan's blue method) compared with the control. The much higher dose of 100 mg/kg only produced a slight increase in mucosal injury as well as producing ulcers in 3/7 animals. As this dose is about 5–10 times that for inhibition of acute or chronic inflammation (Chapter 4) it would seem that the TI of nimesulide is still relatively favourable despite the extreme experimental conditions.

Damage to the small intestine was also observed with nimesulide given at the above-mentioned doses, but this did not seem to be dose-related. The other comparator drugs mentioned above also produced intestinal injury but this was dose-related.

These authors also compared the gastric ulcerogenic effects of nimesulide and some other NSAIDs with their effects on production of mucosal prostaglandins [312]. These experiments were performed in animals in which polyester sponges had been implanted so that the prostaglandin content in the inflammatory exudates could also be determined and compared with that in the gastric mucosa. The gastric mucosal PGE_2 and 6-keto-$PGF_{1\alpha}$ was determined in extracted mucosal scrapings, which are rather crude methods since there can be uncontrolled release of arachidonic acid during the mincing and extraction process. Nimesulide 0.3 mg/kg p.o. given in two doses for 1 and 24 h caused a significant reduction in 6-keto-$PGF_{1\alpha}$ but not PGE_2 in the gastric mucosa. At doses of 3 and 30 mg/kg the concentrations of both these prostaglandins was significantly reduced. These results suggest that in comparison with other studies in rats given nimesulide the inhibition of prostaglandin production occurs at doses which are lower than those required for the development of mucosal injury in the stomach. Similar reduction in prostaglandin concentrations occurring at lower doses than those required for ulceration have been observed with some other NSAIDs [307, 313–315].

Nakatsugi et al. [316] compared the effects of nimesulide 10–100 mg/kg p.o. with indomethacin 0.3–3.0 mg/kg p.o. and ibuprofen 3–30 mg/kg p.o. in the water-stress immersion assay for gastric lesions and gastric mucosal PGE_2 concentrations. In contrast to the studies of Tanaka et al. [312] inhibition of mucosal PGE_2 was only observed after dosage of 100 mg/kg of nimesulide and a few gastric lesions were observed at all doses of this drug although there was no significant difference in the lesion numbers compared with controls (in which there was a low level of lesion development confirming that the animals responded to the effects of the stress). These results are in agreement with the earlier work in cold-stressed animals in which a dose of 100 mg/kg nimesulide was with any injurious effects [303].

In the studies by Nakatsugi and co workers [316], no COX-2 was observed in the gastric mucosa of the rats by Western blotting although COX-1 was detected. This suggests that some inhibition of COX-1 activity was responsible for the reduction in mucosal PGE_2 at the high dose of 100 mg/kg nimesulide. Reduction in mucosal PGE_2 was observed with 1 and 3, but not 0.3 mg/kg indomethacin whereas significant lesion development was observed at the highest dosage of this drug. Similarly, reduction in mucosal PGE_2 was evident with all doses of ibuprofen whereas significant lesion development was only apparent at the two higher doses of this drug. Clearly, lesion development occurs with indomethacin and ibuprofen at doses that are higher than required for reduction in mucosal prostaglandins showing that there is a differentiation between effects on prostaglandin production form lesion development. This differentiation is even more apparent with nimesulide since only high doses of the drug cause reduction in mucosal prostaglandins with any significant lesion development.

Laudanno et al. [317] undertook an extensive examination of the GI mucosal damaging effects in rats of a wide range of 15 NSAIDs with varying COX-2/COX-1 activity and paracetamol [318, 319] administered orally or subcutaneously. The gastric mucosal injury effects were determined in a fasting and fed model that predisposes the development of antral as compared with fundic lesions. Nimesulide 200 mg/kg s.c. produced no lesions in the antrum but did produce a few small lesions in the small intestine and these were relatively few in comparison with diclofenac sodium, etodolac, ibuprofen, ketoprofen, ketorolac, mefenamic acid, naproxen, piroxicam and tenoxicam (all given at 60 or 500 mg/kg s.c.) which produced extensive area of mucosal lesions in the intestine. These are mostly toxic doses of these NSAIDs but it does show that nimesulide given at the high dose of 200 mg/kg s.c produces no gastric lesions and minimal intestinal injury in this model. In 36 h fasted rats (which is an exceptionally long period of food deprivation) slight but non-significant antral and small intestinal injury was apparent with nimesulide 200 mg/kg.

Kataoka and co-workers [320] undertook a sub-chronic study in rats designed to investigate the ulcerogenic effects of orally administered nimesulide 100 mg/kg or indomethacin 5 or 10 mg/kg alone or in combination with either 5 days prior treatment or concomitant treatment for the same period with the corticosteroid, prednisolone 10 mg/kg s.c. This investigation has particular therapeutic interest because of the frequent use of corticosteroids in therapy of rheumatic conditions. In relationship to the pathogenesis of ulcer disease, this steroid is obviously one of the hormones that are involved in mediating stress responses although under single dose conditions it does not always lead to marked increase in ulcerogenic effects of NSAIDs [320].

In the steroid pretreatment protocol 4 days administration with 10 mg/kg prednisolone s.c. followed by overnight fasting and then administration of 100 mg/kg nimesulide p.o. did not lead to any gastric mucosal injury with or without the steroid [320]. However, indomethacin 5 or 10 mg/kg p.o. alone caused dose-related increase in lesions and lead to a marked increase in gastric lesions when given with the steroid. Co-administered daily dosage of nimesulide 30 mg/kg/d p.o. with prednisolone s.c. over 5 days did not lead to any signs of mucosal injury. However, indomethacin 5 mg/kg/d p.o. alone caused substantial mucosal damage alone that was markedly exacerbated by prednisolone.

Measurements of gastric mucosal concentrations of PGE_2 undertaken in either the pretreatment or concurrent treatment protocols did not show any significant changes with nimesulide treatments from control or prednisolone treated animals. However, PGE_2 concentrations were reduced to within 5–10% of control values with indomethacin alone or with the steroid [320].

As noted in Chapter 4 several studies have confirmed the sparing of inhibition of COX-1 in the gastric mucosa by nimesulide both *in vitro* and *ex vivo* [318, 319, 321–325]. This is undoubtedly a major factor together with high pKa of the drug molecule in the low gastric ulcerogenicity observed with nimesulide. As shown in Table 6 and Figure 15 there are appreciable differences in the actions of nimesulide on the gastric mucosa compared with that of other NSAIDs which accounts for its low gastric irritancy. This may in part be related to low uptake of the drug into gastric mucosal cells and mitochondria which is postulated to be lower than that of low pKa carboxylic acid NSAIDs (Fig. 15). Although nimesulide does affect mitochondrial oxidative phosphorylation (see page 364) this effect is only apparent at high drug concentrations.

Hirata and co-workers [326] investigated the effects of nimesulide, indomethacin and NS398 on the transmucosal potential difference (PD), mucosal

Table 6 – Summary of physiopathological and biochemical changes involved in the pathogenesis of gastric mucosal injury by NSAIDs and comparisons with the effects of nimesulide. Based on Rainsford [330] with modifications to include the effects of nimesulide (right hand column) from literature cited in this review

Factor	Principal consequences	Effects of nimesulide
Sloughing of surface mucus, decreased bicarbonate, and altered phospholipid hydrophobicity	Impaired surface mucus and surface membrane protection	–
Immediate (Primary) actions	Breakdown of membrane integrity. Low pKa (3–4.5) carboxylic acid COX-I/COX-2 NSAIDs more damaging than selective COX-2 drugs with high pKa (5.5–6.0)	High pKa of nimesulide (pKa 6.5) associated with less damage to membrane (see Fig. 15)
Back diffusion of acid from acidic drugs	Local decrease in cell pH (promotes drug uptake and local cellular autolysis). Will be pKa dependent, i.e., low pKa carboxylic acids will produce more back diffusion than high pKa NSAIDs	Low potential for back-diffusion of acid due to high pKa
Inhibition of COX-1 leading to (a) decreased PGE_2 and PGI_2 synthesis, and (b) diversion of arachidonate to lipoxygenase products	Altered blood flow ischaemia and anoxia-reperfusion injury $\rightarrow$ oxyradicals I	Oxyradical scavenger, so any tissue damage (e.g., from irritants, *H.pylori*) counteracted

Table 6 – (continued)

Factor	Principal consequences	Effects of nimesulide
–	Platelet-vessel adhesion promotes microvascular injury → bleeding from injured vessels Reduced 'cyto'-protection by decreased mucus production, decreased bicarbonate secretion	Low COX-1 activity leads to less platelet aggregation
–	Promotion of leucocyte accumulation, adhesion (from increased LTB_4 production and/or degradation by chemotactic peptides from local cell injury) contribution to ischaemia	Reduces leucocyte adhesion and activation so reduced inflammatory reactions
Reduced NO	$NO + OH^{\bullet} \rightarrow$ peroxynitrite	Oxyradical scavenging reduces potential for peroxynitrate formation
Later induction of NO	Pro-inflammatory reactions	–
Increase of IL-1 and TNF-alpha	Localised tissue destruction	–
Enhanced oxyradical production Reduced sulphydryl	Possible loss of reductive protection by mucosal bio-molecules against oxyradical damage and perturbed eicosanoid metabolism	–

Table 6 – (continued)

Factor	Principal consequences	Effects of nimesulide
Oxyradicals generation	Activation of NFκB → expression of adhesion molecules on endothelia and leucocytes, increased interleukin I and TNF-α during inflammation	Inhibits TNFα production
Caspase activation	Apoptosis, cell death	Inhibits apoptosis (Fig. 15)
Release of lysosornal hydrolases cholinergic activation	Local cellular autolysis	–
Histamine release from mast cells	Acid/pepsin secretion enhanced in stomach Promotes acid secretion, vasodilation (stomach)	Inhibits acid secretion and mast cell release of histamine
Longer time effects Enhanced motility (amplitude)	Altered G-I transit relation to prostaglandin/NO control of smooth muscle functions	–

Table 6 – (continued)

Factor	Principal consequences	Effects of nimesulide
Inhibition of ATP production	Further caspase activation → apoptosis. Reduced capacity to resist cell injury from mucus and other synthetic reactions	–
COX-2 inhibition	Impaired ulcer healing	–
Altered cAMP levels (? from phosphodiesterase inhibition)	Altered cell metabolism, including effects on acid and mucus secretion (stomach)	–
Inhibition of production of mucus layer and inhibition of mucus biosynthesis (at enzyme level)	Reduced mucus protection	–
Variable apoptosis and adaptation with different NSAIDs	Induction in mucosal injury and bleeding with repeated dosage: varies with different NSAIDs	–

– denotes no effect of nimesulide

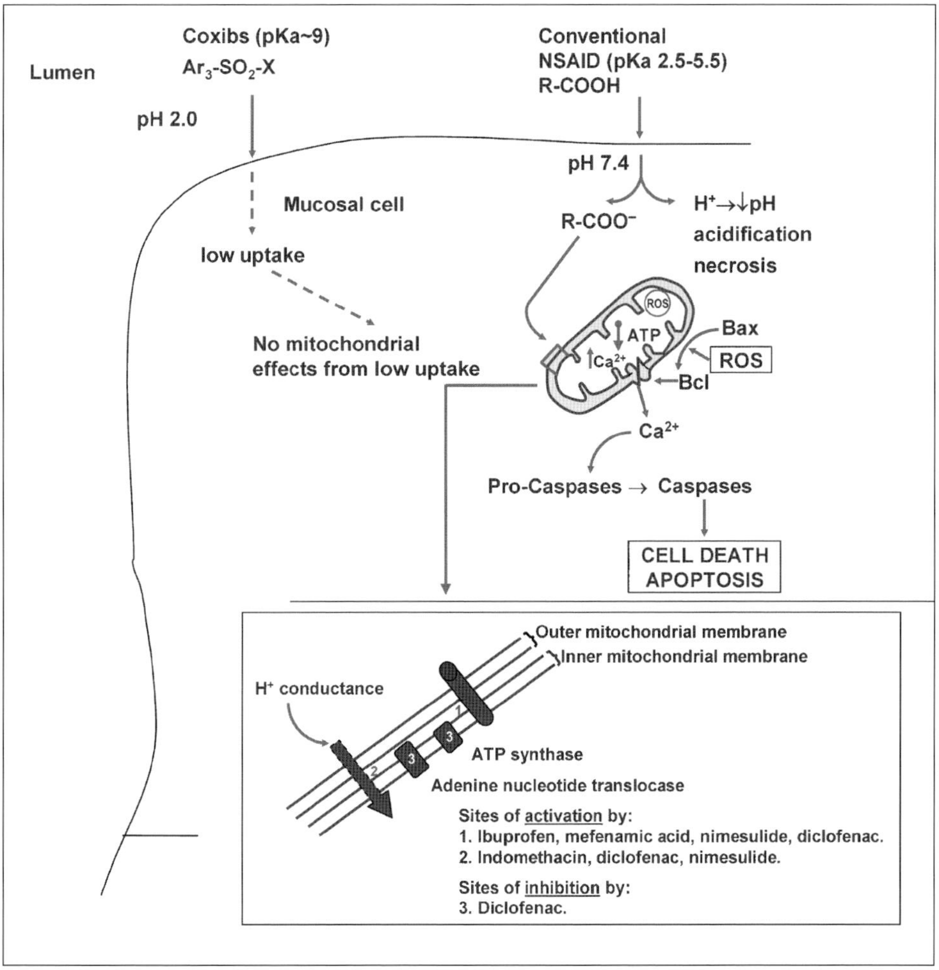

Figure 15
Mechanisms of cellular reactions by NSAIDs and the differing actions of nimesulide in the gastric mucosa focussing on mitochondrial induced apoptosis and the role of pKa of the drugs.

blood flow (MBF), luminal acid loss and luminal PGE_2 before, during and after exposure to the mucosal barrier breaker, taurocholate, of rat stomachs mounted in *ex vivo* in gastric perfusion chambers. Pretreatment with indomethacin 10 mg/kg s.c. attenuated the hyperaemic response to taurocholate and reduced gastric PGE_2 production without affecting PD or acid loss. However, indomethacin caused haemorrhagic lesions in the gastric mucosa. No changes were observed with nimesulide 10 mg/kg s.c. or the same dose of NS398. In the period following exposure of the stomach to taurocholate 10 mM + HCl, indomethacin given at the same dose as in the pretreatment period, reduced PD and MBF in this so-called recovery period but the other two COX-2 inhibitors did not cause any changes in PD, MBF or PGE_2. The authors measured COX-1 and COX-2 mRNA 30 min after exposure to taurocholate and no changes were observed with the former and only slight increase was observed in the expression of the latter. Thus, selective COX-1 inhibition in this model was observed with indomethacin and the lack of effects of nimesulide and NS-398 on gastric PGE_2 production was probably related to little COX-2 and predominant COX-1 being present in the mucosa. The results show that nimesulide had no effects on the mucosal barrier-disrupting effects of taurocholate + acid treatment or influences on the taurocholate-stimulated PGE_2 production and the hyperaemic response. However, indomethacin did show barrier breaking and impaired the hyperaemic response to taurocholate + acid probably as a consequence of inhibition of COX-1-derived PGE_2.

In another study by the same group using the rat gastric and duodenal perfusion models, histamine 8 mg/kg/h stimulated acid production was unaffected by the same doses of NSAIDs as used in the above experiment, and duodenal bicarbonate secretion was reduced by indomethacin or NS-398 [327]. The reduction in bicarbonate secretion coincided with development of duodenal lesions; no mucosal injury was observed with nimesulide or NS-398. These results suggest that COX-1 regulated production of duodenal bicarbonate secretion is important in duodenal mucosal protection. Inhibition of COX-1 by indomethacin reduces bicarbonate secretion and is associated with lesion development. Nimesulide clearly does not impair bicarbonate secretion and does not cause duodenal injury. These differing effects of indomethacin and nimesulide on duodenal secretion do not appear to be related to gastric acid secretion in the rat. Süleyman et al. [328] investigated the effects in 24 h fasted rats of nimesulide 100–500 mg/kg p.o. on the development of gastric mucosal lesions following oral administration 5 min later of indomethacin 25 mg/kg or ibuprofen 400 mg/kg. The rationale behind these experiments was not clear from the author's explanations in their paper except for the possible involvement of COX-1 inhibition being somehow influenced by COX-2, even though there is virtually no COX-2 enzyme in the stomach of fasted rats. The authors found there was complete suppression by nimesulide at all doses of lesions induced by indomethacin or ibuprofen. A control treatment by the H_2-receptor antagonist, ranitidine 150 mg/kg p.o., only resulted to partial suppression of lesion development due to indomethacin or ibuprofen, suggesting that inhibition of acid secretion only partly reduces gastric mucosal lesions induced by these NSAIDs. The authors eliminated the possibility of chemical interactions between nimesulide and indomethacin or ibuprofen by examination of their ^{1}H-NMR and ^{13}C-NMR spectra. Possible explanations for the protective effects of nimesulide in these studies include the mast cell stabilising effects of this drug and inhibitory effects on leucocyte accumulation and activation. In a related study by the same group [329] they showed, using the same experimental design in 24 h fasted rats that nimesulide increased gastric mucosal levels of reduced glutathione from which they suggested this was the reason for the effect of nimesulide in reducing the development of gastric lesions from indomethacin. Unfortunately, the authors did not determine the levels of oxidant species (O_2-, OH$^\bullet$) or the enzymes involved in oxidoreductive reactions by GSH/GSSG or superoxide production and its dismutation which may be important in mucosal damage [330], so it is not possible to conclude that effects on production of GSH (in its reduced state) were a factor in the reduction by nimesulide of indomethacin-induced mucosal injury. Furthermore, the lack of dose- and time-dependent effects of nimesulide in producing its protective actions limits the conclusions that can be drawn from these studies.

An interesting observation that was also made by the authors of this study [329] was that neither celecoxib 10 mg/kg p.o. nor rofecoxib 25 mg/kg p.o. reduced the lesion development from indomethacin 25 mg/kg, and they did not restore mucosal GSH levels that were reduced by indomethacin. These results imply that the effect of nimesulide in stimulating mucosal GSH is specific to this drug and unrelated to its effects as a COX-2 inhibitor. Further studies are, however, needed in order to substantiate these claims.

In another study from the same group [331] using the same general study design as above [328, 329] it was found that nimesulide protected the gastric mucosa against injury from ethanol. It therefore appears that nimesulide has generalised protective effects against gastric mucosal injury by noxious agents. Whether this involves restoring levels of reduced glutathione to normal or near normal or other mechanisms as noted above has still to be resolved.

In a brief report Ramesh et al. [332] undertook a study in eight mongrel dogs half of which were given 2 mg/kg nimesulide twice daily for 4 days, while the others served as controls. It was claimed that the stomachs from dogs given nimesulide "showed multiple ulcers of various sizes and shapes with haemorrhages" but no quantitative data were provided on this group or the controls. A progressive increase in blood urea nitrogen (BUN) was observed up to 96 h after the first dose of nimesulide but again no quantitative data were provided. These studies provide only inconclusive information concerning the GI effects of nimesulide in dogs. It should be noted, however, that this species is notoriously sensitive to the GI effects of NSAIDs. As the studies were performed in mongrel animals there is also the risk of parasitic infection that could complicate the drug effects.

Wilson and co-workers [333] have determined the effects of nimesulide and other NSAIDs on COX-1 and COX-2 activities in canine tissues and it would appear that the degree of selectivity of nimesulide for COX-2 is not unlike that observed in human tissues [318]. The pharmacokinetic and pharmacodynamic studies of Toutain et al. [334, 335] would suggest that optimal dosage for anti-inflammatory and analgesic effects as well as that for showing COX-2 selectivity of nimesulide is about 5 mg/kg. Since the total daily dose of nimesulide in the study by Ramesh at al. [332] was 4 mg/kg this would be within the range for sparing the inhibition of GI mucosal COX-1 and the selectivity for COX-2. Given that the substantial evidence cited above shows that the low gastric ulcerogenicity of nimesulide is related to its lack of inhibitory effects on gastric prostaglandin production [318, 319, 321–325], it is hard to see how these studies by Ramesh et al. [332] can be reconciled with the known pharmacological effects of the drug on the gastric mucosa including that in dogs [334, 335].

In an attempt to establish if inhibitory effects on gastric acid secretion could underlie the low gastro-ulcerogenicity of nimesulide, Tavares et al. examined the effects of this drug on acid secretion in studies in conventional rodent gastric secretory systems [336, 337]. In the isolated and perfused mouse stomach *in vitro* nimesulide 1.0 to 30 µmol/L caused a concentration-related reduction in histamine-induced acid secretion (determined by changes in gastric pH) [336]. Using Hill plots of effects on histamine induced acid production, nimesulide produced a rightward shift in the cumulative agonist concentration-effect (E/A) curve up to 10 µmol/l but at higher concentrations up to 100 µmol/L markedly reduced the maximal response to about 10–15% of the basal level (Fig. 16a–c [336]). In contrast, the H_2-selective antagonist, famotidine, caused a concentration-dependent rightward shift in the E/A curve which yield a pA_2 value of 7.55 consistent with its actions on histamine receptors. In combination with nimesulide 20 µmol/L this H_2-antagonist suggested additive effects (Fig. 16b and 16c), indicative of nimesulide being without effects on H_2-receptors. Similar effects of nimesulide to those observed with histamine were observed with the stable analogue, 5-methylfurmetide. Indomethacin only produced inhibition of acid secretion at high concentration (100 µmol/l). These results suggest that nimesulide may have low ulcerogenicity, in part, from its anti-acid secretory effects.

It is possible that the effects of nimesulide on mitochondrial ATP production (Fig. 15) [338, 339] (discussed in the later section on liver toxicity) could limit the availability of energy available for acid secretion in a manner observed with salicylates [340]. Studies in rat intestinal mucosal mitochondria show that nimesulide is a less potent uncoupler of oxidative phosphorylation than indomethacin [323] so it is possible that the inhibitory effects of nimesulide on ATP production may occur only at higher dosage of the drug than that observed with indomethacin.

In other studies nimesulide has been found to inhibit histamine release [341] from mast cells and subsequent actions in an anaphylactic model in guinea pigs [341, 342]. Thus, nimesulide may, in contrast to that of other NSAIDs, affect both the production and actions of histamine-regulated acid production. No effects appear to have been reported of the actions of nimesulide on other agonist-induced gastric secretagogues (penta-gastrin, acetylcholine) or those due to nerve stimulation.

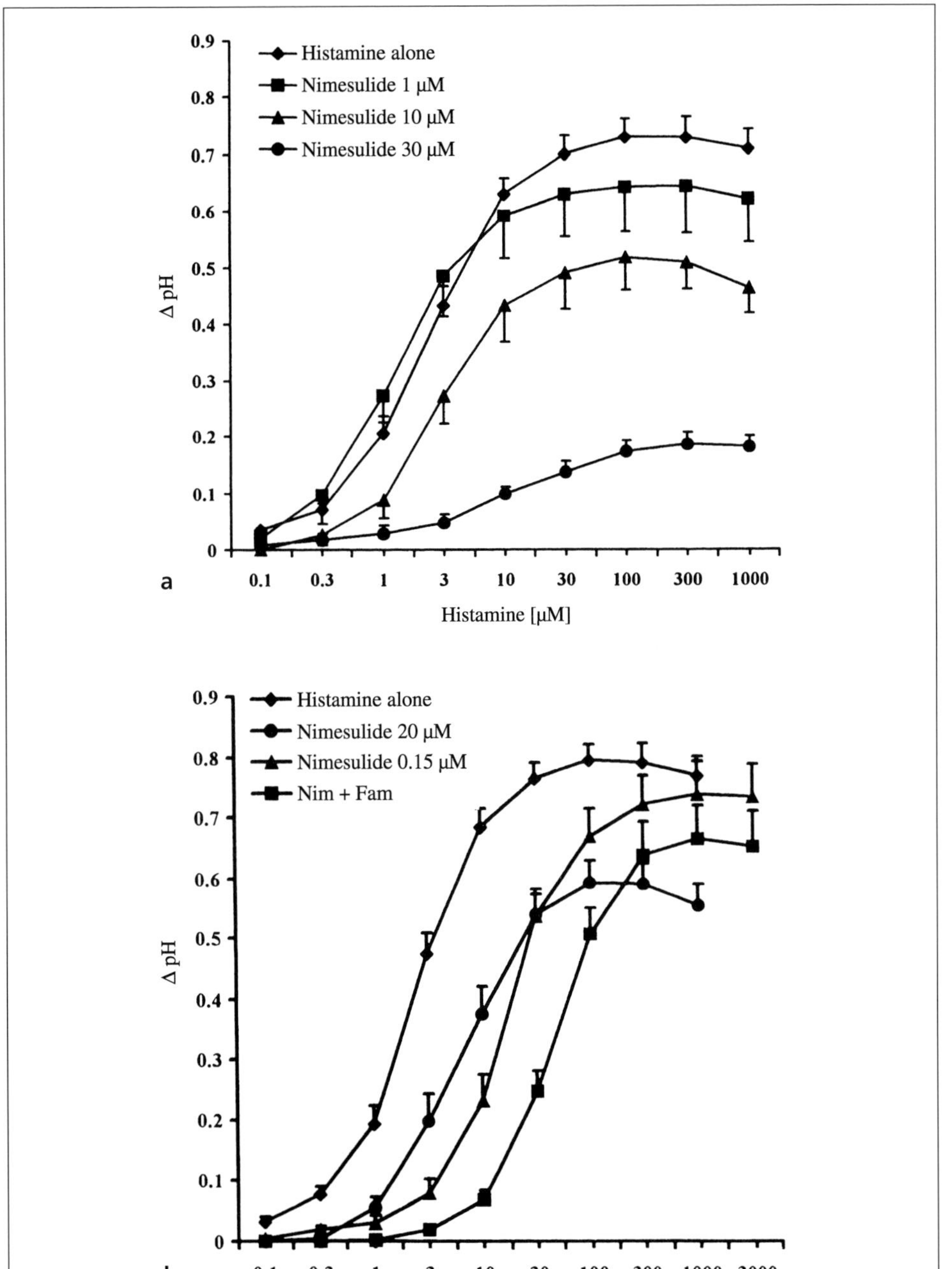
Histamine alone
Nimesulide 1 μM
Nimesulide 10 μM
Nimesulide 30 μM
Δ pH
0.9
0.8
0.7
0.6
0.5
0.4
0.3
0.2
0.1
0
a
0.1 0.3 1 3 10 30 100 300 1000
Histamine [μM]
Histamine alone
Nimesulide 20 μM
Nimesulide 0.15 μM
Nim + Fam
Δ pH
b
0.1 0.3 1 3 10 30 100 300 1000 3000
Histamine [μM]

Figure 16 a, b

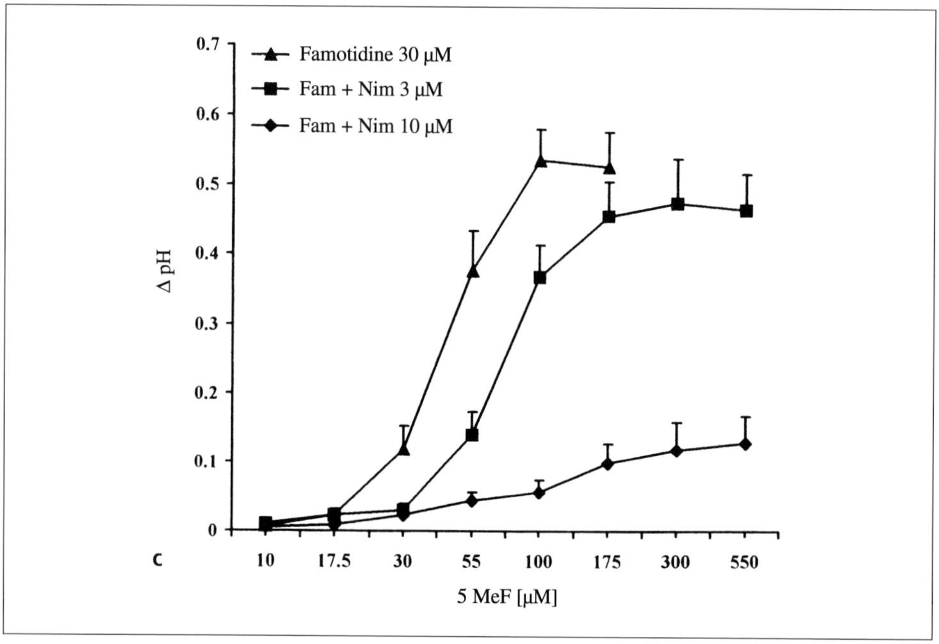

Figure 16c
Effects of nimesulide compared with the H_2-receptor antisecretory agent, famotidine, on acid production stimulated by histamine in the isolated mouse stomach. (a) Acid secretion in the isolated mouse stomach stimulated with histamine in the presence and absence of nimesulide 1.0–30 μmol/L. Nimesulide causes a marked reduction in both the slope and maximum secretion of acid. (b) Acid secretion in the isolated mouse stomach stimulated with histamine in the presence and absence of nimesulide 20 μmol/L, alone or with the H_2-receptor antagonist, famotidine 0.15 μmol/L, or with famotidine alone. Nimesulide and famotidine shift the acid secretion curve to the right.(c) Acid secretion in the isolated mouse stomach stimulated with 5-methyl-furmethide (5-MeF) in the presence of nimesulide 3.0 μmol/L, famotidine 30 μmol/L, or the combination of nimesulide 3.0 or 10 μmol/L with famotidine 30 μmol/L. After Tavares et al. (2001) [336]. Reproduced with permission of the Editor and Publishers of Clinical and Experimental Rheumatology.

Bleeding due to the inhibition of COX-1-derived thromboxane A-2 has been a common feature observed with a wide range of NSAIDs and has been particularly considered to be a factor underlying the development of GI bleeding from aspirin [340]. Saeed and Shah [343] showed that nimesulide could inhibit thromboxane A-2 formation at relatively low concentrations; the IC_{50} being 1.0 mM/L. These authors found that nimesulide inhibited the site of aggregation induced by adrenalin and platelet activating factor with minimum inhibitory effects about 10 mM/L and concentration-related inhibition up to 100 or 200 mM/L. These concentrations are well above those that are encountered during therapy but there may be concentrations that are evident in the gastric mucosa within the focus of dissolution of tablets where there are relatively high concentrations of the drug. Paradoxically however, these authors found that low concentrations of nimesulide 0.01–0.1 mM/L potentiated the aggregatory response to sub-threshold concentrations of adrenalin. The implication from these studies was that using selective inhibitors of calcium channel or activation of nitric oxide that there was a nitric oxide-related and calcium channel effect that was responsible for this potentiation by nimesulide. While the significance of these effects is unclear in relation to whether or not nimesulide may, when the gastric mucosa is damaged, cause bleeding, it does imply that there is a type of biphasic effect on platelet aggregation where low concentrations of nimesulide may promote the aggregation due to adrenalin and maybe other platelet aggregating factors while at high concentrations it inhibits the thromboxane production. Further studies are clearly indicated to establish if, in the event that there is mucosal injury, that nimesulide might or might not potentiate bleeding.

A common feature, which has emerged in the studies of NSAID induced injury, has been the contribution of *Helicobacter pylori* which is often associated with the development of gastroduodenal ulceration, gastritis and mucosal damage [344–346]. Arguments have ranged from the view that *H. pylori* and NSAIDs may be separate factors in ulceration through to the effect of *H. pylori* in stimulating inflammation maybe counteracted by the anti-inflammatory effects of the NSAIDs [345, 346]. Reactions involving increased COX-2 expression from *H.* pylori and in the ulcer crater as well apoptosis promoted by NSAIDs have led to complex interactions between these two groups of ulcerogenic agents [347, 348]. The general consensus is that *H. pylori* certainly is a major factor with the NSAIDs in ulceration and that patients that are *H. pylori* positive have significantly higher rates of GI ulceration and bleeding that those without the infection from *H. pylori* [344]. Most of the endoscopy studies that have been performed with nimesulide have shown that it has either no significant effects compared with placebo or relatively low irritancy depending on the dose and duration of administration of the drug [45, 89, 93]. Certainly, it is of lower observed injury following gastroduodenoscopy in volunteers than observed with most NSAIDs. It was therefore surprising that a study by Kapicioglu and co-workers [349] in which the effects on normal healthy volunteers that had been fasted overnight of nimesulide 100 mg were compared with placebo and aspirin 500 mg, with the endoscopy being undertaken 3 h after intake of the drug. These studies showed that nimesulide had significantly higher scores of gastric mucosal injury that placebo although the average endoscopy lesion score was 1.41 in comparison with that of aspirin that had endoscopy scores of 2.8; the placebo was 0.2. On inspection however, of the data it emerges that about two-thirds of the patients were *H. pylori* positive. Thus it is possible that *H. pylori* played a role in the development of acute mucosal irritancy by nimesulide in these patients. Clearly further investigations are warranted to establish if *H. pylori* is a factor in mucosal irritancy due to nimesulide as this has been shown with some other NSAIDs.

The involvement of blood flow in the development of mucosal injury has been considered a major factor from NSAID injury from NSAIDs [340]. Studies by Guslandi and co-workers [350] further confirmed this but implied that effects of nimesulide were minimal in disturbing blood flow. Blood flow is certainly a feature which emerged from the studies of Hirata and co-workers [326] and the results from the studies in rats indicated that nimesulide had little or any effects on impairing mucosal blood flow and that the normal hyperaemic response to taurocholate was not impaired by nimesulide.

In comparing the actions of nimesulide on the gastric mucosa in relationship to the development of mucosal injury in comparison with other NSAIDs the summary in Table 6 highlights important differences in biochemical and cellular effects of nimesulide compared with that of more ulcerogenic NSAIDs. Major differences include (a) the lack of effects on COX-1, (b) antioxidant activities, (c) inhibitory effects on leucocyte accumulation (and activation and subsequent oxyradical and nitric oxide production), (d) the lack of effects on impairing mucosal blood flow, (e) the inhibition of TNFα, (f) the inhibition of acid secretion and (g) mast cell destabilisation leading to prevention of the release of histamine. A number of other factors are as yet unresolved including the possibility that in inflammatory conditions and cancer cells that normally it would be expected based on studies in isolated tumor cells, there would be an inhibition of proliferation and induction of apoptosis by nimesulide [351]. However, in some inflammatory conditions nimesulide has been shown, particularly in cells like chondrocytes to protect against apoptosis by a mechanism that may involve in part modulating the production of nitric oxide (see Chapter 4) [351]. Thus, the question of whether or not nimesulide produces apoptosis or protects against apoptosis in the stomach where there may be injury or inflammation due to *H. pylori* [347, 348] is as yet unresolved.

Other mechanisms that may be involved in the predisposition to mucosal injury could involve CNS mechanisms including that of acid secretion mediated by the cholinergic pathway. Studies in rats in which the α-2 adrenergic receptor agonist, tizanadeine 0.25 mg/kg p.o., given before administration of nimesulide, naproxen or meloxicam were associated with much less injury to the gastric mucosa and acid production compared with that of the animals that were given the NSAIDs alone [352]. These studies implied that there may be some mediation of α-2 adrenergic receptors in the development in mucosal injury but thesignificance of this in relationship to known mechanisms of mucosal injury is as yet not clear.

Conclusions

Extensive studies in laboratory animal models have shown that nimesulide has relatively low gastric ulcerogenic potential compared with that of other NSAIDs. These studies essentially are confirmed by endoscopic observations in humans. There are clear differences between the actions of nimesulide on the gastric mucosa in comparison with other NSAIDs. It is clear from these observations that nimesulide has a lower potential to impair mucosal defence processes or potentiate the stomach to gastric injury than observed with many NSAIDs.

Intestinal enteropathy

The extensive studies by Bjarnason and co-workers of the effects of NSAIDs on the intestinal mucosa of humans, rats and mice have shown that many of the established NSAIDs produce changes in intestinal permeability, injury and mucosal inflammation in these species [105–107, 110, 139, 354]. Depending on the experimental conditions coxibs may also be without this potential for causing intestinal damage (although notably not in COX-deficient mice [107, 110]). A proposed mechanism of NSAID-induced enteropathy developed by Bjarnason and co-workers [102, 106] is shown in Figure 17. In addition to highlighting the relative role of local mucosal contact with high concentrations of NSAIDs (the "topical effect") in contrast to the systemic effects due to COX-1 inhibition, vascular effects and nitric oxide, this model also emphasises the role of reduction in mitochondrial ATP production, the influence of enteric bacteria, bile and enzymic hydrolytic/proteolytic reactions for combining to cause local tissue reactions [102, 106–108]. To discriminate the effects of nimesulide from that of other NSAIDs in this model the absence of COX-1 inhibitory effects, intestinal permeability changes (probably a function of the high pKa of nimesulide) and possibly anti-proteolytic activities would be expected to reduce the possibility of local as well as systemic components of intestinal injury. What is unclear at this stage is the role played by uncoupling of oxidative phosphorylation with consequent reduction in ATP that has been found to occur with high concentrations or doses of nimesulide [323].

Bjarnason and co-workers [323] compared the effects of nimesulide p.o. with indomethacin 10 mg/kg p.o. on intestinal permeability, mucosal prostanoid concentrations and ATP production in the small intestinal mucosa of rats. Mucosal concentration of PGE, 6-keto $PGF_{1\alpha}$ and TXB_2 were not affected by 10 or 15 mg/kg nimesulide coincident with no effects on mucosal permeability but higher doses of the drug reduced the concentrations of all three prostanoids. Indomethacin likewise reduced the mucosal concentrations of these prostanoids. None of the doses of nimesulide caused inflammatory changes or ulcers although at the highest dose of 60 mg/kg nimesulide there was a transient change of mucosal permeability.

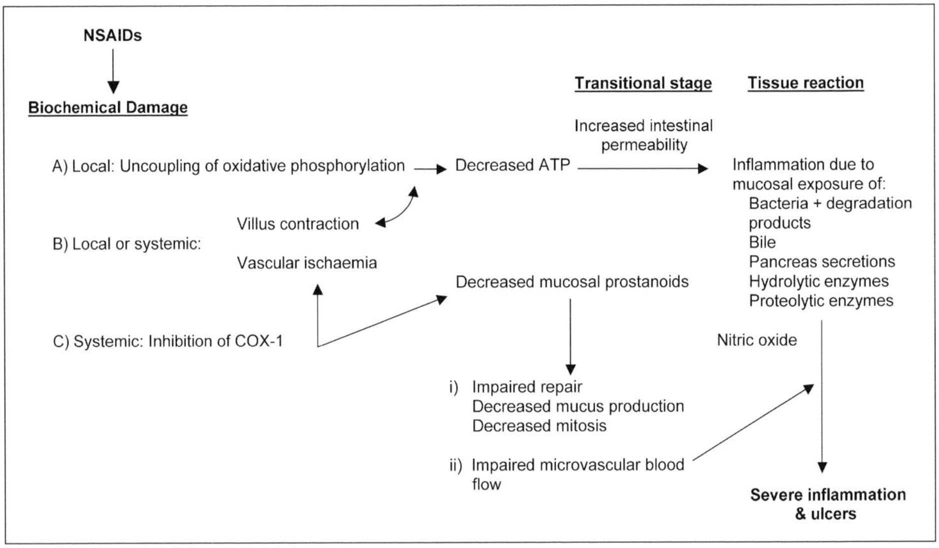

Figure 17
Proposed mechanisms of NSAID-induced enteropathy that involve various local and systemic reactions. The mechanisms differ from those in the stomach due to the added presence of acid, pepsin and Helicobacter pylori *along with unique cellular and physiological (smooth muscle contractile) responses in this organ (see Tab. 5). From Bjarnason and Thodleifsson (1999) [102]. Reproduced with permission of the Editor and Publisher of* Rheumatology (Oxford).

These results show that there is a clear dose-dependent differentiation of effects of nimesulide on the rat small intestinal mucosa. At low–moderate doses there is no inhibition of prostanoid production with any evident mucosal injury, while there are both with indomethacin. At high doses there is inhibition of mucosal prostanoids and no evident injury. The differential effects of nimesulide in contrast to indomethacin have been proposed by Bjarnason and co-workers [108, 323] to be related to the differences in pKa of the drugs (nimesulide pKa 6.5, c.f. indomethacin pKa 3.75).

In addition to having little, if any, effects on the intestinal mucosa, nimesulide 5 mg/kg/d has been shown to have significant protective effects against intestinal inflammation induced by 8-hydroxy-deoxyguanosine and dextran sodium sulphate (DSS)-induced intestinal inflammation in the rat [353]. The effects of nimesulide were shown to be partly related to the reduction in superoxide production and apoptosis that was stimulated by DSS. Protection against the intestinal mucosal lesions induced in rats from burn injury in rats has also been observed with nimesulide and this has been related to the protection against the oxidant stress by the drug [354].

As noted earlier in the section on "NSAIDs and inflammatory bowel disease" most NSAIDs, notably also those with COX-2 selectivity, aggravate the intestinal symptoms in patients with ulcerative colitis and Crohn's disease and this can have important clinical consequences for patients that need these drugs for long-term treatment of arthritic conditions (e.g., HLA-B27 associated ankylosing spondylitis). Nimesulide has been found to be without effects in exacerbating intestinal symptoms in these states [172, 173]. Since COX-2 selective drugs have protective effects against inflammatory reactions in the intestinal tract of rats, in contrast to that of indomethacin or other unselective COX-1/COX-2 inhibitors [355–357] it would appear that nimesulide might also have protective effects through its inhibitory effects on intestinal inflammation.

Overall, nimesulide has been found to be without any appreciable effects on the intestinal tract of rheumatic or normal subjects and in animal models. It also appears that the drug may have little effects in patients with chronic inflammation of the intestinal and may even be protective under some conditions. However, further studies are warranted to fully investigate the clinical and mechanistic aspects of these observations.

Hepatotoxicity

As noted earlier NSAIDs is in general an infrequent cause of hepatotoxicity [47–49, 180–192, 358]. The degree of hepatotoxicity varies from simple elevation of liver transaminases, altered liver function tests, with some minor variations through to more serious manifestations such as cholestatic jaundice, fulminant liver failure and complications an hepato-renal syndrome, manifest from failure of both renal and hepatic detoxifying systems. The development of hepatotoxicity is to some extent an unpredictable phenomenon with NSAIDs and can be regarded as an idiosyncratic event for many of these drugs [43, 359]. It may not be related to the classical pharmacological actions of the NSAIDs, for example as inhibitors of COX-2 and COX-1 or the other actions that are known to underlie anti-inflammatory activity. The link to drug metabolism has however been considered with a number of NSAIDs as well as with paracetamol [359, 360]. In these cases the development of reactive metabolites has been a common underlying feature. To some extent the development of hepatic reactions tends to be screened out during the drug discovery process, particularly with newer NSAIDs as a consequence of long-term toxicity screening in rodents and non-rodent species in the preclinical stage of development and prior to the phase 1 studies which are of course also essentially screening for frequent events [361]. Much information has been derived from *in vitro* studies and *in vivo* investigations in rodents on the mechanisms of hepatotoxic reactions from paracetamol and more commonly hepatotoxic drugs for example diclofenac. These highlight the general concept that there may be reactive metabolites that form under situations where there is evidence of metabolic load from either the drugs of other concomitant medication or agents which lead to the peculiarly high appearance of the reactive metabolite or metabolites under conditions of metabolic or physiopathological stress [359–364].

Complications are frequently seen in the liver in patients who have taken either previously or concomitantly other hepatotoxic agents and this may include statins, oestrogenic steroids, antibiotics or some disease modifying anti-rheumatic agents such as methotrexate [362]. Furthermore, some arthritic conditions predispose hepatic injury from NSAIDs and this is seen in the case of aspirin being taken by patients with systemic lupus erythematosus or in patients with severe hepatic function associated with RA [234–241, 340]. Replication of these kinds of conditions in animal model studies is often difficult although there are indications that salicylate-associated hepatotoxicity is more frequent in rats with adjuvant induced arthritis [340, 363, 364] (see sections on "Clinical aspects of nimesulide-related hepatic reactions from published case reports" and "Hepatic adverse events reported in Finland").

As indicated in the previous section on hepatic adverse events from nimesulide, there has been clear indication of a wide range of concomitant events or intake of hepatotoxic medications that may have predisposed the development of liver toxicity [10, 31, 363, 364]. Analysis of the case reports over the years has highlighted in particular that all of the above mentioned co-factors seem to be common to hepatic injury from nimesulide. The large numbers of case reports that have been published highlight the roles of these concomitant factors [363]. Concomitant intake of antibiotics especially clavulanic acid with amoxycillin or fluoroquinolines has been highlighted [16, 363, 365]. Ranitidine has also been noted as another potentially hepatotoxic agent in cases attributed also to nimesulide [366].

Some of the aspects of the mechanisms of NSAID-induced hepatotoxicity have been reviewed by Boelsterli [43, 364]. This author has made a case for the appearance of a reactive metabolite of nimesulide probably a nitroso or hydroxylamine derivatives that can be postulated from the metabolism of nimesulide [359] by what is actually a very minor route of metabolism of the drug. Regrettably there is virtually little evidence to support the concept that nimesulide produces a reactive metabolite that is the cause of direct toxicity to the liver. The postulated reactions involving the development of the reactive metabolites is shown in Figure 18a and the subsequent reactions focussed on mitochondrial toxicity leading to apoptosis are shown in Figure 18b. Leaving out the question of mitochondrial effects, the development of reactive metabolites has never been demonstrated *in vivo* or even in cells in culture and so this must be regarded as a theoretical concept without any substantive evidence in support of it. It is also questionable in view of the fact that the route for the reduction of the nitro group to form nitroso- or hydroxylamine metabolites is actually a minor pathway probably constituting no more that about 1% of the total metabolites that are excreted [367]. The earlier studies by Swingle

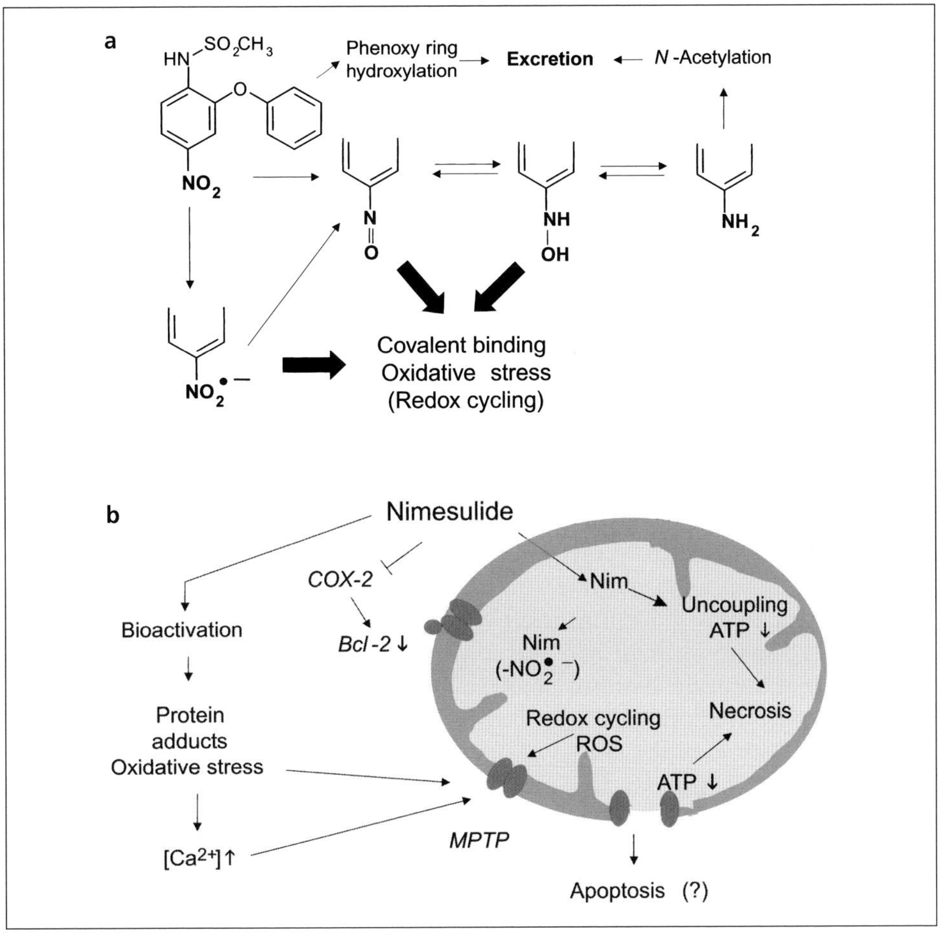

Figure 18
Theoretical formation of reactive metabolites and their effects on mitochondrial uncoupling of oxidative phosphorylation, depletion of ATP and apoptosis via Bcl. (A) Postulated formation of nitroso- and hydroxylamine-metabolites of nimesulide. (B) Suggested actions of metabolites and nimesulide on mitochondria leading to apoptosis. Reproduced with permision of the publishers of International Journal of Clinical Practice. [After Boelsterli (2002) [359].

and Moore [301] (Fig. 19) included the possibility that the 4-nitro group of nimesulide may be activated metabolically. They considered on the basis of known chemistry of the sulphonanilides that they had investigated during the drug discovery process when nimesulide was identified (see Chapter 1) that it was unlikely that the oxidation of nimesulide at the sulpho-amino group was unlikely, but that reduction of the nitro group could involve an electronic transfer process where the

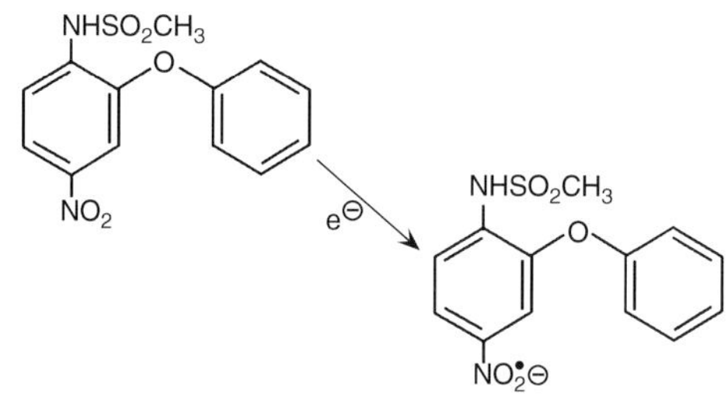

Figure 19
Scheme for formation of the reduction of the nitro-moiety of nimesulide postulated by Swingle and Moore (1984) [301].

nitro group became an electron acceptor (Fig. 19). This poses intriguing possibilities in so much as the involvement of an electron donor process might imply some NADH-reductase or cytochrome P450 reactions, as yet undiscovered. Thus, the possibilities that nimesulide may itself be directly hepatotoxic as a consequence of forming the nitroso- or hydroxylamine metabolites is as yet unproven and probably unlikely at this stage of knowledge. Support for the concept that the drug is unlikely to be directly hepatotoxic is shown from the studies [368] involving an extensive investigation of the effects of nimesulide, its metabolites or manufacturing impurity on the viability and growth of the human hepatoma cell line HepG2 *in vitro*. These studies show that despite intensive investigation of the *in vitro* effects on these human cells, there was evidence for any direct hepatotoxic effects of the drug. Furthermore, investigations using the same cellular system as well as in human primary cells where antibiotics, hormones and various other known hepatotoxic agents were co-incubated with nimesulide or its metabolites show that there is unlikely to be any drug interactions of any major significance with the possible exception of paracetamol (unpublished studies, KD Rainsford).

Another aspect of the mechanisms of liver injury involving radical formation has been highlighted by the studies of Sohi and co-workers [369]. They found that oral administration of 9 mg/kg/d nimesulide twice daily for 1 week followed by intratracheal administration with 2 µg of lipopolysaccharide for 18 h reduced lipid peroxides, stimulated liver glutathione peroxidase (GPO) and surprising reduced superoxide dismutase (SOD) activity. The authors considered that the inhibition of SOD may be a factor in liver injury. However, the fact that there was reduction in lipid peroxides and increased GPO would point to some type of compensation mechanisms on oxidant defence by nimesulide.

Considerable interest has been shown in the effects of nimesulide on liver mitrochondria involving uncoupling reactions linking ATP to mitrochondrial oxido-reductive reactions [338, 339, 370–373]. It should be noted that uncoupling of these oxidative phosphorylation reactions is a common property of NSAIDs with acidic pKa properties [374]. Some indication of the potential for nimesulide to have more pronounced effects in the liver of arthritic subjects was provided by the studies of Caparroz-Assef et al. [370]. They found that supra-therapeutic concentrations of 30–50 μmol/L nimesulide caused stimulation of oxygen consumption in perfused livers and isolated mitochondria, inhibition of gluconeogenesis and stimulation of glycogenolysis, and reduction in the ADP/O and respiratory control ratios in livers from normal but not arthritic rats. Since mitochondria of rats with adjuvant arthritis were found to have high rates of oxygen consumption and altered glucose metabolism, it seems likely that the more pronounced mitochondrial effects of nimesulide in arthritic rats could be related to the arthritic rats being defective along with that of glucose metabolism. This would predispose arthritic animals to mitochondrial effects of nimesulide.

Mingatto et al. [373] observed that incubation of rat hepatocytes with 0.1–1.0 mmol/L nimesulide resulted in a time-dependent decrease in cell viability as measure by the leakage of lactate dehydrogenase, decrease in mitochondrial membrane potential determined by rhodamine 123 retention and reduction in cellular ATP. The concentrations of nimesulide employed by these authors far exceed those encountered therapeutically and thus they must be regarded as high toxic drug concentrations. The amino-metabolite of nimesulide did not elicit these toxic effects.

In a survey of effects of NSAIDs on oxidant stress, Galati and co-workers [375] found that several fenamate drugs and diclofenac that possess a diphenylamine structure along with some sulphonamide drugs were oxidised by tissue peroxidases to form pro-oxidant derivatives. These had the effect of reducing GSH and NADH. Neither nimesulide nor indomethacin had pro-oxidant activity. This observation may with nimesulide be a reflection of its antioxidant effects. Contrasted with diclofenac, a well-known hepatotoxic NSAID, highlights the importance of intrinsic oxidant activity of this NSAID compared with to that of nimesulide and suggests that the mode of action of these dugs is different in the liver.

Of the cholestatic mechanisms that may be associated with liver injury by nimesulide and other NSAIDs bile transporters and multi-drug resistance (MDR) phenotypes may be important [376–378]. The role of MDR1 phenotype in PGE_2 from COX-2, NO from iNOS and cell proliferation was investigated by Fantappe et al. [378]. Both nimesulide and celecoxib inhibited cell proliferation in MDR1 but not normal human liver cells.

A suggestion that the Bcl-2 expression in hepatocytes may be regulated by COX-2 expression in Kupffer cells [379] raises the possibility that nimesulide by inhibiting PGE_2 in the latter cells may regulate Bcl-2 during the development of apoptosis in hepatocytes.

Given that *Helicobacter pylori* infection is associated with cholestatic and other liver reactions [380] and this being frequently found in arthritic patients, it is possible that the cholestatic liver reactions attributed to nimesulide or other NSAIDs may be, in part, a consequence of infections with this organism.

Renal toxicity

As indicated in the section on pharmacoepidemiology nimesulide has been only infrequently associated with renal injury. Since COX-1 as well as COX-2 are involved in renal functions though effects on the renin–angiotensin systems and the renal tubular excretion/re-absorption systems it is more likely that the predominant COX-2 effects of nimesulide would be expected to affect only part of the renal functions [381, 382]. In normal subjects transient effects of nimesulide have been noted on sodium and potassium excretion [383]. Furosemide increase in plasma renin and aldosterone has been found to be blunted by nimesulide 200 mg b.i.d. with concomitant reduction in urinary excretion of PGE_2 in normal healthy subjects [383]. Nimesulide with or without furosemide reduced glomerular filtration rate and renal plasma flow and increased urinary flow and excretion of water [383]. These physiological effects are common to most NSAIDs [384].

Neonatal renal failure has been reported following *in vitro* exposure to NSAIDs [385], but nimesulide has not been found to be associated with renal adverse effects in children [386]. In the newborn rabbit intravenous bolus administration of 2–200 µg/kg nimesulide followed by 0.05–5 µg/kg/min by i.v. infusion caused a dose-dependent increase in renal vascular resistance, decreased glomerular filtration rate, diuresis and renal blood flow [387].

In the isolated perfused rat kidney from normal and diabetic rats indomethacin 10 µmol/L abolished the vasoconstrictor effects of perfused arachidonic acid whereas nimesulide 5 µmol/L only reduced perfusion pressure in diabetic rats coincident with increased expression of renal cortical COX-2. The results imply that nimesulide may have vascular resistance effects in the renal system of diabetic individuals and this may be related to the COX-2 inhibitory effects of the drug.

Renal dynamics in response to nimesulide have been investigated in dogs [388, 389]. In anaesthetised dogs that received an intra-renal infusion of noradrenaline (50–250 ng/kg/min) pretreatment with nimesulide before the noradrenaline, those dogs with the normal sodium intake resulted in decrease in glomerular filtration rate which was greater than that of noradrenaline alone; these effects were not paralleled in changes in renal vascular resistance [389]. Similar effects were noted with meclofenamate. In those dogs that have been on low sodium intake high doses of noradrenaline resulted in decrease of glomerular filtration rate and a rise in renal vascular resistance. Administration of nimesulide resulted in a further fall in glomerular filtration rate and an increase in renal vascular resistance that was greater than that due to noradrenaline alone. These results were paralleled by changes in the renal excretion rates for PGE_2 and 6-keto-$PGF_{1\alpha}$. The results were interpreted by the authors as showing that the inhibition of COX-2 by nimesulide potentiates the renal haemodynamic effects of noradrenaline and that effects on renal haemodynamics are mediated by COX-2 production of prostaglandins.

In a similar model of anaesthetised dogs the same group observed that the concomitant administration of the nitric oxide synthesis inhibitor L-NAME (N^G-nitro-1-arginine-nitro ester) potentiated to a greater extent the noradrenaline-induced vasoconstriction which was evident from either of these drugs alone. The administration of an angiotensin-1 receptor antagonist partially reversed these effects. The authors concluded that there are interactions between nitric oxide and COX-2 derived prostaglandins on the renal vasculature and that angiotensin II partly mediates these effects [390]. In another study the same group [389] investigated the effects of varying sodium load following administration over 8 days of nimesulide but this time the animals were conscious. Sodium excretion was found to be reduced during the first day of administration of nimesulide in animals that were on normal or high sodium load. An increase in plasma potassium levels was evident in those dogs that were on a low sodium intake. These effects were enhanced when nitric oxide was inhibited showing that there was no interaction between nitric oxide inhibition and COX-2 inhibition in mediating these effects on potassium but they did not appear to be related to alterations in plasma aldosterone levels.

These three studies [388–390] probably constitute the most thorough investigation of the interrelationships between COX-2 inhibition by nimesulide, nitric oxide and the renal haemodynamic and excretion mechanisms. The results can be interpreted in similar ways to that observed from the administration of non-selective COX-1/2 NSAIDs [391]. The clinical significance of these observations is that they do show that nimesulide like other NSAIDs can affect normal renal function and that sodium status can have a marked influence on renal excretion and haemodynamics. Studies in streptozotocin-induced diabetic rats suggest that the increase in renal cortical expression of cyclooxygenase leading to the production of not only prostaglandins but also of 20-hydroxyeicosa-tetra-enoic acid which may lead to more profound vasoconstrictor effects in diabetics [392]. Any alterations in production of these arachidonic acid metabolites may have more profound effects in the diabetic compared with the normal state [392]. Factors that may also influence the actions of nimesulide in the kidney could also relate to the uncoupling of oxidative phosphorylation and inhibition of ATP, which has been a known mechanism of action of salicylates, and several other NSAIDs [340, 374].

Further studies are indicated to define the mechanisms of action of nimesulide on renal function in arthritic states and those where there may be elements of renal compromise such as in the elderly.

Cutaneous reactions

Cutaneous reactions have been reported infrequently to nimesulide as noted earlier [70–74]. The more serious manifestations of skin reactions such as Stevens-Johnson and Lyell's syndromes that are common to many NSAIDs [7, 8, 60, 62, 68] have essentially no defined mechanism. Skin rashes and cutaneous eruptions have however been attributable to an immunological cause and may involve T-cell sensitisation. Skin sensitisation assays were undertaken by Kanikkannan and co-workers [393] using the mouth local lymph node assay. Concentrations of nimesulide ranging from 0.5–10% W/V dissolved in acetone-olive oil (4:1) were applied in 20 ml concentrations of volumes to the ear of female mice for 3 days, on day 6 radioactive thymidine was administered intravenously and the uptake of radioactivity in the draining lymph nodes was then determined. It was found that nimesulide did not exhibit any skin sensitisation like that observed with dinitrochlorobenzine.

The production of reactive metabolites of NSAIDs has been postulated as a mechanism of action in causing cutaneous reactions [8]. Obvious candidates as reactive metabolites may include those that have been postulated to be involved in liver reactions such as the nitroso- or hydroxylamine derivatives of the drug. However, no evidence exists for this postulated mechanism and as with liver toxicity that has been postulated to occur with this it is unlikely to be of significance since, in particular, not only it has been pointed out that the formation of these metabolites and the detoxification is essentially a very minor pathway but also in the case of any reactive metabolites accumulating in the skin and leading to activation of resident Langerhans and other immune cells and T-cell activation is probably going to be unlikely.

In a study of 260 patients with a history of recent pseudo-allergic skin reactions induced by NSAIDs, Asero [394] explored the skin reactions occurring from oral intake of either paracetamol or nimesulide; it was found that 19% of the patients reacted to paracetamol and nimesulide and those with a history of aspirin-induced urticaria did not tolerate either of these drugs. Thus, aspirin intolerance represents the major factor in paracetamol or nimesulide induced urticaria. Atopic status was associated with a higher risk of reactions to nimesulide, these being about double that encountered in non-atopic individuals. Interestingly, a history of intolerance to antibacterial drugs was not associated with any increased sensitivity to paracetamol or nimesulide. These results regarding the low sensitivity of NSAID-intolerant patients to nimesulide have been observed by others [395–397] and are indicative of some risks of developing urticaria and other related skin reactions from nimesulide being related largely to a history of sensitivity to other NSAIDs or anaphylaxis. Pastorello and co-workers [397] also observed an increase in intolerance to NSAIDs in patients that had a history of reactions to antimicrobial agents, although this was not so prevalent in patients that received nimesulide.

In another study involving challenge of patients with the history of urticaria or angioedema Sanchez Borges and co-workers [395] noted that there was similar cross-reactions to COX-2 inhibitors with the exception of rofecoxib which had a lower incidence of cross reactivity in patients with urticaria or angioedema. Similar observations were noted in another study by Quiralte and co-workers [396].

Since skin reactions are very difficult to elicit in laboratory animal models very little information can be gleaned from such studies even when skin sensitisers are used since these are in many cases toxic in themselves and elicit a very specific array of reactions. Of the studies that have been done in humans they are only indications of risk factors but no information is available on the mechanism of skin reactions even though these are of relatively minor occurrence in the case of nimesulide compared with that of other NSAIDs which are known to be more likely to induce these conditions [8].

Discussion and conclusions

The clinical epidemiological and experimental evidence reviewed in this chapter has highlighted the relative safety of nimesulide in comparison with that of other NSAIDs including the newer range of coxibs. A recent detailed review at the European Medicines Evaluation Agency of the epidemiological and clinical data supports claims for the drug having a favourable benefit/risk profile in patients with acute pain conditions and those with osteoarthritis and other conditions such as low back pain, dysmenorrhoea.

The clinical data and information from studies in experimental animal models strongly supports the epidemiological data showing that nimesulide has a relatively low risk of serious GI reactions. The sparing of COX-1 combined with its properties of controlling histamine released from mast cells, anti-acid secretory activity, inhibitory effects on leucocyte emigration and activation, antioxidant activity and possibly effects on the production and action of proinflammatory cytokines are all important in protective effects of the drug on the gastric mucosa. Of particular interest is the potential for the drug to have little if any effects on the intestinal mucosa; indeed, it may even have protective effects against intestinal inflammation.

Nimesulide, as with other NSAIDs, has a risk of developing hepatic and renal adverse reactions, though the latter may be of low grade. The hepatic reactions appear to be largely related to problems of concomitant medication with potentially hepatotoxic drugs (paracetamol, diclofenac, oestrogenic steroids, statins) as well as hepatic conditions that predispose development of disease states that may be triggered by nimesulide. The evidence that there may be reactive metabolites produced during the hepatic metabolism of nimesulide (e.g., nitroso- or hydroxylamine- derivatives) has not yet been sufficiently persuasive to enable a mechanism to be proven. The relatively minor route of reductive metabolism of the nitro group of nimesulide, coupled with relatively rapid metabolism to the amino derivative, and the lack of evidence for direct cytotoxicity of the drug, its metabolites and manufacturing impurities does not support adequately the reactive metabolite hypothesis. The effect of nimesulide in uncoupling oxidative phosphorylation leading to apoptosis is a feature observed with many NSAIDs and although these have some risk for developing hepatic toxicity, this alone can hardly be regarded as a substantive mechanism for hepatic injury of either these drugs or nimesulide. The fact that supra-therapeutic concentrations of nimesulide are required *in vitro* to demonstrate effects on oxidative phosphorylation and reduction in ATP suggests that this mechanism may not alone account for hepatocellular damage by nimesulide, as indeed has been shown with the salicylates [340].

The renal effects of nimesulide on haemodynamic, renal tubular and water and electrolyte excretion are similar to those encountered with other NSAIDs. There are no indications of any unique drug interactions between drugs that influence renal tubular excretion or angiotensin inhibitors and nimesulide that may unduly influence the normal physiological functions, except for transient changes in renal excretion and blood flow that are commonly observed with NSAIDs.

Little is known about the mechanisms of cutaneous reactions from nimesulide, but evidence from studies with other NSAIDs implicates effects on T-cells and the skin Langerhans cells in mediating these reactions that may well be due to some as yet unspecified reactive metabolites.

The risks of serious cardiovascular reactions, which have been recently observed with the coxibs, have not been observed with nimesulide. Indeed, there appears to be a relatively low risk of developing myocardial infarction or congestive heart failure from nimesulide as evidenced from spontaneous reporting. Clearly controlled clinical evaluation is required to support the view that there may be a lower risk of nimesulide precipitating myocardial infarction or other cardiovascular events. The pharmacological properties of nimesulide as a weak inhibitor of platelet aggregation may confer on this drug some partial protection against the development of thrombosis that appears to be a problem with the highly selective COX-2 inhibitors (coxibs).

While the principal marketing authorisation holders based in Europe no longer recommend nimesulide for use in children in Europe as well as in Central and South America and Turkey, it is to be noted that nimesulide is widely prescribed in India as generics [398–402]. Recent concerns in India following reports of serious reactions in children, along with some reports received during 1993–1999 by the National Pharmacovigilance Centre in Portugal, of some serious reactions, have led to concerns about the use of nimesulide in children [398, 399]. This is clearly a regulatory issue, possibly unique to India where there appears to be vigorous promotion of the drug by way of advertisements in some of the medical journals in that country for paediatric uses with highly flavoursome formulations. Some studies have been reported from India in paediatric patient populations that have been well below the age deemed to be safe for taking the drug [401]. Use of the drug in children is clearly not recommended, even though the pharmacokinetics in children would indicate that the drug has a relatively favourable biodisposition in this population (see Chapter 2). In the elderly, the drug appears to be well tolerated and aside from interactions with antihypertensive and diuretics would not be expected to result in serious adverse reactions in this patient population.

Summary

Nimesulide, like other NSAIDs, exhibits adverse reactions in the major organ systems comprising the upper GI tract, liver, kidney, skin and immune systems. Noteworthy is the fact that this drug has relatively low occurrences of GI ulcers and bleeding, asthma and respiratory tract reactions and does not appear to have the cardiovascular reactions (congestive heart failure, myocardial infarction) that has been observed recently with the coxibs and some other NSAIDs.

Summary of evidence in major organ systems

Gastrointestinal tract

- Reports of serious GI events from nimesulide are rare. Over the past 5 years, the total of all GI ADRs has averaged 1.1 cases for 10^6 treatment courses (range 0.77–2.01 cases per 10^6 treatment courses) per annum. A total of 315 cases of GI ADRs attributed to or involving nimesulide have been reported since the drug was first introduced in 1985, during which there have been 415 million treatment courses sold. These GI reports comprised 15.7% of all ADR reports and 4.4% were fatal (possibly not due to the drug). In many cases confounding factors were evident (e.g., pre-existing ulcer disease, concomitant ulcer organic drug intake).
- Pharmaco-epidemiological studies have given data on relative risks of serious GI events (haemorrhage, ulcers) with nimesulide in comparison with other NSAIDs. These data derive from case-control, cohort and hospitalisation studies in which the Relative Risks (RR) or Odds Ratios (OR) have ranged from 1.2, 2.0 and 4.4 respectively. In comparison, drugs with the highest risk in these studies were ketorolac (4.2), piroxicam, (4.6) and ketorolac (24.7) or piroxicam (15.5), respectively. When the exposure period is restricted to 15 days the rate ratios for all NSAIDs rose but not those for nimesulide. This highlights the benefit of nimesulide over other NSAIDs for short-term treatments.

  Thus, on average nimesulide has the lowest rate ratio being roughly comparable to that from ibuprofen at anti-rheumatic doses (e.g., 2.4 g/d) or diclofenac, both of which are accepted as low GI risk drugs and much lower than other NSAIDs, e.g., aspirin, naproxen and piroxicam.

- Upper GI endoscopy studies in normal human volunteers showed that the standard therapeutic dose of nimesulide 100 mg b.i.d. taken for 1–2 weeks was markedly less irritant to the stomach than either naproxen 500 mg b.i.d. or indomethacin 50 mg t.i.d. In the first study to be reported demonstrating COX-2 selectivity from an NSAID in humans, nimesulide 100 mg b.i.d taken for 2 weeks did not reduce PGE_2 or 6-keto$PGF_{1\alpha}$ in the gastric mucosa or COX-1-derived TxB_2 in the serum, whereas naproxen 500 mg b.i.d. caused pronounced inhibition of both the gastric mucosal and serum prostanoids. These results were paralleled by relatively little gastroduodenal mucosal damage observed endoscopically from nimesulide but marked damage with the naproxen treatment. Moreover, there was increased intestinal inflammation as observed by faecal calprotectin excretion and intestinal permeability with naproxen, but no significant increase in inflammation or permeability was found with nimesulide. Other endoscopy studies in patients with dyspepsia or osteoarthritis confirmed the low gastric irritancy of nimesulide.
- Nimesulide has low gastric irritancy in rats and other laboratory animal models when given at oral or i.p. doses up to 20–40 times those that are required for acute or chronic anti-inflammatory effects. Even exposure to physical stress or concomitant treatment with the corticosteroid, prednisolone, failed to exacerbate the mucosal effects of nimesulide. In most studies in rats COX-1-derived PGE_2 production by the gastric mucosa was unaffected by nimesulide given orally within the dose-range for anti-inflammatory effects; higher doses could result in inhibition of mucosal PGE_2 or 6-keto $PGF\alpha$. No effects have been observed in rats with nimesulide on gastric mucosal permeability, potential difference mucosal blood flow, or the secretion of gastric acid or duodenal bicarbonate stimulated with histamine. In the isolated perfused mouse stomach nimesulide reduced histamine- or 5-methylfurmetidine-induced acid secretion and exhibited additive inhibitory effects with an H_2-receptor antagonist. Histamine release from mast cells has been shown to be inhibited by nimesulide and this may account for the reduction in acid-secretion by this drug.
- The low propensity for inhibition of COX-1, antioxidant and anti-secretory effects combined with inhibitory effects on leucocyte emigration by nimesulide combined with its high pKa (6.5) may account for the low irritancy on the gastric mucosa of both animals and humans.
- Little if any intestinal injury, permeability or inflammatory changes have been observed in rats or pigs dosed orally with nimesulide, although under some conditions reduction in mucosal PGE_2 may occur in the small intestine. Intestinal inflammation induced by dextran sulphate + 8-hydroxy-deoxyguanosine in rats is reduced by nimesulide. In contrast to the effects of NSAIDs, including coxibs,

nimesulide does not aggravate intestinal symptoms in patients with ulcerative colitis or Crohn's disease.

- Overall, all the evidence shows that nimesulide does not exhibit the marked GI effects seen with many other NSAIDs. This appears to be related to the combined effects of sparing of effects on COX-1 activity, high pKa, antioxidant, anti-secretory, antihistamine and leucocyte inhibitory effects.

Hepatic

- As with most NSAIDs, nimesulide has been associated with the occurrence of hepatic reactions. These include the elevation of plasma levels of the liver transaminase enzymes (ALT, AST, γ-GT), which are observed infrequently, alterations of liver function tests (alkaline phosphatase [ALP], free and conjugated bilirubin) and rarely evidence of cholestatic jaundice. A few cases of liver failure have been reported.
- Hepatobiliary disorders account for 14.3% of all ADRs, and abnormal laboratory findings (6.6%) (principally those involving abnormal liver function tests which accounts for some of the hepatobiliary reactions).
- In most cases withdrawal of the drug has resulted in return of liver function enzymes or liver tests to normal.
- Most cases have confounding factors (other hepatotoxic drugs or liver diseases).
- Epidemiological data show that the occurrence of hepatopathy is at the upper end of the range observed with NSAIDs. The relative risks in comparison with NSAIDs is 1.3 (CI = 0.7–2.3) with an increase in RR to 1.9 where data on ALT above 5 × upper normal limit are included. In a nested case-control study the risk of hepatopathy from nimesulide was estimated to be 1.4.
- Nimesulide does not cause direct cytotoxic damage to liver cells in culture, although there may be increased cell damage when paracetamol or other hepatotoxic drugs were added.
- There has been speculation that nitroso- or hydroxylamine reactive metabolites of nimesulide may be responsible for the liver damage from the drug, in analogy to reactive metabolite injury from diclofenac, paracetamol and other hepatotoxins. To date there is no evidence to support this reactive metabolite hypothesis of cell injury by nimesulide. Reduction in mitochondrial ATP and other functions has been observed with nimesulide following administration of high doses of the drug to rats. This phenomenon is related to uncoupling of oxidative phosphorylation has been observed with a number of acidic NSAIDs and may account for the development of liver injury by these drugs. Reduction in ATP may initiate apoptosis by these drugs.

Renal

- As with other NSAIDs adverse events in the renal system have been rarely observed with nimesulide. These include nephropathies such as tubular or interstitial nephritis and nephrotic syndrome and these along with renal failure are rare.
- Renal and urinary disorders account for 4.7% of all ADRs reported, which is relatively low in comparison with the standard NSAIDs.
- Inhibition of renal prostaglandin production accounts for most of the renal effects (e.g., electrolyte and water impairments) of NSAIDs which are mostly temporary effects. Renal abnormalities are more common in the elderly in whom renal clearance is impaired. The occurrence and development of such symptoms is not evident with nimesulide especially considering the widespread use of the drug.
- Nimesulide blunts the effects of furosemide-induced increase in plasma renin concomitant with reduction in renal PGE_2; this effect is commonly observed with other NSAIDs.
- Renal dynamics have been investigated in normal and diabetic rats as well as in dogs in comparison with standard NSAIDs. In these models nimesulide appears to have lesser effects on PG regulated renal functions compared with other NSAIDs including the transient reduction in renal excretion of sodium which is noted with most NSAIDs. The inhibition of COX-2 appears to potentiate the effects of noradrenaline, but the biological and clinical significance of this is not known.
- Overall, the effects of nimesulide on haemodynamics and renal functions are similar to those observed with other NSAIDs. There is a small effect of COX-1 sparing in the kidney that might account for the drug being less likely to have nephrotoxic effects but the biological and clinical significance of this is not known.

Cutaneous and allergic reactions

- Like other NSAIDs, nimesulide is a frequent cause of minor skin reactions (erythematous rashes, urticaria, etc.), and these are the most frequent of ADRs that have been reported. Mostly cessation of intake of drug causes the symptoms to disappear.
- Rarely, angioedema, Stevens-Johnson and Lyell's syndromes have been reported. The impression is that the occurrence of these may be lower with nimesulide than with other NSAIDs.
- Nimesulide has low intolerance in patients with pseudo-allergic reactions to other NSAIDs. In atopic individuals and in aspirin-intolerant patients the incidence of allergic reactions from nimesulide (19%) is the same as that from paracetamol, a drug which is known to have low intolerance.

- In laboratory animal models skin sensitisation is not apparent with nimesulide as with agents like dinitro-chlorobenzene, which induce such reactions.
- Asthma is a very rare event associated with nimesulide.
- It is possible that the mast cell stabilising and antihistamine effects confer some protection against the allergic reactions in subjects that are predisposed to these conditions.

Cardiovascular system

- Serious cardiovascular reactions (e.g., myocardial infarction, congestive heart failure) which have recently been reported with the coxibs and some other NSAIDs have only very rarely been reported to any appreciable extent with nimesulide.
- The pharmacological properties of nimesulide on COX-2 balanced by weak effects on COX-1 combined with modest antiplatelet effects and short plasma half-life of the drug may confer on it unique properties that may account for nimesulide not being associated with serious cardiovascular complications.

Overall

Nimesulide has been intensively investigated for adverse reactions and their mechanisms in the two decades since the drug was marketed. The low propensity for nimesulide to have GI, renal, cardiovascular and allergic reactions may relate to its novel pharmacological, toxicological and pharmacokinetic properties. Liver reactions are of the same risk as those from other NSAIDs.

The benefit/risk assessment of nimesulide is indicative of it being most favourable for the treatment of exacerbation of chronic conditions, such as osteoarthritis, musculoskeletal and various painful and other inflammatory conditions where the drug is recommended.

References

1. Rainsford KD, Velo GP (Eds) (1983) *Side-Effects of Anti-Inflammatory/Analgesic Drugs*. Raven Press, New York
2. Rainsford KD, Velo GP (Eds) (1987) *Side-Effects of Anti-Inflammatory Drugs. Clinical and Epidemiological Agents*. MTP Press, Lancaster
3. Rainsford KD, Velo GP (Eds) (1992) *Side-Effects of Anti-Inflammatory Drugs*. III. Kluwer Academic Publishers, Lancaster

4. Rainsford KD (1999) Profile and mechanisms of gastrointestinal and other side-effects from NSAIDs. *Am J Med* 107(6A): 27S–36S
5. Simon RA, Namazy J (2003) Adverse reactions to aspirin and nonsteroidal antiinflammatory drugs (NSAIDs) *Clin Rev Allergy Immunol* 24(3): 239–252
6. Day R (2003) COX-2: where are we in 2003? Distinction from NSAIDs becoming blurred. *Arthritis Res Ther* 5(3): 116–119
7. Sanchez-Borges M, Capriles-Hulett A, Caballero-Fonseca F (2003) Cutaneous reactions to aspirin and nonsteroidal antiinflammatory drugs. *Clin Rev Allergy Immunol* 24(2): 125–136
8. Rainsford KD (1992) Mechanisms of rash formation and related skin conditions induced by non-steroidal anti-inflammatory drugs. In: KD Rainsford, GP Velo (Eds): *Side-Effects of Anti-Inflammatory Drugs–3*. Kluwer Academic Publishers, Lancaster, 287–301
9. Sturkenboom M, Nicolosi A, Cantarutti L, Mannino S, Picelli G, Scamarcia A, Giaquinto C; NSAIDs Paediatric Research Group (2005) Incidence of mucocutaneous reactions in children treated with niflumic acid, other nonsteroidal antiinflammatory drugs, or nonopioid analgesics. *Pediatrics* 116:e26–33 [e-print ahead of publication]
10. Rainsford KD (1999) Relationship of nimesulide safety to its pharmacokinetics: Assessment of adverse Reactions. *Rheumatology (Oxford)* 38 (Suppl): 4–10
11. de Abajo F, Montero D, Madurga M, Garcia Rodriguez LA (2004) Acute and clinically relevant drug-induced liver injury: A population-based case-control study. *Br J Clin Pharmacol* 58: 71–80
12. Van Steenbergen W, Peeters P, De Bondt J, Staessen D, Buscher H, Laporta T et al. (1998) Nimesulide-induced acute hepatitis: Evidence from six cases. *J Hepatol* 29: 135–141
13. Schattner A, Sokolovskaya N, Cohen J (2000) Fatal hepatitis and renal failure during treatment with nimesulide. *J Intern Med* 247: 153–155
14. Weiss P, Mouallem M, Bruck R, Hassin D, Tanay A, Brickman CM et al. (1999) Nimesulide-induced hepatitis and acute liver failure. *Isr Med Assoc J* 1: 89–91
15. McCormick PA, Kennedy F, Curry M, Traynor O (1999) COX 2 inhibitor and fulminant hepatic failure. *Lancet* 353: 40–41
16. Elmalem E (2000) Nimesulide, clavulanic acid and hepatitis. *J Intern Med* 248: 168–169
17. Andrade RJ, Lucena MI, Fernandez MC, Gonzalez M (2000) Fatal hepatitis associated with nimesulide. *J Hepatol* 32: 174
18. Romero-Gomez M, Nevado Santos M, Otero Fernandez MA, Fovelo MJ, Suarez-Garcia E, Castro Fernandez M (1999) Acute cholestatic hepatitis induced by nimesulide. *Liver* 19: 164–165
19. Ferreiro C, Vivas S, Jorquera F, Dominguez AB, Espinel J, Munoz F et al. (2000) Toxic hepatitis caused by nimesulide, presentation of a new case and review of the literature. *Gastroenterol Hepatol* 23: 428–430
20. Rodrigues de Oliveira J, Correia J, Silvestre F, Meirelles A, Bernardo A (2000) Severe acute hepatitis probably induced by nimesulide. *Gastroenterol Clin Biol* 24: 592–593
21. Merlani G, Fox M, Oehen HP, Cathomas G, Renner EL, Fattinger K et al. (2001) Fatal hepatotoxicity secondary to nimesulide. *Eur J Clin Pharmacol* 57: 321–326

22. Sbeit W, Krivoy N, Shiller M, Farah R, Cohen HI, Struminger L et al. (2001) Nimesulide-induced acute hepatitis. *Ann Pharmacother* 35: 1049–1052
23. Macia MA, Carvajal A, del Pozo JG, Vera E, del Pino A (2002) Hepatotoxicity associated with nimesulide: Data from the Spanish Pharmacovigilance System. *Clin Pharmacol Ther* 72: 596–597
24. Garcia Rodriguez LA, Perez Gutthann S, Walker AM, Lueck L (1992) The role of nonsteroidal anti-inflammatory drugs in acute liver injury. *Br Med J* 305: 865–868
25. Carson JL, Strom BL, Duff A, Gupta A, Das K (1993) Safety of nonsteroidal anti-inflammatory drugs with respect to acute liver disease. *Arch Intern Med* 153: 1331–1336
26. Perez Gutthann S, Garcia Rodriguez LA (1993) The increased risk of hospitalizations for acute liver injury in a population with exposure to multiple drugs. *Epidemiology* 4: 496–501
27. Garcia Rodriguez LA, Williams R, Derby LE, Dean AD, Jick H (1994) Acute liver injury associated with nonsteroidal anti-inflammatory drugs and the role of risk factors. *Arch Intern Med* 154: 311–316
28. Garcia Rodriguez LA, Ruigomez A, Jick H (1997) A review of epidemiologic research on drug-induced acute liver injury using the general practice research database in the United Kingdom. *Pharmacotherapy* 17: 721–728
29. Traversa G, Bianchi C, Da Cas R, Abraha R, Menniti-Ippolito F, Venegoni M (2003) Cohort study of hepatotoxicity associated with nimesulide and other non-steroidal anti-inflammatory drugs. *Br Med J* 327: 18–22
30. Press Release (2003) European Medicines Evaluation Agency (EMEA) states that nimesulide is safe and effective. www.nimesulide.net (accessed 16 Sept 2004)
31. Rainsford KD (1998) An analysis from clinico-epidemiological data of the principal adverse events from the COX-2 selective NSAID, nimesulide, with particular reference to hepatic injury. *Inflammopharmacology* 6: 203–221
32. Rainsford KD, Shepherd P (2003) Analysis of the adverse reactions in the liver, renal and urinary systems and gastrointestinal tract from the COX-2 NSAID nimesulide. Report to Helsinn Healthcare SA
33. Lewis SC, Langman MJS, Laporte J-R, Matthews JNS, Rawlins MD, Wiholm B-E (2002) Dose-response relationships between individual non-aspirin non-steroidal anti-inflammatory drugs (NANSAIDs) and serious upper gastrointestinal bleeding: a meta-analysis based on individual patient data. *Br J Clin Pharmacol* 54: 320–334
34. García-Rodríguez LA, Hernández-Díaz S (2001) Relative risk of upper gastrointestinal complications among users of acetaminophen and nonsteroidal anti-inflammatory drugs. *Epidemiology* 12: 570–576
35. Henry D, Lim LL-Y, García Rodríguez LA, Pérez Gutthann S, Carson JL, Griffin M, Savage R, Logan R, Moride Y, Hawkey C et al. (1996) Variability in risk of gastrointestinal complications with individual non-steroidal anti-inflammatory drugs: results of a collaborative meta-analysis. *Br Med J* 312: 1563–1566
36. Hernández-Diaz S, García Rodriguez LA (2000) Association between nonsteroidal anti-inflammatory drugs and upper gastrointestinal tract bleeding and perforation: An

overview of epidemiological studies published in the 1990s. *Arch Int Med* 160: 2093–2099

37. Traversa G, Walker AM, Ippolito FM, Caffari B, Capurso L, Dezi A, Koch M, Maggini M, Alegiani SS, Raschetti R (1995) Gastroduodenal toxicity of different nonsteroidal antiinflammatory drugs. *Epidemiology* 6: 49–54
38. Menniti-Ippolito F, Maggini M, Raschetti R, Da Cas R, Traversa G, Walzer AM (1998) Ketorolac use in outpatients and gastrointestinal hospitalization: a comparison with other non-steroidal anti-inflammatory drugs in Italy. *Eur Clin Pharmacol* 54: 393–397
39. García Rodríguez LA, Cattaruzzi C, Troncon MG, Agostinis L (1998) Risk of hospitalization for upper gastrointestinal tract bleeding associated with ketorolac, other nonsteroidal anti-inflammatory drugs, calcium antagonists, and other antihypertensive drugs. *Arch Intern Med* 158: 33–39
40. Laporte JR, Ibanez L, Vidal X, Vendrell L, Leone R (2004) Upper gastrointestinal bleeding associated with the use of NSAIDs: Newer versus older agents. *Drug Saf* 27: 411–420
41. Trechot P, Gillet P, Gay G, Hanesse B, Netter P, Castot A, Larrey D (1996) Incidence of hepatitis induced by non-steroidal anti-inflammatory drugs (NSAID) *Ann Rheum Dis* 55: 936
42. Walker AM (1997) Quantitative studies of the risk of serious hepatic injury in persons using nonsteroidal antiinflammatory drugs. *Arthritis Rheum* 40: 201–208
43. Boelsterli UA (2002) Mechanisms of NSAID-induced hepatotoxicity: Focus on nimesulide. *Drug Saf* 25: 633–648
44. Schattner A, Sokolovskaya N, Cohen J (2000) Fatal hepatitis and renal failure during treatment with nimesulide. *J Intern Med* 247: 153–155
45. Gatti S, Bertazzoli M (2002) Evaluation of isolated case reports on hepatotoxicity. *Eur J Clin Pharmacol* 57: 919–920
46. Merlani G, Fox M, Oehen HP, Cathomas G, Renner EL, Fattinger K, Schneemann M, Kullak-Ublick GA (2001) Fatal hepatoxicity secondary to nimesulide. *Europ J Clin Pharmacol* 57: 321–326
47. Macia MA, Carvajal A, del Pozo JG, Vera E, del Pino A (2002) Hepatotoxicity associated with nimesulide: data from the Spanish Pharmacovigilance System. *Clin Pharmacol Therap* 72: 596–597
48. Conforti A, Leone R, Moretti U, Mozzo F, Velo GP (2001) Adverse drug reactions related to the use of NSAIDs with a focus on nimesulide: Results of spontaneous reporting from a Northern Italian area. *Drug Saf* 24: 1081–1090
49. Agence francaise de securitè sanitarie des produits de Santé. www. Agmed.sante.gouv.fr (16 June 2003)
50. Traversa G, Bianchi C, Da Cas R, Abraha I, Menniti-Ippolito F, Venegoni M (2003) Cohort study of hepatotoxicity associated with nimesulide and other non-steroidal anti-inflammatory drugs. *Br Med J* (5) 327: 18–22
51. Krähenbühl S, Reichen J (2000) Drug hepatotoxicity. In: Bacon BR, Di Bisceglie, AM (Eds): *Liver disease diagnosis and management.* Churchill Livingstone, New York, 294–309

52. Papachristou GI, Demetris AJ, Rabinovitz M (2004) Acute cholestatic hepatitis associated with long-term use of rofecoxib. *Dig Dis Sci* 49: 459–461
53. Grieco A, Miele L, Giorgi A, Civello IM, Gasbarrini G (2002) Acute cholestatic hepatitis associated with celecoxib. *Ann Pharmacother* 36: 1887–1889
54. Svensson CK, Cowen EW, Gaspari AA (2001) Cutaneous drug reactions. *Pharmacol Rev* 53: 357–379
55. Roujeau J-C, Kelly JP, Naldi L, Rzany B, Stern RS, Anderson T, Auquier A, Bastuji-Garin S, Correia O, Locati F et al. (1995) Medication use and the risk of Stevens-Johnson syndrome or toxic epidermal necrolysis. *New Engl J Med* 333: 1600–1607
56. Mockenhaupt M, Kelly JP, Kaufman D, Stern RS; SCAR Study Group (2003) The risk of Stevens-Johnson syndrome and toxic epidermal necrolysis associated with nonsteroidal antiinflammatory drugs: A multinational perspective. *J Rheumatol* 30: 2234–2240
57. Schmutz JL, Barbaud A, Trechot P (2004) Toxic epidermal necrolysis and celecoxib *Annal Dermatol Venerol* 131: 107
58. Perna AG, Woodruff CA, Markus RF, Hsu S (2003) Toxic epidermal necrolysis as a complication of treatment with celecoxib. *Dermatol Online J* 9: 25
59. Giglio P (2003) Toxic epidermal necrolysis due to administration of celecoxib (Celebrex) *South Med J* 96: 320–321
60. Anonymous (1993) Cutaneous reactions to analgesic-antipyretics and nonsteroidal anti-inflammatory drugs. Analysis of reports to the spontaneous reporting system of the Gruppo Italiano Studi Epidemiologici in Dermatologia. *Dermatol* 186: 164–169
61. Nettis E, Di Paola R, Napoli G, Ferrannini A, Tursi A (2002) Benzydamine: An alternative nonsteroidal anti-inflammatory drug in patients with nimesulide-induced urticaria. *Allergy* 57: 442–445
62. Sanchez-Borges M, Capriles-Hulett A, Caballero-Fonseca F (2003) Cutaneous reactions to aspirin and nonsteroidal antiinflammatory drugs. *Clin Rev Immunology* 24: 125–136
63. Strom BL, Carson JL, Morse ML, West SL, Soper KA (1987) The effect of indication on hypersensitivity reactions associated with zomepirac sodium and other nonsteroidal antiinflammatory drugs. *Arthritis Rheum* 30: 1142–1148
64. Jenkins C, Costello J, Hodge L (2004) Systematic review of prevalence of aspirin induced asthma and its implications for clinical practice. *Br Med J* (21) 328: 434
65. Asero R (2000) Leukotriene receptor antagonists may prevent NSAID-induced exacerbations in patients with chronic urticaria. *Ann All Asth Immunol* 85: 156–157
66. Perez C, Sanchez-Borges M, Capriles E (2001) Pretreatment with montelukast blocks NSAID-induced urticaria and angioedema. *J All Asth Clin Immunol* 108: 1060–1061
67. Perrone MR, Artesani MC, Viola M, Gaeta F, Caringi M, Quaratino D, Romano A (2003) Tolerability of rofecoxib in patients with adverse reactions to nonsteroidal anti-inflammatory drugs: A study of 216 patients and literature review. *Int Arch All Immunol* 132: 82–86

68. Gagnon R, Julien M, Gold P (2003) Selective celecoxib-associated anaphylactoid reaction. *J All Asth Clin Immunol* 111: 1404–1405
69. Grob M, Pichler WJ, Wuthrich B (2002) Anaphylaxis to celecoxib. *Allergy,* 57: 264–265
70. Valero Santiago A, Gonzalez-Morales MA, Marti Guadano E (GETNIA) Grupo de Estudio de Tolerancia a Nimesulida en pacientes Intolerantes a AINES (2003) Tolerance of nimesulide in NSAID intolerant patients. *Allergy* 58(4): 367–368
71. Senna GE, Passalacqua G, Andri G, Dama AR, Albano M, Fregonese L, Andri L (1996) Nimesulide in the treatment of patients intolerant of aspirin and other NSAIDs. *Drug Saf* 14: 94–103
72. Bavbek S, Celik G, Ozer F, Mungan D, Misirligil Z (2004) Safety of selective COX-2 inhibitors in aspirin/nonsteroidal anti-inflammatory drug-intolerant patients: comparison of nimesulide, meloxicam, and rofecoxib. *J Asthma* 41: 67–75
73. Nettis E, Marcandrea M, Ferrannini A, Tursi A (2001) Tolerability of nimesulide and paracetamol in patients with NSAID-induced urticaria/angioedema. *Immunopharmacol Immunotoxicol* 23: 343–354
74. Rossoni G, Berti F, Buschi A, Villa LM, Bella DD (1993) New data concerning the antianaphylactic and antihistaminic activity of nimesulide. *Drugs* 46 (Suppl 1): 22–28
75. Griffin MR, Yared A, Ray WA (2000) Nonsteroidal antiinflammatory drugs and acute renal failure in elderly persons. *Am J Epidemiol* (1) 151: 488–496
76. Ravnskov U (1999) Glomerular, tubular and interstitial nephritis associated with nonsteroidal antiinflammatory drugs. Evidence of a common mechanism *Br Clin Pharmacol* 47: 203–210
77. Clinard F, Sgro C, Bardou M, Hillon P, Dumas M, Kreft-Jais C, Escousse A, Bonithon-Kopp C (2004) Association between concomitant use of several systemic NSAIDs and an excess risk of adverse drug reaction. A case/non-case study from the French Pharmacovigilance system database. *Eur J Clin Pharmacol* 60: 279–283
78. Leone R, Conforti A, Ghiotto E, Moretti U, Valvo E, Velo GP (1999) Nimesulide and renal impairment. *Eur J Clin Pharmacol* 55: 151–154
79. Schattner A, Sokolovskaya N, Cohen J (2000) Fatal hepatitis and renal failure during treatment with nimesulide. *J Intern Med* 247: 153–155
80. Van der Niepen P, Janssen van Doorn K, Van den Houte K, Verbeelen D (2002) Nimesulide and acute renal failure caused by oxalate precipitation. *Nephrol Dialysis Transplant* 17: 315–316
81. Stollberger C, Finsterer J (2003) Nonsteroidal anti-inflammatory drugs in patients with cardio- or cerebrovascular disorders. *Z Kardiol* 92: 721–729
82. Garcia Rodriguez LA, Hernandez-Diaz S (2003) Nonsteroidal antiinflammatory drugs as a trigger of clinical heart failure. *Epidemiology* 14: 240–246
83. Hernandez-Diaz S, Garcia-Rodriguez LA (2001) Epidemiologic assessment of the safety of conventional nonsteroidal anti-inflammatory drugs. *Am J Med* 110 (Suppl 3A): 20S–27S

84. Topol EJ (2004) Failing the public health – Rofecoxib, Merck, and the FDA. *N Engl J Med* 351: 1707–1709
85. Mamdani M, Juurlink DN, Lee DS, Rochon PA, Kopp A, Naglie G, Austin PC, Laupacis A, Stukel TA (2004) Cyclo-oxygenase-2 inhibitors *versus* non-selective non-steroidal anti-inflammatory drugs and congestive heart failure outcomes in elderly patients: a population-based cohort study. *Lancet* 363: 1751–1756
86. Anonymous (2005) FDA discloses safety reviews of COX-2's Arcoxia, Prexige and parecoxib and Vioxx cardiovascular risk is unique to molecule, Pfizer and Novartis tell FDA. *"The Pink Sheet" FDC Reports* February: 11–13
87. Drazen JM (2005) COX-2 inhibitors – A lesson in unexpected problems. *N Engl J Med* 352: 1131–1132
88. Psaty BM, Furberg CD (2005) COX-2 inhibitors – Lessons in drug safety. *N Engl J Med* 352: 1133–1135
89. Solomon DH, Schneeweiss S, Glynn RJ, Kiyota Y, Levin R, Mogun H, Avorn J (2004) Relationship between selective cyclooxygenase-2 inhibitors and acute myocardial infarction in older adults. *Circulation* 109: 2068–2073
90. Brinker A, Goldkind L, Bonnel R, Beitz J (2004) Spontaneous reports of hypertension leading to hospitalisation in association with rofecoxib, celecoxib, nabumetone and oxaprozin. *Drugs Aging* 21: 479–484
91. Cho J, Cooke CE, Proveaux W (2003) A retrospective review of the effect of COX-2 inhibitors on blood pressure change. *Am J Ther* 10: 311–317
92. Fitzgerald GA (2004) Coxibs and cardiovascular disease. *N Engl J Med* 351: 1709–1711
93. Hernandez MR, Tonda R, Pino M, Serradell M, Arderiu G, Escolar G (2004) Evaluation of effects of rofecoxib on platelet function in an *in vitro* model of thrombosis with circulating human blood. *Eur J Clin Invest* 34: 297–302
94. Niederberger E, Manderscheid C, Grosch S, Schmidt H, Ehnert C, Geisslinger G (2004) Effects of the selective COX-2 inhibitors celecoxib and rofecoxib on human vascular cells. *Biochem Pharmacol* 68: 341–350
95. Baber SR, Champion HC, Bivalacqua TJ, Hyman AL, Kadowitz PJ (2003) Role of cyclooxygenase-2 in the generation of vasoavtive prostanoids in the pulmonary and systemic vascular beds. *Circulation* 108: 896–901
96. Cesarani R, Carboni L, Germini M, Mainardi P, Passoni A (1993) Antipyretic and platelet antiaggregating effects of nimesulide. *Drugs* 46 (Suppl 1): 48–51
97. Shi W, Wang YM, Cheng NN, Chen BY, Li D (2003) Meta-analysis on the effect and adverse reaction on patients with osteoarthritis and rheumatoid arthritis treated with non-steroidal anti-inflammatory drugs *Zhonghua Liu Xing Bing Xue Za Zhi* 24: 1044–1048
98. Hooper L, Brown TJ, Elliott R, Payne K, Roberts C, Symmons D (2005) The effectiveness of five strategies for the prevention of gastrointestinal toxicity induced by non-steroidal anti-inflammatory drugs: a systematic review. *Br Med J* 329: 948–958
99. Wober W (1999) Comparative efficacy and safety of nimesulide and diclofenac in patients with acute shoulder, and a meta-analysis of controlled studies with nimesulide. *Rheumatology* 38 (suppl 1): 33–38

100. Shah AA, Thjodleifsson B, Murray FE, Sigthorsson G, Oddson E, Gudjonsson H, Price AB, Fitzgerald DJ, Bjarnason I (2001) A randomised, double blind, double dummy, cross-over study of the effects of nimesulide and naproxen on the gastrointestinal tract and an *in vivo* assessment of their selectivity for cyclooxygenase 1 and 2. *Gut* 48: 339–348
101. Shah AA, Murray FE, Thjodleifsson B et al. (1998) Nimesulide, a selective COX-2 inhibitor, causes less gastrointestinal damage compared with naproxen: a prospective study in human volunteers. *Gastroenterology* 115: 101–109
102. Bjarnason I, Thjodleifsson B (1999) Gastrointestinal toxicity of non-steroidal anti-inflammatory drugs: The effect of nimesulide compared with naproxen on the human gastrointestinal tract. *Rheumatology* 38 (suppl 1): 24–32
103. Feldman M, McMahon AT (2000) Do cyclooxygenase-2 inhibitors provide benefits similar to those of traditional nonsteroidal anti-inflammatory drugs, with less gastrointestinal toxicity? *Ann Intern Med* 132: 134–143
104. Whittle BJR (1992) Unwanted effects of aspirin and related agents on the gastrointestinal tract. In: JR Vane, RM Botting (Eds): *Aspirin and other Salicylates*. Chapman & Hall Medical, London, 465–509
105. Bjarnason I, Hayllar J, Macpherson AJ, Russell AS (1993) Side effects of nonsteroidal anti-inflammatory drugs on the small and large intestine. *Gastroenterology* 104: 1832–1847
106. Somasundaram S, Hayllar J, Rafi S, Wrigglesworth J, Macpherson A, Bjarnason I (1995) The biochemical basis of NSAID-induced damage to the gastrointestinal tract: A review and a hypothesis. *Scand J Gastroenterol* 30: 289–299
107. Sigthorsson G, Simpson RJ, Walley M, Anthony A, Foster R, Hotz-Behoftsitz C, Palizban A, Pombo J, Watts J, Morham SG et al. (2002) COX-1 and 2, intestinal integrity and pathogenesis of NSAID-enteropathy in mice. *Gastroenterology* 122: 1913–1923
108 Somasundaram S, Sigthorsson G, Price AB, Tavares IA, Rafi S, Mahmud T, Roseth A, Foster R, Macpherson S, Wrigglesworth JM et al. (2000) The relative importance of inhibition of cyclooxygenase and uncoupling of oxidative phosphorylation in the gastrointestinal toxicity of nonsteroidal anti-inflammatory drugs. *Aliment Pharm Therap* 14: 639–650
109. Wallace JL, McKnight, Reuter BK, Vergnolle N (2000) Dual inhibition of both cyclooxygenase (COX)-1 and COX-2 is required for NSAID-induced erosion formation. *Gastroenterology* 119: 704–714
110. Hotz-Behofsits CM, Walley MJM, Simpson R, Bjarnason IT (2003) COX-1, COX-2 and the topical effect in NSAID-induced enteropathy. *Inflammopharmacology* 11: 363–370
111. Newberry RD, Stenson WF, Lorenz RG (1999) Cyclooxygenase-2-dependent arachidonic acid metabolites are essential modulators of the intestinal immune response to dietary antigen. *Nature Med* 5: 900–906
112. Bjarnason I, Scarpignato C, Takeuchi J, Rainsford KD (2005) NSAID-induced gastric damage: Do physicochemical properties matter? *Submitted*
113. Fenn GC (1994) Controversies in NSAID-induced gastroduodenal damage-do they matter? *Aliment Pharmacol Ther* 8: 15–26

114. Walt R (1992) Drug therapy: Misoprostol for the treatment of peptic ulcer and anti-inflammatory drug-induced gastroduodenal ulceration. *N Engl J Med* 327: 1575–1580
115. Mamdani M, Rochon PA, Juurlink DN, Kopp A, Anderson GM, Naglie G, Austin PC, Laupacis A (2002) Observational study of upper gastrointestinal haemorrhage in elderly patients given selective cyclo-oxygenase-2 inhibitors or conventional non-steroidal anti-inflammatory drugs. *Br Med J* 325: 624
116. Mamdani M, Juurlink DN, Kopp A, Naglie G, Austin PC, Laupacis A (2004) Gastrointestinal bleeding after the introduction of COX 2 inhibitors: ecological study. *Br Med J* 328: 1415–1416
117. Singh G, Ramey DR, Morfeld D, Shi H, Hatoum HT, Fries JF (1996) Gastrointestinal tract complications of nonsteroidal anti-inflammatory drug treatment in rheumatoid arthritis. A prospective observational cohort study. *Arch Intern Med* 156: 1530–1536
118. Silverstein FE, Graham GY, Senior JR, Davies HW, Struthers BJ, Bittman RM, Geis SG (1995) Misoprostol reduces serious gastrointestinal complications in patients with rheumatoid arthritis receiving nonsteroidal anti-inflammatory drugs. *Ann Int Med* 123: 241–249
119. Bombardier C, Laine L, Reicin A, Shapiro D, Burgos-Vargas R, Davis B, Day R, Ferraz MB, Hawkey CJ, Hoshberg MC et al. (2000) Comparison of upper gastrointestinal toxicity of rofecoxib and naproxen in patients with rheumatoid arthritis. *N Engl J Med* 343: 1520–1528
120. Silverstein FE, Faich G, Goldstein JL, Simon LS, Pincus T, Whelton A, Makuch R, Eisen G, Agrawal NM, Stenson WF et al. (2000) Gastrointestinal toxicity with celecoxib *versus* nonsteroidal anti-inflammatory drugs for osteoarthritis and rheumatoid arthritis: the CLASS study: A randomised controlled trial. Celecoxib Long-term Arthritis Study. *JAMA* 284: 1247–1255
121. Chan FKL GD (2004) Prevention of non-steroidal anti-inflammatory drug gastrointestinal complications – Review and recommendations based on risk assessment. *Aliment Pharmacol Ther* 19: 1051–1061
122. Chan FKL, Hung LCT, Suen BY, Wu JCY, Lee KC, Leung VKS, Hui AJ, To KF, Leung WK, Wong VWS et al. (2002) Celecoxib *versus* diclofenac and omeprazole in reducing the risk of recurrent ulcer bleeding in patients with arthritis. *N Engl J Med* 347: 2104–2110
123. Hawkey JC, Jones IJ, Atherton CT, Slelly MM, Bebb JR, Fagerholm U, Jonzon B, Karlson P, Bjarnason IT (2003) Gastrointestinal safety of AZD3582, a cyclooxygenase inhibiting nitric oxide donator: proof of concept study in humans. *Gut* 52: 1537–1542
124. Bjarnason I, Williams P, Smethurst P, Peters TJ, Levi AJ (1986) The effect of NSAIDs and prostaglandins on the permeability of the human small intestine. *Gut* 27: 1292–1297
125. Bjarnason I, Smethurst P, Fenn GC, Lee CF, Menzies IS, Levi AJ (1989) Misoprostol reduces indomethacin induced changes in human small intestinal permeability. *Dig Dis Sci* 34: 407–411

126. Bjarnason I, Fehilly B, Smethurst P, Menzies IS, Levi AJ (1991) The importance of local *versus* systemic effects of non-steroidal anti-inflammatory drugs to increase intestinal permeability in man. *Gut* 32: 275–277
127. Sigthorsson G, Crane R, Simon T, Hoover M, Quan H, Bolognese J, Bjarnason I (2000) COX-2 inhibition with rofecoxib does not increase intestinal permeability in healthy subjects: A double blind crossover study comparing rofecoxib with placebo and indomethacin. *Gut* 47: 527–532
128. Smecuol E, Sugai E, Maurino E, Vazquez H, Niveloni S, Pedreira S, Meddings J (1998) Gastrointestinal permeability to non-steroidal anti-inflammatory drugs: A prospective study comparing meloxicam to slow release indomethacin. Preliminary results. *Gastroenterology* 114: A1087
129. Smecuol E, Bai JC, Sugai E, Vazquez H, Niveloni S, Pedreira S, Maurino E, Meddings J (2001) Acute gastrointestinal permeability responses to different non-steroidal anti-inflammatory drugs. *Gut* 49: 650–655
130. Sigthorsson G, Tibble J, Hayllar J, Menzies I, Macpherson A, Moots R, Scott D, MJ G, Bjarnason I (1998) Intestinal permeability and inflammation in patients on NSAIDs. *Gut* 43: 506–511
131. Bjarnason I, Macpherson AJM, Hollander D (1995) Intestinal permeability: An overview. *Gastroenterology* 108: 1566–1581
132. Bjarnason I, Smethurst P, Clarke P, Menzies IS, Levi AJ, Peters TJ (1989) Effect of prostaglandins on indomethacin induced increased intestinal permeability in man. *Scand J Gastroenterol* 29 (suppl 164): 97–103
133. Goldstein JL, Eisen G, Lewis B, Gralnek I, Fort JG, Zlotnick S (2003) Celecoxib is associated with fewer small bowel lesions than naproxen + omeprazole in healthy subjects as determined by capsule enteroscopy. *Gut* 52 (suppl 4): A16
134. Bjarnason I, Zanelli G, Smith T, Prouse P, Williams P, DeLacey G, Gumpel MJ, Levi AJ (1987) Nonsteroidal antiinflammatory drug induced intestinal inflammation in humans. *Gastroenterology* 93: 480–489
135. Morris AJ, Madhok R, Sturrock RD, Capell HA, Mackenzie JF (1991) Enteroscopic diagnosis of small bowel ulceration in patients receiving non-steroidal antiinflammatory drugs. *Lancet* (1) 337: 520
136. Bjarnason I, Zanelli G, Prouse P, Smethurst P, Levi S, Gumpel MJ, Levi AJ (1987) Blood and protein loss via small intestinal inflammation induced by nonsteroidal anti-inflammatory drugs. *Lancet* 2: 711–714
137. Hayllar J, Price AB, Smith T, Macpherson A, Gumpel MJ, Bjarnason I (1994) Nonsteroidal antiinflammatory drug-induced small intestinal inflammation and blood loss: effect of sulphasalazine and other disease modifying drugs. *Arthritis Rheum* 37: 1146–1150
138. Bjarnason I, Hopkinson N, Zanelli G, Prouse P, Gumpel MJ, Levi AJ (1990) The treatment of non-steroidal anti-inflammatory drug induced enteropathy. *Gut* 31: 777–780
139. Bjarnason I, Hayllar J, Smethurst P, Price AB, Menzies IS, Gumpel MJ (1992) Metronidazole reduces inflammation and blood loss in NSAID enteropathy. *Gut* 33: 1204–1208

140. Morris AJ, Murray L, Sturrock RD, Madhok R, Capell HA, Mackenzie JF (1994) Short report: The effect of misoprostol on the anaemia of NSAID enteropathy. *Aliment Pharmacol Ther* 8: 343–346
141. Langman MJS, Morgan L, Worrall A (1985) Use of anti-inflammatory drugs by patients with small or large bowel perforation and haemorrhage. *Br Med J* 290: 347–349
142. Allison MC, Howatson AG, Torrance CJ, Lee FD, Russell RI (1992) Gastrointestinal damage associated with the use of nonsteroidal anti-inflammatory drugs. *N Engl J Med* 327: 749–754
143. Bjarnason I, Macpherson A (1994) Treatment of nonsteroidal antiinflammatory drug induced damage to the small and large intestine. In: TM Bayless (Ed.): *Current Therapy in Gastroenterology and Liver Disease* (4th Ed.) Mosby, St Louis, 295–298
144. Bjarnason I, Zanelli G, Smethurst P, Burke M, Gumpel MJ, Price AB, Levi AJ (1988) Clinico-pathological features of nonsteroidal antiinflammatory drug induced small intestinal strictures. *Gastroenterology* 94: 1070–1074
145. Lang J, Price AB, Levi AJ, Burk M, Gumpel JM, Bjarnason I (1988) Diaphragm disease: The pathology of non-steroidal anti-inflammatory drug induced small intestinal strictures. *J Clin Path* 41: 516–526
146. Whitcombe DC, Martin SP, Trellis DR, Evans BA, Beicich MJ (1992) "Diaphragmlike" stricture and ulcer of the colon during diclofenac treatment. *Arch Int Med* 152: 2341–2343
147. Speed CA, Bramble MG, Corbett WA, Haslock I (1994) Non-steroidal anti-inflammatory induced diaphragm disease of the small intestine: complexities of diagnosis and management. *Br J Rheumatol* 33: 778–780
148. McCune KH, Allen D, Cranley B (1992) Small bowel diaphragm disease-Strictures associated with non-steroidal anti-inflammatory drugs. *Ulster Med J* 61: 182–184
149. Levi S, DeLacey G, Price AB, Gumpel MJ, Levi AJ, Bjarnason I (1990) 'Diaphragm like' strictures of the small bowel in patients treated with non-steroidal anti-inflammatory drugs. *Br J Radiol* 63: 186–189
150. Hershfield NB (1992) Endoscopic description of diaphragm disease induced by non-steroidal anti-inflammatory drugs. *Gastroint Endosc* 38: 267
151. Halter F, Weber B, Huber T, Eigenmann F, Frey M, Rutchi C (1993) Diaphragm disease of the ascending colon associated with sustained release diclofenac. *J Clin Gastroenterol* 16: 74–80
152. Laine L, Connors LG, Reicin A, Hawkey CJ, Burgos-Vargas R, Schnitzer TJ, Yu Q, Bombardier C (2003) Serious lower gastrointestinal clinical events with nonselective NSAID or coxib use. *Gastroenterology* 124: 288–292
153. Goldstein JL, Bjarnason I, Eisen G, Lewis B (2004) Serious lower bowel complications of NSAIDs in the CLASS study. *Gastroenterology* 126 (Suppl. 2): A1
154. Warner TD, Giuliano F, Vojnovic I, Bukasa A, Mitchell JA, Vane JR (1999) Nonsteroid drug selectivities for cyclo-oxygenase-1 rather than cyclo-oxygenase-2 are associated with human gastrointestinal toxicity: a full *in vitro* analysis. *Proc Natl Acad Sci USA* 96: 7563–7568

155. Blain H, Boileau C, Laqicque F, Nedelec E, Laeuille D, Guillaume C, Gaucher A, Jeandel C, Netter P, Jy J (2002) Limitation of the *in vitro* whole blood assay for predicting the COX selectivity of NSAIDs in clinical use. *Br J Clin Pharmacol* 53: 255–265
156. Brooks P, Emery P, Evans JF, Fenner H, Hawkey CJ, Patrono C, Smolen J, Breedveld F, Day R, Dougados M et al. (1999) Interpreting the clinical significance of the differential inhibition of cyclooxygenase-1 and cyclooxygenase-2. *Rheumatology* 38: 779–788
157. Wight NJ, Gottesdiener K, Garlick NM, Atherton CT, Novak S, Gertz BJ, Calder NA, Cote J, Wong P, Dallob A et al. (2001) Rofecoxib, a COX-2 inhibitor, does not inhibit human gastric mucosal prostaglandin production. *Gastroenterology* 120: 867–873
158. Atherton C, Jones J, KcKaig B, Bebb J, Cunliffe R, Burdsall J, Brough J, Stevenson D, Bonner J, Rodorf C et al. (2004) Pharmacology and gastrointestinal safety of lumiracoxib, a novel syslooxygenase-2 selective inhibitor: An integrated study. *Clin Gastroenterol Hepatol* 2: 113–120
159. Cipollini F, Mecozzi V, Altilia F (1989) Endoscopic assessmsnt of the effects of nimesulide on the gastric mucosa: Comparison with Indomethacin. *Curr Ther Res* 45: 1042–1049
160. Marini U, Spotti D (1993) Gastric tolerability of nimesulide. A double blind comparison of 2 oral damage regimens and placebo. *Drugs* 46 (suppl 1): 249–252
161. Porto A, Reis C, Perdigoto R, Concalves M, Freitas P, Macciocchi A (1998) Gastroduodenal tolerability of nimesulide and diclofenac in patients with osteoarthritis. *Curr Ther Res* 59: 654–665
162. Hunt RH, Harper S, Watson DJ, Yu C, Quan H, Lee M, Evans JK, Oxenius B (2003) The gastrointestinal safety of the COX-2 selective inhibitor etoricoxib assessed by both endoscopy and analysis of upper gastrointestinal events. *Am J Gastroenterol* 98: 1725–1733
163. Hunt RH, Harper S, Callegary P, Yu C, Quan H, Evans J, James C, Bowen B, Rashid F (2003) Complementary studies of the gastrointestinal safety of the cyclo-oxygenase-2-selective inhibitor etoricoxib. *Aliment Pharmacol Ther* 17: 201–210
164. Gleeson M, Ramsey D, Hutchinson S, Spencer D, Montieth G (1994) Colitis associated with non-steroidal anti-inflammatory drugs. *Lancet* 344: 1028
165. Gleeson MH, Davis AJ (2003) Non-steroidal anti-inflammatory drugs, aspirin and newly diagnosed colitis: a case-control study. *Aliment Pharmacol Ther* 17: 817–825
166. Rask-Madsen J, Bukhave K, Laursen LS, Lauritsen K (1990) Eicosanoids in inflammatory bowel disease – physiology and pathology In: TJ Peters (Ed.): *The Cell Biology of Inflammation in the Gastrointestinal Tract.* Corners Publication, Hull, 255–271
167. Campieri M, Franchi LGA, Bazzocchi G, Brignola C, Benatia B, Boccia S, Labo G (1980) Prostaglandins, indomethacin and ulcerative colitis. *Gastroenterology* 78: 193
168. Rampton DS, Sladen GE (1981) Relapse of ulcerative proctocolitis during treatment with NSAID. *Postgrad Med J* 57: 297–299

169. Kaufman HJ, Taubin HL (1987) NSAID activate quiescent inflammatory bowel disease. *Ann Int Med* 107: 513–516
170. British Medical Association & Royal Pharmaceutical Society of Great Britain (2002) *British National Formulary* #43. Pharmaceutical Press, Wallingford
171. Tibble J, Sigthorsson G, Foster R, Scott D, Roseth A, Bjarnason I (1999) Faecal calprotectin: A simple method for the diagnosis of NSAID-induced enteropathy. *Gut* 45: 362–366
172. Tibble J, Teahon K, Thjodleifsson B, Roseth A, Sigthorsson G, Bridger S, Fioster R, Sheerwood R, Fagerhol M, Bjarnason I (2000) A simple method for assessing intestinal inflammation in Crohn's disease. *Gut* 47: 506–513
173. Tibble J, Sigthorsson G, Fagerhol M, Bjarnason I (2000) Surrogate markers of intestinal inflammation are predictive for relapse in patients with inflammatory bowel disease. *Gastroenterology* 119: 15–22
174. Kaufman DW, Kelly JP, Sheehan JE, Laszlo A, Wiholm BE, Alfredsson L, Koft RS, Shapiro S (1993) Nonsteroidal antiinflammatory drug use in relation to major gastrointestinal bleeding. *Clin Pharmacol Ther* 53: 485–494
175. Henry D, Dobson A, Turner C (1993) Variability in the risk of major gastrointestinal complications from nonsteroidal anti-inflammatory drugs. *Gastroenterology* 105: 1078–1088
176. Langman MJS, Weil J, Wainwright P, Lawson DH, Rawlins MD, Logan RFA, Murphy M, Vessey MP, Colin-Jones DG (1994) Risk of bleeding peptic ulcers associated with individual non-steroidal anti-inflammatory drugs. *Lancet* 343: 1075–1078
177. Rodrigues LAG, Jick H (1994) Risk of upper gastrointestinal bleeding and perforation associated with individual non-steroidal anti-inflammatory drugs. *Lancet* 343: 769–772
178. Henry D, Lim LL-Y, Rodrigues LAG, Gutthann PS, Carson JL, Griffin M, Savage R, Logan R, Moride Y, Hawkey C et al. (1996) Variability in risk of gastrointestinal complications with individual non-steroidal anti-inflammatory drugs: results of a collaborative meta-analysis. *Br Med J* 312: 1563–1566
179. MacDonald TM, Morant SV, Robinson GC, Shield MJ, McGilchrist MM, Murray FE, McDevitt DG (1997) Association of upper gastrointestinal toxicity of non-steroidal anti-inflammatory drugs with continued exposure: Cohort study. *Br Med J* 315: 1333–1337
180. Larrey D, Pageaux GP (2005) Drug-induced acute liver failure. *Eur J Gastroenterol Hepatol* 17: 141–143
181. Andrade RJ, Camargo R, Lucena MI, Gonzalez-Grande R (2004) Causality assessment in drug-induced hepatotoxicity. *Expert Opin Drug Saf* 3: 329–344
182. Croft AM, Whitehouse DP, Cook GC, Beer MD (2002) Safety evaluation of the drugs available to prevent malaria. *Expert Opin Drug Saf* 1: 19–27
183. Lucena MI, Carvajal A, Andrade RJ, Velasco A (2003) Antidepressant-induced hepatotoxicity. *Expert Opin Drug Saf* 2: 249–262
184. Liu ZX, Kaplowitz N (2002) Immune-mediated drug-induced liver disease. *Clin Liver Dis* 6: 467–486

185. Hartleb M, Biernat L, Kochel A (2002) Drug-induced liver damage – A three-year study of patients from one gastroenterological department. *Med Sci Monit* 8: CR292–296
186. Marino G, Zimmerman HJ, Lewis JH (2001) Management of drug-induced liver disease. *Curr Gastroenterol Rep* 3: 38–48
187. Degott C (1997) Drug-induced liver injury. Cholestatic injury, acute and chronic. *Pathol Oncol Res* 3: 260–263
188. Biour M, Poupon R, Grangé JD, Chazouilleres O (2000) Hépatotoxicité des médicaments. *Gastroenterol Clin Biol* 11: 1052–1091
189. Lee WM (2003) Drug-induced hepatotoxicity. *N Engl J Med* 349: 474–485
190. Zimmerman HJ (1998) Drug-induced liver injury – clinical. In: *Clinical and Pathological Correlations in Liver Disease: Approaching the Next Millennium.* American Association for the Study of Liver Diseases Postgraduate Course, 252–268.
191. Drug Consults (2004) *Nonsteroidal drug-induced hepatotoxicity.* Micromedex® Healthcare Series, Vol. 121
192. Larrey D (2000) Drug-induced liver diseases. *J Hepato* 32 (Suppl 1): 77–88
193. Vandenbroucke JP (2001) In defence of case reports and case series. *Ann Intern Med* 134: 330–334
194. Striker BHCh, Psaty BM (2004) Detection, verification and quantification of adverse drug reactions *Br Med J* 329: 44–47
195. Bessone F, Fay F, Vorobioff J, Passamonti ME, Godoy A, Tanno H (1997) Nimesulide hepatotoxicity: Evidence from six cases. *Hepatology* 26: 483A
196. Carniato A, Vaglia A (1997) Hepatitis-like syndrome induced by nimesulide? *Le infezioni in Medicina* 4: 265
197. Van Steenberger W, Roskams T, Desmet V (1997) Nimesulide-induced acute hepatitis: Evidence from three cases. *Gastroenterology* 112 (suppl): A1407
198. Grignola JC, Arias L, Rondan M, Sola L, Bagnulo H (1998) Hepatoxicity associated to nimesulide (Revision of five cases) *Arch Med Int* 20: 13–18
199. Selig J, Liberek C, Kondo Oestreicher M, Desmeules J, Stoller R (1998) *Nimesulid hepatotoxicity.* Poster presented at the 6th Annual Meeting of the European Society of Pharmacovigilance – Budapest, September 28–29
200. Bakhle YS (1999) Nimesulide and COX-2 inhibitors. *Lancet* 354: 772
201. Malhotra S, Pandhi P (2000) Analgesics for pediatric use. *Indian J Pediatr* 67: 589–590
202. Carvajal A, Macia MA, del Pozo JG, de Abajo F (2003) Small risk ratios may have strong public health impact. *Br Med J* 327: 1050–1051
203. Romero Gomez M, Nevado Santos M, Fobelo MJ, Castro Fernandez M (1999) Hepatitis aguda por nimesulida: descripcion de tres casos. *Medicina Clinica* 113: 357–358
204. Teoh NC, Farrell GC (2003) Hepatotoxicity associated with nonsteroidal anti-inflammatory drugs. *Clin Liver Dis* 7: 401–413
205. Lewin S (2002) Post-marketing surveillance of nimesulide suspension. *Indian Pediatr* 39: 890–891

206. Bessone F, Tanno H (2000) Hepatotoxicidad inducida por antiinflamatorios no esteroides. *Gastroenterol Hepatol* 23: 200–205
207. Andrade RJ, Lucena MI, Fernandez MC, Gonzalez M (2000) Fatal hepatitis associated with nimesulide. *J Hepatol* 32: 174
208. Perez Moreno, Llerena Guerrero RM, Puertas Montenegro M, Jimenez Arjona MJ (2000) Hepatitis toxica en gestante por nimesulide. *Gastroenterol Hepatol* 23: 498–499
209. Papaioannides D, Korantzopoulos P, Athanassiou E, Sinapidis D (2003) Nimesulide-induced acute hepatotoxicity. *Indian J Gastroenterol* 22: 239
210. Lacroix I, Lapeyre-Mestre M, Bagheri H, Pathak A, Montastruc JL; Club de Reflexion des cabinets de Groupe de Gastro-Enterologie (CREGG); General Practitioner Networks (2004) Nonsteroidal anti-inflammatory drug-induced liver injury: a case-control study in primary care. *Fundam Clin Pharmacol* 18: 201–206
211. Tejos SC, Torrejon NS, Rey H, Meneses MC (2000) Ulceras gástricas sangrantes y hepatitis aguda: dos reacciones adversas simultáneas por nimesulida, en un caso. *Rev Méd Chile* 128: 1349–1353
212. Dourakis SP, Sevastianos VA, Petraki K, Hadziyannis SJ (2001) Nimesulide induced acute icteric hepatitis. *Iatriki* 79: 275–278
213. Castaneda Hernandez G, Barragan Padilla SB (2004) Hepatotoxicity of nimesulide *Gac Med Mex* 140: 679 [Spanish]
214. Montesinos S, Hallal H, Rausell V, Conesa FJ, Lopez A (2001) Hepatitis aguda por nimesulida. *Gastroenterol Hepatol* 24: 219–220
215. Quadranti P (2000) Acute hepatitis after use of nimesulide: drug-induced or is there something more? Mononucleosis. *Schweiz Rundsch Med Prax* 93: 1785–1787 [German]
216. Dumortier J, Borel I, Delafosse B, Vial T, Scoazec JY, Boillot O (2002) Transplantation hépatique pour hépatite subfulminante après prise de Nimesulide. *Gastroentrol Clin Biol* 26: 415–416
217. Gallego Rojo FJ, Fernandez Perez F, Fernandez Perez R, Porcel A, Blas JM, Diez F (2002) Hepatotoxicidad por nimesulida. *Rev Esp Enferm Dig* 94: 41–42
218. Rodrigo L, de Francisco R, Perez-Pariente JM, Cadahia V, Tojo R, Rodriguez M, Lucena MI, Andrade RJ (2002) Nimesulide-induced severe hemolytic anemia and acute liver failure leading to liver transplantation. *Scand J Gastroenterol* 37: 1341–1343
219. Stadlmann S, Zoller H, Vogel W, Offner FA (2002) COX-2 inhibitor (nimesulide) induced acute liver failure. *Virchows Arch* 440: 553–555
220. Ozgur O, Hacihasanoglu A, Karti SS, Ovali E (2003) Nimesulide-induced fulminant hepatitis. *Turk J Gastroenterol* 14: 208–210
221. Danan G, Benichou C (1993) Causality assessment of adverse reactions to drugs – I. A novel method based on the conclusions of International Consensus Meetings: application to drug-induced liver injuries. *J Clin Epidemiol* 46: 1323–1330
222. Larson AM (2004) *Drugs and the liver: patterns of hepatotoxicity.* UpToDate® online, version 12.1

223. Ishak KG (1998) Drug-induced liver injury – Pathology. In: *Clinical and Pathological Correlations in Liver Disease: Approaching the Next Millennium*. American Association for the Study of Liver Diseases Postgraduate Course, 236–251
224. Goodman ZD (2002) Drug hepatotoxicity. *Clin Liver Dis* 6: 381–397
225. Rainsford KD (2003) Analyses of adverse drug reactions attributed to nimesulide worldwide with particular reference to those in the hepatic body system. Helsinn Internal Report TSD No. 8104
226. Weber JCP (1984) Epidemiology of adverse reactions to nonsteroidal antiinflammatory drugs. In: KD Rainsford, GP Velo (Eds): *Side-Effects of Antiinflammatory Drugs*. Raven Press, New York, 1–7
227. Velayudham LS, Farrell GC (2003) Drug-induced cholestasis. *Expert Opin Drug Saf* 2: 287–304
228. Kistner RW (1971) Present status of oral contraceptives: 1. Effectiveness; basis for selection; side effects; metabolic changes. *Drug Ther (NY)* 1: 14–29
229. Dourakis SP, Tolis G (1998) Sex hormonal preparations and the liver. *Eur J Contracept Reprod Health Care* 3: 7–16
230. Lindgren A, Olsson R (1993) Liver damage from low-dose oral contraceptives. *J Intern Med* 234: 287–292
231. Zacur HA, Stewart D (1992) New concepts in oral contraceptive pill use. *Curr Opin Obstet Gynecol* 4: 365–371
232. Adlin EV (1979) Postmenopausal estrogen therapy. *Ann Intern Med* 91: 488–489
233. Blau SP (1977) Study Report R-805-010-02. Helsinn Internal Report TSD No. 7344
234. Russell AJ, Sturge RA, Smith MA (1971) Serum transaminases during salicylate therapy. *Br Med J* 2: 428–429
235. Seaman WE, Ishak KG, Plotz PH (1974) Aspirin-induced hepatotoxicity in patients with systemic lupus erythematosus. *Ann Intern Med* 80: 1–8
236. Seaman WE, Plotz PH (1976) Effect of aspirin on liver tests in patients with RA or SLE and in normal volunteers. *Arthritis Rheum* 19: 155–160
237. Wolfe JD, Metzger AL, Goldstein RC (1974) Aspirin hepatitis. *Ann Intern Med* 30: 74–76
238. Zimmerman HJ (1981) Effect of aspirin and acetaminophen on the liver. *Ann Intern Med* 141: 333–342
239. Prescott LF (1992) The hepatotoxicity of non-steroidal anti-inflammatory drugs. *In:* KD Rainsford, GP Velo (Eds): *Side-effects of Anti-inflammatory Drugs 3*. Kluwer Academic Publishers, Dordrecht, 176–187
240. Banks AT, Zimmerman HJ, Ishak KG, Harter JG (1995) Diclofenac-associated hepatotoxicity: analysis of 180 cases reported to the Food and Drug Administration as adverse reactions. *Hepatology* 22: 820–827
241. Zucker P, Daum F, Cohen MI (1975) Aspirin hepatitis. *Am J Dis Child* 129: 1433–1434
242. Huuponen R, Helin-Salmivaara A, Klaukka T (2003) Heavy users of NSAIDs in Finland. A prescription database study. In: *6th Congress of the European Association for Clinical Pharmacology and Therapeutics*. Istanbul, Turkey, June 24–28, Abstract No. P-133

243. Lewis JH (2002) The rational use of potentially hepatotoxic medications in patients with underlying liver disease. *Expert Opin Drug Saf* 1: 159–172
244. Parra JL, Reddy KR (2003) Hepatotoxicity of hypolipidemic drugs. *Clin Liver Dis* 7: 415–433
245. Heerey A, Barry M, Ryan M, Kelly A (2000) The potential for drug interactions with statin therapy in Ireland. *Ir J Med Sci* 169: 176–179
246. Simpura J, Tigerstedt C, Hanhinen S, Lagerspetz M, Leifman H, Moskalewicz J, Torronen J (1999) Alcohol misuse as a health and social issue in the Baltic Sea region. A summary of findings from the Baltica Study. *Alcohol Alcohol* 34: 805–823
247. Dreyer L, Winther JF, Andersen A, Pukkala E (1997) Avoidable cancers in the Nordic countries. Alcohol consumption. *APMIS* 76 (suppl): 48–67
248. Anonymous (2000) *Alcohol Consumption and Harm in the UK and EU.* IAS Factsheet. Institute of Alcohol Studies, St Ives (Cambridge)
249. Perola M, Vuori E, Penttila A (1994) Abuse of alcohol in sudden out-of-hospital deaths in Finland. *Alcohol Clin Exp Res* 18: 255–260
250. Seppa K, Lof K, Sinclair D, Sillanaukee P (1994) Hidden alcohol abuse among women. *Br J Psychiatry* 164: 544–546
251. Seppa K, Sillanaukee P, Koivula T (1990) The efficiency of a questionnaire in detecting heavy drinkers. *Br J Addict* 85: 1639–1645
252. Agren G, Romelsjo A (1992) Mortality in alcohol-related diseases in Sweden during 1971–80 in relation to occupation, marital status and citizenship in 1970. *Scand J Soc Med* 20: 134–142
253. Hakulinen T, Lehtimaki L, Lehtonen M, Teppo L (1974) Cancer morbidity among two male cohorts with increased alcohol consumption in Finland. *J Natl Cancer Inst* 52: 1711–1714
254. Keso L, Kivisaari A, Salaspuro M (1988) Fractures on chest radiographs in detection of alcoholism. *Alcohol Alcoholism* 23: 53–56
255. Koskenvuo M, Kaprio J, Kesaniemi A, Poikolainen K (1986) Alcohol-related diseases associated with ischaemic heart disease: a three-year follow-up of middle-aged male hospital patients. *Alcohol Alcoholism* 21: 251–256
256. Laitinen K, Valimaki M, Lamberg-Allardt C, Kivisaari L, Lalla M, Karkkainen M, and Ylikahri R (1990) Deranged vitamin D metabolism but normal bone mineral density in Finnish noncirrhotic male alcoholics. *Alcohol Clin Exp Res* 14: 551–556
257. Olkinuora M (1984) Alcoholism and occupation. *Scand J Work Environ Health* 10: 511–515
258. Penttila A (1980) Sudden and unexpected natural deaths of adult males. An analysis of 799 forensic autopsies in 1976. *Forensic Sci Int* 16: 249–259
259. Poikolainen K (1983) Accuracy of hospital discharge data: five alcohol-related diseases. *Drug Alcohol Depend* 12: 315–322
260. Poikolainen K (1982) Seasonality of alcohol-related hospital admissions has implications for prevention. *Drug Alcohol Depend* 10: 65–69

261. Valimaki MJ, Laitinen K, Tiitinen A, Steman UH, and Ylostalo P (1995) Gonadal function and morphology in non-cirrhotic female alcoholics: A controlled study with hormone measurements and ultrasonography. *Acta Obstet Gynecol Scand* 74: 462–466

262 Pikkarainen P (1978) The liver, alcohol and women. *Duodecim* 94: 620–622 [in Finnish]

263. Savolainen VT Penttila A, Karhunen PJ (1992) Delayed increases in liver cirrhosis mortality and frequency of alcoholic liver cirrhosis following an increment and redistribution of alcohol consumption in Finland: evidence from mortality statistics and autopsy survey covering 8533 cases in 1968–1988. *Alcohol Clin Exp Res* 16: 661–664

264. Savander M, Ropponen A, Avela K, Weerasekera N, Cormand B, Hirvioja ML, Riikonen S, Ylikorkala O, Lehesjoki AE, Williamson C et al. (2003) Genetic evidence of heterogeneity in intrahepatic cholestasis of pregnancy. *Gut* 52: 1025–1029

265. Koivurova S, Hartikainen AL, Karinen L, Gissler M, Hemminki E, Martikainen H, Tuomivaara L, Jarvelin MR (2002) The course of pregnancy and delivery and the use of maternal healthcare services after standard IVF in Northern Finland 1990–1995. *Hum Reprod* 17: 2897–2903

266. Eloranta ML, Heiskanen JT, Hiltunen MJ, Mannermaa AJ, Punnonen KR, Heinonen ST (2002) Multidrug resistance 3 gene mutation 1712delT and estrogen receptor alpha gene polymorphisms in Finnish women with obstetric cholestasis. *Eur J Obstet Gynecol Reprod Biol* 105: 132–135

267. Reyes H, Baez ME, Gonzalez MC, Hernandez I, Palma J, Ribalta J, Sandoval L, Zapata R (2000) Selenium, zinc and copper plasma levels in intrahepatic cholestasis of pregnancy, in normal pregnancies and in healthy individuals, in Chile. *J Hepatol* 32: 542–549

268. Pasanen P, Partanen K, Pikkarainen P, Alhava E (1992) Characteristics of jaundice and cholestasis in a Finnish population. *Ann Chir Gynaecol* 81: 284–289

269. Fusi D, Corsello FP, Piacentino R, Marchino GL, Grio R (1993) Cholestasis in pregnancy. *Minerva Ginecol* 45: 307–314 [in Italian]

270. Hirvioja ML, Tuimala R, Vuori J (1992) The treatment of intrahepatic cholestasis of pregnancy by dexamethasone. *Br J Obstet Gynaecol* 99: 109–111

271. Eloranta ML, Hakli T, Hiltunen M, Helisalmi S, Punnonen K, Heinonen S (2003) Association of single nucleotide polymorphisms of the bile salt export pump gene with intrahepatic cholestasis of pregnancy. *Scand J Gastroenterol* 38: 648–652

272. Eloranta ML, Heiskanen JT, Hiltunen MJ, Mannermaa AJ, Punnonen KR, Heinonen ST (2002) Multidrug resistance 3 gene mutation 1712delT and estrogen receptor alpha gene polymorphisms in Finnish women with obstetric cholestasis. *Eur J Obstet Gynecol Reprod Biol* 105: 132–135

273. Eloranta ML, Heiskanen JT, Hiltunen MJ, Mannermaa AJ, Punnonen KR, Heinonen ST (2002) Multidrug resistance 3 gene mutation 1712delT and estrogen receptor alpha gene polymorphisms in Finnish women with obstetric cholestasis. *Eur J Obstet Gynecol Reprod Biol* 104: 109–112

274 Eloranta ML, Heinonen S, Mononen T, Saarikoski S (2001) Risk of obstetric cholestasis in sisters of index patients. *Clin Genet* 60: 42–45

275 Heinonen S, Eloranta ML, Heiskanen J, Punnonen K, Helisalmi S, Mannermaa A, Hiltunen M (2001) Maternal susceptibility locus for obstetric cholestasis maps to chromosome region 2p13 in Finnish patients. *Scand J Gastroenterol* 36: 766–770

276 Eloranta ML, Heiskanen J, Hiltunen M, Helisalmi S, Mannermaa A, Heinonen S (2000) Apolipoprotein E alleles in women with intrahepatic cholestasis of pregnancy. *Scand J Gastroenterol* 35: 966–968

277 Bergmann C, Senderek J, Sedlacek B, Pegiazoglou I, Puglia P, Eggermann T, Rudnik-Schoneborn S, Furu L, Onuchic LF, De Baca M et al. (2003) Spectrum of mutations in the gene for autosomal recessive polycystic kidney disease (ARPKD/PKHD1) *J Am Soc Nephrol* 14: 76–89

278. Tahvanainen E, Tahvanainen P, Kaariainen H, Hockerstedt K (2004) Genetic defects underlying polycystic liver disease are discovered. *Duodecim* 20: 1081–1084

279. Drenth JP, Tahvanainen E, te Morsche RH, Tahvanainen P, Kaariainen H, Hockerstedt K, van de Kamp JM, Breuning MH, Jansen JB (2004) Abnormal hepatocystin caused by truncating PRKCSH mutations leads to autosomal dominant polycystic liver disease. *Hepatology* 39: 924–931

280. Tahvanainen P, Tahvanainen E, Reijonen H, Halme L, Kaariainen H, Hockerstedt K (2003) Polycystic liver disease is genetically heterogeneous: Clinical and linkage studies in eight Finnish families. *J Hepatol* 38: 39–43

281. Todorova B, Kubaszek A, Pihlajamaki J, Lindstrom J, Eriksson J, Valle TT, Hamalainen H, Ilanne-Parikka P, Keinanen-Kiukaanniemi S, Tuomilehto J et al. Finnish Diabetes Prevention Study (2004) The G-250A promoter polymorphism of the hepatic lipase gene predicts the conversion from impaired glucose tolerance to type 2 diabetes mellitus: The Finnish Diabetes Prevention Study. *J Clin Endocrinol Metab* 89: 2019–2023

282. Connelly PW, Hegele RA (1998) Hepatic lipase deficiency. *Crit Rev Clin Lab Sci* 35: 547–572

283. Tahvanainen E, Syvanne M, Frick MH, Murtomaki-Repo S, Antikainen M, Kesaniemi YA, Kauma H, Pasternak A, Taskinen MR, Ehnholm C (1998) Association of variation in hepatic lipase activity with promoter variation in the hepatic lipase gene. The LOCAT Study Investigators. *J Clin Invest* 101: 956–960

284. Miettinen TA, Gylling H, Lindbohm N, Miettinen TE, Rajaratnam RA, Relas H (2003) Finnish Treat-to-Target Study Investigators Serum noncholesterol sterols during inhibition of cholesterol synthesis by statins. *J Lab Clin Med* 141: 131–137

285. Jarvelainen HA, Orpana A, Perola M, Savolainen VT, Karhunen PJ, Lindros KO (2001) Promoter polymorphism of the CD14 endotoxin receptor gene as a risk factor for alcoholic liver disease. *Hepatology* 33: 1148–1153

286. Savolainen VT, Pajarinen J, Perola M, Penttila A, Karhunen PJ (1997) Polymorphism in the cytochrome P450 2E1 gene and the risk of alcoholic liver disease. *J Hepatol* 26: 55–61

287. Liiv I, Teesalu K, Peterson P, Clemente MG, Perheentupa J, Uibo R (2002) Epitope mapping of cytochrome P450 cholesterol side-chain cleavage enzyme by sera from patients with autoimmune polyglandular syndrome type 1. *Eur J Endocrinol* 146: 113–119
288. Mitrunen K, Jourenkova N, Kataja V, Eskelinen M, Kosma VM, Benhamou S, Vainio H, Uusitupa M, Hirvonen A (2000) Steroid metabolism gene CYP17 polymorphism and the development of breast cancer. *Cancer Epidemiol Biomarkers Prev* 9: 1343–1348
289. Levo A, Jaaskelainen J, Sistonen P, Siren MK, Voutilainen R, Partanen J (1999) Tracing past population migrations: genealogy of steroid 21-hydroxylase (CYP21) gene mutations in Finland. *Eur J Hum Genet* 7: 188–196
290. Kupari M, Hautanen A, Lankinen L, Koskinen P, Virolainen J, Nikkila H, White PC (1998) Associations between human aldosterone synthase (CYP11B2) gene polymorphisms and left ventricular size, mass, and function. *Circulation* 97: 569–575
291. Hirvonen A (1999) Polymorphisms of xenobiotic-metabolizing enzymes and susceptibility to cancer. *Environ Health Perspect* 107 (suppl 1): 37–47
292. Hirvonen A (2003) Combinations of susceptible genotypes and individual responses to toxicants. *Environ Health Perspec* 105 (suppl 4): 755–758
293. Levo A, Koski A, Ojanpera I, Vuori E, Sajantila A (2003) Post-mortem SNP analysis of CYP2D6 gene reveals correlation between genotype and opioid drug (tramadol) metabolite ratios in blood. *Forensic Sci Int* 135: 9–15
294. Tiitinen H (1969) Isoniazid and ethionamide serum levels and inactivation in Finnish subjects. *Scand J Respir Dis* 50: 110–124
295. Helske T (1974) Carriers of hepatitis B antigen and transfusion hepatitis in Finland. *Scand J Haematol Suppl* 22: 1–65
296. Hokkanen OT and Sotaniemi EA (1978) Liver injury and multiple drug therapy. *Arch Toxicol Suppl* 173–176
297 Leino T, Leinikki P, Hyypia T, Ristola M, Suni J, Sutinen J, Holopainen A, Haikala O, Valle M, Rostila T (1997) Hepatitis A outbreak amongst intravenous amphetamine abusers in Finland. *Scand J Infect Dis* 29: 213–216
298. Pohjanpelto P (1992) Risk factors connected with hepatitis C infections in Finland. *Scand J Infect Dis* 24: 251–252
299. Sinkkonen S, Rantalainen AL, Paasivirta J, Lahtipera M (2004) Polybrominated methoxy diphenyl ethers (MeO-PBDEs) in fish and guillemot of Baltic, Atlantic and Arctic environments. *Chemosphere* 56: 767–775
300 Lamminpaa A, Pukkala E, Teppo L, Neuvonen PJ (2002) Cancer incidence among patients using antiepileptic drugs: a long-term follow-up of 28,000 patients. *Eur J Clin Pharmacol* 58: 137–141
301. Swingle KF, Moore GGI (1984) Preclinical pharmacological studies with nimesulide. *Drugs Exptl Clin Res* 10: 597–597
302. Swingle KF, Moore GGI, Grant TJ (1976) 4-Nitro-2-phenoxymethanesulfonanilide (R-805): A chemically novel anti-inflammatory agent. *Achiv int Pharmacodyn* 221: 132–139

303. Rainsford KD (1975) A synergistic interaction between aspirin, or other non-steroidal anti-inflammatory drugs, and stress which produces severe gastric mucosal damage in rats and pigs. *Agents Actions* 5: 553–558
304. Rainsford KD (1975) Studies on the effects of R-805 on the gastro-intestinal tract of domestic pigs – A model for the study of human gastro-intestinal functions and ulcer disease. *Unpublished studies*
305. Rainsford KD (1977) Towards assays of gastrointestinal toxicity of non-steroidal anti-inflammatory drugs with improved predictive value in man. *Agents Actions* 7: 245–248
306. Rainsford KD (1981) Comparison of the gastric ulcerogenic activity of new non-steroidal anti-inflammatory drugs in stressed rats. *Br J Pharmacol* 73: 79c–80c
307. Rainsford KD and Willis C (1982) Relationship of gastric mucosal damage induced in pigs by anti-inflammatory drugs to their effects on prostaglandin production. *Dig Dis Sci* 27: 624–635
308. Rainsford KD (1982) An analysis of the gastro-intestinal side-effects of non-steroidal anti-inflammatory drugs, with particular reference to comparative studies in man and laboratory species. *Rheumatol Internat* 2: 1–10
309. Rainsford KD (1985) Relationships of gastric irritancy/ulcerogenicity and anti-oedemic activity of non-steroidal anti-inflammatory drugs. *J Pharm Pharmacol* 37: 678–679
310. Rainsford KD (1989) Animal models for the assay of gastrointestinal toxicity of anti-inflammatory drugs. In: RA Greenwald, HS Diamond (Eds): *CRC Handbook of Animal Models for the Rheumatic Diseases*. CRC Press, Boca Raton, Florida, 181–206
311. Tanaka K, Shimotori T, Makino S, Aikawa Y, Inaba T, Yoshida C, Tanako S (1992) Pharmacological studies of the new anti-inflammatory agent 3-formylamino-7-methylsulfonylamino-6-phenoxy-4H-1-benzopyran-4-one. 1st Communication: Antiinflammatory, analgesic and other related properties. *Arzneim Forsch* 42: 935–944
312. Tanaka K, Makino S, Shimotori T, Aikawa Y, Inaba T, Yoshida C (1992) Pharmacological studies of the new antiinflammatory agent 3-formylamino-7-methylsulfonylamino-6-phenoxy-4′-1-benzopyran-4-one. 2nd communication: Effect on the arachidonic acid cascades. *Arzneim Forsch* 42: 945–950
313. Rainsford KD, Fox SA, Osborne, DJ (1984) Comparative effects of some non-steroidal anti-inflammatory drugs on the ultrastructural integrity and prostaglandin levels in the rat gastric mucosa: Relationship to drug uptake. *Scand J Gastroenterol* 19 (suppl 101): 55–68
314. Rainsford KD (1986) Structural damage and changes in eicosanoid metabolites in the gastric mucosa of rats and pigs induced by anti-inflammatory drugs of varying ulcerogenicity. *Int J Tiss React* 8: 1–14
315. Rainsford KD (1988) Comparative irritancy of oxaprozin on the gastrointestinal tract of rats and mice: Relationship to drug uptake and effects *in vivo* on eicosanoid metabolism. *Aliment Pharmacol Therap* 2: 439–450
316. Nakatsugi S, Terada N, Yoshimura T, Horie Y, Furukawa M (1996) Effects of nimesulide, a preferential cyclooxygenase-2 inhibitor, an carrageenan-induced pleurisy

and stress-induced gastric lesions in rats. *Prost Leuk Essential Fatty Acids* 55: 395–402

317. Laudanno OM, Cesolari JA, Esnarriaga J, San Miguel P, and Bedini OA (2000) *In vivo* selectivity of nonsteroidal antiinflammatory drugs and gastrointestinal ulcers in rats. *Dig Dis Sci* 45: 1359–1365
318. Tavares IA, Bishai PM, Bennett A (1995) Activity of nimesulide on constitutive and inducible cyclooxygenases. *Arzneim Forsch* 45: 1093–1095
319. Cryer B, Feldman M (1998) Cyclooxygenase-1 and cyclooxygenase-2 selectivity of widely used nonsteroidal anti-inflammatory drugs. *Am J Med* 104: 413–421
320. Kataoka H, Horie Y, Koyama R, Nakatsugi S, and Furukawa M (2000) Interaction between NSAIDs and steroid in rat stomach: safety of nimesulide as a preferential COX-2 inhibitor in the stomach. *Dig Dis Sci* 45: 1366–1375
321. Tofanetti O, Casciarri I, Cipolla PV, Cazzlani P, Omini C (1989) Effect of nimesulide on cyclooxygenase activity in rat gastric mucosa and inflammatory exudates. *Med Sci Res* 17: 745–746
322. Ceserani R, Casciarri I, Cavaletti E, Cazzulani P (1991) Action of nimesulide on rat gastric prostaglandins and renal function. *Drug Invest* 3 (suppl 2): 14–21
323. Sigthorsson G, Jacob M, Wrigglesworth J, Somasundaram S, Tavares I, Foster R, Roseth A, Rafi S, Mahmud T, Simpson R et al. (1998) Comparison of indomethacin and nimesulide, a selective cyclooxygenase-2 inhibitor, on key pathophysiologic steps in the pathogenesis of nonsteroidal anti-inflammatory drug enteropathy in the rat. *Scand J Gastroenterol* 33: 728–735
324. Shah AA, Murray FE, Fitzgerald DJ (1999) The *in vivo* assessment of nimesulide cyclooxygenase-2 selectivity. *Rheumatology (Oxford)* 38 (suppl 1): 19–23
325. Ogino K, Hatanaka K, Kawamura M, Ohno T, Harada Y (2000) Meloxicam inhibits prostaglandin E_2 generation via cyclooxygenase 2 in the inflammatory site but not that via cyclooxygenase 1 in the stomach. *Pharmacology* 61: 244–250
326. Hirata T, Ukawa H, Yamakuni H, Kato S, and Takeuchi K (1997) Cyclo-oxygenase isozymes in mucosal ulcerogenic and functional responses following barrier disruption in rat stomachs. *Br J Pharmacol* 122: 447–454
327. Hirata T, Ukawa H, Kitamura M, Takeuchi K (1997) Effects of selective cyclooxygenase-2 inhibitors on alkaline secretory and mucosal ulcerogenic responses in rat duodenum. *Life Sci* 61: 1603–1611
328. Suleyman H, Altinkaynak K, Gocer F, Maras A, Akcay F, Onuk MD, Gepdiremen A (2002) Effect of nimesulide on the indomethacin- and ibuprofen-induced ulcer in rat gastric tissue. *Pol J Pharmacol* 54: 255–259
329. Altinkaynak K, Suleyman H, Akcay F (2003) Effect of nimesulide, rofecoxib and celecoxib on gastric tissue glutathione level in rats with indomethacin-induced gastric ulcerations. *Pol J Pharmacol* 55: 645–648
330. Rainsford KD (2001) The ever-emerging anti-inflammatories. Have there been any real advances? *J Physiol Paris* 95: 11–19

331. Suleyman H, Akcay F, and Altinkaynak K (2002) The effect of nimesulide on the indomethacin- and ethanol-induced gastric ulcer in rats. *Pharmacol Res* 45: 155–158
332. Ramesh N, Jayakumar K, Narayana HK, Vijayasarathi SK (2001) Nimesulide toxicity in dogs. *Indian J Pharmacol* 33: 217–218
333. Wilson JE, Chandrasekharan NV, Westover KD, Eager KB, and Simmons DL (2004) Determination of expression of cyclooxygenase-1 and -2 isozymes in canine tissues and their differential sensitivity to nonsteroidal anti-inflammatory drugs. *Am J Vet Res* 65: 810–818
334. Toutain PL, Cester CC, Haak T, Laroute V (2001) A pharmacokinetic/pharmacodynamic approach *versus* a dose titration for the determination of a dosage regimen: the case of nimesulide, a Cox-2 selective nonsteroidal anti-inflammatory drug in the dog. *J Vet Pharmacol Therap* 24: 43–55
335. Toutain PL, Cester CC, Haak T, Metge S (2001) Pharmacokinetic profile and *in vitro* selective cyclooxygenase-2 inhibition by nimesulide in the dog. *J Vet Pharmacol Therap* 24: 35–42
336. Tavares IA, Borrelli F, Welsh NJ (2001) Inhibition of gastric acid secretion by nimesulide: a possible factor in its gastric tolerability. *Clin Exp Rheumatol* 19: S13–S15
337. Borrelli F, Tavares IA (2003) Effect of nimesulide on gastric acid secretion in the mouse stomach *in vitro*. *Life Sci* 72: 885–896
338. Moreno-Sanchez R, Bravo C, Vasquez C, Ayala G, Silveira LH, Martinez-Lavin M (1999) Inhibition and uncoupling of oxidative phosphorylation by nonsteroidal anti-Inflammatory drugs – Study in mitochondria, submitochondrial particles, cells, and whole heart. *Biochem Pharmacol* 57: 743–752
339. Caparroz-Assef SM, Salgueiro-Pagadigorria CL, Bersani-Amado CA, Bracht A, Kelmer-Bracht AM, Ishii-Iwamoto EL (2001) The uncoupling effect of the nonsteroidal anti-Inflammatory drug nimesulide in liver mitochondria from adjuvant-induced arthritic rats. *Cell Biochem Funct* 19: 117–124
340. Rainsford KD (2004) Side-effects and toxicology of the salicylates. In: KD Rainsford (Ed.): *Aspirin and Related Drugs*. CRC Press, Boca Raton, Florida, 367–554
341. Berti F, Rossoni A, Buschi A, Rebeshi M, Villa LM (1990) Antianaphylactic and antihistaminic activity of the non-steroidal anti-inflammatroy compound, nimesulide, in guinea pig. *Arneim Forsch* 40: 1011–1016
342. Casolaro V, Meliota S, Marino O, Patella V, de Paulis A, Guidi G, and Marone G (1993) Nimesulide, a sulfonanilide nonsteroidal anti-inflammatory drug, inhibits mediator release from human basophils and mast cells. *J Pharmacol Exp Ther* 267: 1375–1385
343. Saeed SA, Shah BH (1998) Dual effects of nimesulide, a COX-2 inhibitor, in human platelets. *Life Sci* 63: 1835–1841
344. Chan FK, Hawkey CJ, Lanas AI (2001) Helicobacter pylori and nonsteroidal anti-inflammatory drugs: A three-way debate. *Am J Med* 110 (1A): 55S–57S
345. Huang JQ, Sridhar S, Hunt RH (2002) Role of Helicobacter pylori infection and non-steroidal anti-inflammatory drugs in peptic-ulcer disease: A meta-analysis. *Lancet* 359: 14–22

346. Hawkey CJ (1999) Personal review: *Helicobacter pylori*, NSAIDs and cognitive dissonance. *Aliment Pharmacol Ther* 13: 695–702
347. Leung WK, To KF, Chan FKL, Lee TL, Chung SCS, Sung JJY (2000) Interaction of *Helicobacter pylori* eradication and non-steroidal anti-inflammatory drugs on gastric epithelial apoptosis and proliferation: implications on ulcerogenesis. *Aliment Pharmacol Ther* 14: 879–885
348. Tatguguchi A, Sakamoto C, Wada K, Akamatsu T, Tsukui T, Miyake K, Futagami S, Kishida T, Fukuda Y, Yamanaka N et al. (2000) Localisation of cyclooxygenase 1 and cyclooxygenase 2 in *Helicobacter pylori* related gastritis and gastric ulcer tissues in humans. *Gut* 46: 782–789
349. Kapicioglu S, Baki AH, Sari M, Ozdemir F, Kavgaci H (2000) Does nimesulide induce gastric mucosal damage? "A double-blind randomized placebo-controlled trial". *Hepatogastroenterology* 47: 1183–1185
350. Guslandi M, Foppa L, Fanti L, Sorghi M (1999) Nonsteroidal anti-inflammatory drugs and gastric mucosal blood flow. *J Clin Gastroenterol* 28: 258–260
351. Li JY, Wang XZ, Chen FL, Yu JP, Luo HS (2003) Nimesulide inhibits proliferation via induction of apoptosis and cell cycle arrest in human gastric adenocarcinoma cell line. *World J Gastroenterol* 9: 915–920
352. Jain NK, Kulkarni SK, Singh A (2002) Modulation of NSAID-induced antinociceptive and anti-inflammatory effects by alpha$_2$-adrenoreceptor agonists with gastroprotective effects. *Life Sci* 70: 2857–2869
353. Tardieu D, Jaeg JP, Deloly A, Corpet DE, Cadet J, Petit CR (2000) The COX-2 inhibitor nimesulide suppresses superoxide and 8-hydroxy-deoxyguanosine formation, and stimulates apoptosis in mucosa during early colonic inflammation in rats. *Carcinogenesis* 21: 973–976
354. Oktar BK, Cakir B, Mutlu N, Celikel C, Alican I (2002) Protective role of cyclooxygenase (COX) inhibitors in burn-induced intestinal and liver damage. *Burns* 28: 209–214
355. Zhang L, Lu YM, Dong XY (2004) Effects and mechanism of the selective COX-2 inhibitor, celecoxib, on rat colitis induced by trinitrobenzene sulfonic acid. *Chin J Dig Dis* 5: 110–114
356. Lugering A, Floer M, Lugering N, Cichon C, Schmidt MA, Domschke W, Kucharzik T (2004) Characterization of M cell formation and associated mononuclear cells during indomethacin-induced intestinal inflammation. *Clin Exp Immunol* 136: 232–238
357. Martin AR, Villegas I, La Casa C, Alarcon de la Lastra C (2003) The cyclo-oxygenase-2 inhibitor, rofecoxib, attenuates mucosal damage due to colitis induced by trinitrobenzene sulphonic acid in rats. *Eur J Pharmacol* 481: 281–291
358. Barbare JC, Imbert A, Benkirane A (2001) Recent developments concerning drug-induced liver toxicity. *Presse Medicale* 30: 673–676
359. Boelsterli UA (2002) Nimesulide and hepatic adverse effects: Roles of reactive metabolites and host factors. *Int J Clin Pract Suppl* 30–36

360. Boess F, Bopst M, Althaus R, Polsky S, Cohen SD, Eugster HP, Boelsterli UA (1998) Acetaminophen hepatotoxicity in tumour necrosis factor-alpha gene knockout mice. *Hepatology* 27: 1021–2019
361. Boelsterli UA (2003) Animal models of human disease in drug safety assessment. *J Toxicol Sci* 28: 109–121
362. Pishvaian AC, Trope BW, Lewis JH (2004) Drug-induced liver disease in 2003. *Curr Opin Gastroenterol* 20: 208–219
363. Rainsford KD (2005) Analysis of confounding factors in non-steroidal anti-inflammatory drug – Associated adverse events. When are the NSAIDs the culprits? Drug Safety, *in preparation*
364. Boelsterli UA (2003) Disease-related determinants of susceptibility to drug-induced idiosyncratic hepatotoxicity. *Curr Opinion Drug Disc Devel* 6: 81–91
365. Lewis JH (2002) Drug-induced liver disease. *Curr Opin Gastroenterol* 18: 307–313
366. Liberopoulos EN, Nonni AB, Tsianos EV, Elisaf MS (2002) Possible ranitidine-induced cholestatic jaundice. *Ann Pharmacother* 36: 172
367. Macpherson D, Best SA, Gedik L, Hewson AT, Rainsford KD, Parisi F (2004) The biotransformation and pharmacokinetics in humans of single dose of ^{14}C-nimesulide. *Submitted*
368. Rainsford KD, Seabrook RW, Spencer S, Hewson AT (2001) Effects of Nimesulide and its metabolites or manufacturing intermediates on the viability and growth of the human hepatoma HepG2 cell line. *Life Sciences* 69: 2965–2973
369. Sohi, KK, Khanduja KL (2003) Nimesulide affects antioxidant status during acute lung inflammation in rats. *Indian J Biochem Biophys* 40: 238–245
370. Caparroz-Assef SM, Bersani-Amado CA, Do Nascimento EA, Kelmer-Bracht AM, Ishii-Iwamoto EL (1998) Effects of the nonsteroidal anti-inflammatory drug nimesulide on energy metabolism in livers from adjuvant-induced arthritic rats. *Res Comm Mol Path Pharmacol* 99: 93–116
371. Mingatto FE, Dos Santos AC, Rodrigues T, Pigoso AA, Uyemura SA, Curti C (2000) Effects of Nimesulide and its reduced metabolite on mitochondria. *Br J Pharmacol* 131: 1154–1160
372. Caparroz-Assef SM, Salgueiro-Pagadigorria CL, Bersani-Amado CA, Bracht A, Kelmer-Bracht AM, Ishii-Iwamoto EL (2001) The uncoupling effect of the nonsteroidal anti-inflammatory drug nimesulide in liver mitochondria from adjuvant-induced arthritic rats. *Cell Biochem Funct* 19: 117–124
373. Mingatto FE, Rodrigues T, Pigoso AA, Uyemura SA, Curti C, Santos AC (2002) The critical role of mitochondrial energetic impairment in the toxicity of nimesulide to hepatocytes. *J Pharmacol Exp Ther* 303: 601–607
374. Whitehouse MW (1965) Some biochemical and pharmacological properties of anti-inflammatory drugs. *Arzneim Forsch–Progress in Drug Research* 8: 321–429
375. Galati G, Tafazoli S, Sabzevari O, Chan TS, O'Brien PJ (2002) Idiosyncratic NSAID drug induced oxidative stress. *Chem Biol Interact* 142: 25–41

376. Ambudkar SV, Dey S, Hrycyna CA, Ramachandra M, Pastan I, Gottesman MM (1999) Biochemical, cellular, and pharmacological aspects of the multidrug transporter. *Annu Rev Pharmacol Toxicol* 39: 361–398
377. Borst P, Evers R, Kool M, Wijnholds J (2000) A family of drug transporters: the multidrug resistance-associated proteins. *J Natl Cancer Inst* 92: 1295–1302
378. Fantappie O, Masini E, Sardi I, Raimondi L, Bani D, Solazzo M, Vannacci A, Mazzanti R (2002) The Mdr phenotype is associated with the expression of Cox-2 and inos in a human hepatocellular carcinoma cell line. *Hepatology* 35: 843–852
379. Souto EO, Miyoshi H, Dubois RN, Gores GJ (2001) Kupffer cell-derived cyclooxygenase-2 regulates hepatocyte Bcl-2 expression in choledocho-venous fistula rats. *Am J Physiol* 280: G805–G811
380. Leong RWL, Sung JJY (2002) Review article: *Helicobacter* species and hepatobiliary diseases. *Aliment Pharmacol Ther* 16: 1037–1045
381. Rodriguez F, Llinás, Moreno C, Salazar FJ (2001) Role of cyclo-oxygenase-2 derived metabolites and NO in renal response to bradykinin. *Hypertension* 37: 129–134
382. Roig F, Llinás MT, Löpez R, Salazar FJ (2002) Role of cyclo-oxygenase-2 in the prolonged regulation of renal function. *Hypertension* 40: 721–728
383. Steinhäuslin F, Munafo A, Buclin T, Macciocchi A, Biollaz J (1993) Renal effects of nimesulide in furosemide-treated subjects. *Drugs* 46 (suppl 1): 257–262
384. Schlondorff D (1993) Renal complications of nonsteroidal anti-inflammatory drugs. *Kidney Intern* 44: 643–653
385. Benini D, Fanos V, Cuzzolin L, Tato L (2004) *In utero* exposure to anti-inflammatory drugs: Neonatal renal failure. *Pediatr Nephrol* 19: 232–234
386. Gupta P, Sachdev HP (2003) Safety of oral use of niemsulide in children: Systematic review of randomised controlled trials. *Indian Pediatr* 40: 518–531
387. Prevot A, Mosig D, Martini S, Guignard JP (2004) Nimesulide, a cyclooxygenase-2 preferential inhibitor, impairs renal function in the newborn rabbit. *Pediatr Res* 55: 254–260
388. Llinas MT, Lopez R, Rodriguez F, Roig F, Salazar FJ (2001) Role of COX-2-derived metabolites in regulation of the renal hemodynamic response norepinephrine. *Am J Physiol Renal Physiol* 281: F975–F982
389. Roig F, Llinas MT, Lopez R, Salazar FJ (2002) Role of cyclooxygenase-2 in the prolonged regulation of renal function. *Hypertension* 40: 721–728
390. Lopez R, Llinas MT, Roif F, Salazar FJ (2003) Role of nitric oxide and cyclooxygenase-2 in regulating the renal hemodynamic response to norepinephrine. *Am J Physiol Regul Integr Comp Physiol* 284(2): R488–R493
391. Eras J, Perazella MA (2001) NSAIDs and the kidney revisited: are selective cyclooxygenase-2 inhibitors safe? *Am J Med Sci* 321: 181–190
392. Quilley J, Chen YJ (2003) Role of COX-2 in the enhanced vasoconstrictor effect of arachidonic acid in the diabetic rat kidney. *Hypertension* 42: 837–843
393. Kanikkannan N, Jackson T, Shaik MS, Singh M (2001) Evaluation of skin sensitisation potential of melatonin and nimesulide by murine local lymph node assay. *Eur J Pharm Sci* 14: 217–220

394. Asero R (1999) Risk factors for acetaminophen and nimesulide intolerance in patients with NSAID-induced skin disorders. *Ann Allergy Asthma Immunol* 82: 554–558
395. Sanchez Borges M, Capriles-Hulett A, Caballero-Fonseca F, Perez CR (2001) Tolerability to new COX-2 inhibitors in NSAID-sensitive patients with cutaneous reactions. *Ann Allergy Asthma Immunol* 87: 201–204
396. Quiralte J, Saenz de San Pedro B, Florido JJ (2002) Safety of selective cyclooxygenase-2 inhibitor rofecoxib in patients with NSAID-induced cutaneous reactions. *Ann Allergy Asthma Immunol* 89: 63–66
397. Pastorello EA, Zara C, Riario-Sforza GG, Pravettoni V, Incorvaia C (1998) Atopy and intolerance of antimicrobial drugs increase the risk of reactions to acetaminophen and nimesulide in patients allergic to nonsteroidal anti-inflammatory drugs. *Allergy* 53: 880–884
398. Thawani V, Sontakke S, Gharpure K, Pimpalkhute S (2003) Nimesulide: The Current Controversy. *Indian J Pharmacol* 35: 121–122
399. Kumar S (2003) Drug link to child deaths is still available in India. *Br Med J* 326: 70
400. Saha K (2003) Use of nimesulide in Indian children must be stopped. *Br Med J* 326: 713
401. Gupta P, Sachdev HP (2003) Safety of oral use of nimesulide in children: systematic review of randomized controlled trials. *Indian Pediatr* 40: 518–531
402. jayashreep@expressindia.com (2004) Paediatric nimesulide: Lack of data leads to continued exposure. www.expresspharmapulse.com/print.php?article=391 (16 Sept)

Index